Red Dot Design Yearbook 2015/2016

Edited by Peter Zec

reddot award
product design

About this book

"Doing" presents more than 400 award-winning products for an active life. All of the products in this book are of outstanding design quality and have been successful in one of the world's largest and most renowned design competitions, the Red Dot Design Award. This book documents the results of the current competition in the field of "Doing", also presenting its most important players – the design team of the year, the designers of the best products and the jury members.

Über dieses Buch

„Doing" präsentiert mehr als 400 ausgezeichnete Produkte für ein aktives Leben. Alle Produkte in diesem Buch sind von herausragender gestalterischer Qualität, ausgezeichnet in einem der größten und renommiertesten Designwettbewerbe der Welt, dem Red Dot Design Award. Dieses Buch dokumentiert die Ergebnisse des aktuellen Wettbewerbs im Bereich „Doing" und stellt zudem seine wichtigsten Akteure vor – das Designteam des Jahres, die Designer der besten Produkte und die Jurymitglieder.

Contents
Inhalt

Professor Dr Peter Zec
Preface of the editor
Vorwort des Herausgebers

Dear Readers,

With the Red Dot Design Yearbook 2015/2016, you are holding a very special book in your hands. In the three volumes, "Living", "Doing" and "Working", it presents to you the winners of this year's Red Dot Award: Product Design, which make up the current state of the art in all areas of product design. At the same time this compendium is a piece of living history, because this year marks the 60th anniversary of the Red Dot Award.

Since 1955, we have set out every year in search of the best that the world of industrial design has to offer. What began 60 years ago as a small industrial show in Essen with a selection of around 100 products, most of which were German, has developed consistently over the past six decades. The competition has grown, ventured out into the world and become cosmopolitan.

A visible sign of this new, international gearing appeared 15 years ago, when the name of the competition was changed to the "Red Dot Design Award" and a new logo, the Red Dot, was introduced. That logo has become synonymous with good design – and the Red Dot Award is now the largest international competition for product design. In this anniversary year alone, 1,994 companies and designers from 56 countries entered 4,928 products in the competition for a Red Dot.

Yet through all the changes, one thing has stayed the same: The heart of the competition is and will always be its independent and fair jury. The following pages will show you this year's selection by our design experts.

I wish you an inspiring read.

Sincerely, Peter Zec

Liebe Leserin, lieber Leser,

mit dem Red Dot Design Yearbook 2015/2016 halten Sie ein besonderes Buch in den Händen: Es präsentiert Ihnen in den drei Bänden „Living", „Doing" und „Working" die Gewinner des diesjährigen Red Dot Award: Product Design und damit den aktuellen State of the Art in allen Bereichen des Produktdesigns. Zugleich ist dieses Kompendium jedoch auch ein Stück Zeitgeschichte, denn der Red Dot Award feiert dieses Jahr sein 60-jähriges Bestehen.

Seit 1955 machen wir uns jedes Jahr aufs Neue auf die Suche nach dem Besten, das die Welt des Industriedesigns zu bieten hat. Was vor 60 Jahren als kleine Industrieschau in Essen mit der Auswahl von rund 100 vornehmlich deutschen Produkten seinen Anfang nahm, hat sich im Laufe der vergangenen sechs Jahrzehnte beständig weiterentwickelt. Der Wettbewerb ist gewachsen, in die Welt hinausgegangen und weltoffen geworden.

Als sichtbares Zeichen für diese neue, internationale Ausrichtung bekam er vor 15 Jahren mit der Änderung zu „Red Dot Design Award" nicht nur einen neuen Namen, sondern auch ein neues Logo, den Red Dot. Dieses Logo ist heute ein Synonym für gutes Design – und der Red Dot Award mittlerweile der größte internationale Wettbewerb für Produktdesign. Alleine in diesem Jubiläumsjahr bewarben sich 1.994 Unternehmen und Designer aus 56 Ländern mit 4.928 Produkten um eine Auszeichnung mit dem Red Dot.

Eine Sache gibt es jedoch, die bei allen Veränderungen gleich geblieben ist: Das Herzstück des Wettbewerbs ist und bleibt seine unabhängige und faire Jury. Welche Auswahl unsere Designexperten in diesem Jahr getroffen haben, das sehen Sie auf den folgenden Seiten.

Ich wünsche Ihnen eine inspirierende Lektüre.

Ihr Peter Zec

The title "Red Dot: Design Team of the Year" is bestowed on a design team that has garnered attention through its outstanding overall design achievements. This year, the title goes to Robert Sachon and the Bosch Home Appliances Design Team. This award is the only one of its kind in the world and is extremely highly regarded even outside of the design scene.

Mit der Auszeichnung „Red Dot: Design Team of the Year" wird ein Designteam geehrt, das durch seine herausragende gestalterische Gesamtleistung auf sich aufmerksam gemacht hat. In diesem Jahr geht sie an Robert Sachon und das Bosch Home Appliances Design Team. Diese Würdigung ist einzigartig auf der Welt und genießt über die Designszene hinaus höchstes Ansehen.

In recognition of its feat, the Red Dot: Design Team of the Year receives the "Radius" trophy. This sculpture was designed and crafted by the Weinstadt-Schnaidt based designer, Simon Peter Eiber.

Als Anerkennung erhält das Red Dot: Design Team of the Year den Wanderpokal „Radius". Die Skulptur wurde entworfen und angefertigt von dem Designer Simon Peter Eiber aus Weinstadt-Schnaidt.

2015	Robert Sachon & Bosch Home Appliances Design Team
2014	Veryday
2013	Lenovo Design & User Experience Team
2012	Michael Mauer & Style Porsche
2011	The Grohe Design Team led by Paul Flowers
2010	Stephan Niehaus & Hilti Design Team
2009	Susan Perkins & Tupperware World Wide Design Team
2008	Michael Laude & Bose Design Team
2007	Chris Bangle & Design Team BMW Group
2006	LG Corporate Design Center
2005	Adidas Design Team
2004	Pininfarina Design Team
2003	Nokia Design Team
2002	Apple Industrial Design Team
2001	Festo Design Team
2000	Sony Design Team
1999	Audi Design Team
1998	Philips Design Team
1997	Michele De Lucchi Design Team
1996	Bill Moggridge & Ideo Design Team
1995	Herbert Schultes & Siemens Design Team
1994	Bruno Sacco & Mercedes-Benz Design Team
1993	Hartmut Esslinger & Frogdesign
1992	Alexander Neumeister & Neumeister Design
1991	Reiner Moll & Partner & Moll Design
1990	Slany Design Team
1989	Braun Design Team
1988	Leybold AG Design Team

Red Dot: Design Team of the Year 2015
Robert Sachon & Bosch Home Appliances Design Team
Technology and Design for a better Quality of Life

For more than 125 years, the name Bosch has stood for pioneering technology and outstanding quality. And for more than 80 years, home appliances made by Bosch have lived up to this name. With the Bosch brand, consumers worldwide associate efficient functionality, reliable quality and internationally acclaimed design. In the Red Dot Award: Product Design alone, the design team led by Robert Sachon has received more than two hundred Red Dots and several Red Dot: Best of the Bests in the past 10 years. To honour this achievement, the design team of Robert Bosch Hausgeräte GmbH is being recognised this year for its consistently high and groundbreaking design performance with the honorary title "Red Dot: Design Team of the Year".

Seit mehr als 125 Jahren steht der Name Bosch für wegweisende Technik und herausragende Qualität. Diesem Anspruch sind auch die Hausgeräte von Bosch seit über 80 Jahren verpflichtet. Mit der Marke Bosch verbinden Konsumenten weltweit effiziente Funktionalität, verlässliche Qualität und ein international ausgezeichnetes Design. Allein im Red Dot Award: Product Design erhielt das Designteam unter der Leitung von Robert Sachon in den vergangenen 10 Jahren mehr als zweihundert Auszeichnungen mit dem Red Dot und darüber hinaus mehrere Auszeichnungen mit dem Red Dot: Best of the Best. In diesem Jahr wird das Designteam der Robert Bosch Hausgeräte GmbH für seine kontinuierlich hohen und wegweisenden Gestaltungsleistungen mit dem Ehrentitel „Red Dot: Design Team of the Year" ausgezeichnet.

Quality of Life

Robert Sachon, Global Design Director of the Bosch brand, summarises the brand values of the company as follows: "Bosch has a long tradition spanning more than 125 years, and has always stood for technical perfection and superior quality." Bosch home appliances always put people first, in keeping with the principles of the company's founder Robert Bosch: technology and quality, credibility and trust, along with a sense of responsibility for people's well-being.

As early as 1932, Robert Bosch wrote that "progress in the development of technology serves, in the broad sense of the term, to afford the largest service to humanity – technology that is intended and able to provide all of humanity with enhanced opportunities and happiness in life." In the wake of the global economic crisis, Robert Bosch anticipated a notion, or value, that was not to enter the public consciousness until the early 1970s: the concept of quality of life.

Bosch develops premium-quality modern home appliances that facilitate people's daily lives and that help them attain a better quality of life. Nevertheless, only few home appliances have a place in the "design hall of fame". Indeed, as a category, home appliances have been largely neglected in the history of design, despite the fact that these products have, due to their technical functionality and everyday use, a high impact on the quality of life.

The Refrigerator

Home appliances from Bosch first saw the light of day at the Leipzig Spring Fair in 1933. The fair visitors must have been impressed when Bosch presented its first refrigerator: a metal barrel with a 60-litre net capacity. At the time, refrigerators were still regarded as "machines" given the obtrusive metal cylinder that gave them more of a tool character. The round form, which we today associate more with a washing machine, served the technical purpose of saving energy and retaining the cold optimally inside the appliance. A refrigerator in the shape of a cabinet, more suitable for daily use, was introduced on the market three years later. Resting on high feet, this newer model looked more like a piece of furniture than a technical machine. In 1949, Bosch then presented the refrigerator that, with its subtle, sober and highly industrialised form, was to become a design classic like no other home appliance.

In 1996, Bosch brought out a new edition of the 1949 design classic, thereby turning the model dating from the time of Germany's post-war economic miracle into a historical icon. Nevertheless, this new edition, introduced in a world that was rapidly changing and becoming faster with the rise of globalisation and the Internet, was designed to remain familiar and recognisable – in other words, to be "cool and calm" at once. That said, the new edition of the classic was much more than a nostalgic reminder of a bygone age. Equipped with modern technology, the Bosch Classic Edition awakened new desires and quickly advanced to become a longtime bestseller on its own terms.

Die Lebensqualität

„Bosch hat eine lange Tradition von mehr als 125 Jahren und steht seit jeher für die Themen ‚Technische Perfektion' und ‚Überlegene Qualität'", fasst Robert Sachon, Global Design Director der Marke Bosch, die Markenwerte des Unternehmens zusammen. Bosch Hausgeräte stellen immer den Nutzen für den Menschen in den Mittelpunkt, getreu den Grundsätzen des Firmengründers Robert Bosch: Technik und Qualität, Glaubwürdigkeit und Vertrauen sowie Verantwortungsbewusstsein zum Wohle der Menschen.

Bereits 1932 schrieb Robert Bosch, „dass die Fortschritte in der Entwicklung der Technik im vollem Umfange des Wortes dazu dienen, der Menschheit die größten Dienste zu leisten. Der Technik, die dazu bestimmt und in der Lage ist, der gesamten Menschheit ein Höchstmaß an Lebensmöglichkeit und Lebensglück zu verschaffen." Unter dem Eindruck der Weltwirtschaftskrise nimmt Robert Bosch einen Gedanken vorweg, der erst zu Beginn der 1970er Jahre in die öffentliche Wahrnehmung treten soll: den Begriff der Lebensqualität.

Bosch entwickelt moderne Haushaltsgeräte von höchster Qualität, die den Alltag der Menschen erleichtern und ihnen zu mehr Lebensqualität verhelfen. Indes ist nur wenigen Hausgeräten ein Platz im Olymp des Designs vorbehalten; wie überhaupt den Hausgeräten nicht immer der gebührende Platz in der Geschichte des Designs eingeräumt wird, obgleich sie aufgrund ihrer technischen Funktionalität und ihres alltäglichen Gebrauchs einen hohen Einfluss auf die Lebensqualität haben, weil sie die Hausarbeit enorm erleichtern und das Leben insgesamt verändern.

Der Kühlschrank

Die Geburtsstunde der Hausgeräte von Bosch schlägt auf der Leipziger Frühjahrsmesse 1933. Die Messebesucher dürften nicht schlecht gestaunt haben, als Bosch seinen ersten Kühlschrank vorstellt: eine Trommel mit 60 Litern Nutzinhalt. Zur damaligen Zeit ist es durchaus üblich, von einer Kältemaschine zu sprechen, da der Metallzylinder unmissverständlich seinen Werkzeugcharakter zum Ausdruck bringt. Die runde Form, die man heute eher mit einer Waschmaschine assoziiert, hat technische Gründe. Es geht darum, Energie zu sparen und die Kälte optimal im Gerät zu halten. Die für den Alltagsgebrauch besser geeignete Schrankform kommt drei Jahre später auf den Markt und erinnert mit ihren hohen Füßen mehr an ein Möbelstück als an eine technische Maschine. 1949 stellt Bosch dann den Kühlschrank vor, der es mit seiner zurückhaltenden, sachlichen und industriell geprägten Form wie kaum ein anderes Hausgerät zum Designklassiker schafft.

1996 bringt Bosch eine Re-Edition der Designikone von 1949 heraus und setzt dem Modell aus der Zeit des deutschen Wirtschaftswunders damit ein Denkmal. In der Zeit der Globalisierung und der sich anbahnenden Veränderung durch das Internet ist die Wiederauflage des Klassikers so etwas wie die eiserne Ration an Vertrautheit in einer sich rasant beschleunigenden Welt: ein Kälte- und Ruhepol zugleich. Doch die erneute Auflage des Klassikers ist weit mehr als die nostalgische Erinnerung an eine verloren gegangene Zeit. Ausgestattet mit moderner Technik weckt die Bosch Classic Edition schnell neue Begehrlichkeiten und avanciert wiederum selbst zum Best- und Longseller.

An evolutionary design approach. In 1933, Bosch unveils a surprise in the form of the first Bosch-manufactured refrigerator. It is compact and round. The appliance is launched on the market in 1949, and its typical shape becomes the archetype for refrigerators and goes on to become an iconic design of the Bosch brand. In 2014, Bosch presents a modern reinterpretation of the design classic in the form of the CoolClassic.
Ein evolutionärer Designansatz. 1933 überrascht Bosch mit dem ersten Kühlschrank aus eigener Produktion. Er ist kompakt und er ist rund. 1949 kommt das Gerät auf den Markt, dessen typische Form den Archetyp des Kühlschranks prägte und der zur Designikone der Marke Bosch avancieren sollte. 2014 stellt Bosch mit dem CoolClassic eine moderne Neuinterpretation des Designklassikers vor.

In 2014, Bosch introduced yet another model, the CoolClassic. In contrast to the 1996 edition, which sought the greatest possible loyalty to the original at the level of form, the CoolClassic aimed for a new interpretation and evolution of the refrigerator. While its rounded corners, sturdy metal housing and large Bosch lettering evoke the original model, the new CoolClassic has a charm of its own. Moreover, behind the façade, touchpad controls and a digital display allow for optimal control.

The CoolClassic is exemplary of a design principle that has been applied to many models of leading car brands. The principle pursues two objectives: the right combination of tradition and innovation, of identity and difference; and an evolutionary approach to design that makes the appliances unmistakably identifiable as "a Bosch" while at the same time underscoring their autonomy. Overall, both the Bosch Classic Edition and the CoolClassic are representative of a certain lifestyle that renders them suitable for use as stand-alone appliances in lofts and living areas as well as in offices, agencies and art studios. Their ultimate design message is that comfort and enjoyment already begin before opening the refrigerator door.

In addition, with its "ColorGlass Edition" of reduced, purist and handle-free models featuring coloured glass fronts, the Bosch brand has once again proved to be a design pioneer, particularly in response to the trend of open-concept kitchen and living spaces that call for a more homelike look of appliances.

2014 kommt mit dem CoolClassic dann eine moderne Neuinterpretation auf den Markt. Im Unterschied zur Re-Edition von 1996, die sich formal um eine möglichst große Originaltreue bemüht, geht es beim Re-Design des neuen CoolClassic um eine Überarbeitung und Weiterentwicklung des Kühlschranks. Die abgerundeten Ecken, das robuste Metallgehäuse und der große Bosch-Schriftzug erinnern zwar noch an das Original, der neue CoolClassic versprüht aber seinen eigenen Charme. Hinter den Fronten verbirgt sich moderne Elektronik, die durch berührungsempfindliche Tasten gesteuert wird und digital ablesbar ist.

Hier kommt ein gestalterisches Prinzip zum Ausdruck, das sich auch bei vielen Modellen erfolgreicher Automarken findet: das gelungene Zusammenspiel von Tradition und Innovation, von Identität und Differenz – ein evolutionärer Ansatz in der Gestaltung, der die Hausgeräte eindeutig der Marke Bosch zuordnet und zugleich ihre Eigenständigkeit unterstreicht. Beiden Geräten, der Bosch Classic Edition wie dem CoolClassic ist der Lifestyle-Charakter gemein, den sie als Solitär im Loft und Wohnbereich, aber auch in Büros, Agenturen und Ateliers unterstreichen. Hier beginnt der Genuss eben schon vor dem Öffnen der Kühlschranktür.

Daneben gelingt es der Marke mit ihrer „ColorGlass Edition" – puristisch reduzierten, grifflosen Modellen mit farbigen Glasfronten – gestalterisch nach vorne zu blicken und der Öffnung der Küche zum Wohnraum und damit einhergehenden wohnlicheren Gerätekonzepten als Trendsetter voranzuschreiten.

The Stove and the Oven

As with the refrigerator, the history of the kitchen stove was initially shaped by engineers. Similar to the "cooling machine", stoves were at first called "cooking machines". And as with refrigerator design, the initial aim of stove design was to emulate furniture and to achieve technical feasibility. The electric stove was not marketed until the 1930s, like the refrigerator, and required the concentration of heat within a narrow space accommodating three or four hotplates of different diameters. In retrospect, the evolution of the cast-iron hotplates of the early years to the modern hobs made of ceramic glass is exemplary of a technological optimisation process during which the energy efficiency of the appliances was continuously improved.

In parallel to the technological development, the design of the control elements and the usage of the appliances evolved. After all, the stove and oven continue to be two of the most used appliances in the kitchen – so much so that Robert Sachon seeks inspiration for his work while cooking, possibly explaining his special focus on these two appliances. As a result, he often prefers to work with them on weekends and enjoys "dealing with appliances that allow one to engage directly and creatively with the product." As head of brand design, Sachon is concerned with the question of which design elements lend themselves to being transferred to other products or categories.

The Series 8 built-in appliances, distinguished with a Red Dot: Best of the Best award, are characterised by a simple, sleek design that prioritises the compatibility of different appliances as well as an easy, intuitive operating concept. The defining element is the flush-surface, centralised control ring with which the user can control the wide range of functions with a simple turn. Moreover, trends occurring in other areas are of relevance for the design of home appliances. For example, open-living concepts have repercussions for the design and integration of home appliances. "The current trend is to design concepts that fit well into the new living environments," explains Robert Sachon. "Integration as a whole is an important topic."

The Kitchen Machine

In 1952, Bosch introduced the first food processor on the market. With a name that says it all – the Neuzeit I (Modern Era I) – this product symbolised the beginning of a new era, since it decreased the number of actions and movements required for preparing food. Neuzeit I was able to stir, knead, cut, chop, mash, grate and grind.

In today's busy world, these small home and kitchen devices are indispensable helpers. As with all other equipment categories, here too the main objective is to improve the usefulness and convenience for the user with good design and innovative technology. Robert Sachon summarises the Bosch corporate philosophy as follows: "We simply try to give the appliances the best that there is."

Der Herd und der Ofen

Die Geschichte des Küchenherds ruht wie die Entwicklung des Kühlschranks zunächst in den Händen von Ingenieuren. Analog zur Kältemaschine spricht man zu Beginn der Entwicklung noch von einer Kochmaschine. Und ebenso wie die Form des Kühlschranks orientiert sich auch die Gestaltung des Herdes zunächst an dem Vorbild des Mobiliars und der technischen Machbarkeit. Der Elektroherd setzt sich parallel zum Kühlschrank erst in den 1930er Jahren durch, schließlich musste zunächst die Wärme auf einen engen Raum konzentriert werden, der drei oder vier Herdplatten mit unterschiedlichen Durchmessern Platz bot. Die Entwicklung von den gusseisernen Herdplatten der Anfangsjahre bis zu den modernen Kochfeldern aus Glaskeramik liest sich rückblickend wie ein technischer Optimierungsprozess, bei dem die Energieeffizienz der Geräte laufend verbessert wurde.

Parallel zur technischen Entwicklung hat sich aber auch die Gestaltung der Bedienelemente und der Gebrauch der Geräte verändert. Herd und Backofen sind nach wie vor zwei der am häufigsten genutzten Geräte in der Küche. Und auch Robert Sachon holt sich nicht zuletzt beim Kochen Inspiration für seine Arbeit. Daher gilt beiden Geräten sein besonderes Augenmerk. Gerade an den Wochenenden nimmt er sich Zeit dafür, „weil man es mit Geräten zu tun hat, die es einem unmittelbar erlauben, kreativ mit dem Produkt umzugehen", sagt Sachon. Als Leiter des Markendesigns beschäftigt ihn dabei auch die Frage, „welche gestalterischen Elemente sich eignen, um sie auf andere Produkte oder Kategorien zu übertragen".

Das mit dem Red Dot: Best of the Best ausgezeichnete Einbauprogramm der Serie 8 zeichnet sich durch eine einfache, klare Linienführung aus, die die perfekte Kombinierbarkeit unterschiedlicher Geräte in den Vordergrund stellt, sowie durch ein einfaches, intuitives Bedienkonzept. Prägendes Element ist der flächenbündig integrierte, zentrale Bedienring, mit dem der Nutzer mit einem Dreh die Vielzahl der Gerätefunktionen im Griff hat. Auch Trends aus anderen Bereichen sind für das Hausgeräte-Design von Bedeutung. Offene Wohnkonzepte wirken sich beispielsweise auch auf das Design und die Integration von Hausgeräten aus. „Der Trend geht aktuell zu Gestaltungskonzepten, die sich gut in die neuen Wohnwelten einfügen", erklärt Robert Sachon. „Integration überhaupt ist ein wichtiges Thema."

Die Küchenmaschine

1952 bringt Bosch die erste Küchenmaschine auf den Markt: die „Neuzeit I". Ihr Name ist Programm. Sie steht für den Beginn einer neuen Zeit, da sie den Hausfrauen viele Handgriffe abnimmt und die Zubereitung des Essens erleichtert. Die „Neuzeit I" kann rühren, kneten, hacken, schnitzeln, pürieren, mahlen und reiben.

Heute sind im modernen Alltag die kleinen Haus- und Küchengeräte unverzichtbare Helfer. Wie bei allen anderen Gerätekategorien gilt auch hier die Prämisse, den Nutzen und den Komfort für den Anwender mit guter Gestaltung und innovativer Technik zu verbessern. „Wir versuchen einfach, den Geräten das Beste mitzugeben", bringt Robert Sachon den Anspruch von Bosch auf den Punkt.

A new design language. The Series 8 range of built-in appliances awarded the Red Dot: Best of the Best sets new standards in the design of user interfaces. The defining element is the flush-surface, centralised control ring with which the user can control the wide range of functions with a simple turn.

Eine neue Designsprache. Das mit dem Red Dot: Best of the Best ausgezeichnete Einbauprogramm der Serie 8 setzt neue Maßstäbe in der Gestaltung von User Interfaces. Prägendes Element ist der flächenbündig integrierte, zentrale Bedienring, mit dem der Nutzer mit einem Dreh die Vielzahl der Gerätefunktionen im Griff hat.

13

Easy to use and featuring high-quality materials and consistent design: the Styline series of small home appliances from Bosch is not only modestly elegant but also uncompromisingly useful.
Hochwertige Materialien, einfache Bedienbarkeit und gestalterisch in einer Linie: Die Styline-Serie der kleinen Hausgeräte von Bosch präsentiert sich nicht nur zurückhaltend elegant, sondern ist auch kompromisslos nützlich.

A good example is the food processor MUM, which received a Red Dot: Best of the Best award as early as 2011. Similar to its predecessor Neuzeit I, its trademark is also versatility and energy-efficient performance. And, with an emphasis on high technological performance, the MUM exhibits the evolutionary design approach unique to Bosch. From one product generation to the next, these appliances affirm and promote the brand identity. Whether large or small home appliances: all are created with the holistic, evolutionary design approach of the Bosch brand, and all follow the uniform design language and uniform design principles derived from the brand values.

The Washing Machine

The first washing machine from Bosch was put into series in 1958, followed in 1960 by the first fully automatic washing machine that allowed clothes to be laundered in a single wash, rinse and spin cycle. In 1972, a new unit was introduced that combined washer and dryer in one. Since then, the washing programme options are being continually refined and expanded. All of these appliances contribute to making the days of hard manual labour a thing of the past. Moreover, the technology and design of the appliances have not only laid the foundation for a better quality of life in the household but also for social change. To the extent that the chore of washing increasingly became a minor matter, women liberated themselves from their role as housewife, and working women were no longer the exception to the rule.

Ein gelungenes Beispiel ist die Küchenmaschine MUM, die bereits 2011 mit dem Red Dot: Best of the Best ausgezeichnet wird. Analog zu ihrem historischen Vorgänger, der „Neuzeit I", ist auch ihr Markenzeichen die Vielseitigkeit und die energieeffiziente Leistung. Im Einklang mit der hohen technischen Leistung spiegelt sich in der Formensprache der Küchenmaschine MUM aber auch der evolutionäre Gestaltungsansatz mustergültig wider. Von Generation zu Generation kommt im Produktdesign auch die Anbindung an die Marke Bosch zum Ausdruck. Ob große oder kleine Hausgeräte – allen wird der ganzheitliche, evolutionäre Designansatz der Marke Bosch zuteil, folgen sie doch sämtlich einer den Markenwerten entspringenden, einheitlichen Designsprache und einheitlichen Gestaltungsprinzipien.

Die Waschmaschine

Die erste Waschmaschine aus dem Hause Bosch geht 1958 in Serie. Es ist noch kein Waschvollautomat. Dieser kommt 1960 auf den Markt und ermöglicht es, die Wäsche erstmals in einem einzigen Durchlauf waschen, spülen und schleudern zu lassen. 1972 werden die Waschmaschine und der Trockner dann eins. Seither wird das Geräteprogramm immer weiter verfeinert und ausgebaut. All diese Geräte sorgen dafür, dass die Zeiten der mühevollen Handarbeit langsam zu Ende gehen. Die Technik und die Gestaltung der Geräte legen nicht nur den Grundstein für mehr Lebensqualität im Haushalt, sondern auch für eine gesellschaftliche Veränderung. Das Thema Waschen wird mehr und mehr zur Nebensache. Die Frauen emanzipieren sich von ihrer Rolle als Hausfrau. Berufstätige Frauen sind keine Ausnahme mehr.

Although women's liberation is usually discussed from the standpoint of a social and political movement, the topic merits further examination from the perspective of home appliance design. Indeed, the changing effect of technology and design extends beyond the material manifestation of the products. On the one hand, design concepts in the home appliance sector are always oriented towards long-term societal trends. Yet on the other hand, the designed products themselves have an impact on society, in particular through their technical function and their user-oriented utility. Here, at the latest, it becomes clear that quality of life is far more than the sum of the products that we own. The importance and impact of design go beyond what is materialised in a product, becoming discernible only in the changing conditions of use and their impact.

The Dishwasher

In 1964, the first dishwasher from Bosch went into serial production. Thanks to technological developments, dishwashers have since become very quiet. So much so that the remaining time of the programme cycles is now explicitly indicated and of late even projected onto the floor so that the user knows when the machine has finished the washing and drying process.

"It's possible to tolerate certain sounds, but it's fair to say that we're probably better off without them," says Robert Sachon. "Essentially, a common priority of all product categories is to ensure that appliances are quiet, so as not to disturb users. At the same time, we have to generate and design the sounds that consumers are supposed to hear which is not always easy." This shows that the team led by Robert Sachon values not only the quality of the product and its exterior design, but also the quality of the experience, from the operation of and interaction with the appliance to the sound design and haptics. Finally, Bosch wants design perfection to be experienceable with all the senses.

The Internet of Things – Design 4.0

Bosch undoubtedly invests heavily in the development and design of new technologies. And Bosch wouldn't be Bosch if it didn't capitalise on emerging technological opportunities to achieve more convenience and ease of use in home appliances.

"Under the slogan the 'Internet of Things', visions have been presented at trade fairs for years," explains Robert Sachon. "With the digitisation and networking of appliances, these visions are now becoming reality and taking on concrete forms," says the chief designer. With the "Home Connect" app, Bosch's digitally connected home appliances can be controlled using a smartphone or tablet PC. Among these is the Series 8 oven, distinguished with a Red Dot: Best of the Best. "'Home Connect' is a good example of how we generate real added value for our customers with new technologies," says Jörg Gieselmann, Executive Vice President Corporate Brand Bosch at BSH Hausgeräte GmbH. He adds: "With the app, setting up the appliance, making or changing basic settings or operating it when away from home becomes child's play. In this way we're creating wholly new possibilities and freedoms in the household."

Während die Emanzipation der Frauen häufig unter dem Gesichtspunkt einer sozialen und politischen Bewegung verhandelt wird, kann es sich durchaus lohnen, das Thema auch aus der Perspektive des Hausgeräte-Designs zu beleuchten. Über die materialisierte Form der Produkte hinaus zeigt sich die verändernde Wirkung von Technik und Design. Auf der einen Seite orientieren sich Gestaltungskonzepte im Hausgerätebereich immer auch an den langfristigen gesellschaftlichen Trends. Auf der anderen Seite nehmen die gestalteten Produkte wiederum selbst Einfluss auf die Gesellschaft, insbesondere über ihre technische Funktion und ihren am Nutzer orientierten Gebrauch. Spätestens hier wird deutlich, dass Lebensqualität weit mehr ist als die Summe der Produkte, die wir besitzen. Die Bedeutung und die Wirkung des Designs gehen über das hinaus, was sich in einem Produkt materialisiert. Sie werden erst in den sich verändernden Bedingungen der Nutzung und des Gebrauchs und deren Wirkung ablesbar.

Der Geschirrspüler

1964 geht der erste Geschirrspüler von Bosch in Serie. Aufgrund der technischen Entwicklung sind die Geschirrspüler inzwischen so leise und geräuscharm geworden, dass die Gestalter über zusätzliche Projektionsflächen die Restlaufzeit der Geräte auf den Boden projizieren, damit der Benutzer weiß, wann das Gerät den Spül- und Trocknungsvorgang beendet hat.

„Man kann mit bestimmten Geräuschen gut leben, man kann ohne sie vielleicht noch etwas besser leben", sagt Robert Sachon. „Es zieht sich im Grunde durch alle Produktkategorien hindurch, dass die Geräte immer angenehm leise arbeiten, damit man eben nicht gestört wird. Und die Geräusche, die der Konsument wahrnehmen soll, versuchen wir bewusst zu gestalten, was nicht immer einfach ist", so Sachon. Hier zeigt sich, dass das Team um Robert Sachon stets versucht, über die Produktqualität hinaus eine Erlebnisqualität zu vermitteln, bei der neben der äußeren Gestaltung auch die Bedienung und Interaktion mit den Geräten, das Sounddesign und die Haptik eine große Rolle spielen. Schließlich will Bosch im Design Perfektion mit allen Sinnen erfahrbar machen.

Das Internet der Dinge – Design 4.0

Bosch gehört ohne Zweifel zu den Unternehmen, die viel in die Entwicklung und Gestaltung neuer Technologien investieren. Und Bosch wäre nicht Bosch, wenn das Unternehmen die sich bietenden technischen Möglichkeiten nicht auch für eine einfachere Handhabung und leichtere Bedienbarkeit der Hausgeräte nutzen würde.

„Unter dem Schlagwort ‚Internet der Dinge' gibt es ja bereits seit Jahren Visionen, die beispielsweise auf Messen vorgestellt werden", erläutert Robert Sachon. „Mit der Digitalisierung und der Vernetzung der Geräte werden diese Visionen heute Realität und nehmen ganz konkrete Formen an", so der Chefdesigner. Mit der „Home Connect"-App können vernetzte Hausgeräte von Bosch mithilfe eines Smartphones oder Tablets gesteuert werden, zum Beispiel der Serie 8 Backofen, der mit dem Red Dot: Best of the Best ausgezeichnet wurde. „Home Connect ist ein gutes Beispiel dafür, wie wir mit den neuen Technologien echte Mehrwerte für unsere Kunden

With the effortless implementation of digitally networked home appliances, Bosch is once again setting a new standard in the design and use of such appliances. And, as in the early twentieth century, the kitchen is today again becoming a new focal point of communication, even if under the changed information and communication environment of the twenty-first century. Of course, these developments also affect the daily work and professional identity of designers. As Robert Sachon explains, "In the past our team was very heavily influenced by industrial designers, while nowadays we are focusing more on the topic of user interface design." Yet this is only to be expected since the perceptions and communication patterns of consumers are evolving. Essentially, it means that technological changes always lead to changes in the design of the future.

generieren", erklärt Jörg Gieselmann, Executive Vice President Corporate Brand Bosch der BSH Hausgeräte GmbH. „Die Verbraucher können mithilfe der App kinderleicht und einfach ihr Gerät in Betrieb nehmen, Grundeinstellungen vornehmen und verändern oder es von unterwegs aus bedienen. So schaffen wir ganz neue Möglichkeiten und Freiheiten im Haushalt."

Mit der mühelosen Benutzung digital vernetzter Hausgeräte setzt Bosch abermals einen neuen Standard in der Gestaltung und im Gebrauch von Hausgeräten. Und wie bereits zu Beginn des 20. Jahrhunderts wird die Küche heute wieder zu einem neuen Mittelpunkt der Kommunikation, wenn auch unter den veränderten Informations- und Kommunikationsbedingungen des 21. Jahrhunderts. Natürlich beeinflussen diese Entwicklungen auch die tägliche Arbeit und das Selbstverständnis der Designer, wie Robert Sachon erläutert: „Wurde unser Team früher sehr stark von Industriedesignern geprägt, so verstärken wir uns heute beim Thema ‚User Interface Design'." Das kann auch nicht anders sein, denn auch die Wahrnehmung und die Kommunikation der Konsumenten entwickeln sich weiter. Insofern ergeben sich grundsätzlich aus den technologischen Veränderungen immer auch Veränderungen für das Design der Zukunft.

The Internet of Things: with the "Home Connect" app, Bosch's digitally connected home appliances can be controlled using a smartphone or tablet PC, true to the Bosch philosophy "Invented for Life".
Das Internet der Dinge: Mit der „Home Connect"-App können vernetzte Hausgeräte von Bosch mithilfe eines Smartphones oder Tablets gesteuert werden, getreu der Bosch-Philosophie „Technik fürs Leben".

Values

When Robert Bosch founded his company in 1886 – a workshop for precision mechanics and electrical engineering in Stuttgart – housework still meant hard work. Hobs were heated with an open fire. Dishes and laundry had to be laboriously cleaned by hand. Food had to be salted or boiled. Housecleaning and the weekly washing day certainly consumed a lot of time and energy.

Robert Bosch recognised this problem and developed technical solutions which others hadn't even imagined. Some 47 years after the founding of the company, in 1933, Robert Bosch moved into the production of home appliances, a division dedicated to making people's everyday lives easier. And still during the Great Depression, he spearheaded a massive overhaul of the Bosch Group, by then globally active in the field of automotive and industrial technology, in order to modernise and diversify its activities.

From the outset it was one of Robert Bosch's principles to produce the best possible quality at all times. As early as 1918, he wrote that "it has always been an unbearable thought that someone might prove, upon examining one of my products, that my performance is inferior in some way. That's why I've always made a point of releasing only products that have passed all quality tests, in other words, that were the best of the best." This maxim essentially guides the corporate philosophy to this day: technology and quality, credibility and trust, along with a sense of responsibility for people's well-being.

And in 1921 he wrote: "I've always acted according to the principle that I would rather lose money than trust. The integrity of my promises, the belief in the value of my products and in my word have always meant more to me than temporary gain." To this day, Robert Bosch and the example he set have a significant impact on the company. The founder is still the reference point that he has always been, and which he will still be tomorrow.

Throughout the years, the design team of Robert Bosch Hausgeräte GmbH has succeeded in finding a modern and appropriate interpretation of the values which its founder embodied and which manifest in the company's brand to this day. It has set pioneering standards in home appliance design and made a significant contribution to improving the quality of life. To acknowledge the design performance of the entire team, which has garnered attention over the years with its high-quality products and consistently outstanding design, the honorary title "Red Dot: Design Team of the Year" for 2015 is bestowed on Robert Sachon and the Bosch Home Appliances Design Team.

Die Werte

Als Robert Bosch im Jahr 1886 sein Unternehmen gründete – eine Werkstatt für Feinmechanik und Elektrotechnik in Stuttgart –, bedeutete Hausarbeit noch Schwerstarbeit. Kochstellen wurden mit offenem Feuer beheizt. Geschirr und Wäsche mussten aufwendig von Hand gereinigt werden. Lebensmittel mussten gepökelt oder eingekocht werden. Der Hausputz und der wöchentliche Waschtag kosteten viel Zeit und Kraft.

Robert Bosch erkannte dieses Problem und entwickelte dafür technische Lösungen, wo andere noch nicht einmal Möglichkeiten erahnten. 47 Jahre nach der Gründung des Unternehmens stieg Robert Bosch im Jahr 1933 in die Produktion von Hausgeräten ein, um mit seinen Produkten den Alltag der Menschen zu erleichtern. Und unter dem Eindruck der Weltwirtschaftskrise verordnete er dem inzwischen weltweit im Bereich der Kraftfahrzeug- und Industrietechnik agierenden Bosch-Konzern einen konsequenten Modernisierungs- und Diversifizierungskurs.

Von Anfang an ist es einer der Grundsätze von Robert Bosch, immer bestmögliche Qualität zu produzieren. „Es war mir immer ein unerträglicher Gedanke, es könne jemand bei der Prüfung eines meiner Erzeugnisse nachweisen, dass ich irgendwie Minderwertiges leiste. Deshalb habe ich stets versucht, nur Arbeit hinauszugeben, die jeder sachlichen Prüfung standhielt, also sozusagen vom Besten das Beste war", schreibt Robert Bosch bereits im Jahr 1918. An dieser Haltung orientiert sich das Unternehmen bis heute: Technik und Qualität, Glaubwürdigkeit und Vertrauen sowie Verantwortungsbewusstsein zum Wohle der Menschen.

„Immer habe ich nach dem Grundsatz gehandelt: Lieber Geld verlieren als Vertrauen. Die Unantastbarkeit meiner Versprechungen, der Glaube an den Wert meiner Ware und an mein Wort standen mir stets höher als ein vorübergehender Gewinn", schreibt Robert Bosch im Jahr 1921. Vieles von dem, was der Firmengründer gedacht und vorgelebt hat, übt bis heute eine große Anziehungskraft aus. Robert Bosch ist der Bezugspunkt des Unternehmens, der er bereits früher war und der er auch morgen noch sein wird.

Dem Designteam der Robert Bosch Hausgeräte GmbH gelingt es seit vielen Jahren, die Werte, die bereits ihr Firmengründer verkörperte und die bis heute in den Markenwerten des Unternehmens zum Ausdruck kommen, auf zeitgemäße Art zu interpretieren. Es hat im Hausgeräte-Design wegbereitende Standards gesetzt und einen wesentlichen Beitrag zu mehr Lebensqualität geleistet. Mit Blick auf die gestalterische Leistung des gesamten Teams, das über Jahre hinweg mit qualitativ hochwertigen Produkten und einem kontinuierlich hohen Gestaltungsniveau auf sich aufmerksam gemacht hat, wird der Ehrentitel „Red Dot: Design Team of the Year" im Jahr 2015 an Robert Sachon und das Bosch Home Appliances Design Team verliehen.

"We simply try to give the appliances the best that there is."
„Wir versuchen einfach, den Geräten das Beste mitzugeben."

Robert Sachon, Global Design Director Bosch

Red Dot: Design Team of the Year 2015
Interview: Robert Sachon
Global Design Director
Robert Bosch Hausgeräte GmbH

For generations, the Bosch name has stood for groundbreaking technology and outstanding quality. Home appliances from Bosch have been committed to these standards for over 80 years now. Consumers around the world associate the Bosch brand with efficient functionality, reliable quality and design that has won awards at an international level. The brand has won more than 500 awards in the past 10 years alone. The man behind this design success is Robert Sachon. He started his career in 1999 at Siemens-Electrogeräte GmbH, at the time under the leadership of Gerd Wilsdorf. In 2005 he moved to Robert Bosch Hausgeräte GmbH, where he took over from his predecessor Roland Vetter. Together with a team of roughly 40 employees worldwide, he has shaped the design of the Bosch brand in his role as Global Design Director. Burkhard Jacob met with him for an interview in the Red Dot Design Museum Essen.

Mr Sachon, you have been Global Design Director for the Bosch brand for 10 years now. How would you describe your role as head designer?

The term "head designer" is pretty accurate. It sounds a little like a head chef. Similar to the chef de cuisine, who is in charge of the kitchen crew in fine dining establishments, I lead a team of employees who are responsible for the design of the home appliances.

Do you also get stuck in?

I am one of those designers who not only manage other people but are also happy to get their own hands dirty in order to set out the design direction. We have lots of different product categories, but at the end of the day the point is of course to shape the face of the Bosch brand in a similar way to a signature or a common design language. And that's something I like to stay involved in.

Mr Sachon, you exude a calmness that makes me curious as to what your star sign is?

Aries – a healthy mix of diplomacy and stubbornness.

Der Name Bosch steht seit Generationen für wegweisende Technik und herausragende Qualität. Diesem Anspruch sind die Hausgeräte von Bosch seit über 80 Jahren verpflichtet: Mit der Marke Bosch verbinden Konsumenten weltweit effiziente Funktionalität, verlässliche Qualität und ein international ausgezeichnetes Design. Allein in den letzten 10 Jahren wurden mehr als 500 Auszeichnungen gewonnen. Der Mann hinter diesen Design-Erfolgen ist Robert Sachon. Er begann seine Karriere 1999 bei der Siemens-Electrogeräte GmbH, damals unter Gerd Wilsdorf. 2005 wechselte er zur Robert Bosch Hausgeräte GmbH und beerbte seinen Vorgänger Roland Vetter. Gemeinsam mit einem Team von weltweit rund 40 Mitarbeitern hat er als Global Design Director das Design der Marke Bosch geprägt. Burkhard Jacob traf ihn im Red Dot Design Museum Essen zum Interview.

Herr Sachon, seit 10 Jahren sind Sie nun Global Design Director der Marke Bosch. Wie würden Sie Ihre Tätigkeit als Chefdesigner beschreiben?

Der Begriff „Chefdesigner" trifft es schon ganz gut. Es klingt ein wenig wie Chefkoch. Ähnlich dem Chef de Cuisine, der in der gehobenen Gastronomie die Küchenbrigade leitet, führe ich ein Team von Mitarbeitern, die für das Design der Haushaltsgeräte verantwortlich sind.

Greifen Sie auch selbst zum Kochlöffel?

Ich bin einer der Gestalter, die nicht nur managen, sondern auch selbst zum Stift greifen, um die Gestaltungsrichtung vorzugeben. Wir haben viele unterschiedliche Produktkategorien, aber am Ende des Tages geht es natürlich darum, das Gesicht der Marke Bosch im Sinne einer Art Handschrift, einer gemeinsamen Designsprache zu prägen. Und das möchte ich mir gerne erhalten.

Herr Sachon, Sie strahlen eine Ruhe aus, die die Frage provoziert, welches Sternzeichen Sie sein könnten?

Widder – eine gesunde Mischung aus Diplomatie und Dickköpfigkeit.

What characteristics of an Aries are helpful for the role of designer?

There are two characteristics that help me greatly in my work: a tendency to be a perfectionist and a certain amount of tenacity. Both of these things help me to work on different projects of differing durations and scope in order to bring long-term topics to fruition for the Bosch brand.

Do you have a design role model?

That's a difficult question, because I suppose we ultimately always come back to the heroes of design history. I have to admit I have huge respect for Dieter Rams, even though that's probably something that lots of designers say. His design language resonates with me: its clarity, order and meaningfulness. I am of a similar mindset. Without wanting to overstate their importance, his ten principles of good design are still valid today. I am fascinated by this clear design language, which has become very popular again nowadays in particular, making it all the more fitting for Bosch brand values.

What does the Bosch brand stand for?

Bosch has a long tradition spanning more than 125 years, and has always stood for technical perfection and superior quality. In our design, we take these rational values and make them tangible at an emotional level by means of clear design that showcases high-quality materials and their uncompromising finish in a precise manner down to the smallest detail. This is a holistic approach to design which we apply to all of our products

Welche Eigenschaften des Widders sind denn hilfreich für die Tätigkeit als Designer?

Es gibt zwei Eigenschaften, die mir bei der Tätigkeit sehr entgegenkommen: ein gewisser Hang zum Perfektionismus und eine gewisse Beharrlichkeit. Beides hilft mir, zeitgleich an unterschiedlichen Projekten mit unterschiedlicher Laufzeit und Tragweite zu arbeiten, um langfristige Themen für die Marke Bosch durchzusetzen.

Haben Sie ein Vorbild in Fragen der Gestaltung?

Das ist eine schwierige Frage, weil man vermutlich immer bei den Heroen der Designgeschichte landet. Trotzdem – ich habe einen wahnsinnigen Respekt vor Dieter Rams, auch wenn das wahrscheinlich viele Designer sagen. Aber seine Gestaltungssprache liegt mir schon sehr nahe: die Klarheit, die Ordnung, die Sinnhaftigkeit. Da sehe ich eine ähnliche Geisteshaltung. Seine zehn Gebote des Designs haben ja heute noch ihre Gültigkeit, ohne sie religiös überhöhen zu wollen. Mich fasziniert diese klare Gestaltungssprache, die gerade heute wieder hoch im Kurs steht – und die umso mehr zu den Markenwerten von Bosch passt.

Wofür steht die Marke Bosch?

Bosch hat eine lange Tradition von mehr als 125 Jahren und steht seit jeher für die Themen „Technische Perfektion" und „Überlegene Qualität". Diese rationalen Werte machen wir in unserem Design emotional erfahrbar durch eine klare Gestaltung, welche hochwertige Materialien und deren kompromisslose Verarbeitung präzise bis ins kleinste Detail in Szene setzt.

worldwide. In this regard, it is fair to speak of a uniform design language, or DNA, of the Bosch brand.

So you are a brand manager as well as a designer?

Most definitely. Unlike other companies, where design is part of technical development, design at Bosch benefits from the fact that it is a key part of brand management. Our design team plays an important role and has a clear remit, as it gets involved with the development of product concepts at a very early stage – long before any thoughts of marketing for the products or of an advertising campaign.

Do you base your design decisions on the Bosch brand values?

The aim is always to bring the brand values to life. In the area of home appliances, we see a lot of products that pass on certain design features to the next generation. Consequently there is an underlying evolutionary thought process involved. This goes without saying with a brand like Bosch, which has been conveying the same values for over 125 years.

Basing design language on the brand values is one side of the coin. The other, which you have just described, relates to a product's use and benefit for the consumer.

Absolutely. We pursue a user-centred design approach where consumer monitoring plays a major role. We benefit from the fact that we too are all users of home appliances. We therefore can observe ourselves as well as others. And when observing ourselves, it's important to always be aware of our blind spot.

Who is it that ultimately decides whether or not a product goes into production?

As a designer and as Global Design Director, I don't make lonely decisions, relying instead on my team of specialists. Maybe that is one reason why we were awarded the Design Team of the Year title. But the decision of whether a product goes into production is not made by the design team alone. That is a joint decision made by top management. Our task is to convince all of those involved in the process to take a proposed course of action.

What role does market observation play for design?

The competitive environment for home appliances is very tough. There are over 1,000 home appliances brands worldwide. We know some of those brands very well. After all, we meet our competitors regularly at trade fairs. So it is in our mutual interest to set ourselves apart very clearly from our competitors. Obviously we don't just look at the market from the perspective of the competition, but also always with a view to understanding long-term developments. For example, connectivity is one keyword that shows where the journey is headed. As a consequence, the

Das ist ein ganzheitlicher Gestaltungsansatz, den wir auf all unsere Produkte weltweit übertragen. Insofern kann man durchaus von einer einheitlichen Designsprache, einer DNA der Marke Bosch reden.

Sie sind also nicht nur Designer, sondern auch Markenmanager?

Definitiv. Im Unterschied zu anderen Unternehmen, in denen das Design Teil der technischen Entwicklung ist, profitiert das Design bei uns davon, ein wesentlicher Teil des Markenmanagements zu sein. Unser Designteam hat eine wichtige Rolle und eine klare Aufgabenstellung, da es sich bereits sehr früh mit der Entwicklung von Produktkonzepten befasst; und zwar lange bevor über deren Vermarktung oder eine Werbekampagne nachgedacht wird.

Orientieren Sie Ihre gestalterischen Entscheidungen an den Markenwerten von Bosch?

Es geht immer darum, die Markenwerte erfahrbar zu machen. Im Bereich der Hausgeräte finden wir viele Produkte, die bestimmte gestalterische Merkmale an die nächste Generation weitergeben, denen also ein evolutionärer Gedanke zugrunde liegt. Das kann ja auch nicht anders sein bei einer Marke wie Bosch, die seit mehr als 125 Jahren dieselben Werte vermittelt.

Die Designsprache an den Markenwerten zu orientieren, ist eine Seite der Medaille. Sie beschreiben auch noch eine andere Seite: die Seite des Gebrauchs und des Nutzens für den Konsumenten.

Absolut. Wir verfolgen einen nutzerzentrierten Gestaltungsansatz, bei dem die Beobachtung der Konsumenten eine wichtige Rolle spielt. Dabei kommt uns entgegen, dass wir auch alle selbst Nutzer von Hausgeräten sind. Wir haben also die Ebene der Fremd- und der Selbstbeobachtung. Bei der Selbstbeobachtung muss man aber immer auch seinen blinden Fleck im Visier haben.

Wer entscheidet letztlich, ob ein Produkt in Serie geht?

Als Designer und als Global Design Director treffe ich keine einsamen Entscheidungen, sondern vertraue auf mein Team, das aus Spezialisten besteht. Vielleicht ist das auch ein Grund dafür, warum wir mit der Auszeichnung zum Designteam des Jahres bedacht worden sind. Die Entscheidung, ob ein Produkt in Serie geht, trifft das Designteam aber nicht allein. Das ist eine gemeinsame Entscheidung des Topmanagements. Unsere Aufgabe ist es, alle Prozessbeteiligten davon zu überzeugen, einen vorgeschlagenen Weg zu gehen.

Welche Rolle spielt die Marktbeobachtung für das Design?

Im Hausgerätemarkt finden wir ein sehr starkes Wettbewerbsumfeld vor. Es gibt über 1.000 Hausgerätemarken weltweit. Einige davon kennen wir auch sehr gut. Wir treffen unsere Wettbewerber ja regelmäßig auf Messen. Da liegt eine gute Differenzierung im wechselseitigen Interesse.

competition is never the only source of inspiration. We regularly attend trade fairs to scout out new trends. In doing so, we also learn from other industries such as the automotive industry, where the products and innovation cycles are similar in length to those on the home appliances market. In addition, we observe the developments in interior design, architecture and in consumer electronics. We get an insight into a range of vastly different industries, filtering innovations according to whether they constitute short-term or more long-term trends and how they impact on technical and design development on the home appliances market. Ultimately we want to develop products that are attractive not only on the day they are purchased but for many years afterwards.

What topics will most influence the industry for home appliances in the coming years?

There are some developments that have been influencing the home appliances industry for quite a while. For example, visions in relation to the "Internet of Things" have featured at trade fairs for some years now. With the connectivity of the appliances and the digital transmission of data, these visions are now becoming a reality and are taking shape in a very real way.

To what extent does that also change the work within your design team?

Naturally, developments like these also affect our daily work as designers. In the past our team was very heavily influenced by industrial designers, while nowadays we are focusing more on the topic of user interface design. Although the design of home appliances has always involved the design of functions for use as well as operating elements, digitalisation and connectivity mean that we are also pursuing independent design concepts which in turn result in new forms of operation and user guidance.

Is this also a general indication of how the design of home appliances will develop?

Yes, and that is something which ultimately makes a lot of sense. Coming back to the general developments and influencing factors again, it is fair to say that the change in how we use our living space has a significant role to play. Open-plan living concepts have resulted in kitchens themselves becoming a part of the living area. This of course is also reflected in the concepts behind the appliances. The current trend is to design concepts that fit well into the new living environments. Integration as a whole is an important topic.

To what extent are materials a general topic in the design of home appliances?

The choice and quality of materials are very important to Bosch. These topics run through all categories of appliances and can be found in all product groups. The materials used in home appliances often also have to fulfil technical product characteristics, as they sometimes come into contact with food or are exposed to high temperatures.

Man schaut sich den Markt natürlich nicht nur aus der Perspektive des Wettbewerbs an, sondern immer auch mit Blick auf die langfristigen Entwicklungen. Vernetzung ist beispielsweise ein Stichwort, das zeigt, wo die Reise hingeht. Und insofern ist die Konkurrenz niemals die einzige Quelle der Inspiration. Wir sind regelmäßig als Trendscouts auf Messen. Dabei lernen wir auch von anderen Branchen wie beispielsweise der Automobilindustrie, wo wir es mit ähnlich langlebigen Produkten und Innovationszyklen zu tun haben wie im Hausgerätemarkt. Daneben beobachten wir die Entwicklungen in Interior Design, Architektur und dem Bereich der Consumer Electronics. Wir nehmen Einblick in die unterschiedlichsten Branchen und filtern die Innovationen danach, ob es sich um kurzfristige oder eher langfristige Trends handelt und wie sie die technische und gestalterische Entwicklung im Hausgerätemarkt beeinflussen. Wir wollen ja Produkte entwickeln, die nicht nur im Moment des Kaufs, sondern noch viele Jahre danach attraktiv sind.

Welche Themen werden die Branche der Hausgeräte in den kommenden Jahren besonders beeinflussen?

Es gibt einige Entwicklungen, die auch nicht erst seit gestern Einfluss auf die Hausgeräte-Branche nehmen. Unter dem Schlagwort „Internet der Dinge" gibt es ja bereits seit Jahren Visionen, die auf Messen vorgestellt wurden. Mit der Vernetzung der Geräte und der digitalen Übertragung von Daten werden diese Visionen heute Realität und nehmen ganz konkrete Formen an.

Inwieweit verändert das auch die Arbeit innerhalb Ihres Designteams?

Natürlich beeinflussen solche Entwicklungen auch unsere tägliche Arbeit als Gestalter. Wurde unser Team früher sehr stark von Industriedesignern geprägt, so verstärken wir uns heute beim Thema „User Interface Design". Die Gestaltung von Hausgeräten befasst sich zwar seit jeher mit der Gestaltung von Gebrauchsfunktionen und Bedienelementen, durch die Digitalisierung und Vernetzung verfolgen wir aber auch eigenständige Designkonzepte, die wiederum neue Formen der Bedienung und Benutzerführung mit sich bringen.

Zeigt sich auch hier eine generelle Entwicklung im Design von Hausgeräten?

Ja, was auch letztlich konsequent ist. Wenn wir noch einmal auf die generellen Entwicklungen und Einflussfaktoren zurückkommen, dann spielt die Veränderung der Wohnwelten eine wichtige Rolle. Die Küche hat sich durch offene Wohnkonzepte selbst zu einem Teil des Wohnraums entwickelt. Und das spiegelt sich natürlich auch in den Gerätekonzepten wider. Der Trend geht aktuell zu Gestaltungskonzepten, die sich gut in die neuen Wohnwelten einfügen. Integration überhaupt ist ein wichtiges Thema.

Inwieweit sind Materialien ein generelles Thema im Design von Hausgeräten?

Für Bosch haben Materialien und Materialqualität einen hohen Stellenwert. Sie ziehen sich durch alle Gerätekategorien und finden

As Global Design Director at Bosch, Robert Sachon likes to get stuck in. He is a designer and brand manager in one.

Als Global Design Director Bosch greift Robert Sachon auch selbst zum Zeichenstift. Er ist Designer und Markenmanager in einer Person.

As a result, part of our day-to-day work as designers when dealing with these materials also involves familiarising ourselves with the corresponding technologies for processing the materials.

In order to also express the values of the brand through the quality and processing of the materials?

Most definitely. In some cases, we take processing to the very limits of what is technically feasible, even though many consumers may not be aware of that at first. But even if it is not necessarily the first thing they see, they will notice it when using the products.

Design as a non-verbal means of communication?

Yes, consumer perception is also evolving. For example, this is where influences from the field of consumer electronics and mobile communication come to bear. The materials used in smartphones and tablets as well as their finishing quality change how consumers perceive products. After all, they hold the devices in their hands day after day, and that also makes them more discerning of quality in other product segments.

sich bei allen Produktgruppen. Die verwendeten Materialien müssen im Bereich der Hausgeräte vielfach auch technische Produkteigenschaften erfüllen, da sie teilweise mit Lebensmitteln in Kontakt kommen oder hohen Temperaturen ausgesetzt sind. Insofern gehört es im Umgang mit diesen Materialien auch zur täglichen Arbeit des Gestalters, sich mit entsprechenden Verarbeitungstechnologien auseinanderzusetzen.

Um über die Qualität und die Verarbeitung der Materialien auch die Werte der Marke zum Ausdruck zu bringen?

Definitiv. Dabei gehen wir bei der Verarbeitung teilweise bis an die Grenzen der technischen Machbarkeit, auch wenn es vielen Konsumenten im ersten Moment nicht bewusst sein mag. Aber selbst wenn sie es vielleicht nicht vordergründig wahrnehmen, spüren sie es doch im Umgang mit den Produkten.

Design als nonverbales Mittel der Kommunikation?

Ja, auch die Wahrnehmung der Konsumenten entwickelt sich weiter. Hier kommen beispielsweise die Einflüsse aus dem Bereich der Consumer Electronics und der mobilen Kommunikation zum Tragen. Die verwendeten Materialien im Bereich der Smartphones und Tablets sowie deren Verarbeitungsqualität verändern die Wahrnehmung der Konsumenten. Sie haben die Geräte ja täglich in der Hand, und das macht sie auch sensibler für Qualitäten in anderen Produktsegmenten.

So Bosch is also being forced to be innovative with its home appliances by other industries?

You could see it like that. As a team, we have to ask ourselves the question every day of how we can constantly tweak the quality ethos and the technical perfection of the Bosch brand in order to make the brand visible and tangible again and again. Such efforts include gaps and bending radii that have a major effect on development, production and quality management.

As head designer, are you allowed to have one topic that is particularly close to your heart?

For me that topic is cooking, because it involves appliances that make it possible to be creative with the product in a very immediate way.

And what are the questions in relation to cooking that interest you as a designer?

For example how to combine appliances, and the question of how these appliances relate to each other. What codes and what design elements are suitable to be transferred to other products or categories? That can be very helpful when developing a design DNA for the brand.

Do we even need to be able to cook nowadays?

While our mothers knew exactly how to handle their home appliances, the younger generation is perhaps more likely to use automatic programmes. As a result, today's appliances are geared to meet different user requirements. It was not until sensor technology and new digital display and operating technologies were developed that these options for use and operation became possible. The very question of how the new displays are designed and programmed has become a very exciting field for designers. Our team developed dedicated style guides for images and animations for this purpose that had not existed in that form beforehand.

Don't the technical possibilities automatically lead to a complexity of products that is maybe not even desirable?

The new operating and display technologies have simultaneously given rise to more possibilities when using the appliances. Part of our work is also to prevent the appliances from becoming too complex as a result of more operating possibilities. The whole point is that they should be simple and intuitive to use. Even complex technology must remain manageable. In a technology-based company like Bosch, we as designers also have to act as a control instance vis-à-vis the marketing or technical product development departments, because we keep the user in sight and design an important interface informed by user-focused concepts.

Bosch wird also auch durch andere Branchen zur Innovation im Bereich der Hausgeräte gezwungen?

Wenn man das so sehen will. Als Team müssen wir uns täglich mit der Frage auseinandersetzen, wie wir den Qualitätsgedanken und die technische Perfektion der Marke Bosch stetig nachschärfen können, um diese immer wieder sichtbar oder erfahrbar zu machen. Dazu gehören Spaltmaße und Biegeradien, die sich erheblich auf Entwicklung, Produktion und Qualitätsmanagement auswirken.

Darf man als Chefdesigner auch ein Thema haben, dass einem besonders am Herzen liegt?

Für mich ist es das Thema Kochen, weil man es mit Geräten zu tun hat, die es einem unmittelbar erlauben, kreativ mit dem Produkt umzugehen.

Und welche Fragen interessieren Sie als Designer beim Thema Kochen?

Da geht es beispielsweise um die Kombinierbarkeit von Geräten und die Frage, wie diese Geräte miteinander in Beziehung stehen. Welche Codes und welche gestalterischen Elemente eignen sich, um sie auch auf andere Produkte oder Kategorien zu übertragen? Das kann sehr hilfreich sein, um eine Design-DNA für die Marke zu entwickeln.

Muss man denn heute überhaupt noch kochen können?

Während unsere Mütter noch sehr genau wussten, wie sie mit ihren Hausgeräten umzugehen hatten, greift die jüngere Generation vielleicht eher auf Automatik-Programme zurück. Die Geräte werden also inzwischen unterschiedlichen Nutzeranforderungen gerecht. Diese Möglichkeiten im Gebrauch und in der Bedienung wurden erst durch Sensortechnik und neue digitale Anzeige- und Bedientechnologien eröffnet. Allein die Frage, wie die neuen Anzeigen und Displays gestaltet und bespielt werden, ist ein sehr spannendes Betätigungsfeld für Designer geworden. Unser Team entwickelte dafür eigene Style Guides für Bildwelten und Animationen, die es so vorher nicht gab.

Führen die technischen Möglichkeiten nicht automatisch zu einer Komplexität von Produkten, die vielleicht gar nicht wünschenswert ist?

Mit den neuen Bedien- und Anzeigetechniken wachsen zugleich die Möglichkeiten des Gebrauchs von Geräten. Ein Teil unserer Arbeit besteht auch darin zu verhindern, dass durch mehr Bedienmöglichkeiten die Geräte zu komplex werden. Sie sollen ja gerade einfach und intuitiv zu bedienen sein. Selbst komplexe Technik muss beherrschbar bleiben. In einem technisch geprägten Unternehmen wie Bosch haben wir als Designer also auch die Rolle eines Korrektivs gegenüber dem Marketing oder der technischen Produktentwicklung, weil wir den Benutzer im Blick behalten und durch nutzerorientierte Konzepte eine wichtige Schnittstelle gestalten.

As a designer, are you not always destined to have one foot in the present and one in the future?

We simply try to give the appliances the best that there is. And, depending on the product, we have to look far into the future. To this end, we have developed a dedicated process within the company which is known as "Vision Range". It is roughly comparable with the show cars and concept studies used in the automotive industry. We design an ideal future scenario in order to gear our brand, our products and our design to that scenario from a strategic perspective. This guarantees us a competitive lead, as the content can flow directly into future projects. Maybe that is one of the major advantages of being able to work for one company and with one team on a long-term basis. Because it gives us the freedom to look to the future, quite separately from the specific product.

Thank you for speaking with us, Mr Sachon.

Steht man als Designer nicht permanent mit einem Bein in der Gegenwart und mit dem anderen Bein in der Zukunft?

Wir versuchen einfach, den Geräten das Beste mitzugeben. Und je nach Produkt müssen wir weit vorausschauen. Wir haben dafür in unserem Hause einen eigenen Prozess entwickelt, den wir „Vision Range" nennen. Man kann das in etwa mit den Showcars und Konzeptstudien in der Automobilindustrie vergleichen. Wir entwerfen ein zukünftiges Idealbild, um unsere Marke, unsere Produkte und unser Design strategisch danach auszurichten. Das sichert uns einen Vorsprung, da die Inhalte direkt in künftige Projekte einfließen können. Vielleicht ist es einer der großen Vorzüge, langfristig in einem Unternehmen und mit einem Team arbeiten zu können. Denn es gibt uns die Freiheit, losgelöst vom konkreten Produkt, einen Blick in die Zukunft zu werfen.

Vielen Dank für das Gespräch, Herr Sachon.

The Red Dot: Design Team of the Year 2015 around Robert Sachon, Global Design Director Bosch and Helmut Kaiser, Head of Consumer Products Design Bosch.
Das Red Dot: Design Team of the Year 2015 um Robert Sachon, Global Design Director Bosch und Helmut Kaiser, Head of Consumer Products Design Bosch.

Red Dot: Best of the Best
The best designers of their category
Die besten Designer ihrer Kategorie

The designers of the Red Dot: Best of the Best
Only a few products in the Red Dot Design Award receive the "Red Dot: Best of the Best" accolade. In each category, the jury can assign this award to products of outstanding design quality and innovative achievement. Exploring new paths, these products are all exemplary in their design and oriented towards the future.

The following chapter introduces the people who have received one of these prestigious awards. It features the best designers and design teams of the year 2015 together with their products, revealing in interviews and statements what drives these designers and what design means to them.

Die Designer der Red Dot: Best of the Best
Nur sehr wenige Produkte im Red Dot Design Award erhalten die Auszeichnung „Red Dot: Best of the Best". Die Jury kann mit dieser Auszeichnung in jeder Kategorie Design von außerordentlicher Qualität und Innovationsleistung besonders hervorheben. In jeder Hinsicht vorbildlich gestaltet, beschreiten diese Produkte neue Wege und sind zukunftsweisend.

Das folgende Kapitel stellt die Menschen vor, die diese besondere Auszeichnung erhalten haben. Es zeigt die besten Designer und Designteams des Jahres 2015 zusammen mit ihren Produkten. In Interviews und Statements wird deutlich, was diese Designer bewegt und was ihnen Design bedeutet.

Matt Ling, Christian Koepf
Bosch Lawn and Garden Ltd.

"To integrate leading technology into established products providing tangible user benefits."

„Führende Technik in etablierte Produkte zu integrieren, um spürbare Vorteile für den Nutzer zu erzielen."

What do you particularly like about your own award-winning product?
The design and performance of this cordless product is a game changer versus petrol products, and this surprises users.

Is there a certain design approach that you pursue?
We started from a blank sheet of paper and had an extensive European user research programme the results of which we were able to build into our design. We used the design language of existing product segments in an established market and adapted to new technology in an evolutionary way.

How do you define design quality?
Simplicity of design – especially when components can perform more than one function.

Was gefällt Ihnen an Ihrem eigenen, ausgezeichneten Produkt besonders gut?
Die Gestaltung und Leistung dieses kabellosen Produkts stellen eine grundlegende Alternative zu benzinbetriebenen Geräten dar und das erstaunt Nutzer.

Gibt es einen bestimmten Gestaltungsansatz, den Sie verfolgen?
Wir haben mit einem leeren Blatt Papier angefangen und hatten ein umfassendes, europäisches Forschungsprogramm, dessen Ergebnisse wir dann in unser Gestaltungskonzept einbauen konnten. Wir haben die Gestaltungssprache bestehender Produktsegmente in einem etablierten Markt verwendet und uns der neuen Technik auf eine evolutionäre Weise angepasst.

Wie definieren Sie Designqualität?
Schlichtheit der Gestaltung, besonders, wenn Komponenten mehr als eine Funktion erfüllen können.

reddot award 2015
best of the best

Manufacturer
Bosch Lawn and Garden Ltd.,
Stowmarket, Great Britain

GRA 53
Professional Lawnmower
Professioneller Rasenmäher

See page 96
Siehe Seite 96

MIZSEI Design Team

"Contribute to an affluent lifestyle through water-related products."
„Mithilfe von Produkten, die mit Wasser zu tun haben, zu einem gehobenen
Lebensstil beitragen."

What do you particularly like about your own award-winning product?
That we removed all irregularities on body and handle and gave it a smart appearance.

Do you have a motto for life?
Take positive action and remain humble.

How do you define design quality?
The beautiful expression of something that is unconsciously felt, such as by touch, and the long-term durability of a safely handled manufacturing.

What do you see as being the biggest challenges in your industry at present?
Updating material and surface treatment methods due to the tightening of regulations relating to water quality standards. The resinification of a component in order to keep costs down and reduce weight.

Was gefällt Ihnen an Ihrem eigenen, ausgezeichneten Produkt besonders gut?
Dass wir alle Unebenheiten vom Körper und vom Griff beseitigt haben und dem Produkt eine elegante Erscheinung haben geben können.

Haben Sie ein Lebensmotto?
Positiv tätig sein und bescheiden bleiben.

Wie definieren Sie Designqualität?
Als Ausdruck der Schönheit von etwas unbewusst Gefühltem, etwa durch Berührung, und auch als langfristige Haltbarkeit durch eine sicher gehandhabte Fertigung.

Worin sehen Sie aktuell die größten Herausforderungen in Ihrer Branche?
Die Verarbeitungsmethoden für Materialien und Oberflächen auf den neuesten Stand zu bringen aufgrund der immer strenger werdenden Gesetze in Bezug auf Wasserqualitätsnormen. Das Verharzen von Bestandteilen, um Kosten zu senken und das Gewicht zu reduzieren.

reddot award 2015
best of the best

Manufacturer
MIZSEI MFG CO., LTD.,
Yamagata City, Gifu Prefecture, Japan

Sprinkle3D
Outdoor Tap
Außenwasserhahn

See page 102
Siehe Seite 102

Andreu Carulla
Andreu Carulla Studio

"We understand our design as a reflection of the Mediterranean way of life, creating products that are lively, bright and emotional by combining art and technology."

„Wir verstehen unser Design als Ausdruck des mediterranen Lebensstils und schaffen Produkte, die lebendig, fröhlich und emotional sind, indem wir Kunst mit Technik kombinieren."

What do you particularly like about your own award-winning product?
The simplicity and elegance of the solution; the fact that we have innovated such a well-known and highly evolved product as a sunshade, and its ease of use.

Is there a certain design approach that you pursue?
Not really, we try to reinvent ourselves in every project. Our goal is to find an adequate solution for our clients and for user needs.

What inspires you?
I am inspired by constant observation. I am a curious person, but overall, I am inspired by hard work. To quote Pablo Picasso, "Inspiration exists, but it has to find us working."

Do you have a motto for life?
Enjoy and don't waste your time.

Was gefällt Ihnen an Ihrem eigenen, ausgezeichneten Produkt besonders gut?
Die Einfachheit und Eleganz der Lösung. Die Tatsache, dass wir ein so bekanntes und hochentwickeltes Produkt wie den Sonnenschirm und seine Benutzerfreundlichkeit erneuert haben.

Gibt es einen bestimmten Gestaltungsansatz, den Sie verfolgen?
Nicht wirklich. Wir versuchen, uns in jedem Projekt neu zu erfinden. Unser Ziel ist es, für unsere Kunden und für die Bedürfnisse der Verbraucher eine passende Lösung zu finden.

Was inspiriert Sie?
Mich inspiriert die ständige Beobachtung. Ich bin neugierig. Aber im Allgemeinen inspiriert mich harte Arbeit. Wenn ich Pablo Picasso zitieren darf: „Inspiration existiert, aber sie muss dich bei der Arbeit finden."

Haben Sie ein Lebensmotto?
Genieße, aber vergeude nicht deine Zeit.

reddot award 2015
best of the best

Manufacturer
Calma, Iniciativa Exterior 3i,
Vilamalla, Spain

OM
Sunshade
Sonnenschirm

See page 118
Siehe Seite 118

Marec Hase, Paulo Mesquita
HASE BIKES

"Enjoy what you do, every day."
„Genieße jeden Tag das, was du tust."

What do you particularly like about your own award-winning product?
It feels like a cross between a mountain bike and an off-road go-kart – it is impossible not to feel a rush of excitement.

Is there a certain design approach that you pursue?
Exploring dynamic design inspired by organic forms. We like to take advantage of modern production processes to create technical parts with innovative designs.

Is there a designer that you particularly admire?
Raymond Loewy. He initiated a major revolution in product and graphic design and is still an inspiration for designers.

How do you define design quality?
Good design is the perfect balance between what people need and how people feel about the products.

Was gefällt Ihnen an Ihrem eigenen, ausgezeichneten Produkt besonders gut?
Es fühlt sich wie eine Mischung aus Mountainbike und Gelände-Gokart an – man kann gar nicht anders als begeistert sein.

Gibt es einen bestimmten Gestaltungsansatz, den Sie verfolgen?
Dynamische Gestaltung auszuloten, inspiriert von organischen Formen. Wir nutzen gerne moderne Fertigungsverfahren, um neu gestaltete technische Komponenten herzustellen.

Gibt es einen Designer, den Sie besonders schätzen?
Raymond Loewy. Er löste eine bedeutende Revolution im Produkt- und im Grafikdesign aus und ist noch immer eine Inspiration für Designer.

Wie definieren Sie Designqualität?
Gutes Design ist das perfekte Gleichgewicht zwischen dem, was Menschen brauchen, und dem, wie sie über die jeweiligen Produkte denken.

reddot award 2015
best of the best

Manufacturer
HASE BIKES, Waltrop, Germany

Kross
Trike with E-Engine
Trike mit E-Motor

See page 124
Siehe Seite 124

Valentin Vodev

"I tried to integrate my own intelligent solutions for details and innovations in the VELLO bike without neglecting design and lifestyle."

„Beim VELLO bike habe ich versucht, eigene intelligente Detaillösungen und
Innovationen zu integrieren, ohne dabei auf Design und Lifestyle zu verzichten."

Is there a certain design approach that you pursue?
I always try to take a novel, previously unexplored approach in my designs. This mostly results in unconventional solutions, which I can then turn into innovations. In my opinion, design is a mixture of logic, aesthetics and art.

What inspires you?
The world around me. For example, I developed the first prototypes of the VELLO bike for a trip to Cuba. I wanted to design a light bicycle that would make it easy to travel in buses or cars whilst discovering a country and its people at the same time.

Do you have a motto for life?
I try to go through life with my eyes wide open to recognise the potential of nature and its beauty. That is why I like opposites and seeing things in different ways.

Gibt es einen bestimmten Gestaltungs-ansatz, den Sie verfolgen?
In meinen Designs versuche ich immer, einen neuen und bis jetzt unerforschten Ansatz zu finden, so komme ich meistens auf unkonventionelle Lösungen, die ich dann in Innovationen umwandeln kann. Für mich ist Design eine Mischung aus Logik, Ästhetik und Kunst.

Was inspiriert Sie?
Die Umwelt um mich herum. Den ersten Prototypen des VELLO bikes habe ich z. B. für eine Reise nach Kuba entwickelt. Ich wollte ein leichtes Fahrrad gestalten, das es ermöglicht, problemlos in Bus oder Auto mitgenommen zu werden, während man gleichzeitig Land und Leute kennenlernt.

Haben Sie ein Lebensmotto?
Ich versuche, mit offenen Augen durchs Leben zu gehen und das Potenzial der Umwelt und ihre Schönheit wahrzunehmen. Deswegen mag ich Gegensätze und verschiedene Blickwinkel.

reddot award 2015
best of the best

Manufacturer
VELLO bike, Vienna, Austria

VELLO bike
Folding Bicycle
Faltrad

See page 132
Siehe Seite 132

R&D Campagnolo

"Pure design. Pure performance."

„Pures Design. Pure Leistung."

Is there a designer that you particularly admire?
Luigi Colani for his aerodynamic shapes and Marc Newson for the sense of freedom he is able to express in his works.

How do you define design quality?
Quality is the ability to produce while respecting project features. Quality in design is the ability to produce different thoughts – all of them innovative.

What do you see as being the biggest challenges in your industry at present?
Reducing the weight of components while increasing their performance.

What does winning the Red Dot: Best of the Best mean to you?
It means that we have been able to successfully express ourselves through our product. Don't wake me up, I'm dreaming!

Gibt es einen Designer, den Sie besonders schätzen?
Luigi Colani für seine aerodynamischen Formen und Marc Newson für den Freiheitssinn, den er in seiner Arbeit zum Ausdruck bringt.

Wie definieren Sie Designqualität?
Qualität ist die Fähigkeit, unter Berücksichtigung der Projekteigenschaften etwas herzustellen. Designqualität ist die Fähigkeit, verschiedene Gedanken hervorzubringen – allesamt neuartig.

Worin sehen Sie aktuell die größten Herausforderungen in Ihrer Branche?
Das Gewicht von Komponenten zu reduzieren und gleichzeitig ihre Leistung zu steigern.

Was bedeutet die Auszeichnung mit dem Red Dot: Best of the Best für Sie?
Es bedeutet, dass wir uns erfolgreich durch unser Produkt zum Ausdruck gebracht haben. Weck' mich nicht. Ich träume!

reddot award 2015
best of the best

Manufacturer
Campagnolo S.r.l., Vicenza, Italy

Super Record EPS 2015
Groupset for Road Bikes
Schaltgruppe für Rennräder

See page 140
Siehe Seite 140

Markus Hachmeyer – Schwalbe
Oliver Zuther – Syntace GmbH

"No gimmicks!"

What do you particularly like about your own award-winning product?
Procore is a product that has a significantly positive influence on the performance of a bike. This is combined with a tremendous increase in fun riding the bike – in our opinion, the most important point about this product.

Is there a certain design approach that you pursue?
The priority is always function, performance and durability of the product. That is why we make no or only marginal compromises in this respect.

What trends are currently having a particularly large influence on design?
Currently, the e-bike exerts the greatest influence of all on the bicycle industry. That not only includes the associated technology, but also the design of the wheels and components.

Was gefällt Ihnen an Ihrem eigenen, ausgezeichneten Produkt besonders gut?
Procore ist ein Produkt, das die Performance eines Bikes in hohem Maße positiv beeinflusst. Dies geht mit enormer Steigerung des Fahrspaßes einher, was uns bei diesem Produkt am allerwichtigsten ist.

Gibt es einen bestimmten Gestaltungsansatz, den Sie verfolgen?
Im Vordergrund stehen immer die Funktion, Performance und Langlebigkeit des Produktes. Daher gehen wir hier keine oder nur unwesentliche Kompromisse ein.

Welche Trends beeinflussen das Design zurzeit besonders stark?
Momentan ist es vor allem das E-Bike, welches die Fahrradbranche am stärksten prägt. Dies gilt für die entsprechende Technik wie auch das Design der Räder und Komponenten.

reddot award 2015
best of the best

Manufacturer
Schwalbe – Ralf Bohle GmbH,
Reichshof, Germany

Schwalbe Procore
Dual Chamber System
Doppelkammersystem

See page 142
Siehe Seite 142

Jasper Den Dekker
REV'IT! Sport International

"Challenge the status quo!"

„Hinterfrage den Status quo!"

Is there a certain design approach that you pursue?
Experience and knowledge are important tools but as a designer it is my duty to challenge myself and avoid repetition. This approach stimulates innovative thinking throughout the whole design process whilst allowing me to keep a keen eye on basic design principles.

What do you see as being the biggest challenges in your industry at present?
Within the motorcycle industry, safety is an ongoing theme. Strict regulations often interfere with the freedom we would like to have. The desire to innovate the core values that define a motorcycle product, while at the same time taking account of the strict regulations, make it a challenge to come up with something that is innovative.

Gibt es einen bestimmten Gestaltungs-ansatz, den Sie verfolgen?
Erfahrung und Wissen sind wichtige Hilfs-mittel, aber als Designer ist es meine Aufgabe, mich ständig kritisch zu hinter-fragen und Wiederholungen zu vermeiden. Dieser Ansatz regt während des gesamten Designprozesses innovatives Denken an und erlaubt mir, ein wachsames Auge auf die Grundregeln des Designs zu haben.

Worin sehen Sie aktuell die größten Herausforderungen in Ihrer Branche?
In der Motorradbranche ist Sicherheit ein ständiges Thema. Strenge Vorschriften be-einträchtigen unsere Gestaltungsfreiheit. Wenn wir sowohl den Kernwerten, die das Produkt Motorrad ausmachen, als auch den strengen Vorschriften Rechnung tragen wollen, ist es eine Herausforderung, sich etwas Neues einfallen zu lassen, das innovativ ist.

reddot award 2015
best of the best

Manufacturer
REV'IT! Sport International,
Oss, Netherlands

REV'IT! Seeflex
Motorcycle Limb Protection
Gliedmaßenprotektoren für Motorradfahrer

See page 166
Siehe Seite 166

Gyudeog Kim
Innus Korea

"Design by walking, think with your hands!"

„Gestalte beim Gehen. Denke mit den Händen!"

Is there a certain design approach that you pursue?
I continuously study humans, cultures and minutely observe objects so that I can turn an idea into design.

What inspires you?
Korean traditional architecture and furniture inspire me and provide solutions.

Is there a product that you have always dreamed about realising someday?
I'm dreaming of making high-tech shoes using Internet of Things technologies such as self-lighting shoes. Smart shoes that can stop children going missing.

How do you define quality?
Quality is a present for the consumer, similar to a gift for someone you love.

Gibt es einen bestimmten Gestaltungs-ansatz, den Sie verfolgen?
Ich beobachte kontinuierlich Menschen und Kulturen, und studiere Objekte bis ins Detail, um eine Idee in Design umwandeln zu können.

Was inspiriert Sie?
Traditionelle koreanische Architektur und Möbel dienen mir als Inspiration und bieten Lösungen.

Welches Produkt würden Sie gerne einmal realisieren?
Ich träume davon, Hightech-Schuhe mit Internet of Things-Technologie herzustellen, z. B. selbstleuchtende Schuhe. Clevere Schuhe, die verhindern können, dass Kinder verlorengehen.

Wie definieren Sie Qualität?
Qualität ist ein Geschenk an den Verbraucher, ähnlich einem Präsent für einen geliebten Menschen.

reddot award 2015
best of the best

Manufacturer
Innus Korea, Busan, South Korea

KI xtrap neo
Shoes
Schuhe

See page 174
Siehe Seite 174

45

Jake Lah
Helinox Inc

"Our guiding principle is that every design aspect should serve a clear functional purpose and should be conveyed intuitively without any unnecessary features."

„Unser Leitprinzip ist, dass jeder Designaspekt einen klaren, funktionalen Zweck erfüllt und dass dies ohne unnötige Extras intuitiv vermittelt werden sollte."

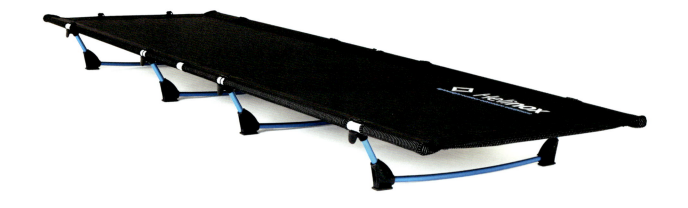

Is there a certain design approach that you pursue?
I try to listen to the voices of users first and build up wish lists for the respective product. Then I use a process of elimination, deciding what needs to be given up in order to achieve a certain key objective. In the end, I come up with the solution that I believe will fulfil the highest number of things on the wish list.

What inspires you?
The satisfied comments of people when they use products that I have designed.

What do you see as being the biggest challenges in your industry at present?
Products in the outdoor industry have become a mixture of fashion and outdoor gear, but the fashion element has become more and more dominant. However, outdoor kit cannot be rebuilt from scratch every single season and can only continue to improve by applying refinements year by year.

Gibt es einen bestimmten Gestaltungsansatz, den Sie verfolgen?
Ich versuche zuallererst auf die Anwender zu hören und stelle eine Wunschliste für das betreffende Produkt zusammen. Dann wende ich ein Ausschlussverfahren an und entscheide, was aufgegeben werden muss, um ein bestimmtes Ziel zu erreichen. Am Ende denke ich mir eine Lösung aus, von der ich glaube, dass sie die meisten Aspekte der Wunschliste erfüllt.

Was inspiriert Sie?
Die zufriedenen Kommentare von Menschen, wenn sie Produkte benutzen, die ich gestaltet habe.

Worin sehen Sie aktuell die größten Herausforderungen in Ihrer Branche?
Produkte für die Outdoor-Branche sind eine Mischung aus Mode- und Outdoor-Zubehör geworden, doch das Mode-Element wird immer wichtiger. Outdoor-Ausrüstung kann allerdings nicht jede Saison von Grund auf neu entworfen werden.

reddot award 2015
best of the best

Manufacturer
Helinox Inc, Incheon, South Korea

Lite Cot
Lightweight Foldable Cot
Leichte faltbare Liege

See page 190
Siehe Seite 190

RØDE New Product Development Team

"Design that works."
„Design, das funktioniert."

Is there a certain design approach that you pursue?
The RØDE innovation and design process begins with a statement; "what if…?". We aim to deliver a product with superlative performance that enhances the user experience with a sophisticated mechanical design combined with the iconic "RØDE Design DNA".

How do you define design quality?
Product quality ultimately comes down to fit for purpose and reliability. The actual and implicit functionality, combined with appropriate visual language for the product category, should convey and reinforce its nature and functional aspects.

What do you see as being the biggest challenges in your industry at present?
Working in business in a world without borders is a big challenge, as established rules of business have to be adapted at an ever quicker pace.

Gibt es einen bestimmten Gestaltungs-ansatz, den Sie verfolgen?
Der RØDE Innovations- und Gestaltungs-prozess beginnt mit dem Satz: „Was wäre, wenn …?" Unsere Absicht ist es, ein Produkt von überragender Qualität zu liefern, die das Anwendererlebnis durch eine raffinierte mechanische Gestaltung kombiniert mit der ikonischen „RØDE Design DNA" verbessert.

Wie definieren Sie Designqualität?
Letztlich geht es bei Produktqualität darum, ob das Produkt zweckmäßig und zuverläs-sig ist. Die tatsächliche und implizite Funk-tionalität in Verbindung mit der für die Produktkategorie passenden Formenspra-che sollte die Natur des Produkts und seine funktionalen Aspekte vermitteln und bekräftigen.

Welche Trends beeinflussen das Design zurzeit besonders stark?
In einer Welt ohne Grenzen zu arbeiten, ist eine große Herausforderung, da etablierte Geschäftsregeln immer schneller angepasst werden müssen.

reddot award 2015
best of the best

Manufacturer
RØDE Microphones, Sydney, Australia

NTR
Microphone
Mikrofon

See page 208
Siehe Seite 208

See page 208
Siehe Seite 208

Stan Spangenberg, Emile Kuenen
Milk Design B.V.

"Products designed around the human body, that of the parent and of the child."
„Produkte um den menschlichen Körper herum entwickelt, um den der Eltern und
 den des Kindes."

What inspires you?
We strongly believe the world needs to move towards a healthy and more sustainable way of living, and that corporations and business have the responsibility to lead the way. We are inspired to look for innovative solutions that contribute to this transition.

Is there a designer that you particularly admire?
We admire the Italian designer Alberto Meda, who is a design engineer. He is the type of designer who finds no pleasure in a project that doesn't contain something new. To this, we would like to add: to achieve something impossible that has never been done before.

Do you have a motto for life?
Never ever take no for an answer: always try to go beyond the limits.

Was inspiriert Sie?
Wir sind fest davon überzeugt, dass die Welt sich in Richtung einer gesünderen, nachhaltigeren Lebensweise entwickeln muss und dass Unternehmen und Geschäftswelt die Verantwortung haben, mit gutem Beispiel voranzugehen. Das animiert uns, neuartige Lösungen zu suchen, die etwas zu diesem Übergang beitragen können.

Gibt es einen Designer, den Sie besonders schätzen?
Wir bewundern den italienischen Designer Alberto Meda, der ein Designingenieur ist. Er ist die Art Designer, dem ein Projekt, das nichts Neues enthält, keine Freude macht. Wir würden dem hinzufügen: etwas Unmögliches zu erreichen, das nie vorher gemacht wurde!

Haben Sie ein Lebensmotto?
Nie Nein als Antwort zu akzeptieren. Immer zu versuchen, über die Grenzen hinwegzugehen.

reddot award 2015
best of the best

Manufacturer
Milk Design B.V., Amsterdam, Netherlands

Joolz Geo
Stroller
Kinderwagen

See page 222
Siehe Seite 222

Tore Vinje Brustad – Permafrost AS
Anders August Kittilsen, Hilde Angelfoss – Stokke AS
Andreas Murray – Permafrost AS

"We live by our vision: In the best interest of the child."

„Wir leben nach unserer Vision: Im besten Interesse des Kindes."

What do you particularly like about your own award-winning product?
Longevity! Our concept is a modular system of seating solutions that accommodates children right from birth up to about ten years. We have therefore given it a timeless aesthetic quality.

Is there a project that you have always dreamed about realising someday?
We are particularly fond of projects that bridge the gap between technology and art. Designers have an obligation to turn all the world's technological achievements into meaningful innovations in people's lives.

Is there a designer that you particularly admire?
One particular favourite is Naoto Fukasawa. We admire how he keeps simplifying and refining the shape and construction until he un-covers the bare essence of the product.

Was gefällt Ihnen an Ihrem eigenen, ausgezeichneten Produkt besonders gut?
Seine Langlebigkeit! Unser Konzept ist ein modulares System aus Sitzlösungen, das sich Kindern von Geburt an bis zum Alter von etwa zehn Jahren anpasst. Wir haben ihm deshalb eine zeitlose Ästhetik gegeben.

Welches Projekt würden Sie gerne einmal realisieren?
Wir sind sehr von Projekten angetan, die eine Brücke zwischen Technologie und Kunst schlagen. Designer sind dazu verpflichtet, all die technischen Errungenschaften der Welt in sinnvolle Innovationen für das tägliche Leben der Menschen zu verwandeln.

Gibt es einen Designer, den Sie besonders schätzen?
Einer unserer wirklichen Lieblingsdesigner ist Naoto Fukasawa. Wir bewundern, wie er die Form und die Struktur immer weiter vereinfacht und verfeinert, bis er die Quintessenz eines Produktes freilegt.

reddot award 2015
best of the best

Manufacturer
Stokke AS, Ålesund, Norway

Stokke® Steps™
High Chair
Hochstuhl

See page 236
Siehe Seite 236

Diego Recchia,
Samsonite Design & Development Team

"There's no poetry without method."

„Ohne Methodik gibt es keine Poesie."

What do you particularly like about your own award-winning product?
Its simple functional aesthetics.

Is there a certain design approach that you pursue?
Our aim is to design products that communicate something, that make things better, that make people think or laugh instead of designing products with a nice shape but without a soul.

What inspires you?
The greatest source of inspiration are humans and their behaviour.

Is there a designer that you particularly admire?
There is no one in particular, but in general, we admire people who design without technical skills just for necessity. That is the real soul of design: a design for a need.

Was gefällt Ihnen an Ihrem eigenen, ausgezeichneten Produkt besonders gut?
Seine einfache, funktionale Ästhetik.

Gibt es einen bestimmten Gestaltungsansatz, den Sie verfolgen?
Unser Ziel ist es, Produkte zu gestalten, die etwas kommunizieren, die Dinge verbessern, die Menschen zum Lachen bringen oder zum Denken anregen, statt Produkte, die zwar eine nette Form, aber keine Seele haben.

Was inspiriert Sie?
Die wichtigste Inspirationsquelle ist der Mensch und sein Verhalten.

Gibt es einen Designer, den Sie besonders schätzen?
Keinen bestimmten, aber im Allgemeinen bewundern wir Menschen, die aus einer Notwendigkeit heraus ohne technische Fachkenntnisse gestalten. Das ist die wirkliche Seele des Designs: gestalten, um ein Bedürfnis zu erfüllen.

reddot award 2015
best of the best

Manufacturer
Samsonite Europe NV,
Oudenaarde, Belgium

Lite-Biz
Cabin-Size Luggage
Kabinengepäck

See page 248
Siehe Seite 248

Chun-Chieh Wang

"Try for the Earth, try for a better life."
„Bemühe dich um die Erde, bemühe dich um ein besseres Leben."

Is there a certain design approach that you pursue?
To explore new possibilities of materials. The idea of Pháin is transforming a heavy industrial item into a hipster accessory. At the onset of the design, my main intention was to explore the sturdy materials of kraft paper and PP-woven fabric to find a novelty use for the bag other than its purpose of holding cement or chemical and industrial materials.

How do you define design quality?
The quality of design is not just about the perfect rendering of the craftsmanship, it is also all about the consistency shown right through – from the concept within to the form outside.

Gibt es einen bestimmten Gestaltungsansatz, den Sie verfolgen?
Das Erkunden neuer Möglichkeiten von Materialien. Die Idee von Pháin ist es, ein Objekt aus der Schwerindustrie in ein hippes Accessoire zu verwandeln. Zu Beginn der Gestaltung war es meine Hauptintention, die robusten Materialien Kraftpapier und PP-Gewebestoff zu erforschen, um eine neuartige Verwendung für den Sack zu finden – etwas anderes als seinen eigentlichen Zweck, Zement oder Chemikalien und Industriematerialien zu verwahren.

Wie definieren Sie Designqualität?
Bei der Qualität von Design geht es nicht nur darum, eine Handwerkskunst perfekt auszuüben, sondern auch darum, Konsequenz zu beweisen – vom inneren Konzept bis zur äußeren Form.

reddot award 2015
best of the best

Manufacturer
Ideoso Design Inc., Taipei, Taiwan

Pháin
Backpack Bag
Rucksacktasche

See page 260
Siehe Seite 260

Klemens Moeslinger, Sian Thomas, Horst Huebner, Claudia Roos, Kathrin Merz
Triumph International AG

"All women deserve a good bra."

„Alle Frauen verdienen einen guten BH."

Is there a certain design approach that you pursue?
Always push the boundaries and be first, but keep consumer excitement uncompromisingly at the heart of all our designs.

Do you have a motto for life?
To improve is to change; to be perfect is to change more often.

How do you define design quality?
Quality design is the aesthetic delivery of a function and a touch of personal luxury that is ultimately measured by fulfilling the consumer's aspirations.

What trends are currently having a particularly large influence on design?
The growth of social media gives us more access to customer feedback and new ways of manufacturing, such as 3D printing and fabric engineering, are challenging us to rethink our design process.

Gibt es einen bestimmten Gestaltungsansatz, den Sie verfolgen?
Immer die Grenzen auszudehnen und der Erste zu sein. Dabei aber stets die Begeisterung der Verbraucher im Mittelpunkt unserer Gestaltung zu bewahren.

Haben Sie ein Lebensmotto?
Verbesserung bedeutet Veränderung. Perfektion bedeutet häufige Veränderung.

Wie definieren Sie Designqualität?
Gestaltung von Qualität ist die ästhetische Erfüllung einer Funktion und ein Hauch von persönlichem Luxus. Letztendlich wird das an der Erfüllung der Erwartungen von Verbrauchern gemessen.

Welche Trends beeinflussen das Design zurzeit besonders stark?
Das Wachstum der sozialen Medien gibt uns einen besseren Zugang zu Kundenfeedback, und neue Herstellungsmöglichkeiten wie der 3D-Druck oder die Gewebetechnik fordern uns dazu auf, den Gestaltungsprozess zu überdenken.

red**dot** award 2015
best of the best

Manufacturer
Triumph Intertrade AG,
Bad Zurzach, Switzerland

Triumph Magic Wire
Bra
BH

See page 270
Siehe Seite 270

Raymond Lao
Raymond-L International Co., Ltd.

"Efficient without rushing."

„Effizienz ohne Hektik."

Is there a certain design approach that you pursue?
We aim at combining the natural, the environmentally friendly with our cultural legacy and with being innovative.

What inspires you?
Our inspiration comes from traditional oriental bamboo knitting skills. This inspiration challenges us to improve these skills through design and innovative technology.

Is there a designer that you particularly admire?
Master Alain Mikli. He is the leading eyewear designer in this industry.

How do you define design quality?
Being both designer and producer, quality is our first priority and also our reputation. We have standards and procedures to make sure only the best quality products are released in the market.

Gibt es einen bestimmten Gestaltungsansatz, den Sie verfolgen?
Wir bemühen uns, das Natürliche, das Umweltfreundliche mit unserem kulturellen Erbe und mit Innovation zu verbinden.

Was inspiriert Sie?
Unsere Inspiration waren traditionelle asiatische Bambus-Verbindungstechniken. Diese Inspiration stellte uns vor die Herausforderung, diese Kenntnisse durch Gestaltungskonzepte und neue Techniken zu vertiefen.

Gibt es einen Designer, den Sie besonders schätzen?
Den meisterhaften Alain Mikli. Er ist der führende Brillendesigner der Branche.

Wie definieren Sie Designqualität?
Sowohl Designer als auch Hersteller zu sein. Qualität ist dabei unsere oberste Priorität und macht unsere Reputation aus. Wir haben Standards und Verfahrensweisen, die sicherstellen, dass nur die hochwertigsten Produkte auf den Markt kommen.

reddot award 2015
best of the best

Manufacturer
Raymond-L International Co., Ltd.,
Taipei, Taiwan

Bamboo Eyewear
Brillen aus Bambus

See page 274
Siehe Seite 274

Maximilian Büsser
MB&F

"A creative adult is a child who survived."

„Ein kreativer Erwachsener ist ein Kind, das überlebt hat."

Is there a certain design approach that you pursue?
My designs are as much my autobiography as my psychotherapy.

What inspires you?
Whatever makes my heart beat faster. In a world of pre-packaged, hyper-marketed goods, it does not happen that often.

Is there a designer that you particularly admire?
All those who had the guts to break free from their era.

What do you see as being the biggest challenges in your industry at present?
Resisting the dictatorship of shareholder value and marketers.

What does winning the Red Dot: Best of the Best mean to you?
An enormous surprise, a great honour and acknowledgement of our team's formidable work.

Gibt es einen bestimmten Gestaltungsansatz, den Sie verfolgen?
Meine Designs sind genauso autobiographisch wie sie psychotherapeutisch sind.

Was inspiriert Sie?
Alles, was mein Herz schneller schlagen lässt. In einer Welt voller abgepackter, übervermarkteter Waren kommt das nicht häufig vor.

Gibt es einen Designer, den Sie besonders schätzen?
Alle, die den Mut hatten, aus ihrer Epoche auszubrechen.

Worin sehen Sie aktuell die größten Herausforderungen in Ihrer Branche?
Der Diktatur von Shareholder-Value und Vermarktern zu widerstehen.

Was bedeutet die Auszeichnung mit dem Red Dot: Best of the Best für Sie?
Eine enorme Überraschung, eine große Ehre und Anerkennung für die beeindruckende Arbeit, die unser Team geleistet hat.

reddot award 2015
best of the best

Manufacturer
MB&F, Geneva, Switzerland

Horological Machine No. 6 "Space Pirate"
Watch
Uhr

See page 292
Siehe Seite 292

Martin Kaufmann, Björn Harms, Ulla Kaufmann
Ulla + Martin Kaufmann

"To develop something so simple that it attains a self-evident ease of use."

„Leichtigkeit zu entwickeln, die im Gebrauch eine selbstverständliche Einfachheit erreicht."

What do you particularly like about your own award-winning product?
The translation of gold as a material into tangible flexibility and the taming of this interplay by means of a simple button.

Is there a certain design approach that you pursue?
To keep exploring the simple golden band with all its versatility and to bring it into harmony with its wearer.

Is there a designer that you particularly admire?
Donald Judd. He managed to express great things with limited means.

What trends are currently having a particularly large influence on design?
The willingness of society to change and exchange things is having a massive influence on the short shelf-life of products.

Was gefällt Ihnen an Ihrem eigenen, ausgezeichneten Produkt besonders gut?
Das Umwandeln des Materials Gold in erfahrbare Flexibilität und das Bändigen dieses Spiels durch einen simplen Knopf.

Gibt es einen bestimmten Gestaltungsansatz, den Sie verfolgen?
Das schlichte goldene Band mit all seiner Vielfältigkeit immer wieder neu auszuloten und es in Harmonie mit dem Träger zu bringen.

Gibt es einen Designer, den Sie besonders schätzen?
Donald Judd. Er hat es geschafft, Großes mit wenigen Mitteln auszudrücken.

Welche Trends beeinflussen das Design zurzeit besonders stark?
Die Bereitschaft der Gesellschaft zum Wechseln und Austauschen beeinflusst die Schnelllebigkeit der Produkte massiv.

reddot award 2015
best of the best

Manufacturer
Ulla + Martin Kaufmann,
Hildesheim, Germany

Tira
Bangle
Armreif

See page 302
Siehe Seite 302

Gorden Wagener
Daimler AG

"Work hard, play hard."

Is there a certain design approach that you pursue?
Our work centres on sensual purity as the definition of modern luxury.

What inspires you?
Inspiration can be found everywhere and the best ideas generally come spontaneously. I gather a lot of ideas while travelling, but also find them in everyday life – in nature, architecture, art and fashion. As a designer, I think about my work all the time. I live design, because as designers we are shaping part of the future.

What trends are currently having a particularly large influence on design?
We are experiencing a revolution in digitalisation and networking which will affect the significance of cars. They will grow beyond their role as a means of transport and will evolve into a private area of seclusion.

Gibt es einen bestimmten Gestaltungsansatz, den Sie verfolgen?
Im Mittelpunkt unseres Schaffens steht sinnliche Klarheit als Definition eines modernen Luxus.

Was inspiriert Sie?
Inspiration ist überall und die besten Ideen kommen meist spontan. Ich sammle viele Impulse auf Reisen, aber auch im Alltag – in der Natur, Architektur, Kunst oder Mode. Als Designer denke ich rund um die Uhr über meine Arbeit nach, ich lebe Design. Denn wir als Designer gestalten ein Stück der Zukunft.

Welche Trends beeinflussen das Design zurzeit besonders stark?
Wir erleben eine Revolution aus Digitalisierung und Vernetzung, wodurch sich die Bedeutung des Autofahrens verändern wird. Das Auto wächst über seine Rolle als Transportmittel hinaus und wandelt sich zum vernetzten Mobilort.

reddot award 2015
best of the best

Manufacturer
Daimler AG, Stuttgart, Germany

Mercedes-AMG GT
Sports Car
Sportwagen

See page 312
Siehe Seite 312

Kevin Rice
Mazda Motor Europe

"I have always been inspired by pure beauty that tells a clear visual story."

„Reine Schönheit, die eine eindeutige visuelle Geschichte erzählt, hat mich schon immer inspiriert."

Is there a certain design approach that you pursue?
At Mazda, our KODO design philosophy guides us as we capture the visual energy of an animal or athlete about to spring into action and then release that energy into the body of the car.

What trends are currently having a particularly large influence on design?
Connecting the driver to the car is currently the major influence on car design. This focuses primarily on electronics, the human-machine interface and social connectivity. In addition, we believe there should be a seamless physical connection between the driver and the car. Jinba Ittai, "horse and rider as one body", describes this connection and the way the driver feels like he or she is actually part of the car as it moves.

Gibt es einen bestimmten Gestaltungsansatz, den Sie verfolgen?
Bei Mazda leitet uns unsere KODO Design-Philosophie. Wir fangen die visuelle Energie eines zum Sprung ansetzenden Tieres oder Athleten ein und übertragen diese Energie auf die Karosserie.

Welche Trends beeinflussen das Design zurzeit besonders stark?
Den Fahrer mit dem Auto zu verbinden, ist derzeit der Haupttrend im Automobildesign. Der Schwerpunkt liegt hauptsächlich auf Elektronik, der Mensch-Maschine-Schnittstelle und auf sozialer Konnektivität. Außerdem glauben wir, dass es eine nahtlose physische Verbindung zwischen Fahrer und Fahrzeug geben sollte. Jinba Ittai, die „Einheit von Pferd und Reiter", beschreibt diese Verbindung und das Gefühl des Fahrers, eins mit dem Fahrzeug zu sein.

reddot award 2015
best of the best

Manufacturer
Mazda Motor Corporation,
Hiroshima, Japan

Mazda MX-5
Roadster

See page 314
Siehe Seite 314

Flavio Manzoni
Ferrari Design

"The future is always in the hands of those who can foresee it."

„Die Zukunft liegt immer in den Händen derer, die sie vorhersehen können."

Enzo Ferrari

Is there a certain design approach that you pursue?
Designing a new Ferrari calls for indispensable elements such as audacity, sensitivity, and the visionary ability that is paramount in anticipating natural evolution. Another fundamental prerequisite is the relentless quest for a design that should translate the technical value of the product into an artistic shape.

Is there a project that you have always dreamed about realising someday?
As an architect, there are many objects that inspire me, "from the spoon to the city", which one day I would like to interpret. But time is a limit with which we have to measure every day... Today, I am fortunate enough to be able to design Ferraris, the most beautiful and technologically advanced cars in the world. And that has been my dream ever since I was a child.

Gibt es einen bestimmten Gestaltungsansatz, den Sie verfolgen?
Die Gestaltung eines neuen Ferraris erfordert unabdingbare Elemente wie z. B. Wagemut und Feingefühl sowie das visionäre Geschick, das vorrangig ist, wenn es darum geht, eine natürliche Entwicklung zu antizipieren.

Welches Projekt würden Sie gerne einmal realisieren?
Als Architekt gibt es viele Objekte, die mich inspirieren, „vom Löffel bis hin zur Stadt", die ich eines Tages gerne einmal interpretieren würde. Aber Zeit ist eine Grenze, die wir jeden Tag vermessen ... Heute habe ich das Glück, Ferraris – die schönsten und technisch fortschrittlichsten Automobile der Welt – gestalten zu dürfen. Und als Kind war das mein Traum.

reddot award 2015
best of the best

Manufacturer
Ferrari SpA, Maranello (Modena), Italy

Ferrari FXX K
Sports Car
Sportwagen

See page 316
Siehe Seite 316

Thomas Ingenlath and Volvo Design Team

"Don't jump around. Put all the effort into the idea, the sketch, the subject you selected. Don't stop till it is perfect."

„Sei nicht sprunghaft, sondern konzentriere deine Bemühungen auf die Idee, die Skizze, das Thema, für das du dich entschieden hast. Hör nicht auf, bis es perfekt ist."

What do you particularly like about your own award-winning product?
Its stately presence on the road and the interior that delivers a true Scandinavian flair.

Do you have a motto for life?
The advice of Hartmut Warkuß: When you are at the top, remember the next thing is the way down. When you are down there, remember the next thing is the way back up again.

How do you define quality?
Quality is true to its purpose, function and brand. And it carries an element of imagination in it that goes beyond the before mentioned.

What do you see as being the biggest challenges in your industry at present?
Cutting carbon dioxide emissions.

Was gefällt Ihnen an Ihrem eigenen, ausgezeichneten Produkt besonders gut?
Seine imposante Ausstrahlung auf der Straße und seine Innenausstattung, die echtes skandinavisches Flair vermittelt.

Haben Sie ein Lebensmotto?
Den Rat von Hartmut Warkuß: Wenn Du ganz oben bist, vergiss nicht, dass der nächste Schritt bergab geht. Wenn Du unten bist, vergiss nicht, dass der nächste Schritt wieder nach oben führt.

Wie definieren Sie Qualität?
Qualität ist ihrem Zweck, ihrer Funktion und ihrer Marke treu. Und sie enthält einen Funken Phantasie, der über das zuvor Genannte hinausgeht.

Worin sehen Sie aktuell die größten Herausforderungen in Ihrer Branche?
Kohlendioxid-Emissionen reduzieren.

reddot award 2015
best of the best

Manufacturer
Volvo Car Group, Gothenburg, Sweden

Volvo XC90
SUV

See page 334
Siehe Seite 334

Toshiyuki Yasunaga – Yamaha Motor Co., Ltd.
Kazumasa Sasanami – GK Dynamics Incorporated

"Enrich people's lives with beauty. Beauty makes sense and enthuses people."

„Das Leben von Menschen mit Schönheit bereichern. Schönheit ist sinnhaft und begeistert Menschen."

What do you particularly like about your own award-winning product?
It has a totally new value, created by fusing simple yet aggressive styling with functionality. Besides, it is well balanced in terms of performance, usability and eco-friendliness. As for the styling, it has an attractively organic and sculptural expression, as if inviting you to ride.

What inspires you?
We always try to keep our eyes wide open for new events, things and information, trying to create a unique trend, capturing the zeitgeist and predicting the future.

How do you define design quality?
Firstly, beauty and user-friendliness; secondly, offering a new value; thirdly, longevity and fourthly, creating great excitement.

Was gefällt Ihnen an Ihrem eigenen, ausgezeichneten Produkt besonders gut?
Es stellt einen völlig neuen Wert dar, der auf der Fusion eines einfachen, aber aggressiven Stylings mit Funktionalität beruht. Außerdem ist es gut ausgewogen in Bezug auf Leistung, Brauchbarkeit und Umweltfreundlichkeit. Was das Styling angeht, ist es ansprechend organisch und plastisch, sodass es auf den Fahrer sehr einladend wirkt.

Was inspiriert Sie?
Wir versuchen immer, für neue Events, Dinge und Informationen die Augen offen zu halten, und bemühen uns, einen einzigartigen Trend ins Leben zu rufen, der den Zeitgeist erfasst und die Zukunft voraussagt.

Wie definieren Sie Designqualität?
Erstens als Schönheit und Nutzerfreundlichkeit. Zweitens als das Angebot eines neuen Wertes. Drittens als langlebig und viertens als etwas, das große Begeisterung auslöst.

red**dot** award 2015
best of the best

Manufacturer
Yamaha Motor Co., Ltd., Shizuoka, Japan

MT-07
Motorcycle
Motorrad

See page 344
Siehe Seite 344

David Bowler – DB Industrial Design
Leonard Huissoon – Diverto Technologies BV

"Keep an open mind!"

„Sei stets für alles offen!"

Is there a certain design approach that you pursue?
We always try to find the right balance between functional and visual dynamics. At the same time, it's important to give a product personality.

What trends are currently having a particularly large influence on design?
The tightening of legislative requirements for noise, emissions and safety. The demand for more powerful and more productive solutions enforces bigger jumps in product evolution.

What does winning the Red Dot: Best of the Best mean to you?
It's like we were flying low, trying to stealth over conservative industry mountains. Then the Red Dot jury flies by, gives us a cocktail of biokerosine and Red Bull and hits our booster button!

Gibt es einen bestimmten Gestaltungs-ansatz, den Sie verfolgen?
Wir versuchen immer, das richtige Gleichgewicht zwischen der funktionalen und der visuellen Dynamik zu finden. Gleichzeitig ist es wichtig, einem Produkt Persönlichkeit zu geben.

Welche Trends beeinflussen das Design zurzeit besonders stark?
Die Verschärfung von gesetzlichen Auflagen in Bezug auf Lärm, Emissionen und Sicherheit. Die Nachfrage nach leistungsstärkeren und produktiveren Lösungen erzwingt größere Sprünge in der Produktentwicklung.

Was bedeutet die Auszeichnung mit dem Red Dot: Best of the Best für Sie?
Es ist so, als ob wir tief flögen und versuchten, heimlich über konservative Industrieberge zu kommen. Dann fliegt plötzlich die Red Dot-Jury vorbei, gibt uns einen Cocktail aus Biokerosin und Red Bull und drückt auf den Booster-Knopf!

reddot award 2015
best of the best

Manufacturer
Diverto Technologies BV,
Wemeldinge, Netherlands

Diverto QS 100
Tractor, Excavator, Loader, Mower
Traktor, Bagger, Radlader, Mäher

See page 356
Siehe Seite 356

Ponsse Oyj
LINK Design and Development Oy

"The development of forest machines is increasingly moving towards putting operator comfort at the centre of design alongside productivity."

„Bei der Entwicklung von Forstmaschinen steht zunehmend der Fahrerkomfort, neben der Produktivität, im Mittelpunkt der Gestaltung."

reddot award 2015
best of the best

Manufacturer
Ponsse Oyj, Vieremä, Finland

Ponsse Scorpion
Forest Machine
Forstmaschine

See page 358
Siehe Seite 358

Guido Kellermann, Johannes Leugers
Kellermann GmbH

"To create timeless design."

„Zeitloses Design schaffen."

Is there a certain design approach that you pursue?
Our products stand for simple, clear and unfussy design. And light is meant to appear as part of the design.

What inspires you?
Almost always organic shapes found in nature. We are convinced that the use of natural forms should also find expression in our products.

How do you define design quality?
When design creates new values and, in doing so, also acts as a counterpoint to today's disposable society.

What do you see as being the biggest challenges in your industry at present?
It is quite exciting to create something new despite the large number of familiar shapes. To recall traditions, but nonetheless dare to work on developing a consistent design – we believe that is the way to go.

Gibt es einen bestimmten Gestaltungsansatz, den Sie verfolgen?
Unsere Produkte stehen für eine einfache, klare und schnörkellose Gestaltung. Und das Licht soll dabei als Bestandteil des Designs wirken.

Was inspiriert Sie?
Fast immer organische Formen aus der Natur. Wir sind davon überzeugt, dass die natürliche Formensprache auch in unseren Produkten zum Ausdruck kommen muss.

Wie definieren Sie Designqualität?
Wenn Design neue Werte schafft und damit auch als Gegenpol zur gegenwärtigen Wegwerfgesellschaft fungiert.

Worin sehen Sie aktuell die größten Herausforderungen in Ihrer Branche?
Es ist durchaus spannend, bei der Vielzahl der bekannten Formen, immer wieder etwas Neues zu schaffen. Sich auf die Tradition besinnen und trotzdem mutig an der Designsprache arbeiten – von diesem Weg sind wir überzeugt.

reddot award 2015
best of the best

Manufacturer
Kellermann GmbH, Aachen, Germany

Bullet 1000
Motorcycle Lamps
Motorradleuchten

See page 366
Siehe Seite 366

Audi Light Design Team
Automotive Lighting Reutlingen GmbH

"Light has no weight, but an incredible amount of importance."
„Licht hat kein Gewicht, aber eine sehr große Bedeutung."

Is there a certain design approach that you pursue?
With long lines and sharp accents, an Audi light has to be associated with an Audi at first glance, even from a distance. When getting closer to the car, it should become clear right away which Audi model it is.

Is there a designer that you particularly admire?
Bruno Munari because of his eye for balanced proportions and his sensibility for creating beauty.

What does winning the Red Dot: Best of the Best mean to you?
The Audi Light Design Team and Automotive Lighting are very proud to have created a headlamp that summarises our aesthetic and innovative brand values in a manner that has found approval by one of the most important design awards in the world.

Gibt es einen bestimmten Gestaltungsansatz, den Sie verfolgen?
Mit langen Linien und scharfen Akzenten muss eine Audi-Leuchte auf den ersten Blick auch auf Entfernung mit einem Audi in Verbindung gebracht werden. Wenn man näher an den Wagen herankommt, sollte sofort klar sein, welches Audi-Modell es ist.

Gibt es einen Designer, den Sie besonders schätzen?
Bruno Munari, weil er ein Auge für ausgewogene Proportionen hat und mit Sensibilität Schönheit schafft.

Was bedeutet die Auszeichnung mit dem Red Dot: Best of the Best für Sie?
Das Team des Audi Lichtdesigns und Automotive Lighting sind sehr stolz darauf, einen Scheinwerfer entwickelt zu haben, der unsere ästhetischen und innovativen Markenwerte auf eine Weise zusammenfasst, die bei einer der bedeutendsten Designauszeichnungen der Welt auf Anerkennung gestoßen ist.

reddot award 2015
best of the best

Manufacturer
Automotive Lighting Reutlingen GmbH,
Reutlingen, Germany

Headlamp for Audi TT3
Scheinwerfer für den Audi TT3

See page 368
Siehe Seite 368

Ray Chen International

"As a loyal minimalist, my design philosophy is: If one line is sufficient, never use two."

„Als getreuer Minimalist lautet meine Designphilosophie: Wenn ein Strich genügt, verwende niemals zwei."

Is there a certain design approach that you pursue?
I pursue the aesthetics of the Song Dynasty.

What inspires you?
The ancient implements at the National Palace Museum, Taipei.

Is there a project that you have always dreamed about realising someday?
To create a stage design for an opera.

Is there a designer that you particularly admire?
Le Corbusier. Full of creativity, he was an artisan more than an architect.

What do you see as being the biggest challenges in your industry at present?
To provide this high-tech age with a human touch.

What trends are currently having a particularly large influence on design?
Eco-friendliness.

Gibt es einen bestimmten Gestaltungsansatz, den Sie verfolgen?
Ich folge der Ästhetik der Song-Dynastie.

Was inspiriert Sie?
Die antiken Werkzeuge im National Palace Museum in Taipei.

Welches Projekt würden Sie gerne einmal realisieren?
Das Bühnenbild für eine Oper.

Gibt es einen Designer, den Sie besonders schätzen?
Le Corbusier. Erfüllt von Kreativität, war er mehr Kunsthandwerker als ein Architekt.

Worin sehen Sie aktuell die größten Herausforderungen in Ihrer Branche?
Darin, diesem Hightech-Zeitalter eine menschliche Note zu verleihen.

Welche Trends beeinflussen das Design zurzeit besonders stark?
Umweltfreundlichkeit.

reddot award 2015
best of the best

Manufacturer
B/E Aerospace, Winston-Salem,
North Carolina, USA

China Airlines Premium Business Class
Aircraft Interior
Flugzeug-Inneneinrichtung

See page 384
Siehe Seite 384

Audi Industrial Design München

"Rendering 'Vorsprung durch Technik' visible."
„Vorsprung durch Technik sichtbar machen."

What do you particularly like about your own award-winning product?
The DNA of both brands of Leica and Audi, becomes equally apparent through reduced design, precise workmanship, high-quality materials and ease of use.

Is there a certain design approach that you pursue?
We enter into a symbiotic cooperation with our partners and find a way to combine their approach with that of Audi.

What do you see as being the biggest challenges in your industry at present?
There is such an incredible amount of design. What matters therefore is the differentiation to other products and brands in the market.

What trends are currently having a particularly large influence on design?
We prefer consistency to so-called trends. That gives customers a clear direction.

Was gefällt Ihnen an Ihrem eigenen, ausgezeichneten Produkt besonders gut?
Die Genetik der Marken Leica und Audi ist durch reduziertes Design, präzise Verarbeitung, hochwertige Materialien und eine einfache Bedienung deutlich sichtbar.

Gibt es einen bestimmten Gestaltungsansatz, den Sie verfolgen?
Wir gehen mit unseren Partnern eine symbiotische Kooperation ein und finden einen Weg, ihren Ansatz mit dem von Audi zu verbinden.

Worin sehen Sie aktuell die größten Herausforderungen in Ihrer Branche?
Es gibt so unglaublich viel Design. Entscheidend ist deshalb die Differenzierung zu anderen Produkten und Marken am Markt.

Welche Trends beeinflussen das Design zurzeit besonders stark?
Wir ziehen Konstanz den sogenannten Trends vor. Das gibt dem Kunden Orientierung.

reddot award 2015
best of the best

Manufacturer
Leica Camera AG, Wetzlar, Germany

Leica T
Camera
Kamera

See page 394
Siehe Seite 394

Byung-Mu Huh, Min-Ji Seo,
Young Kyoung Kim, Sang-Ik Lee
LG Electronics Inc.

"Design is a process that only makes sense when it is based on humanity
and returns to the user."

„Gestaltung ist ein Prozess, der nur Sinn ergibt, wenn er auf Menschlichkeit
beruht und zum Nutzer zurückkehrt."

Is there a certain design approach that you pursue?
We observe people and their behaviour consciously and unconsciously. Through close observation, design becomes based on an understanding of human kind and is able to create products that take user needs into account.

How do you define design quality?
The quality of design lies not only in an aesthetic outward appearance but also in a proper and elaborate finish that enhances the quality and value of a product.

What do you see as being the biggest challenges in your industry at present?
It is all about endless change. Rather than insisting on things of the past and developing a tolerance, one must keep changing constantly, in order to elicit new paradigms.

Gibt es einen bestimmten Gestaltungs-ansatz, den Sie verfolgen?
Wir beobachten Menschen und ihre Verhaltensweisen bewusst und unbewusst. Dadurch kann Gestaltung auf einem Verständnis der Menschen aufbauen und in Produkten zum Vorschein kommen, deren Form die Nutzerbedürfnisse berücksichtigt.

Wie definieren Sie Designqualität?
Die Qualität von Design besteht nicht nur aus einer ästhetischen äußeren Erscheinung, sondern auch aus der richtigen und aufwendigen Ausführung, welche die Qualität und den Wert des Produkts hervorhebt.

Worin sehen Sie aktuell die größten Herausforderungen in Ihrer Branche?
Es dreht sich alles um Veränderung. Anstatt auf Dinge aus der Vergangenheit zu bestehen und Toleranz dafür zu entwickeln, muss man sich ständig verändern, um neue Paradigmen auszulösen.

reddot award 2015
best of the best

Manufacturer
LG Electronics Inc., Seoul, South Korea

77EG9900
OLED TV

See page 422
Siehe Seite 422

Yusuke Tsujita
Sony Corporation

"Have passion and confidence."

„Leidenschaft und Zuversicht haben."

What do you particularly like about your own award-winning product?
It is functional but looks like furniture and harmonises with its surroundings.

Is there a certain design approach that you pursue?
Pursuing ideas that are pure, simple and compelling, I push the limit by leaving out unnecessary elements.

Is there a designer that you particularly admire?
I particularly appreciate the work of Konstantin Grcic, because it is based on clear thinking, diligent testing and careful refinement.

How do you define design quality?
For me, design quality means the way in which design presents a beautiful, optimal solution.

Was gefällt Ihnen an Ihrem eigenen, ausgezeichneten Produkt besonders gut?
Es ist funktional, sieht aber wie ein Möbelstück aus und harmoniert mit seiner Umgebung.

Gibt es einen bestimmten Gestaltungsansatz, den Sie verfolgen?
Indem ich Ideen verfolge, die rein, einfach und fesselnd sind, gehe ich beim Weglassen unnötiger Elemente bis an die Grenzen des Möglichen.

Gibt es einen Designer, den Sie besonders schätzen?
Ich mag die Arbeit von Konstantin Grcic sehr, da sie auf klarem Denken, gewissenhaften Untersuchungen und vorsichtigen Verbesserungen beruht.

Wie definieren Sie Designqualität?
Für mich bedeutet Designqualität die Art, in der Design eine schöne, optimale Lösung liefert.

reddot award 2015
best of the best

Manufacturer
Sony Corporation, Tokyo, Japan

LSPX-W1S
4K Ultra Short Throw Projector
4K-Ultrakurzdistanz-Projektor

See page 432
Siehe Seite 432

Erwin Weitgasser – Zeug Design Ges.m.b.H.
Ferdinand Maier – ruwido austria gmbh
Detlev Magerer – Zeug Design Ges.m.b.H.

"Design is soul."
„Design ist Seele."

How do you define design quality?
If the design's attitude and emotional statement are fully respected during the manufacturing process, and the resulting product comes out as planned without compromises.

Is there a designer that you particularly admire?
Luigi Colani. Less for his aesthetic abilities than for his principle of open communication, which we all value.

What does winning the Red Dot: Best of the Best mean to you?
We are tremendously proud of this award. It is a confirmation of the consistent way in which the guiding principle is applied: "Design is soul." Receiving the Red Dot: Best of the Best for a remote control shows how important this input device will continue to be in the future for interactive television as a technology and brand carrier.

Wie definieren Sie Designqualität?
Wenn die Haltung und emotionale Aussage des Designs bei der Umsetzung in allen Details respektiert wird und sich damit das Produkt ohne Kompromisse so darstellt wie geplant.

Gibt es einen Designer, den Sie besonders schätzen?
Luigi Colani, weniger wegen seiner ästhetischen Kompetenz, vielmehr schätzen wir sein offenes Kommunikationsprinzip.

Was bedeutet die Auszeichnung mit dem Red Dot: Best of the Best für Sie?
Wir sind ausgesprochen stolz auf diese Auszeichnung. Es ist die Bestätigung der Konsequenz in der Leitbildumsetzung „Design ist Seele". Die Auszeichnung „Red Dot: Best of the Best" für eine Fernbedienung zu erhalten zeigt, wie wichtig dieses Eingabegerät auch in Zukunft als Technologie- und Markenträger für interaktives Fernsehen ist.

reddot award 2015
best of the best

Manufacturer
ruwido austria gmbh, Neumarkt, Austria

leaf
Remote Control
Fernbedienung

See page 434
Siehe Seite 434

Garden
Garten

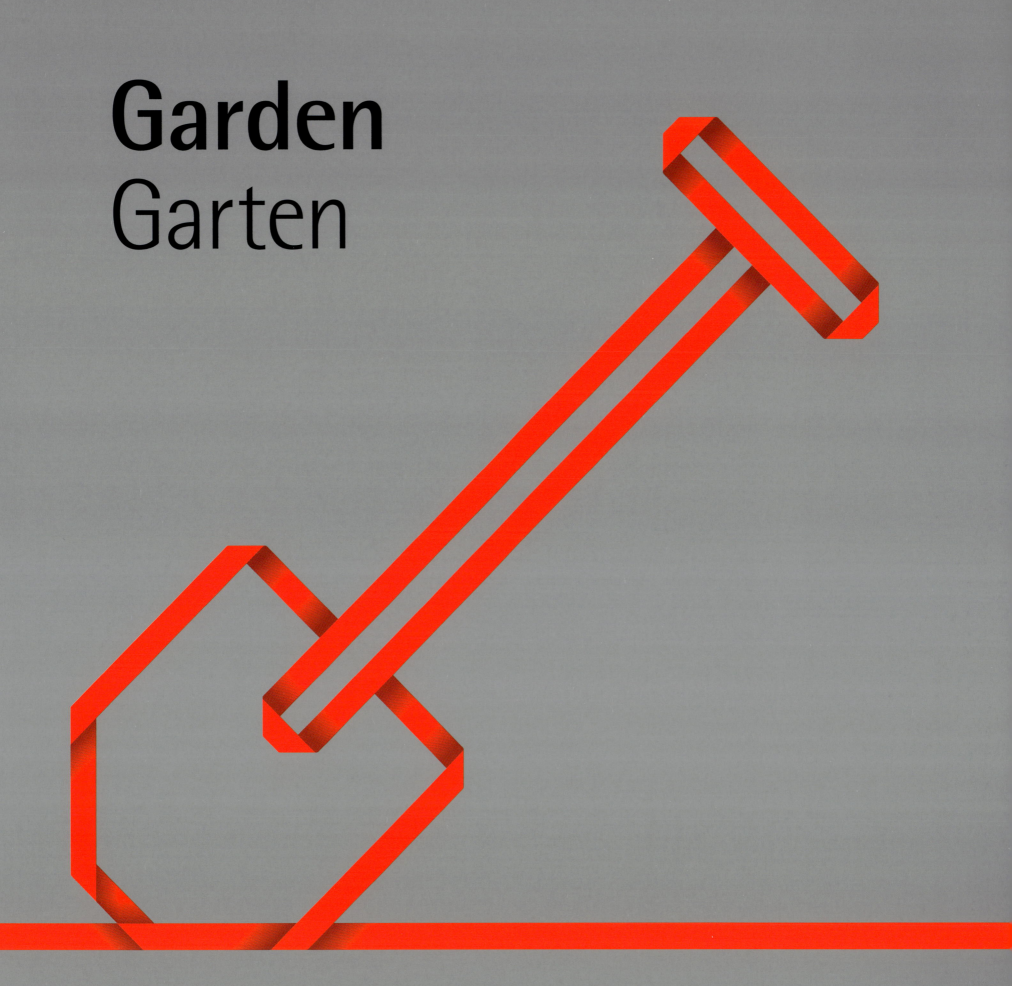

GRA 53
Professional Lawnmower
Professioneller Rasenmäher

Manufacturer
Bosch Lawn and Garden Ltd.,
Stowmarket, Great Britain

Design
TEAMS Design GmbH, Esslingen,
Germany

Web
www.bosch-professional.com
www.teamsdesign.com

reddot award 2015
best of the best

Silent power

Mowing the lawn, particularly of public spaces, is not to be seen as an easy task; not least because people living in the immediate surroundings might feel irritated by the noise. The GRA 53 is a professional lawnmower offering many benefits for this type of work. It showcases a design of sporting appearance as well as highly user-friendly elements for operation. The 53 cm lawnmower is powered by two 6.0 Ah 36 volt batteries that deliver a long run time and being weatherproof facilitate working without any interruption. It is equipped with two brushless EC motors that deliver optimum power and efficiency entirely without gas emission. An important aspect, especially for professional use, is that this motor does away with the need for maintenance even when in permanent use. The independent drive unit features variable speed controls, allowing users to cut lawns neatly in accordance to the respective lawn condition. The design with optimised blades ensures a good constant cutting performance. The air flow and grass bag design deliver efficient cutting as well as an equally efficient collection of the cut grass. Compared to gasoline driven lawnmowers, the GRA 53 emits a noise level of just 92 dB(A) and thus lends itself for use particularly in noise-sensitive areas, where it not only aestheticises the task of lawnmowing with its appearance but also makes the task easier thanks to its uncompromising user-friendliness.

Leise Kraft

Das Rasenmähen ist besonders im öffentlichen Bereich eine nicht zu unterschätzende Aufgabe. Nicht zuletzt können sich auch Anwohner in unmittelbarer Umgebung durch die Mäharbeiten belästigt fühlen. Der GRA 53 ist ein professioneller Rasenmäher, der für eine derartige Arbeit viele Vorteile bietet. Gestaltet ist er mit einer sportiv anmutenden Formensprache sowie sehr nutzerfreundlichen Bedienelementen. Der 53-cm-Rasenmäher wird von zwei Akkus mit je 6.0 Ah und 36 Volt angetrieben, die eine lange Laufzeit und, aufgrund ihrer Witterungsbeständigkeit, ein Arbeiten ohne Unterbrechungen ermöglichen. Er verfügt über zwei bürstenlose EC-Motoren, welche keinerlei Emissionen erzeugen und dabei eine optimale Leistung wie Effizienz bieten. Ein wichtiger Aspekt, insbesondere für professionelle Nutzer, ist, dass dieser Motor auch im Dauerbetrieb keinerlei Wartung erfordert. Das unabhängige Antriebssystem weist eine variable Drehzahlsteuerung auf, weshalb das Gras entsprechend den Rasenbedingungen gemäht werden kann. Die Gestaltung mit optimierten Schneidemessern gewährleistet eine konstant gute Schneidleistung. Der Luftstrom sowie der Grasauffangbehälter sorgen dabei für einen effizienten Schnitt und ein ebenso effizientes Sammeln des Schnittguts. Im Vergleich zu einem Benzinrasenmäher ist der GRA 53 mit einem garantierten Schallleistungspegel von 92 dB(A) zudem gut für den Einsatz in geräuschempfindlichen Bereichen geeignet. Er ästhetisiert dort die Tätigkeit des Rasenmähens durch seine Formensprache ebenso wie er sie durch seine kompromisslose Nutzerfreundlichkeit erleichtert.

Statement by the jury

With its outstandingly user-oriented design, the GRA 53 lawnmower satisfies the demands and needs of especially professional users. All user elements follow an overly clear layout structure and thanks to a well thought-out colour concept also ensure outstandingly comfortable operation. Equipped with brushless EC motors, this lawnmower delivers optimum power whilst being highly efficient with no need for servicing. Since it produces no exhaust fumes, the lawnmower is easy on both the environment and its user.

Begründung der Jury

Mit seiner ausgesprochen nutzerorientierten Gestaltung wird der Rasenmäher GRA 53 gerade professionellen Ansprüchen gerecht. Alle Bedienelemente sind sehr übersichtlich angeordnet. Sie erlauben auch durch ein wohl durchdachtes Farbkonzept eine intuitive und überaus komfortable Handhabung. Ausgestattet mit bürstenlosen EC-Motoren ist dieser Rasenmäher kraftvoll, robust und sehr effizient. Da er ohne Emissionen arbeitet, werden sowohl der Nutzer wie auch die Umwelt geschont.

Designer portrait
See page 28
Siehe Seite 28

VIKING 7 Series
Lawn Mower
Rasenmäher

Manufacturer
VIKING GmbH, Langkampfen, Austria
Design
Busse Design+Engineering GmbH, Elchingen, Germany
Web
www.viking-garden.com
www.busse-design.com

The 7 Series professional lawn mowers comprise a range of machines for continuous use. Their innovative magnesium housing in combination with a plastic insert makes the lawn mowers particularly lightweight. The MB 756 YC features a mono-comfort-handlebar, providing easy access to the 80-litre grass catcher bag. The cutting height can be adjusted to six steps. Further features include powerful engines, steel side protection, solid rubber front bumpers, thick rubber tyres and metal rims.

Die Profi-Rasenmäher der Serie 7 umfassen Geräte für den Dauereinsatz. Ihre innovativen Magnesiumgehäuse, in Kombination mit einer Kunststoffeinlage, machen die Rasenmäher besonders leicht. Das Modell MB 756 YC ist mit einem Mono-Komfortlenker ausgestattet, der einen freien Zugriff auf den 80-Liter-Grasfangkorb erlaubt. Die Schnitthöhe lässt sich in sechs wählbaren Stufen justieren. Zu den weiteren Merkmalen gehören starke Motoren, ein Seitenschutz aus Stahl, eine Vollgummi-Frontstoßstange, dicke Gummireifen und Metall-Felgen.

Stiga Park Pro 740 IOX 4WD
Ride-on Lawn Mower
Aufsitzmäher

Manufacturer
GGP Italy S.p.A., Castelfranco Veneto (Treviso), Italy
Design
Designit, Copenhagen, Denmark
Web
www.stiga.com
www.ggp-group.com
www.designit.com

The Stiga Park Pro 740 IOX 4WD is a ride-on lawn mower designed to combine good usability and driving comfort. Its four-wheel drive, an exclusive articulated steering system and the hydrostatic transmission, ensure safe driving control. The driving comfort is further supported by an adjustable seat. The front-mounted cutting deck allows for enhanced visibility of the cutting area while versatile mowing techniques guarantee accurate cutting. In addition, an LED screen informs on the mower's performance.

Der Stiga Park Pro 740 IOX 4WD ist ein Aufsitzmäher, der eine gute Bedienbarkeit mit Fahrkomfort verbindet. Sein Allrad-antrieb, die exklusive Knicklenkung und das Hydrostatgetriebe sorgen für eine sichere Kontrolle beim Fahren. Der Fahrkomfort wird durch einen verstellbaren Sitz zusätz-lich unterstützt. Das vorne platzierte Mähwerk ermöglicht eine bessere Sichtbar-keit des Schnittbereichs, während die Kombination verschiedener Mähtechniken ein exaktes Schnittergebnis garantiert. Zusätzlich informiert ein LED-Bildschirm über die Leistung des Rasenmähers.

Statement by the jury
This ride-on lawn mower impresses at a formally as well as functionally high level. Technically sophisticated details allow for a high degree of working comfort.

Begründung der Jury
Formal wie funktional überzeugt dieser Aufsitzmäher auf hohem Niveau. Technisch raffinierte Details ermöglichen einen hohen Arbeitskomfort.

Verve Cutting Tool
Verve Schneidwerkzeug

Manufacturer
Kingfisher, Lille, France
In-house design
Web
www.kingfisher.com

Made of fiberglass, with a nylon body and TPR handles, these loppers make gardening more comfortable. The design concept follows the objective of offering a product range with fiberglass technology that is nevertheless affordable for hobby gardeners. A multi-level gearing system was integrated to improve the cutting ability of the carbon steel blades. The shape of the handles helps with difficult work while the non-skid material increases safety.

Statement by the jury
These cutting tools establish themselves as a functionally sophisticated product solution. The high-grade materials ensure efficient and comfortable work in the garden.

Gefertigt aus Fiberglas, einem Nylon-Körper sowie einem TPR-Griff erleichtern diese Astscheren die Gartenarbeit. Das Gestaltungskonzept folgt dabei der Zielsetzung, trotz Nutzung der Fiberglas-Technologie eine kostengünstige Produktserie für den Hobbygärtner anzubieten. Um die Schneidfähigkeit der Klingen aus Carbonstahl zu verbessern, wurde ein mehrstufiges Verzahnungssystem integriert. Die Konturgestaltung der Griffe hilft bei schwierigen Arbeiten, während das griffige Material die Sicherheit erhöht.

Begründung der Jury
Als funktional durchdachte Produktlösung überzeugen diese Schneidwerkzeuge, ihre hochwertigen Materialien gewähren ein effektives und komfortables Arbeiten.

Edyn
Garden Sensor
Gartensensor

Manufacturer
fuseproject, San Francisco, USA
In-house design
Web
www.fuseproject.com

Edyn is an innovative system to support the cultivation of crop plants by measuring soil conditions and by recording these data. The solar-powered Wifi garden sensor is put into the ground to measure, among other factors, nutrient levels, temperature and sunlight exposure, and sends the information in real-time to the Edyn app. Based on the derived data, suitable plants and plant pairings are suggested using a database with information about more than 5,000 plants.

Statement by the jury
The Edyn garden sensor opens up an innovative way for soil-specific plant cultivation and furthermore impresses as a formally concise design.

Edyn ist ein innovatives System zur Unterstützung des Anbaus von Nutzpflanzen, welches die Bodenbeschaffenheit erfasst. Der solarbetriebene, Wi-Fi-fähige Gartensensor wird in den Boden gesteckt und misst unter anderem die dort vorhandenen Nährstoffe, die Temperatur sowie die Sonneneinstrahlung und sendet diese Daten in Echtzeit an die Edyn-App. Passend zu den ermittelten Daten werden geeignete Pflanzenarten und -kombinationen empfohlen, wobei auf eine Datenbank mit über 5.000 Pflanzen zurückgegriffen wird.

Begründung der Jury
Der Gartensensor Edyn eröffnet auf innovative Weise einen bodenspezifischen Pflanzenanbau und überzeugt zudem als formal prägnanter Entwurf.

Sprinkle3D
Outdoor Tap
Außenwasserhahn

Manufacturer
MIZSEI MFG CO., LTD.,
Yamagata City Gifu Prefecture,
Japan

In-house design
MIZSEI Design Team

Web
www.mizsei.co.jp

reddot award 2015
best of the best

Functional garden sculpture

Be it for watering the garden or washing the car, people need water outdoors for the most various activities. The Sprinkle3D outdoor tap presents itself as an excellent functional solution that comes with an impressively new aesthetic. It possesses a clear language of form with the appearance of an elegant sculpture. This outdoor tap was designed against the backdrop that traditional watering taps of the box type buried in the ground or mounted onto the wall are not only unsanitary but also complicated to use. Therefore this water tap is buried into the ground in a way that it sticks out only 30 cm, blending in well with almost any environment. Since it is made of high-quality materials it is robust and durable. An additional innovative feature is presented by the construction of the tap with a spout that is able to turn 360 degrees into an upright or horizontal position. This solves the problem of connected water hoses bending or even twisting, hence stopping the flow of water. Instead, hoses connected to the Sprinkle3D water tap can move and turn freely in all directions preventing any problems of this kind. The shape of the spout has been designed to suit the connecting of a hose; however, it can also be used without a hose as a regular water outlet. With its plain and clear design as well as its fascinatingly logical functionality, the Sprinkle3D water tap presents itself as a highly convincing reinterpretation of a water tap.

Funktionale Gartenskulptur

Ob für die Bewässerung des Gartens oder das Autowaschen, im Freien benötigt man Wasser für die unterschiedlichsten Tätigkeiten. Der Außenwasserhahn Sprinkle3D stellt dafür eine exzellente funktionale Lösung dar, die mit einer beeindruckend neuen Ästhetik einhergeht. Er besitzt eine klare Formensprache mit der Anmutung einer eleganten Skulptur. Gestaltet wurde dieser Außenwasserhahn vor dem Hintergrund, dass traditionelle Bewässerungshähne mit in die Erde oder an der Wand montiertem Kasten unhygienisch und zudem umständlich zu bedienen sind. Er wird deshalb so in den Boden eingelassen, dass er nur ca. 30 cm herausragt und sich dabei harmonisch in das umgebende Ambiente einfügt. Da er aus hochwertigen Materialien gefertigt wird, ist er auch robust und langlebig. Ein zusätzlicher innovativer Aspekt ist die Konstruktion dieses Wasserhahns mit einem vertikal und horizontal um 360 Grad drehbaren Auslauf. Auf diese Weise wird das Problem gelöst, dass sich angeschlossene Wasserschläuche verdrehen können und plötzlich kein Wasser mehr fließt. Der an den Sprinkle3D angeschlossene Schlauch lässt sich dagegen ohne derartige Probleme völlig frei bewegen. Der Wasserauslass ist dabei zwar für den Anschluss eines Schlauches geformt, kann aber auch gut ohne Schlauch benutzt werden. Mit seiner klaren, einfachen Formensprache und einer verblüffend logischen Funktionalität stellt der Außenwasserhahn Sprinkle3D eine überaus gelungene Neuinterpretation dar.

Statement by the jury

A fascinatingly simple solution that was realised to elegant effect. Thanks to a spout able to be rotated 360 degrees vertically and horizontally, a watering hose connected to the Sprinkle3D water tap can move around freely in all directions without bending or twisting; an approach that facilitates highly versatile and comfortable operation. This water tap can be integrated easily into many different environments – and it always cuts quite a figure.

Begründung der Jury

Eine bestechend einfache Lösung, die auf elegante Art und Weise umgesetzt wurde. Dank eines vertikal und horizontal um 360 Grad drehbaren Auslaufs kann beim Außenwasserhahn Sprinkle3D ein angeschlossener Bewässerungsschlauch völlig frei bewegt werden, ohne sich zu verdrehen. Dadurch wird eine äußerst vielseitige und komfortable Handhabung ermöglicht. Dieser Außenwasserhahn lässt sich leicht in die unterschiedlichsten Umgebungen integrieren – und macht dort stets eine gute Figur.

Designer portrait
See page 30
Siehe Seite 30

ERGO LINE
Watering System
Bewässerungssystem

Manufacturer
Cellfast Sp. z o.o., Stalowa Wola, Poland
Design
KABO & PYDO (Tomasz Pydo,
Katarzyna Borkowska), Warsaw, Poland
Web
www.cellfast.com.pl
http://kabo-pydo.com

Ergo Line is a watering system that offers innovative solutions: the Safetouch system with its reliable grip protects the product against accidental damage, the Hang Up system allows for the comfortable storage of the sprinklers and the extended clench mechanism makes sure that everything can be effectively connected with the hose. The watering system offers a long service life and is designed for efficient ergonomics.

Statement by the jury
A functionally well thought-through design characterises Ergo Line. The watering system impresses with its technical details.

Ergo Line ist ein Bewässerungssystem, das innovative Lösungen bietet: Sein Safetouch-System schützt vor Beschädigungen und garantiert einen sicheren Griff. Zudem kann der Sprenger mithilfe des Hang-up-Systems an jedem beliebigen Ort hängend aufbewahrt werden. Die verlängerte Klemmmechanik ermöglicht zudem eine präzise Verbindung mit dem Schlauch. Das Bewässerungssystem ist für eine langlebige und ergonomisch sinnvolle Nutzung ausgelegt.

Begründung der Jury
Ein funktional durchdachtes Gestaltungskonzept charakterisiert Ergo Line. Das Bewässerungssystem überzeugt aufgrund seiner technischen Details.

Neo
Rainwater Tank
Regenwasser-Flachtank

Manufacturer
Premier Tech Aqua GmbH,
Hamburg, Germany
In-house design
Design
Lormp Universallaboratorium
(Heinrich B. Oeker), Lenzen, Germany
Web
www.premiertechaqua.de

Neo was developed with the objective of designing a large, durable water tank with beneficial built-in properties. With its basic geometry of a cylinder, the flat rainwater tank offers high stability. In addition, less material is required as, due to the tank's bagel-shaped contour, front sides are no longer necessary. The thickness of its walls could be reduced as well as a result of optimised statics. The rainwater tank is easy to install and only takes up a small amount of space in the ground of the garden.

Statement by the jury
This flat rainwater tank offers an unconventional design language and convincing functionality – a demand-oriented design idea.

Mit der Zielsetzung, einen großen, langlebigen Wasserspeicher mit vorteilhaften Einbaueigenschaften zu konzipieren, entstand Neo. Der Regenwasser-Flachtank mit der Grundgeometrie eines zylindrischen Behälters bietet eine hohe Stabilität. Zudem benötigt er weniger Material, da die Stirnseiten aufgrund der bagelförmigen Kontur entfallen. Die Wanddicke konnte mithilfe einer optimierten Statik reduziert werden. Der Regenwassertank ist leicht einzubauen und nimmt nur wenig Platz im Erdreich des Gartens ein.

Begründung der Jury
Der Regenwasser-Flachtank bietet eine unkonventionelle Formensprache und eine überzeugende Funktionalität – eine bedarfsorientierte Gestaltungsidee.

Nimbus
Self-Watering Planter
Selbstbewässerungs-Pflanzgefäß

Manufacturer
Houseworks Group, Nimbus, Atlanta, USA
In-house design
Web
www.nimbuspot.com
Honourable Mention

Nimbus is an innovative self-watering planter that does not require any electronic sensors and control systems. Nimbus adjusts the amount of water the plants get, depending on the specific environment they are in. As the plants grow larger, or as humidity and evaporation levels change, the water supply will change accordingly. Alternating natural dry and wet phases prevent fungal infestation. The two-part reservoir holds up to four litres of water, sufficient for up to four months of watering per fill.

Statement by the jury
Thanks to its functionally well thought-through design, Nimbus allows the users to water their plants automatically for several months.

Nimbus ist ein innovatives Selbstbewässerungs-Pflanzgefäß, das ohne elektronische Sensoren und Steuerungssysteme auskommt. Abhängig von dem jeweiligen Umfeld bekommt die Pflanze so viel Wasser wie sie braucht. Die Wasserzufuhr ändert sich, sobald die Pflanze wächst oder sich die Luftfeuchtigkeit oder die Wasserverdunstung ändern. Indem sich natürliche Trocken- und Nassphasen abwechseln, wird ein Pilzbefall vermieden. Das zweiteilige Reservoir fasst bis zu vier Liter Wasser und reicht für mehr als vier Monate.

Begründung der Jury
Aufgrund seiner funktional durchdachten Gestaltung erlaubt Nimbus dem Nutzer über Monate eine automatische Pflanzenbewässerung.

Voltasol
Planter
Pflanzgefäß

Manufacturer
Livingthings, Andorra la Vella, Andorra
Design
Studio BAG Disseny (Sandra Compte, F. Xavier Mora),
Andorra la Vella, Andorra
Web
www.wearelivingthings.com
www.bagdisseny.com

The Voltasol flowerpot has a conical base that allows the plant to face the sun at any time of the day by means of slight movement. A simple rotation on the axis of the pot increases the amount of sunlight and thus promotes the growth of the plant. The terracotta planter comes in two sizes and is available in white, yellow or salmon in its lower part. The matching bottom tray comes in the respective colour of the pot.

Der Blumentopf Voltasol hat einen konischen Boden, wodurch die Pflanze mittels einer leichten Bewegung zu jeder Tageszeit der Sonne zugewandt werden kann. Die einfache Drehung um die Achse des Topfes vergrößert den Lichteinfall und fördert somit das Wachstum der Pflanze. Erhältlich ist das Pflanzgefäß in zwei unterschiedlichen Größen, wobei der untere Bereich des Terrakotta-Topfes in den Farben Weiß, Gelb und Lachs gehalten ist. Der passende Unterteller nimmt die jeweilige Farbgestaltung des Topfes auf.

Statement by the jury
The coherent implementation of an original design idea impresses the beholder. Voltasol achieves an autonomous overall appearance.

Begründung der Jury
Die stimmige Umsetzung einer originellen Gestaltungsidee beeindruckt den Betrachter. Voltasol erreicht ein eigenständiges Gesamtbild.

Wave
Planter
Pflanzgefäß

Manufacturer
Kasper Design s.r.o., Trutnov, Czech Republic
Design
Petr Novague, Prague, Czech Republic
Web
www.kasper.cz
www.novague.com

The Wave planter features a design of curved lines and is both suitable for outdoor and indoor use. Integrated LEDs illuminate the lower part of the planter. Its long sides are made of stainless steel while the short sides are available with an optional black, red or white coating. The plant containers are designed to create an architecturally impressive room division when placed in a row. The riveted accessory features two handles for convenient handling. Wave is produced with the dimensions 800 x 330 x 605 mm.

Das Pflanzgefäß Wave zeigt eine geschwungene Linienführung und eignet sich sowohl für den Einsatz draußen als auch drinnen. Integrierte LEDs beleuchten die Unterseite. Seine Längsseiten sind aus geschliffenem Edelstahl, während die Querseiten optional in den Farben schwarz-, rot- oder weißlackiert erhältlich sind. Die Pflanzgefäße sind so konzipiert, dass sie hintereinander aufgestellt eine architektonisch imposante Raumtrennung schaffen. Die genietete Konstruktion wurde für eine komfortable Handhabung mit Griffen versehen. Wave wird in den Abmessungen 800 x 330 x 605 mm gefertigt.

Statement by the jury
The Wave planter owes its independent aesthetics to minimalist lines in combination with high-grade, glossy materials.

Begründung der Jury
Einer minimalistischen Linienführung in Kombination mit hochwertig glänzenden Materialien verdankt das Pflanzgefäß Wave seine eigenständige Ästhetik.

FireGlobe
Fire Bowl
Feuerschale

Manufacturer
Eva Solo A/S, Måløv, Denmark
Design
Tools Design, Copenhagen, Denmark
Web
www.evasolo.com
www.toolsdesign.com

The FireGlobe fire bowl enhances the beauty of the flames with its sculptural shape and can be safely used in the garden or on the patio. It provides a cosy warmth, allowing users to enjoy even chilly summer nights outdoors. When not in use, the device can easily be moved to a different place by the integrated handle on the top. The bowl is made of black, matt enamelled steel while its legs are made of aluminium.

Die Feuerschale FireGlobe hebt mit ihrer skulptural anmutenden Kontur die Flammen wirkungsvoll hervor und ist im Garten oder auf der Terrasse sicher einsetzbar. Sie verströmt eine behagliche Wärme, wodurch auch kühle Sommerabende draußen genossen werden können. Wenn die Feuerschale nicht im Gebrauch ist, kann sie mithilfe des oben integrierten Griffs leicht an einen gewünschten Ort versetzt werden. Hergestellt ist die Feuerschale aus schwarzem, matt emailliertem Stahl, die Beine bestehen hingegen aus Aluminium.

Statement by the jury
This fire bowl conveys a minimalist aesthetic. FireGlobe impresses with its robust materials and harmonious proportions.

Begründung der Jury
Eine minimalistische Ästhetik vermittelt diese Feuerschale. FireGlobe überzeugt dank ihrer robusten Materialien und ihrer ausgewogenen Proportionen.

CONE
Charcoal Grill
Holzkohlegrill

Manufacturer
höfats GmbH, Kraftisried, Germany
In-house design
Christian Wassermann, Thomas Kaiser
Web
www.hoefats.com

Cone is an innovative charcoal grill that allows for efficient heat regulation: when high heat is required, users simply lift the coal grid and thus the heat source. To reduce the heat, the coal grid is lowered and thereby the air supply decreased. At the lowest and coldest position, the air ducts are closed. Since the coal grid can be lifted up to the edge of the cone after the barbecue, the grill can be transformed into a fireplace that only needs few logs of wood to create a great ambience.

Cone ist ein innovativer Holzkohlegrill, der eine effektive Hitzeregulierung ermöglicht: Sobald große Hitze zum Angrillen erwünscht ist, kann der Kohlerost und somit die Hitzequelle einfach angehoben werden. Um die Wärme zu reduzieren, wird der Kohlerost abgesenkt und damit die Luftzufuhr reduziert. In der tiefsten und zugleich kältesten Position sind die Luftöffnungen geschlossen. Da der Kohlerost bis zur Trichterkante angehoben werden kann, lässt sich Cone nach dem Grillen in eine Feuerstelle wandeln, die mit nur wenig Holz eine schöne Stimmung erzeugt.

Statement by the jury
Thanks to its sculptural appearance, Cone achieves a powerful visual presence. Furthermore, its multifunctional use is impressive.

Begründung der Jury
Dank seiner skulpturalen Anmutung erreicht Cone eine starke visuelle Präsenz. Darüber hinaus überzeugt seine multifunktionale Nutzung.

Xtreme
Sunlounger
Sonnenliege

Manufacturer
Boxmark Leather d.o.o., Kidricevo, Slovenia
Kreativni aluminij d.o.o., Kidricevo, Slovenia
Design
Interartro d.o.o. (Sabina Zerezghi), Ankaran, Slovenia
Web
www.boxmark.com
www.kreal.si
www.sabinazerezghi.com

The comfortable Xtreme sunlounger features an elegant combination of an aluminium frame with an upholstery made of a patented, highly durable genuine leather. The innovative material is equally suited for indoor and outdoor use. The surface of the aluminium is protected by a powder coating durable even against seaside conditions. The aluminium is furthermore treated with a micro-structured paint and a matt metallic gloss, ensuring that no fingerprints are visible. The sunloungers are stackable and take up little space during storage.

Die komfortable Sonnenliege Xtreme zeigt die elegante Kombination eines Aluminium-Rahmens mit einer Bespannung aus einem patentierten, stark belastbaren Echtleder. Das innovative Material ist für den In- und Outdoor-Bereich gleichermaßen geeignet. Die Oberfläche des Aluminiums ist mit einer Pulverbeschichtung selbst vor Einflüssen des Meeresklimas geschützt. Das Aluminium wurde außerdem mit einer Feinstrukturfarbe und einem eleganten, matt metallischen Glanz veredelt, weshalb Fingerabdrücke nicht sichtbar sind. Die Liegen sind stapelbar und benötigen daher wenig Platz während der Lagerung.

Statement by the jury
An ergonomically sophisticated design approach characterises this sunlounger. Its high-grade materials are visually and haptically outstanding.

Begründung der Jury
Ein ergonomisch durchdachter Gestaltungsansatz zeichnet diese Sonnenliege aus, wobei die edlen Materialien visuell und haptisch auffallen.

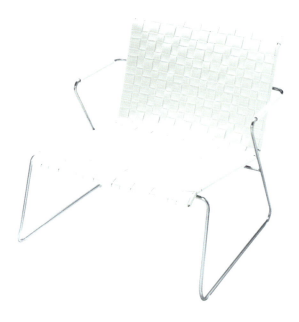

Slim Belt
Lounge Chair
Lounge Sessel

Manufacturer
VITEO GmbH, Graz, Austria
In-house design
Wolfgang Pichler
Web
www.viteo.com

The Slim Belt lounge chair was developed for use in public and business environments and is characterised by stackability, low maintenance and high comfort. The seat consists of special straps made of spun-dyed acrylic yarn, offering reliable protection against weathering. A marine grade steel frame embraces the seat with dynamically appealing lines and, simultaneously, serves as the chair's armrests. The material of the straps recurs in the delicate armrest padding and thus creates an attractive accent.

Statement by the jury
With its harmonious proportions, the Slim Belt lounge chair conveys an elegant lightness. The high level of comfort is another convincing quality.

Der Lounge Sessel Slim Belt wurde für den Objekteinsatz konzipiert und zeichnet sich durch seine Stapelbarkeit, seine Pflegeleichtigkeit und einen hohen Komfort aus. Die Sitzfläche besteht aus speziellen Gurten aus spinndüsengefärbten Acrylgarnen, welche zuverlässig witterungsbeständig sind. Ein salzwasserbeständiger Edelstahldraht umrahmt die Sitzschale in einer dynamisch anmutenden Linienführung und stellt zugleich Armlehnen dar. Die Wiederholung des Gurtmaterials als grazile Armauflage setzt einen ansprechenden Akzent.

Begründung der Jury
Aufgrund seiner ausgewogenen Proportionen vermittelt der Lounge Sessel Slim Belt eine elegante Leichtigkeit. Zudem überzeugt der hohe Komfort.

Slim Wood
Lounge Chair
Lounge Sessel

Manufacturer
VITEO GmbH, Graz, Austria
In-house design
Wolfgang Pichler
Web
www.viteo.com

The design concept of the Slim Wood lounge chair combines the purist-looking material Corian with a wood frame in a warm colour tone. This results in an exciting contrast effect. The chair offers a versatile array of possible applications as both materials are suitable for indoor and outdoor use. The delicate seat provides impressive stability and seems to float above the Iroko substructure, a common African type of wood. The furniture is part of the Slim Wood collection, which unites an international design approach with local craftsmanship.

Statement by the jury
A material mix rich in contrast characterises the elegant appeal of this chair, which, in addition, allows for high seating comfort.

Das Gestaltungskonzept des Lounge Sessels Slim Wood kombiniert das puristisch anmutende Material Corian mit einem Holzgestell in einem warmen Farbton. So entsteht ein spannender Kontrasteffekt. Der Stuhl bietet flexible Nutzungsmöglichkeiten, da sich beide Materialien für draußen und drinnen eignen. Die filigran wirkende Sitzschale bietet eine beeindruckende Stabilität. Sie scheint über der Unterkonstruktion aus massivem Iroko, einer in Afrika weit verbreiteten Holzart, zu schweben. Das Möbel ist Bestandteil der Slim Wood Kollektion, welche einen internationalen Gestaltungsansatz mit regionaler Handwerkskunst vereint.

Begründung der Jury
Ein kontrastreicher Materialmix charakterisiert diesen elegant anmutenden Lounge Sessel, der darüber hinaus einen hohen Sitzkomfort ermöglicht.

Yard
Armchair
Armlehnstuhl

Manufacturer
Emu Group S.p.A., Marsciano (Perugia), Italy
Design
Stefan Diez Office, Munich, Germany
Web
www.emu.it
www.stefan-diez.com

Yard is an armchair with a lightweight frame made of powder-coated aluminium. The elastic straps of the seat and the back guarantee a pleasant comfort even without additional cushioning. In an elegant way, the plaited straps are inserted flush into the frame. After years of use, the potentially worn straps can be exchanged if required. Yard is available in various colour combinations including black/grey, red/red, blue/blue or mint green/beige.

Statement by the jury
This armchair is convincing because of its elastic seat, while colour concept and lines convey a distinctive appearance.

Yard ist ein Armlehnstuhl, dessen leichtes Gestell aus pulverbeschichtetem Aluminium besteht. Der Sitz und die Rückenlehne aus elastischen Gurten sorgen auch ohne zusätzliche Polster für einen angenehmen Komfort. Auf elegante Weise sind die geflochtenen Gurten flächenbündig in das Gestell eingeschoben. Bei Bedarf können die nach Jahren eventuell abgenutzten Gurte ausgewechselt werden. Yard ist in diversen Farbkombinationen, unter anderem in Schwarz/Grau, Rot/Rot, Blau/Blau und Minzgrün/Beige, erhältlich.

Begründung der Jury
Dieser Armlehnstuhl überzeugt aufgrund seiner elastischen Sitzfläche, wobei Farbkonzept und Linienführung eine charakteristische Anmutung vermitteln.

urban
Garden Furniture
Gartenmöbel

Manufacturer
acamp by Dandler GmbH & Co. KG, Stockach, Germany
Design
guggenbichlerdesign…
(Harald Guggenbichler), Vienna, Austria
Web
www.acamp.de
www.dandler.de
www.guggenbichler.at

urban is a stackable garden furniture collection which was developed with the objective of creating an aluminium-textilene construction that is fit for daily use and reduced to the essentials. The filigree design is based on a triangle-shaped profile with a tapering structure, which does not reduce the functionality and stability. This design principle is consistent in all items of the collection, including a stacking chair, folding chair, relax chair, side table and folding lounger. The densely woven structure of the textile fabric provides a high level of comfort.

Statement by the jury
A high quality of materials and usability characterises this garden furniture collection. The conclusive design yields timeless aesthetics.

Die stapelbare Gartenmöbel-Serie urban entstand mit der Zielsetzung, eine alltagstaugliche und auf das Wesentliche reduzierte Aluminium-Textilgewebe-Konstruktion zu kreieren. Die filigran anmutende Formensprache beruht auf einem Triangel-Profil mit verjüngter Struktur, ohne dabei Abstriche bei der Funktionalität und Stabilität zu machen. Dieses Gestaltungsprinzip zeigt sich durchgängig bei dem Stapelsessel, dem Klappsessel, dem Relaxsessel, dem Beistelltisch und der Klappliege. Für einen guten Sitzkomfort sorgt die dichte Webstruktur des Textilgewebes.

Begründung der Jury
Eine hohe Material- und Gebrauchsqualität zeichnet die Gartenmöbel-Serie urban aus. Der stimmige Entwurf erreicht eine zeitlose Ästhetik.

Grid
Outdoor Sofa

Manufacturer
Gloster Furniture Ltd, Bristol, Great Britain
Design
365° (Henrik Pedersen), Aarhus, Denmark
Web
www.gloster.com
www.365.dk

The Grid modular outdoor sofa can be flexibly assembled to meet specific needs. The shapes of the various elements allow the combination of the seating units in multiple ways and different directions. The wooden table elements are adjustable in height. Their graphic lines harmoniously blend with the comfortable cushioning. Grid creates a cosy lounge atmosphere of distinctive character in both private and public spaces.

Statement by the jury
With its generous dimensions, the Grid outdoor sofa provides a high level of comfort. Its mix of materials underlines the elegant appeal.

Das modulare Outdoor-Sofa Grid lässt sich flexibel und bedarfsgerecht zusammenstellen: Die Struktur der unterschiedlichen Module erlaubt es, die einzelnen Sitzeinheiten in vielfältiger Weise und verschiedenen Richtungen aufzubauen. Tischelemente aus Holz können in unterschiedlichen Höhen positioniert werden. Deren graphische Linienführung fügt sich harmonisch zur komfortablen Polsterung. Grid schafft sowohl im öffentlichen wie privaten Bereich eine gemütliche Lounge-Atmosphäre mit einem unverwechselbaren Charakter.

Begründung der Jury
Mit seinen großzügigen Ausmaßen erreicht das Outdoor-Sofa Grid ein hohes Maß an Komfort. Seine Materialität unterstreicht die elegante Anmutung.

Pantagruel
Outdoor Table
Outdoor-Tisch

Manufacturer
Extremis, Poperinge, Belgium
Design
Dirk Wynants Design Works, Poperinge, Belgium
Web
www.extremis.be
www.dwdw.be

This outdoor table has been named after the giant Pantagruel, a character of a novel, who has a huge appetite. For increased comfort, the round tabletop is fitted with a "Lazy Susan", a central revolving tray. The wood surface of the table matches the seat bench with its four separate elements. The four legs connect table and bench into one formal unit and are optionally available with a powder coating. The overall impression is both sturdy and elegant.

Dieser Outdoor-Tisch wurde nach der Romanfigur Pantagruel benannt, einem Riesen mit großem Appetit. Um den Komfort zu erhöhen, ist die runde Tischplatte mittig mit einer „Lazy Susan" ausgestattet, einer drehbaren Servierplatte. Die Holzfläche des Tisches korrespondiert mit der viergeteilten Sitzbank. Vier Beine verbinden Tisch und Bänke zu einer formalen Einheit und sind optional mit einer Pulverbeschichtung erhältlich. Der Gesamteindruck ist zugleich robust und elegant.

Statement by the jury
The Pantagruel outdoor table is marked by an autonomous appearance. The compact design achieves a convincing functionality.

Begründung der Jury
Eine eigenständige Formensprache prägt den Outdoor-Tisch Pantagruel. Die kompakte Konstruktion erreicht eine überzeugende Funktionalität.

Mountain
Outdoor Furniture
Outdoor-Möbel

Manufacturer
Guangzhou Gardenart Furniture Co., Ltd., Guangzhou, China
In-house design
Roger Cao
Web
www.gardenart.cn

The generous Mountain outdoor furniture was conceived for both private and public gardens. The purist design unites a large-scale table with two seat benches as one visual entity. The ergonomically sophisticated arrangement provides good seating comfort without the need for adjusting the position of the benches. The aluminium material minimises the weight while a weather-proof powder coating for all surfaces protects the table from weathering.

Das großzügige Outdoor-Möbel Mountain wurde sowohl für private wie auch öffentliche Gärten konzipiert. Eine puristische Formensprache vereint den großflächigen Tisch mit den beiden Sitzbänken zu einer visuellen Einheit. Die ergonomisch durchdachte Anordnung ermöglicht einen guten Sitzkomfort, ohne dass dafür die Position der Bänke verändert werden muss. Die Fertigung aus Aluminium minimiert das Gewicht, wobei eine wetterfeste Pulverbeschichtung alle Oberflächen dauerhaft vor Verwitterung schützt.

115

Pool Lounge
Cover
Abdeckung

Manufacturer
A-Spas by Armstark GmbH, Schärding, Austria
In-house design
Reinhard Helmig
Web
www.a-spas.com
Honourable Mention

Pool Lounge is a walkable, fully resilient and automated cover for in-ground pools, hot tubs and swim spas, able to withstand 320 kg/m² of pressure. Heated rails allow for a smooth opening, even at freezing temperatures. Thanks to its dirt preventive properties, Pool Lounge reduces the use of water cleaning agents, and, furthermore, brings down water heating costs to a considerable degree and minimises the evaporation of water. The innovative insulation is based on a nanotechnology coating.

Pool Lounge ist eine begehbare, mit 320 kg/m² belastbare und automatisierte Abdeckung für in den Boden eingelassene Pools, Whirlpools und Swim-Spas. Beheizte Laufschienen ermöglichen ein problemloses Öffnen selbst bei Frosttemperaturen. Aufgrund des Schutzes vor Verschmutzung verringert die Pool Lounge den Einsatz von Wasserhygienemitteln, zudem werden die Wasserheizkosten und die Wasserverdunstung deutlich minimiert. Die innovative Isolierung beruht auf einer auf Nanotechnologie basierenden Beschichtung.

Statement by the jury
The Pool Lounge cover represents a functionally sophisticated product solution that harmoniously blends in with the ambience of the patio.

Begründung der Jury
Eine funktional durchdachte Produktlösung stellt die Abdeckung Pool Lounge dar, die sich harmonisch in das Terrassenambiente einfügt.

ek mini for2
Wooden House
Holzhaus

Manufacturer
Ekokoncept d.o.o., prefabricated buildings, Ljubljana, Slovenia
In-house design
Web
www.ekokoncept.com
Honourable Mention

ek mini for2 is an energy-efficient wooden house that can be used, for instance, as a garden office or holiday home. Produced from local spruce wood, it is well insulated and equipped with high-grade aluminium windows and doors. Its shape is reminiscent of a classical Slovene steep roof house and blends harmoniously with urban or rural environments. The glass façade on the side of the house offers plenty of light. The minimalist interior comprises wooden floors, a small kitchenette and a toilet.

ek mini for2 ist ein kleines, energie-effizientes Holzhaus, das beispielsweise als Gartenbüro oder Ferienwohnung genutzt werden kann. Gefertigt aus regionalem Fichtenholz, ist es gut isoliert und mit hochwertigen Aluminiumfenstern und Türen ausgestattet. Seine Kontur erinnert an ein klassisches slowenisches Steildach-Haus und fügt sich gleichermaßen gut in eine städtische oder ländliche Umgebung ein. Die seitliche Glasfront bietet viel Licht. Das minimalistische Interieur umfasst einen Holzboden, eine kleine Küchenzeile und ein WC.

Statement by the jury
This wooden mini-house with its single pitch roof and glass façade impresses as an aesthetically appealing and energy-saving structure.

Begründung der Jury
Als ästhetisch ansprechende und energiesparsame Baukonstruktion überzeugt dieses Mini-Holzhaus mit Pultdach und Glasfassade.

OM
Sunshade
Sonnenschirm

Manufacturer
Calma, Iniciativa Exterior 3i,
Vilamalla, Spain

Design
Andreu Carulla Studio,
Banyoles, Spain

Web
www.calma.cat
www.andreucarulla.com

reddot award 2015
best of the best

Form for a lifestyle

Sunshades have always stood symbolically for a sense of lightness while being outdoors. They create a zone of peacefulness and protect against the sun. The OM reinterprets the classic form of a sunshade as an aesthetic object that catches the eye and possesses high functionality and solidity. Made of high-quality materials, it fascinates viewers with a clear and elegant language of form. In addition, it also surprises users with a refined rotating system that not only facilitates very easy and self-explanatory opening and closing of the umbrella, but also allows users to lock the opening in different configurations, with each lending the sunshade a different appearance and function. Thus, when fully opened, it presents itself with an overtly modern vision, and when loosely expanded, is reminiscent of Ibiza-style sunshades with their typical wave-like appearance. The innovative design concept thus offers great versatility as it allows the use of the umbrella standing or put to the wall at different angles, such as 90 degrees for an inside corner, 180 degrees for plain walls and 270 degrees around corners. The 180-degree configuration can even be used vertically as a light windscreen or space partition. Showcasing a clear and elegant use of forms as well as a sophisticated functionality, the OM sunshade embodies a perfect harmony of lifestyle and design at all times.

Form für ein Lebensgefühl

Der Sonnenschirm ist seit jeher ein Symbol für die Leichtigkeit des Lebens im Freien. Er schafft Ruhezonen und schützt vor der Sonne. Der OM interpretiert die klassische Form eines Sonnenschirms neu als ein die Blicke auf sich ziehendes ästhetisches Objekt, das zugleich von hoher Funktionalität und Solidität ist. Er wird aus hochwertigen Materialien gefertigt und begeistert mit seiner eleganten und klaren Formensprache. Er verblüfft zudem mit einem sorgfältig durchdachten Rotationsmechanismus, der dem Nutzer ein sehr einfaches und auch selbsterklärendes Öffnen und Schließen erlaubt. Intuitiv lässt sich dieser Sonnenschirm in verschiedene Positionen einstellen und verändert dadurch immer wieder Anmutung und Funktion. So hat er vollständig aufgespannt eine betont moderne Ausstrahlung und erinnert locker gespannt mit einem wellenartigen Erscheinungsbild an die Markisen auf Ibiza. Das innovative Gestaltungskonzept begründet seine große Vielfältigkeit, da er als Stand- oder Wandschirm in verschiedenen Winkeln aufgestellt werden kann, etwa mit 90 Grad für den Einsatz in Ecken, mit 180 Grad an geraden Wänden oder mit 270 Grad über Eck. Der 180-Grad-Schirm kann vertikal zudem auch als leichter Windschutz oder Raumtrenner genutzt werden. Mit seiner eleganten Formgebung sowie seiner ausgeklügelten Funktionalität verkörpert der Sonnenschirm OM stets perfekt die Einheit von Design und Lifestyle.

Statement by the jury

The OM sunshade merges a clear and elegant design with highly refined functionality. The innovative, outstandingly smart rotating system is highly fascinating as it allows for comfortable opening and closing. The sunshade can be used in many different configurations and enhances almost any ambiance. It can serve as standing umbrella or attached to the wall and also as space partition. Moreover, the use of high-quality materials gives it high durability.

Begründung der Jury

Der Sonnenschirm OM vereint eine klare und elegante Formensprache mit einer ausgefeilten Funktionalität. Beeindruckend ist der innovative, ausgesprochen intelligent konstruierte Rotationsmechanismus, der ein komfortables Öffnen und Schließen ermöglicht. Dieser Sonnenschirm kann sehr vielseitig eingesetzt werden und bereichert jederzeit das Ambiente. So dient er als Stand- oder Wandschirm ebenso wie als Raumtrenner. Die Fertigung aus hochwertigen Materialien verleiht ihm überdies Langlebigkeit.

Designer portrait
See page 32
Siehe Seite 32

China
Parasol
Sonnenschirm

Manufacturer
Zhejiang Zhengte Group Co., Ltd.,
Linhai, China
In-house design
Jianfeng He
Web
www.zhengte.com.cn
Honourable Mention

This parasol is reminiscent of traditional parasols from the ancient Chinese Qin dynasty. Its 36 ribs form a distinctive visual contrast with the fabric covering. The opening mechanism is easy to handle, maintenance-free and reliable. Especially the proportions of the centre part and the upper part are well balanced while the radius of the parasol has been deliberately limited. The high-quality anodised pole lends the overall appearance a distinctive character and is durably protected against rust.

Statement by the jury
Characterised by an elegant appearance, this parasol stands out as a space-saving product solution.

Dieser Sonnenschirm erinnert an traditionelle Schirme der chinesischen Qin-Dynastie. Charakteristisch heben sich seine 36 Streben visuell von der textilen Bespannung ab. Der Öffnungsmechanismus ist einfach zu handhaben, wartungsfrei und zuverlässig. Die Proportionen des mittleren und oberen Teils sind besonders ausgewogen, wobei der Radius des Sonnenschirms bewusst begrenzt wurde. Die hochwertig eloxierte Schirmstange verleiht dem Gesamtbild eine unverwechselbare Note und ist langfristig vor Rost geschützt.

Begründung der Jury
Eine elegante Anmutung charakterisiert diesen Sonnenschirm, der sich als platzsparende Produktlösung auszeichnet.

Knirps Silver
Parasol
Sonnenschirm

Manufacturer
doppler, e. doppler & Co. GmbH,
Braunau-Ranshofen, Austria
In-house design
Herbert Forthuber
Design
Synowation GmbH (Bernhard Rothbucher),
Salzburg, Austria
Web
www.dopplerschirme.com
www.synowaytion.com

The Knirps Silver parasol combines elegant design with user-friendly functionality. Poles made of solid aluminium minimise the weight and facilitate handling. A simple turn of the runner/opening handle tilts the umbrella cover by up to 30 degrees, providing excellent protection even when the sun is low. The high-quality lightfast fabric with stain-repellant finish comes in different colours and offers a high UV-protection factor.

Statement by the jury
As an expression of a high demand on quality, this parasol showcases a purist design. Sophisticated details facilitate comfortable handling.

Der Sonnenschirm Knirps Silver kombiniert eine elegante Gestaltung mit einer nutzerfreundlichen Funktionalität. Schirmrohre aus solidem Aluminium minimieren das Gewicht und unterstützen so die komfortable Handhabung. Durch einfaches Drehen am Öffnungs-Schiebegriff neigt sich das Schirmdach um bis zu 30 Grad und bietet selbst bei tief stehender Sonne einen guten Sonnenschutz. Der lichtechte Qualitätsstoff mit schmutzabweisender Imprägnierung bietet einen hohen UV-Schutzfaktor und ist in verschiedenen Farben erhältlich.

Begründung der Jury
Als Ausdruck eines hohen Qualitätsanspruchs zeigt dieser Sonnenschirm eine puristische Formensprache. Durchdachte Details erleichtern die Handhabung.

Bags, baskets and containers	Antrieb und Schaltung
Bicycle tyre inflators	Beleuchtung
Bike computers	Campinggeschirr
Bike equipment	Campingmöbel
Camping furnishings	Cityräder
Camping tableware	DJ und VJ Equipment
Citybikes	E-Bikes
DJ and VJ equipment	Fahrradcomputer
E-bikes	Fahrradzubehör
Functional clothing	Funktionsbekleidung
Games	Helme
Gear and gear shift	Hobby
Helmets	Luftpumpen
Hobbies	Matratzen
Lighting	Mikrofone
Locks	Mountainbikes
Mattresses	Musikinstrumente
Microphones	Outdoorausrüstung
Mountain bikes	Outdoorbekleidung
Musical instruments	Recording und Computer
Outdoor clothing	Schlafsäcke
Outdoor equipment	Schlösser
Pet supplies	Sextoys
Recording and computers	Spiele
Sex toys	Spielzeuge
Sleeping bags	Sportschuhe
Sports shoes	Taschen
Tents	Tierartikel
Toys	Zelte

Leisure, sport and games
Freizeit, Sport und Spiel

Kross
Trike with E-Engine
Trike mit E-Motor

Manufacturer
HASE BIKES, Waltrop, Germany

In-house design
Marec Hase, Paulo Mesquita

Web
www.hasebikes.com

Three-wheeled riding pleasure

Trikes are a logical further development of the tricycle and thus a construction of high stability that has long inspired new design ideas. The Kross presents itself as a contemporary and highly exciting interpretation of a trike. Its harmoniously balanced design projects the properties of safety and vitality at a single glance. It facilitates a new form of trike mobility by integrating a mid-mounted engine directly into the frame: the high-performance Shimano Steps pedelec motor can be switched in when needed and thus enhances the trike's operating range. The integration is formally well conceived and made possible by a unique geometry, leading to an optimal weight distribution. That makes the use of the pedelec motor possible, supporting the rear wheel drive – the Kross thus features a rear drive and all the advantages this brings with. In addition, the outstanding riding comfort of the Kross is achieved through the well-balanced combination of an equally innovative 80 mm travel full-suspension system with independent wheel suspension of the rear axle, complemented by a differential gear. This allows users to ride the trike even on curvy, hilly and unsurfaced roads, without compromising on riding comfort and operability. Furthermore, safety is ensured by a high-performance disc brake system. Last but not least, the high riding pleasure with the Kross is further enhanced by the wide, knobby tyres. Thanks to an extraordinary design, the trike thus turns into a means of transport offering a new, highly exciting experience.

Fahrspaß auf drei Rädern

Das Trike ist eine Weiterentwicklung des Dreirades und damit einer Konstruktion, die wegen ihrer Stabilität schon immer die Ideen der Designer beflügelt hat. Das Kross ist nun eine zeitgemäße und überaus spannende Interpretation des Trikes. Mit einer ausgewogenen Formensprache visualisiert es auf den ersten Blick die Attribute Sicherheit und Vitalität. Eine neue Form der Mobilität mit dem Trike erlaubt ein direkt in den Rahmen integrierter Mittelmotor: Der leistungsfähige Shimano Steps-Pedelecmotor lässt sich bei Bedarf zuschalten und vergrößert so den Einsatzbereich dieses Trikes. Die Integration ist formal sehr gut gelungen, da sie auf einer einzigartigen Geometrie basiert, die zu einer optimalen Gewichtsverteilung führt. Der so ermöglichte Einsatz des Pedelecmotors unterstützt den Antrieb der Hinterräder – das Kross hat also einen Heckantrieb und die damit verbundenen Vorteile. Verantwortlich für den erstaunlichen Komfort des Kross ist zudem eine gut abgestimmte Kombination aus einer ebenso innovativen Vollfederung mit 80 mm Federweg, einer Einzelradaufhängung der Hinterachse sowie einem Differentialgetriebe. Der Fahrer kann mit diesem Trike deshalb problemlos auch auf kurvigen, hügeligen und unbefestigten Straßen fahren, ohne dass es unbequem wird. Sicherheit bieten ihm überdies hochentwickelte Scheibenbremsen. Für ein hohes Maß an Fahrspaß sorgt nicht zuletzt auch die Ausstattung des Kross mit dicken Stollenreifen. Das Fortbewegungsmittel Trike lässt sich so dank einer außergewöhnlichen Gestaltung sehr spannungsreich neu erleben.

Statement by the jury

The Kross delivers users a feeling of instant riding comfort and fun. Representing a well-conceived innovative combination of a full-suspension trike with an e-engine, it possesses excellent handling characteristics. It offers outstanding shock absorption on the road and, equipped with wide knobby tyres and three disc brakes, is also suitable for off-road riding. This new type of trike is refreshingly different and ultimately convinces with its safety concept.

Begründung der Jury

Das Kross gibt dem Fahrer auf Anhieb ein Gefühl von Komfort und Spaß. Als gelungene innovative Kombination aus einem vollgefederten Trike mit einem E-Motor bietet es exzellente Fahreigenschaften. Auf der Straße federt es ausgezeichnet ab, und, ausgestattet mit dicken Stollenreifen und drei Bremsscheiben, ist es auch gut im Offroadbereich einsetzbar. Diese neue Art von Trike ist erfrischend anders und überzeugt außerdem mit seinem Sicherheitskonzept.

Designer portrait
See page 34
Siehe Seite 34

Bergamont E-Ville C XT
E-Bike

Manufacturer
Bergamont Fahrrad Vertrieb GmbH,
Hamburg, Germany
Design
TEAMS Design GmbH (Oliver Keller),
Hamburg, Germany
Web
www.bergamont.de
www.teamsdesign.com

Highly practical, comfortable to ride and easy to mount are the key benefits of the E-Ville C XT bike. Fixing the battery to the seat tube has resulted in a low centre of gravity that improves the riding and handling of the e-bike. The smooth integration of the motor and battery, sophisticated two-tone finish and use of high quality components offer users the possibility to connect lifestyle with urban mobility.

Statement by the jury
The classic design of the E-Ville C XT conveys comfort and stability – two characteristics that define it as an all-rounder for both everyday life and leisure activities.

Eine hohe Alltagstauglichkeit, angenehme Fahreigenschaften und ein bequemer Aufstieg zählen zu den Vorzügen des E-Ville C XT. Durch seinen am Sattelrohr befestigten Akku ergibt sich ein tiefer Schwerpunkt, der den Fahr- und Tragekomfort verbessert. Die optisch ansprechende Integration von Motor und Akku, die aufwendige Bicolor-Lackierung und hochwertige Komponenten bieten so die Möglichkeit, Lifestyle und urbane Mobilität miteinander zu verbinden.

Begründung der Jury
Die klassische Gestaltung des E-Ville C XT vermittelt Komfort und Stabilität – zwei Eigenschaften, die es in Alltag und Freizeit zum Allrounder machen.

BESV Panther PS1
E-Bike

Manufacturer
Darfon Innovation Corp.,
Taoyuan, Taiwan
In-house design
Jonathan Yin
Web
www.besv.com

Lightweight and equipped with a proprietary Algorhythm power drive system; the BESV Panther PS1 is ideal for urban cycling. Taking into account both cycling conditions and the user's individual pedalling power, the Smart Mode calculates and automatically adjusts the drive power to the level of assistance needed for optimal comfort whilst cycling. The stiff frame is constructed in one piece and consists of more than 10 layers of carbon fibre.

Statement by the jury
Quick, manoeuvrable, stylish – these characteristics of the BESV Panther PS1 are complemented by a smart drive system in order to excellently meet the requirements of contemporary urban mobility.

Das leichte, mit geschütztem Algorithmus-Antriebssystem ausgestattete BESV Panther PS1 ist sehr gut für die Stadt geeignet. Unter Berücksichtigung der äußeren Bedingungen und der individuellen Tretleistung errechnet die Smart-Einstellung die für die aktuelle Fahrt benötigte Stromkapazität und stimmt die Motorleistung automatisch darauf ab. Der aus einem Stück gefertigte stabile Rahmen besteht aus mehr als zehn Carbonfaser-Schichten.

Begründung der Jury
Flink, wendig, stilvoll – so präsentiert sich das BESV Panther PS1 und wird, ergänzt von einem intelligenten Antrieb, einer zeitgemäß urbanen Mobilität hervorragend gerecht.

Mando Footloose IM
E-Bike

Manufacturer
Mando, Seoul, South Korea
Design
MAS Design Products Ltd. (Mark Sanders),
Windsor, Great Britain
Web
www.mandofootloose.com
www.mas-design.com

Mando Footloose IM is a new concept in e-mobility using the latest automotive technology. It features a built-in Series Hybrid System motor to power the vehicle without chains. The vehicle operation is fully managed by an automotive-level Electronic Control Unit paired with CAN communication. The unit generates power with an alternator and offers a choice between throttle and pedal-power.

Statement by the jury
Footloose IM represents a new interpretation of a bicycle: the hybrid system hidden in the frame does without chains thus creating an aesthetic that draws attention.

Mando Footloose IM ist ein neues Konzept der E-Mobility und mit neuster Automobiltechnik ausgestattet. Es verfügt über einen eingebauten Series-Hybrid-System-Motor, der das Gefährt kettenlos antreibt. Der Betrieb erfolgt vollständig über eine per CAN-Kommunikation gelenkte elektronische Steuereinheit, wie sie auch im Automobilbereich zum Einsatz kommt. Die Einheit produziert über einen Wechselgenerator Strom und ermöglicht die Wahl zwischen Motor- und Pedalantrieb.

Begründung der Jury
Das Footlose IM stellt eine Neudefinition des Fahrrads dar: Das im Rahmen verborgene Hybridsystem kommt ohne Kette aus und erzielt dadurch eine aufsehenerregende Ästhetik.

BESV Lion LX1
E-Bike

Manufacturer
Darfon Innovation Corp., Taoyuan, Taiwan
In-house design
Jonathan Yin
Web
www.besv.com

The BESV Lion LX1 captivates with elegant contouring, a sturdy, reliable aluminium frame and a power drive system featuring Algorhythm technology. Taking into account the rider's personal pedaling power, the intelligent power supply calculates the exact level of assistance needed for optimal riding comfort with the help of peak sensors and the Algorhythm software. This allows it to automatically adjust the drive power to the individual rider's needs.

Statement by the jury
This e-bike impresses with its powerful profile that does away with a rear fork. The technology used for the specially developed drive system is outstanding.

Das BESV Lion LX1 besticht durch eine elegante Formgebung, einen stabilen, zuverlässigen Aluminiumrahmen und ein Antriebssystem mit Algorithmus-Technologie. Auf Basis der persönlichen Tretleistung berechnet eine intelligente Stromversorgung mithilfe von Spitzensensoren und der Algorithmus-Software immer genau, wie viel Strom für den optimalen Fahrkomfort gerade benötigt wird, und passt die Motorleistung automatisch den individuellen Bedürfnissen an.

Begründung der Jury
Dieses E-Bike beeindruckt mit seiner kraftvollen Silhouette, die ohne Hinterradgabel auskommt. Technisch herausragend ist sein speziell entwickeltes Antriebssystem.

Ökovänlig
E-Bike

Manufacturer
Protanium/Sandwichbike,
Apeldoorn, Netherlands
In-house design
Basten Leigh, Brian Hoehl, Lars Munksø
Web
www.protanium.com
Honourable Mention

Ökovänlig, meaning "environmentally friendly", is elegantly presented with a plywood frame. The two plywood panels made of PEFC certified wood harvested in Germany, along with all other specially designed aluminium parts for the frame, are to be assembled by the buyers themselves into the finished bike. The easy-to-remove lithium battery is integrated into the plywood sandwich construction. The bike complies with EN and ISO standards.

Statement by the jury
This electric bike is the realisation of a sustainable approach and attracts attention with its innovative plywood frame design.

Ökovänlig, zu Deutsch umweltfreundlich, präsentiert sich mit einem aus Sperrholz bestehenden Rahmen. Die beiden Sperrholzplatten, deren PEFC-zertifiziertes Holz aus Deutschland stammt, und die übrigen aus Spezialaluminium bestehenden Rahmenelemente setzt der Käufer selbst zum fertigen Rad zusammen. Der leicht abnehmbare Lithium-Akku ist in die Sperrholz-Schichtkonstruktion integriert. Das Rad erfüllt die entsprechenden EN- und ISO-Normen.

Begründung der Jury
Dieses E-Bike verwirklicht einen nachhaltigen Ansatz und zieht mit seinem innovativen Sperrholzrahmen die Blicke auf sich.

SENGO 275
E-Mountain Bike

Manufacturer
Simplon Fahrrad GmbH, Hard, Austria
In-house design
Web
www.simplon.com

The Sengo 275 showcases a light full-carbon frame with integrated drive unit and battery for both protection and elegant appearance. This reduces the bike weight to approx. 17.5 kg. The Vibrex+ Advanced Damping technology and sturdy tyres with their high traction ensure an outstandingly comfortable ride even in difficult terrain. All cables and tubes are completely routed inside the frame for protection. The installation points for kickstand and luggage carrier are integrated unobtrusively into the frame.

Das Sengo 275 besitzt einen leichten Vollcarbon-Rahmen, in den Mittelmotor und Akku geschützt und formschön integriert sind. Sein Gesamtgewicht beträgt nur ca. 17,5 kg. Die Vibrex+ Advanced-Damping-Technologie sowie robuste Reifen mit Kurvengrip sorgen für hervorragenden Fahrkomfort auch im Gelände. Leitungen und Seile sind komplett im Rahmen verlegt und dadurch gut geschützt. Die Montagepunkte für Seitenständer oder Gepäckträger wurden unauffällig am Rahmen angebracht.

Statement by the jury
The clear and high quality design of the Sengo 275 has created optimal technological sophistication and outstanding user experience.

Begründung der Jury
In dem klar und hochwertig gestalteten Sengo 275 gehen technologische Raffinesse und außergwöhnlich gute Fahreigenschaften eine optimale Verbindung ein.

Elite Hybrid HPC SLT 29
E-Mountain Bike

Manufacturer
CUBE Bikes, Waldershof, Germany
In-house design
Maximilian Fricke
Web
www.cube.eu

The Elite Hybrid HPC 29 signifies high performance and sophisticated design. The tuned drive position helps to skilfully integrate the Bosch drive system into the newly developed high-end carbon frame. Sunk into the down tube, the battery is additionally protected from impact by a protection shield, and thanks to an innovative fastening mechanism it can be charged while attached to the bike. Additional reinforcements in the chain stays lend high stiffness to the frame.

Statement by the jury
High-quality components and a dynamic design vocabulary predestine this mountain e-bike for use in town and out in nature, turning it into a true eye-catcher.

Das Elite Hybrid HPC 29 steht für hohe Performance und ein ausgereiftes Design. Mithilfe der Tuned Drive Position wurde das Bosch-Antriebssystem geschickt in den neu entwickelten High-End-Carbonrahmen integriert. Der Akku ist im Unterrohr versenkt, wird durch das Protection Shield zusätzlich vor Schlägen geschützt und durch ein innovatives Verschlusssystem im Rad geladen. Zusätzliche Verstärkungen in den Kettenstreben sorgen für hohe Steifigkeit des Rahmens.

Begründung der Jury
Hochwertige Komponenten und eine dynamische Formensprache prädestinieren dieses E-Mountainbike für Stadt und Natur und machen es zum auffälligen Blickfang.

LEAOS
E-Bike

Manufacturer
Leaos, Bolzano, Italy
In-house design
Francesco Sommacal
Web
www.leaos.com

The Leaos solar E-Bike catches the eye with its distinctive design. Instead of familiar tubes, it showcases a compact carbon body that integrates all technical components, including a display and a light system by Supernova. Equipped with a stepless NuVinci Harmony gear system, the rider can choose between manual and automatic gear modes. Leaos is designed as a unisex product ideally suited for everyday use.

Statement by the jury
This e-bike convinces not only with its eye-catching, formal independence, but also with a highly pleasant, smooth-running riding experience.

Das Solar-E-Bike Leaos fällt durch seine markante Formgebung ins Auge. Anstelle der üblichen Röhren präsentiert es einen kompakten Carbonkörper, in den alle technischen Komponenten wie ein Display und eine Lichtanlage von Supernova integriert sind. Ausgestattet mit dem stufenlosen Schaltsystem NuVinci Harmony kann der Fahrer zwischen manuellem und automatischem Schaltmodus wählen. Leaos ist als Unisexausführung gestaltet und für den Alltag hervorragend geeignet.

Begründung der Jury
Das E-Bike überzeugt nicht nur mit seiner ins Auge springenden formalen Eigenständigkeit, sondern auch mit einem angenehm leichtgängigen Fahrgefühl.

Gocycle G2
E-Bike

Manufacturer
Karbon Kinetics Ltd, London, Great Britain
In-house design
Richard Thorpe
Web
www.gocycle.com

The Gocycle G2 e-bike merges the joy of cycling with stylish, high-quality technical components. It is manufactured with a lightweight injection-moulded magnesium alloy construction. An integrated dash display monitors the lithium battery status, gear position, speed and driving mode. Other features include pedal torque sensing, electronic predictive gear shift, as well as proprietary side mounted quick-detach PitstopWheels®. The front-hub motor, also a proprietary feature, separates the battery-powered drive from the rear pedal drive. The automotive inspired frame design adjusts to different body types and folds into a compact storage size. Gocycle's appealing appearance reflects a smooth, fun ride.

Das Gocycle G2 verbindet die Freude am Radfahren mit stilvollen und hochwertig technischen Komponenten. Es ist aus einer leichten Spritzgussmagnesiumlegierung gefertigt. Eine im Lenker integrierte Anzeige zeigt den Lithium-Batterieladezustand, die Gangschaltung, die Geschwindigkeit und den Fahrmodus an. Weitere Features sind die Pedaldrehmomenterfassung, die elektronische Gangschaltung sowie die seitlich montierten und schnell abnehmbaren PitstopWheel®-Laufräder. Der Frontnabenmotor trennt den Batterieantrieb von dem rückseitigen Pedalantrieb. Die von Automobilen inspirierte Rahmengestaltung passt sich optimal an unterschiedliche Körpergrößen an und lässt sich einfach zu kompakten Dimensionen falten. Das ansprechende Erscheinungsbild von Gocycle spiegelt sich in dem entspannten und unterhaltsamen Fahrvergnügen wider.

VELLO bike
Folding Bicycle
Faltrad

Manufacturer
VELLO bike, Vienna, Austria

Design
Valentin Vodev, Vienna, Austria

Web
www.vello.bike
www.valentinvodev.com

reddot award 2015
best of the best

Compact mobility

The VELLO bike combines both sophisticated function-ality and aesthetic appeal. It is a compact and man-oeuvrable folding bicycle developed for urban mobility and designed to master short and long distances easily. The 20" wheels allow for speedy handling, perfect for the city where many stop-and-go moves are part of one's everyday commute. With its unique folding mechanism, it can be easily folded to half its size with-in seconds and then conveniently wheeled along – perfect for bringing on public transportation or fitting into small elevators. Once at destination, the VELLO bike can be fully folded for storage or transportation without the use of any additional tools. Its robust, unisex chrome-moly steel frame, as well as the spe-cially designed magnet- and rear suspension system that also includes the folding mechanism, helps to ideally smoothen the ride on bumpy streets. Another well thought-out detail are the foldable fenders de-signed to protect the rider's clothing, especially in wet weather conditions. All bicycles are manufactured individually by hand and offer users astonishing riding comfort. The handlebar and seat post height are easily adjustable to individual user requirements. Emerging as a contemporary and highly sophisticated reinter-pretation of a folding bike, it presents the user with a functional and graceful design that leaves a lasting impression.

Kompakte Mobilität

Das VELLO bike beeindruckt durch eine ausgeklügelte Funktionalität ebenso wie durch seine Ästhetik. Es ist ein kompaktes und wendiges 20"-Faltrad, das speziell für die urbane Mobilität entwickelt wurde und das jede Kurz- und Langstrecke mit Leichtigkeit bewältigt. Dank seines einzigartigen Faltmechanismus lässt sich seine Gesamtlänge sekundenschnell und unkompliziert um etwa die Hälfte falten und anschließend bequem wei-terrollen. Es ist deshalb ideal zum Mitnehmen, zum Beispiel in engen Liften oder öffentlichen Verkehrs-mitteln. Für den Transport oder zum Verstauen kann das VELLO bike ohne Verwendung von Werkzeugen komplett gefaltet werden. Der stabile Unisex-Chrom-Molybdän-Stahlrahmen sowie das eigens angefertigte Magnet- und Stoßdämpfersystem, das auch den Falt-mechanismus beinhaltet, federn die Unebenheiten der Straße ideal ab. Ein gut durchdachtes Detail ist zu-dem, dass sogar das Schutzblech faltbar ist, damit die Kleidung des Fahrers auch bei Regenwetter sauber bleibt. Dieses einzeln von Hand gefertigte Faltrad bietet den Nutzern darüber hinaus einen erstaunlichen Fahr-komfort, da es wendig ist und sich mit seiner Mehr-gangschaltung im Verkehr gut anpassen kann. Lenker-höhe und Sattelstütze lassen sich mit Leichtigkeit individuell auf den Fahrer abstimmen. Als eine zeitge-mäße und sehr durchdacht gestaltete Neuinterpreta-tion des Faltrades erfreut es den Fahrer jeden Tag mit seiner Funktionalität und Grazie.

Statement by the jury

This folding bike fascinates with sophisticated func-tionality perfectly adapted to everyday use in urban areas. It captivates users with its well-balanced de-tails and allows intuitive folding to a compact size. Equipped with multiple gears and specially designed magnetic shock absorbers, the bike delivers outstand-ing riding comfort. A beautiful bike that offers itself as an ideal companion for everyday use.

Begründung der Jury

Dieses Faltrad begeistert mit einer ausgereiften Funk-tionalität, die perfekt auf das Leben in der Stadt eingestellt ist. Es verblüfft durch gut aufeinander ab-gestimmte Details und kann auf intuitive Weise kom-pakt zusammengefaltet werden. Ausgestattet mit einer Mehrgangschaltung und speziellen magnetischen Stoß-dämpfern zeichnet es sich durch hohen Fahrkomfort aus. Das schöne Faltrad ist damit der ideale Begleiter für jeden Tag.

Designer portrait
See page 36
Siehe Seite 36

Lector ULC World Cup
Mountain Bike

Manufacturer
GHOST-Bikes GmbH, Waldsassen, Germany
In-house design
Web
www.ghost-bikes.de

With the Lector ULC World Cup, this bicycle manufacturer introduces a new model in the series of race-oriented hardtails. The model features a Tune 29" wheelset and a light carbon frame that keeps the overall weight low. Innovative technical components such as a full Shimano XTR fit-out, Ritchey WCS carbon mounting parts, as well as Fox forks with Kashima coating in combination with needle bearings, all lend this bike outstanding ride characteristics that make the Lector ULC World Cup with its angular yet dynamic design emerge as a high-end hardtail.

Mit dem Lector ULC World Cup präsentiert der Hersteller ein neues Modell dieser Reihe, die auf hohe Geschwindigkeiten ausgerichtet ist. Es verfügt über 29"-Tune-Laufräder und einen leichten Carbon-Rahmen, der das Gewicht niedrig hält. Innovative technische Komponenten wie die komplette Shimano-XTR-Ausstattung, spezielle Carbon-Anbauteile und ein Fox-Fahrwerk mit Kashima-Beschichtung in Verbindung mit Nadellagern sorgen für hervorragende Fahreigenschaften und machen das dynamisch und kantig gestaltete Lector ULC World Cup zu einem anspruchsvollen Hardtail.

Statement by the jury
The high-performance technology and flowing design of the distinctive Lector model deliver optimum power transmission therefore creating an exciting all-mountain biking experience.

Begründung der Jury
Die leistungsfähige Technik und fließende Formgebung des ausdrucksstarken Lector-Modells sorgen für einen optimalen Kraftfluss und damit auch für eine packende All-Mountain-Erfahrung.

Commuter
Bicycle
Fahrrad

Manufacturer
Canyon Bicycles GmbH, Koblenz, Germany
In-house design
Web
www.canyon.com

The Commuter urban bicycle combines purist elegance with athletic functionality. Ascending clear lines blend together seamlessly around the cockpit, forming a cohesive unity with the frame and illumination. Internal cable and line routing, the light built into the cockpit and an integrated dynamo in the front dropout further contribute towards the clean design of this light and agile bicycle. Low-maintenance parts, such as the belt drive, internal hub shifting and hydraulic disc brakes, ensure high reliability, while leather contact points underline a high-quality finish.

Das Urban Bike Commuter verbindet puristische Eleganz mit sportlicher Funktionalität. Aufsteigende klare Linien treffen am Cockpit zusammen und bilden mit Rahmen und Beleuchtung eine formale Einheit. Innenverlegte Züge und Leitungen, die im Cockpit integrierte Beleuchtung sowie der in den Ausfallenden kontaktierte Dynamo unterstreichen die aufgeräumte Gestaltung des leichten, agilen Fahrrads. Wartungsarme Komponenten wie Riemenantrieb, Nabenschaltung und hydraulische Scheibenbremsen bieten hohe Zuverlässigkeit und die Kontaktflächen aus Leder betonen den Qualitätsanspruch.

Ubike Smart
Folding Bicycle
Faltrad

Manufacturer
Ubike Co., Ltd., Changhua, Taiwan
In-house design
Web
www.ubike-tech.com

The Ubike Smart folding bicycle showcases a streamlined frame that stands out with its original two-tone paint finish. Instead of being founded on a trouble-prone magnetic mechanism, the easy-to-operate folding system follows the idea of "push and fold" based on the connection between the x, y and z axes. This smart stay fold system is highly user-friendly; even when folded, the bicycle can be rolled along and therefore need not be carried by hand.

Das Faltrad Ubike Smart besitzt einen stromlinienförmigen Rahmen, der durch eine originelle Lackierung in zwei Farben hervorsticht. Sein einfach zu bedienendes Faltsystem nach dem Motto „Drücken und Falten" basiert statt auf dem anfälligen Magnetmechanismus auf der Verbindung von x-, y- und z-Achse. Dieses „Smart-Stay-Faltsystem" ist sehr benutzerfreundlich; das Rad lässt sich auch in gefaltetem Zustand rollen und muss daher nicht getragen werden.

Statement by the jury
The gently curved, distinctively coloured frame lends this bicycle a sporty appearance. The sophisticated folding mechanism is also an outstanding feature.

Begründung der Jury
Sein sanft gebogener und prägnant lackierter Rahmen verleiht dem Faltrad eine sportive Anmutung. Bemerkenswert ist auch sein ausgeklügelter Faltmechanismus.

Strive CF
Mountain Bike

Manufacturer
Canyon Bicycles GmbH, Koblenz, Germany
In-house design
Web
www.canyon.com

Focusing on dynamics and design, the full-suspension Strive CF merges a purist, straightforward language of form with innovative technology. At the core is Shapeshifter, a system that allows the rider to transform the frame geometry, kinematics and suspension travel of the bicycle at the push of a button, thus perfectly adapting it to the terrain while riding. Complementing the vision of a perfect design, the gear and brake cable routing has been fully integrated into the frame.

Das auf Dynamik und Design fokussierte vollgefederte Strive CF kombiniert eine puristische, geradlinige Formensprache mit innovativer Technik. Der Shapeshifter, das Herzstück des Rahmens, ermöglicht es dem Fahrer, die Geometrie, Kinematik und den Federweg des Bikes per Knopfdruck zu verändern und es so während der Fahrt genau an die Umgebung anzupassen. Im Sinne der Vision einer perfekten Gestaltung wurden die Schalt- und Bremszüge komplett in den Rahmen verlegt.

Statement by the jury
A distinctive frame design and components that adapt flexibly at the push of a button define the Strive CF as a high-performance mountain bike.

Begründung der Jury
Eine markante Rahmengestaltung und flexibel per Knopfdruck veränderbare Komponenten kennzeichnen das Strive CF als sehr leistungsstarkes Mountainbike.

Super Record EPS 2015
Groupset for Road Bikes
Schaltgruppe für Rennräder

Manufacturer
Campagnolo S.r.l., Vicenza, Italy

In-house design
R&D Campagnolo

Web
www.campagnolo.com

reddot award 2015
best of the best

Comfortable precision

In cycling sports shifting gears quickly is of utmost importance. Even the slightest irregularity can affect the riding flow and translate into a delay in time. The Super Record EPS 2015 is a groupset for road bikes that delivers a drivetrain for highly effective gear shifting performance. A central aspect of its design is that it merges the know-how of a long-established company with highly advanced technology. Specifically, the rear and front derailleurs are equipped with electric motors that allow the rider to shift gears in a quick and highly precise manner. A single action via the shift control makes it possible to intuitively and seamlessly shift up or down for a multiple number of sprockets (multi-shifting). The shifting itself offers a high degree of quality and comfort, since the "click feel" and position of the controls have been optimised in order to prevent any risk of unintentional shifting. Another innovation in the field of gear shifts that further enhances the safety of the rider has been realised with the possibility of releasing the rear derailleur from the motor manually for emergency functionality in the event of a fault. Carbon fibre and titanium have been brought together to make the groupset very light and durable. The seamless merging of tradition with modern technology has thus led to an innovative product of elegant appearance that signifies high efficiency and safety for the rider.

Komfortable Präzision

Im Radsport ist das richtige Schalten von enormer Bedeutung. Schon kleinste Verzögerungen stören den Fahrfluss und bedeuten sofort einen Zeitverlust. Die Super Record EPS 2015 ist eine Schaltgruppe für Rennräder, die dem Fahrer ein sehr effektives Schaltverhalten ermöglicht. Ein zentraler Aspekt ihrer Gestaltung ist, dass sich hier das Wissen eines traditionsreichen Unternehmens gekonnt mit hochentwickelter Technologie vereint. So sind bei diesen Schaltungen im hinteren Schaltwerk sowie im vorderen Umwerfer Motoren integriert, die dem Fahrer einen schnellen wie auch präzisen Gangwechsel erlauben. Mit nur einer Bedienung des Schalthebels kann er intuitiv mehrere Gänge gleichzeitig und ohne Unterbrechung hinauf- oder herunterschalten (Multi-Shifting). Der Vorgang des Schaltens selbst bietet ein hohes Maß an Qualität und Komfort, denn das Feedback des Einrastens wurde so optimiert, dass ein unabsichtliches Schalten ausgeschlossen ist. Sicherheit gibt dem Fahrer zusätzlich die im Bereich der Schalttechnik innovative Möglichkeit, den Motor des hinteren Schaltwerks von der Mechanik zu entkoppeln, um in Notsituationen immer noch manuell schalten zu können. Da sie aus den Materialien Carbon und Titan gefertigt wird, ist diese Schaltgruppe langlebig und sehr leicht. Indem Tradition und moderne Technologie nahtlos ineinandergreifen, entstand so ein innovatives, elegant anmutendes Produkt, das für den Radfahrer hohe Effizienz und Sicherheit bedeutet.

Statement by the jury

Thanks to an elegant design, the Super Record EPS 2015 merges the advantages of mechanical and electronic bicycle technology into an innovative gear shift groupset of the highest quality. The approach of integrating electric motors into both the rear and front derailleurs offers riders maximum gear shifting precision and effectiveness. It facilitates safe and seamless shifting up or down several sprockets with one single action.

Begründung der Jury

Mittels einer eleganten Formensprache vereint die Gestaltung der Super Record EPS 2015 die Vorteile mechanischer und elektronischer Fahrradtechnik zu einer innovativen Schaltung auf höchstem Niveau. Dem Fahrer bietet sich durch die Integration von Motoren im hinteren Schaltwerk wie auch im vorderen Umwerfer ein Höchstmaß an Präzision und Effektivität des Schaltens. Mit einer einzigen Schaltaktion lassen sich mehrere Gänge zugleich und ohne Unterbrechung sicher betätigen.

Designer portrait
See page 38
Siehe Seite 38

Schwalbe Procore
Dual Chamber System
Doppelkammersystem

Manufacturer
Schwalbe – Ralf Bohle GmbH,
Reichshof, Germany

In-house design
Markus Hachmeyer

Design
Syntace GmbH (Oliver Zuther),
Tacherting, Germany

Web
www.schwalbe.com
www.syntace.de

reddot award 2015
best of the best

Safe grip

In mountain bike sports the tyres play a central role.
The quality of the tyres is decisive, particularly in
demanding terrain and on difficult trails, and whether
the rider has good control over the bike or not. The
innovative Schwalbe Procore dual chamber system
ensures maximum performance of and control over
tyre pressure. The well-conceived idea of the dual
chamber system implies differentiated distribution of
air into the tyres according to preference and terrain
condition. This is achieved by the core chamber of the
tyre being fillable with high pressure of around four
to six bars for extremely high shock absorption and
excellent emergency running properties, and the outer
chamber allowing the pressure to be reduced to only
one bar. In combination, this means that the outer
chamber can be driven with extremely low pressure
for enormously improved tyre performance. A central
role in this system of optimised pressure is played by
a patented dual valve that distributes the air into both
chambers. The upper part of the valve is a selector
which is simply screwed in or out to individually fill
the desired chamber. An additional "air-guide" integrated over the valve then allows air to enter into the
outer chamber. The result of this new tyre concept
is that it allows mountain bikers to ride much lighter
tyres even in hard use conditions, thanks to a sophisticated design and user-friendly functionality.

Sicher im Griff

Im Mountainbikesport spielen die Reifen eine zentrale Rolle. Besonders in anspruchsvollem Gelände oder
auf schwierigen Trails entscheidet die Qualität der
Reifen darüber, ob der Fahrer das Rad gut im Griff hat
oder nicht. Mit dem innovativen Doppelkammersystem Schwalbe Procore wird eine sichere Kontrolle des
Reifenluftdrucks ermöglicht. Eine gut zu Ende gedachte
Zweiteilung impliziert eine differenzierte Verteilung
der Luft in den Reifen, je nach Bedarf oder Gelände-
beschaffenheit. Erreicht wird dies durch die Tatsache,
dass im Inneren der Reifen eine Hochdruckkammer mit
etwa vier bis sechs Bar für eine extrem hohe Durchschlagsicherheit und exzellente Notlaufeigenschaften
sorgt. In der Außenkammer dagegen lässt sich der
Druck bis auf ein Bar reduzieren. Dadurch kann man
den äußeren Reifen mit extrem geringem Luftdruck
fahren, was zu einer enorm verbesserten Reifenper-
formance führt. Eine zentrale Rolle in diesem System
der Optimierung des Luftdrucks spielt ein patentiertes
Dualventil, welches die Luft in beide Kammern verteilt.
Der obere Teil des Ventils ist ein Selektor, über den
durch einfaches Ein- und Ausdrehen die ausgewählte
Kammer befüllt werden kann. Ein hier zusätzlich über
dem Ventil integrierter Air-Guide lenkt die Luft dann in
die Außenkammer. Das Ergebnis dieses neuen Reifen-
konzepts ist, dass der Mountainbiker damit selbst im
harten Einsatz deutlich leichtere Reifen fahren kann –
dank einer durchdachten Gestaltung und nutzerfreundlichen Funktionalität.

Statement by the jury

Embodying an innovative principle of a dual chamber
system, the Schwalbe Procore also sets a new aesthetic standard in the field of tyre technology for
mountain bikes. The pressure in the tyre can be controlled via a patented dual valve, with the high-
pressure chamber in the tyre's core ensuring high
shock absorption. The result is a clearly improved performance, and the possibility to ride a much lighter
tyre even under hard use.

Begründung der Jury

Schwalbe Procore setzt mit dem innovativen Prinzip
des Doppelkammersystems auch ästhetisch neue
Standards im Bereich der Reifentechnik für Mountain-
bikes. Der Luftdruck im Reifen kann über ein patentiertes Dualventil gesteuert werden, wobei eine
Hochdruckkammer im Inneren eine äußerst hohe
Durchschlagsicherheit gewährleistet. Das Ergebnis ist
eine deutlich verbesserte Performance, und auch
unter extremen Bedingungen können sehr viel leichtere
Reifen gefahren werden.

Designer portrait
See page 40
Siehe Seite 40

PROCORE

SCHWALBE DUAL CHAMBER SYSTEM | 27.5" | 4-6 BAR | 55-75 PSI

iGauge MINI VELOCE ROA
Mini Pump
Minipumpe

Manufacturer
Airace Enterprise Co., Ltd.,
Taichung, Taiwan
In-house design
Web
www.airace.com.tw

The mini pump developed for racing bicycles is equipped with Bluetooth 4.0 technology. The pump pressure is no longer shown on the pump itself but is sent directly to the iGauge app and thus directly to the user's smart mobile device. Made of CNC-milled aluminium, the iGauge Mini Veloce Roa features a clever thumb-lock twin valve. Presta valves do not separate when the pump head is taken off after use.

Statement by the jury
Transmitting the pump pressure via an app onto a smart phone turns the mini pump into a smart high-tech tool. The pump also boasts an elegant appearance.

Diese für Rennräder entwickelte Minipumpe ist mit der Bluetooth-4.0-Technologie ausgestattet. Sie zeigt den gemessenen Pumpendruck nicht mehr selbst an, sondern schickt ihn an die App iGauge und damit direkt auf das Smartphone des Benutzers. Die aus CNC-gefrästem Aluminium gefertigte iGauge Mini Veloce Roa besitzt ein Doppelventil mit Daumenarretierung. Nach der Anwendung können Presta-Ventile nicht mehr abgetrennt werden, wenn der Pumpenkopf vom Rad entfernt wird.

Begründung der Jury
Der via App auf das Smartphone übertragene Pumpendruck macht die Minipumpe zu einem smarten Hightech-Tool. Darüber hinaus sieht sie auch noch edel aus.

Transformer X
Pump
Pumpe

Manufacturer
Topeak, Inc., Taichung, Taiwan
In-house design
Bill Liao, Louis Chuang
Web
www.topeak.com

The Transformer X stand/floor pump comes with an integrated bike stand. The pump features a hose long enough to easily reach the front wheel, as well as extendable support legs for greater stability when pumping. Adjustable frame hooks that can adapt to any frame size, without the need for tools, provide support for maintenance work and storage needs. The gauge is conveniently located at the top of the stand, while the automatic SmartHead pump accurately regulates the pressure.

Statement by the jury
This stand/floor pump captivates with original practical details and a functionality that is oriented towards user comfort and ease of operation.

Die Standpumpe Transformer X wartet mit integriertem Bikestand auf. Der Pumpschlauch ist lang genug, um auch das Vorderrad zu erreichen, wobei ausziehbare Standbeine zusätzliche Stabilität gewährleisten. Die einstellbaren Rahmenaufnahmen können ohne Werkzeug an das Fahrrad angepasst werden und geben Halt bei Montagearbeiten. Das Manometer ist gut sichtbar am oberen Teil der Standpumpe angebracht. Ein automatischer SmartHead-Kopf reguliert den Luftdruck.

Begründung der Jury
Diese Standpumpe besticht durch originelle, praktische Details und eine Funktionalität, die sich an einer einfach zu handhabenden Bedienung orientiert.

Tool Monster
Bike Tool
Fahrradwerkzeug

Manufacturer
Topeak, Inc., Taichung, Taiwan
In-house design
Dennis Chiang, Louis Chuang
Web
www.topeak.com

Tool Monster is an all-in-one tool with 22 integrated functions suitable for almost all bike fixing tasks including fixing a broken chain, installing and removing pedals, tightening a loose crank bolt and adjusting the derailleur or seat – be it mountain, trekking or racing bikes. The two-piece design delivers higher leverage than comparable mini tools of this type and thus makes dealing with bike maintenance issues much easier.

Statement by the jury
The demands on mini tools are high: they have to be both compact and all-rounders. The cleverly designed Tool Monster sets a new high standard in this respect.

Das Tool Monster ist ein All-in-One-Tool und kombiniert 22 Werkzeuge, mit denen sich nahezu alle Fahrradreparaturen wie die Behebung von Kettenschäden, Pedalwechsel, Anziehen loser Kurbelschrauben, Einstellen des Schaltwerks oder Feinjustierung des Sattels durchführen lassen – am Mountainbike, Trekking- oder Rennrad. Die zweiteilige Konstruktion ermöglicht deutlich größere Hebelkräfte als herkömmliche Minitools und erleichtert anfallende Ausbesserungen dadurch maßgeblich.

Begründung der Jury
Die Anforderungen an Multitools sind hoch: Sie müssen zugleich Alleskönner und kompakt sein. Das klug entwickelte Tool Monster erfüllt diese Ansprüche vorbildlich.

KS LEV Ci
Dropper Seat Post
Sattelstütze

Manufacturer
Kind Shock Hi-Tech Co., Ltd.,
Tainan, Taiwan
In-house design
Grace Chen
Web
www.kssuspension.com

The KS LEV Ci, with its 65 mm adjustment range, aims to complement any bike. The unidirectional carbon mast routes the Recourse Ultralight cable system through the frame to a KGSL remote, thus saving over 50 grams. The new cable interface system at the lower end of the seat post is easy to set up and service, and the saddle is held by a high compression-moulded head. Thanks to a special roller clutch bearing and seal collar, the KS LEV Ci runs smoothly and free of side movement.

Statement by the jury
The design of the KS LEV Ci convinces with its sophisticated technical construction that also keeps the weight low.

Mit dem 65-mm-Verstellbereich möchte die KS LEV Ci ein Bike sinnvoll ergänzen. Der unidirektionale Schaft leitet das Recourse-Ultralight-Kabel durch den Rahmen zum KGSL-Remotehebel, sodass sich über 50 Gramm Gewicht sparen lassen. Das neue Verbindungsstück am unteren Ende der Stütze ist leicht zu installieren. Den Sattel hält ein hoch komprimierter Carbon-Kopf. Dank der Führung mit dem speziellen Walzenlager und der gedichteten Überwurfmutter läuft die KS LEV Ci sanft und ohne Spiel.

Begründung der Jury
Die Gestaltung der KS LEV Ci überzeugt durch ihre ausgeklügelte technische Konstruktion, die gleichzeitig für Gewichtsersparnis sorgt.

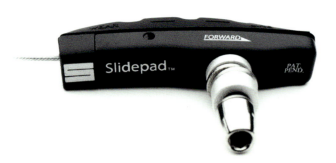

SureStop
Bicycle Braking System
Fahrrad-Bremssystem

Manufacturer
SureStop, Irvine, USA
In-house design
Web
www.surestop.bike

SureStop is a patented, mechanical solution to make bicycle brakes easier to operate and safer by preventing wheel lock while braking. Unlike traditional two-lever brake systems, this system prevents the wheels from locking so riders can maintain more control during braking. The simple one-lever, all-wheel braking mode integrates seamlessly into a common bicycle brake system with its clear design.

Statement by the jury
Safety was put first in this braking system with its deliberately simple, functional design. It is easy to operate and suitable for almost all bicycle types.

SureStop ist eine patentierte, mechanische Lösung, um Fahrradbremsen leichter bedienbar und durch Antiblockiersystem sicherer zu machen. Anders als bei herkömmlichen Zweihebelsystemen verhindert sie, dass die Räder blockieren und ermöglicht so eine bessere Kontrolle beim Bremsen. Der schlichte, über einen Hebel gesteuerte Allradbremsansatz passt sich mit seiner klaren Gestaltung nahtlos in gängige Radbremsen ein.

Begründung der Jury
Die Sicherheit stand bei diesem schlicht und funktional gestalteten Bremssystem im Vordergrund. Einfach zu bedienen, ist es für nahezu alle Fahrräder verwendbar.

AX-009DG & AX-007GR
Bike Light
Fahrradleuchte

Manufacturer
Gentos Co., Ltd., Tokyo, Japan
In-house design
Masahiro Saito
Web
www.gentos.jp

This aerodynamic bike light focuses on brightness and, thanks to the use of lateral emitters, is also visible from the left and right sides. It offers five different light patterns: normal, eco, high-power, rapid-flashing and slow-flashing. A built-in battery power indicator allows the energy status of the battery to be checked at a glance. The integrated battery is rechargeable via a Micro USB cable.

Statement by the jury
The sleek, compact design of the AX-009DG & AX-007GR emphasises the high functionality of this bike light, which provides illumination not only to the front but also to the sides.

Diese aerodynamische Fahrradleuchte legt Wert auf Helligkeit und ermöglicht mittels Seitenemittern, dass sie auch von links und rechts sichtbar ist. Sie bietet fünf verschiedene Lichtmuster an: Normal, Eco, High-Power, Schnelles Blinken und Langsames Blinken. Ein eingebauter Batteriestandsanzeiger macht die verbleibende Leistung der Batterie auf einen Blick sichtbar. Mit einem Micro-USB-Kabel lässt sich der integrierte Akku wieder aufladen.

Begründung der Jury
Die schlanke, kompakte Gestaltung der AX-009DG & AX-007GR betont die hohe Funktionalität dieser Leuchte, die nicht nur nach vorne, sondern auch zur Seite Licht abgibt.

GA2
Mountain Bike Grip
Mountainbike-Griff

Manufacturer
RTI Sports GmbH, Koblenz, Germany
In-house design
Ergon Werksdesign
Design
Veryday, Bromma, Sweden
Web
www.rtisports.de
www.ergon-bike.com
www.veryday.com

GA2 is an ergonomic all-round mountain bike grip designed to require less gripping force thanks to an optimised shape. It is suitable for extensive mountain biking, enduro racing and ambitious gravity riding adventures. Along with the internally butted thickness of the tubing, the soft and UV-resistant rubber compound delivers outstanding cushioning comfort. The internal aluminium clamp ensures proper upper-body positioning and twisting resistance.

Statement by the jury
The GA2 stands out by being pleasant to hold, and in particular for providing an extremely good grip, therefore proving itself as a high quality product for sporting adventures on the bike.

Der GA2 ist ein ergonomischer Allround-Mountainbike-Griff, der durch die optimierte Form geringe Greifkräfte erfordert und sich für ausgedehnte Mountainbike-Touren, für Enduro-Rennen oder ambitionierte Gravity-Abenteuer eignet. Die weiche, UV-stabile Gummimischung bietet zusammen mit der konifizierten Wandstärke des Innenkerns hervorragenden Dämpfungskomfort. Die innenliegende Klemme aus Aluminium hält auch auf Carbonlenkern mit höchster Verdrehfestigkeit.

Begründung der Jury
Die angenehme Haptik und vor allem die sehr gute Griffigkeit sind es, die den GA2 besonders auszeichnen und seine hohe Qualität bei sportlichem Fahren belegen.

Saddle with Modula System
Fahrradsattel-Modulsystem

Manufacturer
Selle Royal SpA, Pozzoleone (Vicenza), Italy
In-house design
Web
www.selleroyal.com

This new Selle Royal System is an innovative bicycle saddle construction: a saddle shell that can be combined with a variety of tops. Thanks to a special attachment system, it can accommodate a wide variety of coloured tops, as well as gel or memory-foam tops, and allows accessories to be attached between the base and the top of the saddle. This concept therefore also prevents theft, as the top can easily be taken off via a clip.

Das neue Selle Royal System ist eine innovative Konstruktion für Fahrradsättel: eine mit verschiedenen Oberteilen kombinierbare Sattelschale. Mithilfe eines speziellen Befestigungssystems lassen sich je nach Vorliebe verschiedenfarbige Oberteile sowie solche mit Gel oder Memory-Schaum verwenden und weitere Elemente zwischen Unter- und Oberteil des Sattels anbringen. Durch diese Kombination lässt sich die Oberseite auch einfach abnehmen und so Diebstahl vorbeugen.

Statement by the jury
The innovation of this modular saddle system convinces with two benefits: it can be customised according to individual taste and it prevents vandalism.

Begründung der Jury
Die Innovation dieses modularen Sattelsystems besticht durch zwei Vorzüge: Es lässt sich nach individuellem Geschmack bestücken und verhindert Vandalismus.

Saly Wheel
Cable Lock
Kabelschloss

Manufacturer
Knollan Ltd., Ramat Yishai, Israel
Design
micamino design (David Darki),
Ramot Menashe, Israel
Web
www.knollan.com
www.micamino-design.com

Instead using of wheels or dialers, this cable lock opens via a joystick-like button that can be moved in four directions – up, down, left, right – in any desired sequence. The numbers and letters around the button enable password selection in any length and easy-to-memorise by meaning. The unlocking is straightforward and can be done with one hand, making the lock suitable for operation in the dark and also catering to the needs of users with disabilities.

Statement by the jury
The Saly Wheel cable lock convinces with its innovative locking mechanism that is both safe and easy to operate.

Statt Zahlenrad oder Wahlvorrichtung besitzt dieses Kabelschloss einen joystick-artigen Knopf, der in jeder gewünschten Reihenfolge in vier Richtungen – nach oben, unten, links, rechts – bewegt werden kann. Über die Nummern und Buchstaben rund um den Knopf kann ein Passwort in jeder Länge und mit leicht zu erinnernder Bedeutung gewählt werden. Die Entsperrung ist unkompliziert und kann mit nur einer Hand erfolgen, sodass das Schloss auch im Dunkeln sowie von Menschen mit Behinderung benutzt werden kann.

Begründung der Jury
Das Kabelschloss Saly Wheel überzeugt durch seinen innovativen Schließmechanismus, der zugleich sicher ist und sich rasch öffnen lässt.

8960 Phantom
Bicycle Lock
Fahrradschloss

Manufacturer
ABUS August Bremicker Söhne KG,
Wetter, Germany
Design
Tom Ayton // aytonlab, Wuppertal, Germany
Web
www.abus.com
www.aytonlab.com

The armoured cable lock boasts a compact carrying size and a versatile adjustable length. Thanks to an innovative Fidlock magnetic holder, the lock can be quickly and easily attached to the bicycle frame and effortlessly removed by using just one hand. The 8960 Phantom offers a high degree of security and simplifies bicycle mobility. The PVC exterior finish protects against paintwork damage and lends the lock a simple yet stylish matt black appearance.

Statement by the jury
Thanks to its special length, the armoured cable lock offers high flexibility. The easy-to-use magnetic holder simplifies its operation.

Das Panzerkabelschloss zeigt ein kompaktes Transportmaß und eine vielseitig anwendbare Länge. Mithilfe eines innovativen Fidlock-Magnethalters lässt es sich leicht am Fahrradrahmen befestigen und mit einer Hand völlig mühelos wieder abnehmen. Das 8960 Phantom bietet ein hohes Maß an Sicherheit und vereinfacht die Mobilität mit dem Rad. Die PVC-Ummantelung beugt Lackschäden vor und erhält durch die mattschwarze Optik eine schlichte Anmutung.

Begründung der Jury
Dank seiner besonderen Länge bietet das Panzerkabelschloss eine hohe Flexibilität. Sein einfach zu handhabendes Magnethaltersystem erleichtert den Gebrauch.

POWER PLUS MOUNT
Mobile Phone Holder
Halter für Mobiltelefon

Manufacturer
Yiwu Welloo Sports Goods Co., Ltd.,
Yiwu, China
Design
Shenzhen Hanov Design Co., Ltd. (Ellen Su),
Shenzhen, China
Web
www.welloo.com.cn
www.hanov.com.cn

Power Plus Mount is a rechargeable mobile phone holder with a built-in 6,000 mAh lithium-polymer battery. The holder serves as a power source when using GPS, a stopwatch or WeChat, and it can also charge the headlight and tail light of a bicycle at the same time. By adopting aluminium as the connecting structure, it can be installed easily on the handlebars. The product is small, light and impact-resistant and provides a safe grip on the mobile phone.

Power Plus Mount ist ein wiederaufladbarer Halter für Mobiltelefone mit einer 6.000-mAh-Lithium-Polymer-Batterie. Während der Nutzung von GPS, Stoppuhr oder WeChat fungiert er als Stromquelle und kann gleichzeitig Vorder- und Rücklicht des Fahrrads aufladen. Mithilfe eines Verbindungsstücks aus Aluminium lässt er sich ganz einfach auf das Steuerrohr montieren. Das Produkt ist handlich, leicht und stoßfest und bietet dem Mobiltelefon einen sicheren Halt.

Statement by the jury
This compact mobile phone holder gains attention through its functional performance: it provides permanent power for the phone, but also for other bicycle components.

Begründung der Jury
Der kompakte Mobiltelefonhalter macht durch seine funktionale Leistung auf sich aufmerksam: Er liefert permanent Strom für das Telefon, aber auch für weitere Fahrradkomponenten.

Polar V650
Cycling Computer
Fahrradcomputer

Manufacturer
Polar Electro Oy, Kempele, Finland
In-house design
Lauri Lumme
Web
www.polar.com

The streamlined and 120 gram Polar V650 can be readily attached to a mount on a bike. Its large 2.8" colour touchscreen displays the latest training data, which can be easily switched by touching or swiping the screen. Users can arrange the display according to their priorities and therefore choose which details are shown during the ride. The front-mounted LED light enhances visibility when cycling in the dark.

Der aerodynamisch gestaltete, 120 Gramm leichte V650 lässt sich mit einer Halterung problemlos an das Rad montieren. Sein großes 2,8"-Farb-Touchdisplay visualisiert die aktuellen Trainingsdaten und ermöglicht schnelles Umschalten durch Berühren oder Wischen. Das Display lässt sich beliebig anordnen und zeigt die gewünschten Informationen während der Fahrt an. Das stirnseitig integrierte LED-Licht erhöht die Sichtbarkeit im Dunkeln.

Statement by the jury
This cycling computer catches the eye with its neat design and straightforward operation that allows cyclists to see their individual training data at just a glance.

Begründung der Jury
Der Fahrradcomputer springt aufgrund seiner handlichen Formgebung und einfachen Bedienung ins Auge, mit der sich die individuellen Daten leicht im Blick behalten lassen.

iGlow Cage B
Bottle Holder
Flaschenhalter

Manufacturer
Topeak, Inc., Taichung, Taiwan
In-house design
Louis Chuang, Terry Lee
Web
www.topeak.com

The integration of a bright RBG LED into the iGlow Cage B water bottle holder means a variety of colours illuminate the clear bottle, creating 360 degree visibility, thus increasing safety for cyclists. Alongside always on and blinking modes, variety is added by allowing easy switching between purple, blue, green, yellow and white light. The liquid in the bottle both enlarges the illumination and ensures an even glow of light.

Eine in den iGlow Cage B Flaschenhalter integrierte helle RGB-LED erleuchtet die durchsichtige Trinkflasche in brillanten Farben und erhöht durch die 360-Grad-Sichtbarkeit die Sicherheit im Straßenverkehr. Neben einem konstanten und einem blinkenden Leuchtmodus lassen sich per Knopfdruck die Farben Lila, Blau, Grün, Gelb und Weiß einstellen. Die in die Flasche gefüllte Flüssigkeit streut das Licht gleichmäßig und vergrößert die Leuchtfläche zugleich.

Statement by the jury
The iGlow Cage B stands out in the truest sense of the word through its innovative function of making the water bottle glow, thus enhancing the cyclist's safety.

Begründung der Jury
Der iGlow Cage B sticht im wahrsten Sinne durch die innovative Funktion hervor, dass er die Trinkflasche zum Leuchten bringt und so die Sicherheit erhöht.

Hyban
Bicycle Helmet
Fahrradhelm

Manufacturer
ABUS August Bremicker Söhne KG,
Wetter, Germany
Design
FWD Sailing, Rolle, Switzerland
Web
www.abus.com
www.forward-sailing.com

The Hyban helmet with an LED rear light offers reliable mobile safety for city cycling, commuting by bike, BMXing and skating. The ABS hard-shell design guarantees extra resistance, high shock-absorption and protection in the event of impact. Including 13 air inlets and 5 air outlets, the in-mould helmet possesses enhanced ventilation properties. The precise "Zoom Evo" adjustment system makes it easier to adjust the head size for both adults and young people.

Der Hyban-Helm mit LED-Rücklicht bietet zuverlässige mobile Sicherheit auf dem Skateboard, City-, Commuting- oder BMX-Rad. Seine ABS-Hartschalenkonstruktion ist besonders widerstandsfähig und gewährleistet hohen Aufprallschutz. Mit 13 Luftein- und fünf Luftauslässen verfügt der InMold-Helm über sehr gute Belüftungseigenschaften. Das fein justierbare Verstellsystem „Zoom Evo" erleichtert die Anpassung an den Kopfumfang von Erwachsenen und Jugendlichen.

Statement by the jury
Safety and functionality – the design of the Hyban helmet focuses on these factors to emerge as a companion that guarantees protection for everyday mobility.

Begründung der Jury
Sicherheit und Funktionalität – darauf legt die Gestaltung des Hyban-Helms großen Wert und macht ihn so zum zuverlässigen Schutz für den mobilen Alltag.

Overtake
Bicycle Helmet
Fahrradhelm

Manufacturer
Smith Optics, Ketchum, Idaho, USA
In-house design
Drew Chilson, Mike Aaskov
Design
Koroyd (Piers Storey), Monaco
Web
www.smithoptics.com
www.koroyd.com

The Overtake bicycle helmet uses a unique construction that absorbs 30 per cent more energy than traditional EPS foam, making it exceptionally impact-resistant. At the same time, the helmet is very light, weighing in at only 250 grams, and includes 21 air vents for balanced aerodynamic performance. Available in three sizes and 12 colour variations, the helmet is adjustable for a personalised and secure fit and can be partially customised with MIPS linings.

Der Fahrradhelm Overtake verwendet eine Aerocore-Konstruktion mit Koroyd, die 30 Prozent mehr Energie als herkömmlicher EPS-Schaum absorbiert und dadurch außergewöhnlich stoßsicher ist. Zugleich ist der Helm mit 250 Gramm sehr leicht und bietet mit 21 Lüftungsöffnungen eine ausgewogene aerodynamische Leistung. Er ist einstellbar auf eine personalisierte und sichere Passform und in drei Größen sowie zwölf Farbvarianten erhältlich, die teils zusätzlich mit MIPS ausgekleidet werden können.

Statement by the jury
The helmet merges extremely light, robust materials with a design that is focused on areodynamics to create a well-balanced symbiosis.

Begründung der Jury
Besonders leichte, stabile Materialien und eine auf Aerodynamik ausgerichtete Gestaltung gehen bei diesem Helm eine stimmige Symbiose ein.

CLOSCA fuga
Bicycle Helmet
Fahrradhelm

Manufacturer
CLOSCA Design, Ollería, Spain
Design
CuldeSac, Valencia, Spain
Web
www.closca.co
www.culdesac.es

The fuga bicycle helmet intelligently solves two flaws in common helmet design. Firstly, it is comprised of mobile rings that allow it to be folded flat therefore reducing its volume and making it easy to transport and store. The helmet is safety-certified and can fit into a handbag, briefcase or backpack. Secondly, fuga is stylish and is available in a variety of colours, therefore offering a more fashion conscious cyclist choice.

Der Fahrradhelm fuga umgeht zwei Nachteile üblicher Helme auf intelligente Weise: Er besteht aus beweglichen Ringen, die sich zusammenschieben lassen, sodass sich das sonst große Volumen verringert und er einfach zu transportieren oder zu verstauen ist. Der zertifizierte Helm passt in eine Handtasche, Aktentasche oder einen Rucksack. Zudem ist fuga stilvoll und lässt sich dank verschiedener wählbarer Farben auf den Geschmack des Benutzers abstimmen.

Statement by the jury
The design of fuga rethinks the bicycle helmet, and solves the impractical bulkiness experienced with other helmets through a smart, and above all, aesthetic folding mechanism.

Begründung der Jury
Die Gestaltung des fuga denkt den Fahrradhelm neu und löst seine oft unpraktische Sperrigkeit mit einem intelligenten und insbesondere ästhetischen Klappmechanismus.

BIKE TRAVEL BAG PRO
Bike Travel Bag
Fahrradreisetasche

Manufacturer
EVOC Sports GmbH, Munich, Germany
In-house design
Holger Feist, Marcus Kern
Web
www.evocsports.com

The Bike Travel Bag Pro protects bikes when travelling, allowing any bike type to be stowed within minutes. The glass fibre reinforced rods blend the padded substructure with the stable bottom structure. The frame is stowed upright and the wheels placed in separate side compartments. Most of the weight is located in the rear of the bag where it can be carried on stable running gear with smooth rolling wheels.

Statement by the jury
The Bike Travel Bag Pro provides an ideal solution for travelling with a bike. Its sophisticated construction offers optimum protection for the bike.

Die Bike Travel Bag Pro schützt Fahrräder beim Verreisen und verstaut verschiedenste Radtypen in wenigen Minuten. Der Oberbau ist durch glasfaserverstärkte Kunststoffstreben mit der stabilen Bodenkonstruktion verbunden und mit einer Polsterung verstärkt. Der Rahmen wird aufrecht verstaut, die Laufräder in separaten Seitentaschen. Das Hauptgewicht liegt im hinteren Teil der Tasche auf dem kippstabilen Fahrwerk mit seinen leicht laufenden Skaterollen.

Begründung der Jury
Die Bike Travel Bag Pro stellt ein ideales Behältnis dar, um ein Rad auf Reisen sicher zu transportieren. Ihre durchdachte Konstruktion bietet dem Fahrrad sehr guten Schutz.

Norco Urban Bag
Bicycle Bag
Fahrradtasche

Manufacturer
ASISTA Teile fürs Rad GmbH & Co. KG, Leutkirch, Germany
Design
Paula-D Design (Martina Stauber), Amtzell, Germany
Web
www.norco-bags.de
www.paula-d.de

Conceived for mobility on a bicycle, the Urban Bag features a design that is aligned to allow seamless use between office, family and everyday life. The bag has a clear, elegant appearance, is made of robust, water-resistant nylon material and attaches to all standard bicycle racks with self-explanatory vario hooks. Thanks to a tablet compartment with a direct-access window integrated into the top and a padded notebook compartment, it is highly versatile.

Statement by the jury
The Urban Bag lends itself to flexible use thanks to its convenient tablet and notebook compartments. It also epitomises the zeitgeist as a formally appealing piece of equipment.

Die für die Mobilität mit dem Fahrrad konzipierte Urban Bag ist auf den fließenden Einsatz zwischen Büro und Freizeit abgestimmt. Sie ist elegant und klar gestaltet, aus robustem, wasserabweisendem Nylonmaterial gefertigt und lässt sich mithilfe der Vario-Haken an allen gängigen Fahrradgepäckträgern befestigen. Dank des in den Deckel integrierten Tablet-Fachs mit Bedienfenster und eines gepolsterten Notebook-Fachs ist sie vielseitig nutzbar.

Begründung der Jury
Mit ihren praktischen Tablet- und Notebook-Fächern lässt sich die Urban Bag flexibel einsetzen und präsentiert sich zudem als dem Zeitgeist gemäßes, formschönes Gebrauchsutensil.

Urbanite ST 7600
Bicycle Backpack
Fahrradrucksack

Manufacturer
ABUS August Bremicker Söhne KG,
Wetter, Germany
Design
Culture Form, Berlin, Germany
Web
www.abus.com
www.culture-form.com

The Urbanite ST 7600 cycling backpack is optimally tailored to the needs of urban bikers offering, for instance, practical internal compartments and an innovative system for attaching a helmet. With its stable shape, 25 litre capacity and padded laptop compartment, it meets the important criteria for everyday use, commuting, as well as for the office. Water-resistant material and welded outer seams guarantee that the contents of the backpack are kept dry.

Statement by the jury
Featuring highly resistant materials and innovative details such as a helmet holder and large storage capacity, the Urbanite ST 7600 yields high practical utility.

Der Fahrradrucksack Urbanite ST 7600 ist ganz auf die Bedürfnisse urbaner Biker ausgerichtet und bietet z. B. eine praktische Innenaufteilung und ein innovatives System zur Befestigung eines Helms. Mit seiner stabilen Form, 25 Litern Stauraum und einem gepolsterten Laptop-Fach erfüllt er wichtige Kriterien für den Alltags-, Pendler- und Büroeinsatz. Das wasserdichte Material und verschweißte Außennähte gewährleisten den trockenen Transport des Inhalts.

Begründung der Jury
Mit widerstandsfähigen Materialien und innovativen Details wie Helmhalter und großem Stauraum erzielt der Urbanite ST 7600 eine hohe Praxistauglichkeit.

BE1 Enduro Protect
Mountain Bike Backpack
Mountainbike-Rucksack

Manufacturer
RTI Sports GmbH, Koblenz, Germany
In-house design
Ergon Werksdesign
Web
www.rtisports.de
www.ergon-bike.com

The design of the BE1 Enduro Protect combines a hydration bladder and spine protector into a functional and ergonomic pack that weighs less than 1 kg. The four-position backrest, self-adjusting shoulder straps and elastic waistband, for one-hand operation, guarantee a perfect fit. An outstanding feature is a moveable central part made of EVA lightweight moulded foam, which allows flexibility along the back and a holder for full face helmets.

Statement by the jury
The design of this mountain bike backpack provides an ideal fit to the body and is additionally supported by a sophisticated combination of materials.

Die Gestaltung des BE1 Enduro Protect, der mit Trinkblase und Protektor unter 1 kg wiegt, verbindet Funktionalität und Ergonomie. Eine 4-fache Rückenlängeneinstellung, selbstjustierende Schultergurte und ein elastischer Hüftgurt mit Einhand-Bedienung stellen eine perfekte Passform sicher. Hervorstechendes Merkmal ist ein bewegliches Mittelteil aus EVA-Leichtbauformschaum, das Flexibilität im Rückenbereich schafft und einen Halter für Integralhelme bietet.

Begründung der Jury
Die Formgebung dieses Mountainbike-Rucksacks ermöglicht einen idealen Sitz am Körper, der durch ausgeklügelte Materialkombinationen zusätzlich unterstützt wird.

Fitbit Surge
Fitness Watch
Fitnessuhr

Manufacturer
Fitbit Inc., San Francisco, USA
In-house design
Design
NewDealDesign LLC., San Francisco, USA
Web
www.newdealdesign.com
www.fitbit.com

Fitbit Surge is a compactly designed fitness watch with GPS, 8-sensor technology and continuous heart rate activity tracker. Equipped with a bright backlit LCD touch-screen display, it tracks and displays daily activity statistics in real time and allows monitoring of fitness sessions as well as all steps and calories burned throughout the day. The watch features automatic sleep recognition and synchronises all data with the Fitbit Dashboard and mobile app.

Statement by the jury
The fitness watch convinces with its broad range of functions, which allow for constant tracking, and through its clear and seamless design, which conveys high quality and reliability.

Fitbit Surge ist eine kompakt ausgeführte Fitnessuhr mit GPS, 8-Sensoren-Technologie und kontinuierlicher PurePulse-Herzfrequenzmessung. Mit einem hintergrundbeleuchteten LCD-Touchscreen-Display lassen sich Aktivitätsstatistiken in Echtzeit einsehen und Fitnesseinheiten, Schritte und Kalorienverbrauch überblicken. Die Uhr besitzt eine automatische Schlaferkennung und synchronisiert alle Daten mit dem Fitbit Dashboard und der mobilen App.

Begründung der Jury
Überzeugend an der Fitnessuhr ist ihr hoher Funktionsumfang mit ständigen Messungen sowie die klare, nahtlose Gestaltung, die Qualität und Zuverlässigkeit vermittelt.

Lohas Watch
Heart Rate Watch
Pulsfrequenzuhr

Manufacturer
Qisda Corporation, Taipei, Taiwan
Design
BenQ Lifestyle Design Center, Taipei, Taiwan
Web
www.qisda.com
www.benq.com

The Lohas Watch is a Bluetooth heart rate monitor watch, which monitors heart rate and burned calories. LED light beams and an electro-optical cell detect the volume of blood flow under the skin and display the current heart rate. An LED indicator notifies users of their heart rate zone; blue for low, green for optimal and orange for high intensity. The metallic surface and the silicon strap lend the watch its stylish look.

Statement by the jury
The design of the Lohas watch convinces with its concise multi-colour LED indicator and overall sporty look.

Die Lohas Watch ist eine Bluetooth-Pulsfrequenzuhr, die den Puls und den Kalorienverbrauch überwacht. LED-Lichtbalken und eine elektrooptische Zelle können das Blutvolumen unter der Haut ermitteln und die aktuelle Pulsfrequenz anzeigen. Die LED-Anzeige informiert mit drei Farben über den Frequenzbereich: Blau bedeutet niedrige, Grün optimale und Orange hohe Intensität. Die metallische Oberfläche und das Silikonarmband verleihen der Uhr einen stilvollen Look.

Begründung der Jury
Die Gestaltung der Lohas Watch überzeugt durch ihre prägnante LED-Anzeige in verschiedenen Farben sowie eine insgesamt sportive Anmutung.

POLAR A300
Activity Monitor
Aktivitätstracker

Manufacturer
Polar Electro Oy, Kempele, Finland
In-house design
Anna-Marja Suvilaakso
Web
www.polar.com

The A300 fitness and activity monitor combines 24/7 activity tracking with Smart Coaching features based on heart rate. It showcases a comfortable shape with rounded-off edges, and is available in six trendsetting colours. Connection with the accurate Polar H7 heart rate sensor allows for heart rate monitored training. Users thus gain insights into whether their training is improving fitness or burning fat. In addition, it also shows the total burned calories from both training and daily activity.

Statement by the jury
The A300 monitors the activities of the wearer around the clock and displays the collected data in a contemporary and sporting design.

Der A300 Aktivitätstracker kombiniert 24-Stunden-Tracking mit herzfrequenzbasierten Smart-Coaching-Funktionen. Er zeigt eine komfortable Formgebung mit abgerundeten Kanten, ist in sechs Trendfarben erhältlich und lässt sich für ein herzfrequenzorientiertes Training mit dem EKG-genauen Polar-Herzfrequenz-Sensor H7 verbinden. So erfährt der User etwa, ob er eher die Fitness oder den Fettstoffwechsel optimiert. Zudem wird der Kalorienverbrauch in Sport und Alltag angezeigt.

Begründung der Jury
Rund um die Uhr verfolgt der A300 die Aktivität des Nutzers und übermittelt ihm seine Werte in einem zeitgemäß und sportlich gehaltenen Design.

UP MOVE
Wearable Fitness Tracker
Tragbarer Fitness-Tracker

Manufacturer
Jawbone, San Francisco, USA
Design
fuseproject, San Francisco, USA
Web
www.fuseproject.com

Up Move is a modular designed fitness tracker that allows the discreet integration of its technology into the life of the user. At the core is a compact TR90 nylon pod that acts as both a pedometer and a sleep tracker. The pod can be used on its own either in the user's pocket or inserted into a clip or wristband. It is available in different colours, and simultaneously acts as a watch. Featuring a sleek design, the tracker is both stylish and easy to operate.

Statement by the jury
The fitness tracker embodies a symbiosis of urban form design and versatile functionality. It thus presents itself as an aesthetic and useful companion to everyday life.

Up Move ist ein modular gestalteter Fitness-Tracker, mit dem sich die technologischen Elemente diskret in den Alltag integrieren lassen. Herzstück ist ein kompakter TR90-Nylon-Pod, der Pedometer wie auch Schlaf-Tracker ist. Der Pod kann eigenständig in der Tasche, einem Clip oder in dem in verschiedenen Farben erhältlichen Armband mitgeführt werden und fungiert zugleich als Uhr. Mit seinem dezenten Design ist der Tracker stilvoll und einfach in der Handhabung.

Begründung der Jury
Der Fitness-Tracker stellt eine Symbiose aus urban anmutender Formgebung und vielseitiger Funktionalität dar. Er präsentiert sich so als ästhetischer wie nützlicher Alltagsbegleiter.

SEECARE HAT03 PC
Heart Rate Monitor Watch
Pulsuhr

Manufacturer
Guangzhou Shirui Electronics Co., Ltd., Guangzhou, China
In-house design
Chu Pengcheng, Tu Jianping
Web
www.cvte.com

The Seecare Smart heart rate monitor watch provides an innovative mechanical structure that aims to enhance the effectiveness of heart rate capturing during exercise. Made of metal and glass, the product conveys an elegant and sophisticated appeal. Turning the dial plate clockwise will tighten the wristband to activate the monitoring function, which continuously collects data on the heartbeat wave and the precise heart rate.

Statement by the jury
A strictly minimalist design and high-quality materials distinguish this heart rate monitor watch, a training accessory that is both functional and up to date.

Die Pulsuhr Seecare Smart verfügt über eine innovative mechanische Struktur, die die Herzfrequenz und damit die Leistungsfähigkeit während des Trainings misst. Das Produkt aus Metall und Glas vermittelt einen eleganten und hochwertigen Eindruck. Beim Drehen des Zifferblatts im Uhrzeigersinn verengt sich das Armband und aktiviert die Überwachungsfunktion, mit der laufend Daten zur Herzstromkurve und zur genauen Herzfrequenz erfasst werden.

Begründung der Jury
Eine streng minimalistische Gestaltung und hochwertige Materialien zeichnen diese Pulsuhr als nicht nur funktionales, sondern auch zeitgemäßes Trainings-Accessoire aus.

InLab3
Tracker

Manufacturer
InBody, Seoul, South Korea
In-house design
Seungmin Jun
Web
www.inbody.com

With its discreet size and lightweight quality, InLab3 is a stylish bracelet that can complement any look. An LED pattern shows the time and displays the daily status of the user's activities. Activated in the dark, it will shine the "sun" at night creating an atmospheric user experience. It is an activity tracker that at first glance looks more like a fashion statement.

Statement by the jury
The design of the InLab 3 tracker merges formal elegance with functional detail, whilst monitoring the fitness of the user, thus creating a stylish lifestyle product.

InLab3 präsentiert sich mit seiner dezenten Größe und Leichtigkeit als modebewusstes Armband, das zu jedem Stil passt. Ein LED-Muster zeigt die Zeit an und gibt Auskunft über den täglichen Status der Benutzeraktivitäten. Bei Nacht zeigt es im Dunkeln aktiviert die „Sonne" als stimmungsvolles Erlebnis. Das Armband ist ein Aktivitätstracker, der auf den ersten Blick wie ein modisches Statement erscheint.

Begründung der Jury
In der Gestaltung des Trackers InLab3 verschmelzen formale Eleganz und funktionale Details, die die Fitness des Benutzers messen, zum trendigen Lifestyleprodukt.

Misfit Flash
Fitness Monitor
Fitness-Tracker

Manufacturer
Misfit, Burlingame, USA
In-house design
Timothy Golnik, James Toggweiler
Web
www.misfit.com

The Misfit Flash is a sleek fitness tracker that measures the activity level and sleep patterns of the user. Thanks to its modular design, it can be worn flexibly as a wristwatch, key ring or it can be clipped to the user's clothes. The front face is a button, which, when pressed, displays a hidden array of LED lights that indicate fitness progress and time of day. The button can also be used to play and stop music, control connected lights or send quick messages to friends.

Der Misfit Flash ist ein schicker Fitness-Tracker, der das Aktivitätsniveau und Schlafverhalten des Nutzers misst. Dank seines modularen Designs lässt er sich flexibel als Armbanduhr, Schlüsselbund oder an Kleidung geklemmt tragen. Die Vorderseite besteht aus einem Knopf, der auf Druck eine sonst unsichtbare Reihe von LED einblendet. Diese zeigen den Fitness-status und die Uhrzeit an. Mit dem Knopf lassen sich auch Musik und Leuchtmittel steuern oder Nachrichten an Freunde verschicken.

Statement by the jury
The flexible usability and versatile functions define the Misfit Flash as an unobtrusive accessory that allows monitoring and the gaining of insights into one's fitness and sleep patterns at any time.

Begründung der Jury
Der flexible Einsatz und vielseitige Leistungsumfang definieren den Misfit Flash als jederzeit passendes Accessoire, mit dem sich die persönlichen Messwerte beständig im Blick behalten lassen.

iHealth Edge
Activity and Sleep Tracker
Aktivitäts- und Schlaf-Tracker

Manufacturer
Andon Health Co., Ltd., Tianjin, China
In-house design
Design
Beijing FromD Design Consultancy Ltd.,
Beijing, China
Web
www.ihealthlabs.com
www.fromd.net

The iHealth Edge is an activity and sleep tracker that can switch between four different modes, tracking steps and distances, calories burned and sleep efficiency. A fun activity level feature motivates the user to be more active and therefore reach higher levels. It can be wirelessly connected to a mobile device via Bluetooth 4.0 and stores up to 14 days of data. Powered by a rechargeable built-in battery, it typically lasts for five to seven days between charges.

Statement by the jury
This wearable tracker scores highly with an impressive array of functions, complemented by the display of motivational icons, while also serving as a classic digital watch.

Der Aktivitäts- und Schlaf-Tracker iHealth Edge erlaubt die Wahl zwischen vier verschiedenen Modi und überwacht sowohl Schritte bzw. Distanzen, Kalorienverbrauch als auch die Schlafeffizienz. Die spielerische Aktivitätsgrad-Funktion motiviert Träger dazu, aktiver weitere Stufengrade zu erreichen. Der Tracker kann kabellos über Bluetooth 4.0 mit anderen Geräten kommunizieren und speichert die Daten von bis zu 14 Tagen. Die eingebaute, wiederaufladbare Lithiumbatterie liefert Strom für fünf bis sieben Tage.

Begründung der Jury
Dieses Wearable imponiert mit einem ansehnlichen Funktionsumfang, ergänzt durch eine eingeblendete Motivationshilfe, und dient zugleich als klassische Digitaluhr.

RunPhones® Wireless Performance Headphones
Hochleistungskopfhörer

Manufacturer
AcousticSheep LLC, Erie, USA
In-house design
Dr Wei-Shin Lai
Web
www.runphones.com
Honourable Mention

The RunPhones Wireless Performance Headphones keep up with an active lifestyle. The patented "headphones in a headband" design secures ultra-thin traditional stereo headphones inside a comfortable, wicking, sport-style headband. The headphones pair with Bluetooth-enabled devices for hours of enjoyment and are rechargeable via micro-USB. They let in a small amount of outside noise for greater awareness, keep sweat at bay and are machine washable.

Statement by the jury
The RunPhones are securely fitting headphones presented in a stylishly designed appearance.

Der Hochleistungskopfhörer RunPhones Wireless passt sich dem aktiven Lebensstil an. Die patentierte Gestaltung als „Kopfhörer im Stirnband" bietet ultradünne traditionelle Stereokopfhörer in einem bequemen, feuchtigkeitsabsorbierenden Stirnband im Sportlook. Die Kopfhörer verbinden sich für langen Musikgenuss mit bluetoothfähigen Geräten und können über Micro-USB geladen werden. Sie lassen einen Teil der Umgebungsgeräusche zur Umweltwahrnehmung durchdringen, verhindern Schweißlauf und sind maschinenwaschbar.

Begründung der Jury
Das RunPhones sorgt für einen sicheren Sitz der Kopfhörer und präsentiert sich zugleich in stilvollem Look.

JuCad Phantom
Golf Caddy

Manufacturer
JUTEC Biegesysteme GmbH,
Limburg/Lahn, Germany
In-house design
Web
www.jucad.de
Honourable Mention

The JuCad Phantom is a golf caddy with a novel appearance distinguished by single spoke full-carbon wheels and a slender frame. Manufactured in titanium or carbon, it offers minimal weight with maximum stability. It conveniently folds into a flat packing size, is equipped with a powerful 48 volt propulsion system and two invisible high-tech motors, and uses safe magnet technology for the charging of its eco-friendly lithium batteries.

Der JuCad Phantom ist ein Golfcaddy, der mit Ein-Speichen-Felgen aus Vollcarbon und dem schlanken Caddyrahmen eine innovative Erscheinung darstellt. Gefertigt aus Titan oder Carbon, gewährleistet er hohe Stabilität bei minimalem Gewicht. Er lässt sich bequem auf ein flaches Packmaß zusammenfalten, besitzt ein kraftvolles 48-Volt-Antriebssystem, zwei unsichtbare Motoren und nutzt beim Ladevorgang der umweltfreundlichen Lithium-Hochleistungsakkus eine sichere Magnettechnik.

Statement by the jury
The JuCad Phantom merges design sophistication with modern technical detail into a golf caddy of a highly distinctive appearance.

Begründung der Jury
Im JuCad Phantom verschmelzen gestalterische Raffinesse und moderne technische Details zu einem markant aussehenden Golfcaddy.

IMAX
Electric Scooter
Elektroroller

Manufacturer
Yongkang Hulong Electric Vehicle Co., Ltd.,
Yongkang, China
In-house design
Can Xiang
Web
www.chinahulong.com

The Imax electric scooter combines convenience, speed and manoeuvrability. Its design, based on the smart integration of functional parts in the frame, underlines the urban and fun aspects of cruising around the city. Further benefits include: a wide deck for the users' feet to stand side by side, a one-hand folding mechanism ensuring user-friendliness, a silent and emission-free e-motor and a two-way adjustable disc brake.

Der Elektroroller Imax verbindet Manövrierfähigkeit mit Geschwindigkeit und Komfort. Seine Gestaltung basiert auf der intelligenten Integration funktionaler Teile in den Rahmen und unterstreicht so den urbanen Aspekt und den Spaßfaktor beim Fahren durch die Stadt. Weitere Pluspunkte sind die breite Ablage, auf der beide Füße genügend Platz finden, ein mit einer Hand bedienbarer, anwenderfreundlicher Klappmechanismus, ein leiser, emissionsfreier E-Motor sowie eine zweifach verstellbare Scheibenbremse.

Statement by the jury
Imax captivates not only through its user-friendly operation and manoeuvrability, but it also stands out in terms of its compact design.

Begründung der Jury
Der Imax besticht nicht nur mit seiner Wendigkeit und benutzerfreundlichen Bedienung, sondern macht auch durch eine kompakte Gestaltung auf sich aufmerksam.

SSE-TN1
Smart Tennis Sensor
Intelligenter Tennis-Sensor

Manufacturer
Sony Corporation, Tokyo, Japan
In-house design
Matthew Forrest, Yasuhide Hosoda,
Nobuhiro Jogano, Katsuji Miyazawa
Web
www.sony.net

Attached to a tennis racket, the SSE-TN1
sensor monitors swing speed, ball spin
and speed, as well as the impact position and other statistics in real time on
a mobile device. Using a specific sound
technology, the results are presented
visually to help users improve their skills.
Encased in a rubbery pod, the sensor
is very light, dust- and water-resistant,
and can be attached easily to the end of
the grip.

Der am Tennisschläger befestigte Sensor
SSE-TN1 kontrolliert die Geschwindigkeit
des Schwingens, Drall und Geschwindigkeit
des Balls, die Aufprallposition und andere
Daten in Echtzeit auf einem mobilen Gerät.
Mithilfe spezifischer Klangtechnologie
übertragen, werden die Ergebnisse visuell
angezeigt und so für den Anwender nach-
vollziehbar. Der mit Gummi verkleidete
kompakte Sensor ist sehr leicht, staub- und
wasserbeständig und lässt sich bequem in
das Griffende drehen.

Statement by the jury
This sensor merges sophisticated
technology and functional design into
a product that provides players with
important performance data.

Begründung der Jury
In diesem Sensor verbinden sich ausgeklügelte Technologie und funktionale Gestaltung zu einem Produkt, das dem Sportler
wichtige Erkenntnisse liefert.

Teqball
Sports Equipment/Sport
Sportgerät/Sportart

Manufacturer
Teqball Ltd., Budapest, Hungary
In-house design
Gábor Borsanyi, Viktor Huszár
Design
Co&Co Designcommunication Ltd.,
Budapest, Hungary
Máté Tóth
Web
www.teqball.com
www.coandco.cc

Teqball is based on football and it is a name given to both a piece of sports equipment and a unique new sport. The equipment, consisting of a football and a curved table similar to a tennis table, is a training tool for both professional and amateur sports people to develop their concentration, agility and stamina. The sport can be practised anywhere including outdoors, on the street or in a park, and by teams of various sizes. The rules are easy to understand and there is a low risk of injury. Teqball has been scientifically developed based on mathematical and physical calculations as well as empirical tests.

Teqball beruht auf Fußball und bezeichnet sowohl ein Sportgerät als auch eine Sportart. Die Ausrüstung, bestehend aus einer Art gebogener Tischtennisplatte und einem Fußball, ermöglicht Profis wie Amateuren, ihre Reaktionsfähigkeit, Beweglichkeit und Ausdauer zu trainieren. Die Besonderheit ist das Spielerische: So kann der Sport überall im Freien, auf der Straße oder der Wiese von unterschiedlich großen Teams ausgeübt werden. Die Regeln sind leicht verständlich und die Verletzungsgefahr ist gering. Teqball wurde auf Basis mathematischer und physikalischer Berechnungen sowie empirischer Tests entwickelt.

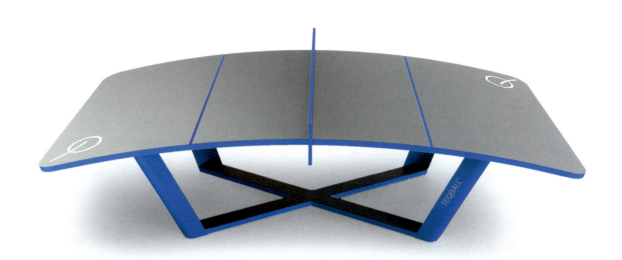

163

Adidas Smart Ball
Football
Fußball

Manufacturer
adidas, Portland, USA
In-house design
Scott Tomlinson
Web
www.adidas.com

The Adidas Smart Ball allows players to improve their striking technique by giving them feedback on the parameters of ball speed, spin and trajectory, as well as strike accuracy. Sensors in the ball analyse the forces acting on the ball, and send the data via Bluetooth to the Smart Ball app on the player's paired iOS device. The app can then calculate the ball's entire trajectory and display it as a precise animation. The Smart Ball is the same size as a regular football and feels like a premium match ball, so that players can't tell the difference in training. The charger evokes the culture of the game through a trophy metaphor.

Der Adidas Smart Ball ermöglicht es dem Spieler, seine Schusstechnik gezielt zu verbessern, denn er gibt ihm Feedback zu Geschwindigkeit, Flugbahn, Spin und Treffergenauigkeit. Sensoren im Inneren analysieren die Kräfte, die während des Flugs auf den Ball einwirken, und senden diese Daten per Bluetooth an die Smart-Ball-App auf dem iOS-Gerät des Spielers. Diese App kann die komplette Flugbahn berechnen und genaue Animationen der Schüsse aufzeichnen. Der Smart Ball hat die Größe eines normalen Fußballs und fühlt sich an wie ein Premium-Spielball, sodass kein Unterschied wahrnehmbar ist. In der Anmutung eines Pokals ruft das Ladegerät die Kultur des Spiels wach.

Equipped with sensors, the innovative Smart Ball opens up an entirely new dimension of football training; the accurate analysis of kicking technique promises high-precision training.

Der mit Sensoren ausgestattete, innovative Smart Ball kündet von neuen Wegen im Fußballtraining: Die präzise Analyse der Schusstechnik verspricht punktgenaues Trainieren.

REV'IT! Seeflex
Motorcycle Limb Protection
Gliedmaßenprotektoren für Motorradfahrer

Manufacturer
REV'IT! Sport International,
Oss, Netherlands

In-house design
Jasper Den Dekker

Web
www.revit.eu

Sleek protection

Since motorcyclists are directly exposed to the road, they have to protect their limbs well with special protectors. In particular, accidents with other vehicles and poor motorcycle handling around curves can easily lead to severe consequences. The REV'IT! Seeflex protectors have been designed using state-of-the-art materials and an innovative construction concept that lend new properties to indispensable protective goods. The result are highly functional protectors with the styling of an aesthetic accessory. Showcasing a highly sporting appearance, they blend well with the individual fashion style worn by motorcyclists. This novel aesthetic is complemented by a particularly high wear comfort, even on longer rides. Thanks to the innovative material choice paired with a sophisticated ergonomic design, these protectors can easily adapt to and fit most body sizes. Another aspect is their protective performance, as these protectors comply with the highest possible protection level requirements (EN1621-1:2012 CE Level 2) defined for these types of products. Featuring outstanding ventilation properties that prevent sweating, these protectors are also highly suitable for all kinds of weather conditions, including very high and very low temperatures. Based on a sophisticated design concept, this product has emerged as a reinterpretation of both the safety requirements and aesthetics of motorcycle limb protection. With their sleek design form, they ensure high wearing pleasure.

Schnittig geschützt

Da Motorradfahrer der Straße unmittelbar ausgesetzt sind, müssen sie ihre Gliedmaßen gut mit speziellen Protektoren schützen. Dies vor allem auch deshalb, weil Unfälle oder ein falsches Kurvenverhalten rasch schlimme Folgen haben können. Bei der Gestaltung von REV'IT! Seeflex führen der Einsatz hochentwickelter Materialien sowie ein innovativer Aufbau zu neuen Eigenschaften dieser unverzichtbaren Schutzmaßnahmen. Das Ergebnis sind hochfunktionale Schützer mit der Anmutung ästhetischer Accessoires. Sie haben eine sehr sportive Ausstrahlung und passen sich damit leicht dem Kleidungsstil des Fahrers an. Diese neue Ästhetik geht einher mit einem besonders großen Komfort auch bei längerem Tragen. Aufgrund ihrer innovativen Materialgebung im Einklang mit einer ausgeklügelten ergonomischen Gestaltung können sich diese Protektoren problemlos den unterschiedlichsten Körpermaßen anpassen. Ein weiterer Aspekt ist ein hoher Grad an Sicherheit, da sie dem höchsten erreichbaren Schutzniveau (EN1621-1:2012 CE Level 2) solcher Produkte entsprechen. Sie bieten dabei sehr gute Bedingungen etwa bei ausgesprochen niedrigen oder auch hohen Umgebungstemperaturen, denn ihre ausgezeichneten Ventilationseigenschaften verhindern starkes Schwitzen. Auf der Grundlage eines überaus durchdachten Gestaltungskonzepts wurden hier die Ästhetik und die Sicherheit von Gliedmaßenprotektoren neu interpretiert. Mit ihrer schnittigen Formensprache trägt und zeigt man sie gerne.

Statement by the jury

The REV'IT! Seeflex motorcycle limb protectors deliver the essence of sporting design. An innovative and consistently implemented combination of advanced materials and smart construction has led to a product with outstandingly ergonomic as well as aesthetic properties. These limb protectors satisfy the highest safety standard even under extreme conditions. They offer good ventilation and impress users with their wearing comfort.

Begründung der Jury

Die Gliedmaßenprotektoren REV'IT! Seeflex wirken sportiv und auf den Punkt gestaltet. Eine innovative und stringent umgesetzte Kombination aus fortschrittlichen Materialien und intelligentem Aufbau führt hier zu einem Produkt mit bestechenden ergonomischen und ästhetischen Eigenschaften. Diese Gliedmaßenprotektoren erfüllen höchste Sicherheitsstandards auch unter extremen Bedingungen. Sie bieten dabei eine gute Ventilation und beeindrucken mit ihrem Tragekomfort.

Designer portrait
See page 42
Siehe Seite 42

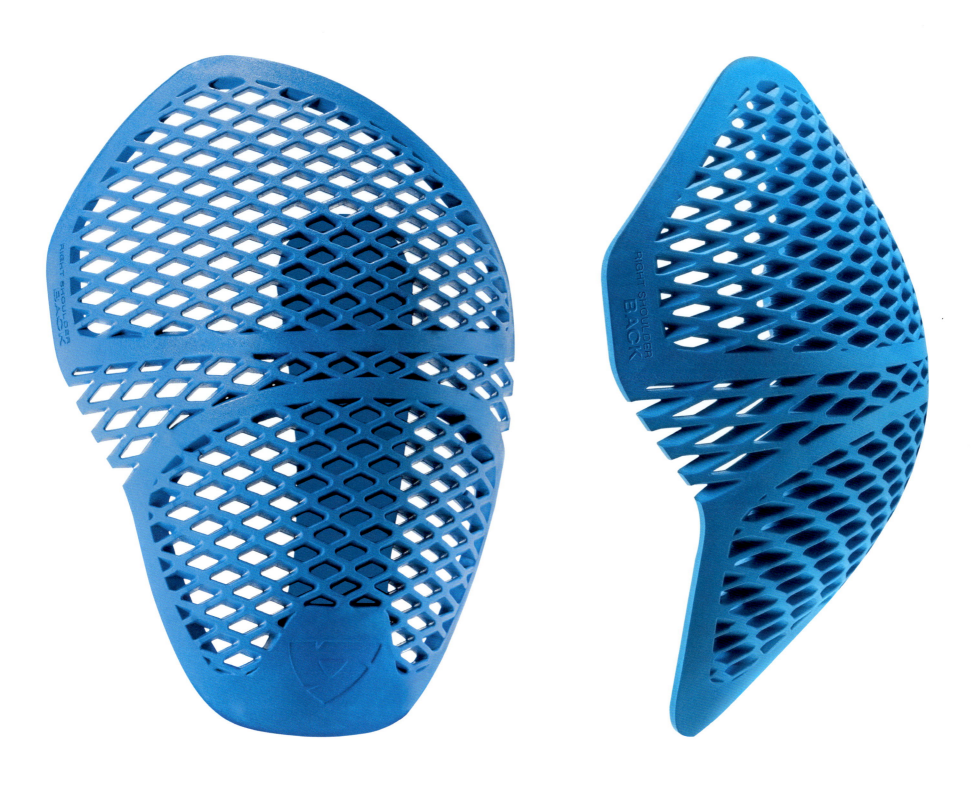

X-BIONIC® Ski Touring SphereWind Jacket
Sportswear
Sportbekleidung

Manufacturer
X-BIONIC®, Asola, Italy
Design
X-Technology Swiss R&D AG,
Wollerau, Switzerland
Web
www.xb4al.com
www.x-bionic.com
www.x-technology.com

The Ski Touring SphereWind Jacket combines a specially developed soft-shell bi-stretch material with a structured 3D mesh fabric to maintain an optimum temperature level. X-Bionic features an advanced air exchange: it removes damp warm air and provides climate control. The jacket also possesses a high warmth-to-weight ratio. The ExtractorPorts, i.e highly stable pads on the shoulders, protect even under high strap load of a backpack and skis.

Statement by the jury
The jacket fully satisfies the key mountain activity requirements of climate control and low weight. The shoulder pads alleviate shoulder stress when carrying a backpack.

Das Ski Touring SphereWind Jacket kombiniert das eigens entwickelte Soft-Shell-Bistretch-Material mit 3-D-Meshgewebe, um ein optimales Temperaturniveau zu erreichen. X-Bionic ermöglicht einen Luftaustausch, der feuchtwarme Luft sicher ableitet und das Klima reguliert. Die Jacke bietet zudem ein gutes Wärme-Gewicht-Verhältnis. ExtractorPorts, formstabile Pads auf den Schultern, schützen bei hoher Gurtbelastung durch Rucksack inklusive Ski.

Begründung der Jury
Den entscheidenden Anforderungen im Gebirge, Klimaregulierung und wenig Gewicht, wird diese Jacke voll gerecht. Schulterpads erleichtern zudem das Tragen von Gepäck.

X-BIONIC® For Automobili Lamborghini Formula Ski Jacket
Functional Jacket
Funktionsjacke

Manufacturer
X-BIONIC®, Asola, Italy
Design
X-Technology Swiss R&D AG,
Wollerau, Switzerland
Web
www.xb4al.com
www.x-bionic.com
www.x-technology.com

Formula Ski XITANIT for Automobili Lamborghini is a highly complex functional jacket: it is handmade and consists of 486 sections. To compensate for frequent changes between demanding ski runs and waiting at the lift, where wearers are in danger of cooling down, functional elements that adapt interactively have been developed to suit all body zones. This innovative interplay enables wearers to achieve the best of their performance under any conditions.

Statement by the jury
The functional jacket fascinates with its elaborate fabrication. Be it hot or cold, it perfectly adapts to climatic conditions, offering optimal usability for the wearer.

Die Formula Ski XITANIT 2.0 for Automobili Lamborghini ist eine hochkomplexe Funktionsjacke: Sie besteht aus 486 Teilen und wird in Handarbeit gefertigt. Um den häufigen Wechsel von anstrengender Abfahrt und dem Warten am Lift, das den Träger gefährlich auskühlen kann, zu kompensieren, wurden an allen Körperzonen Funktionselemente platziert, die adaptiv interagieren. Dieses innovative Zusammenspiel verleiht dem Träger die Fähigkeit, seine Leistung unter allen Bedingungen abzurufen.

Begründung der Jury
Die Funktionsjacke fasziniert durch ihre aufwendige Fertigung. Ob warm oder kalt, sie passt sich an die klimatischen Gegebenheiten an und bietet ihrem Träger hohen Nutzen.

X-BIONIC® For Automobili Lamborghini Golf Jacket
Golf Jacket
Golfjacke

Manufacturer
X-BIONIC®, Asola, Italy
Design
X-Technology Swiss R&D AG,
Wollerau, Switzerland
Web
www.xb4al.com
www.x-bionic.com
www.x-technology.com

The development of this golf jacket aims to combine unrestricted freedom of movement and optimum weather protection. The solution comes in the form of a high-tech, two-piece jacket comprising a symbionic membrane outer jacket that provides protection from wind and rain, and an inner fabric that guarantees high elasticity. Therefore every movement of a golf swing can be carried out without hindrance, and should protection from wind and rain be needed the Shelter Cover can also be worn.

Statement by the jury
The concept of the golf jacket presents a two-layer model that fulfils in both the necessity of freedom of movement and protection against wind and rain.

Die Entwicklung der Golfjacke verfolgte das Ziel, uneingeschränkte Bewegungs-freiheit mit bestmöglichem Wetterschutz zu verbinden. Die Lösung ist ein Hightech-Zweiteiler, bestehend aus einer Überjacke aus Symbionic-Membran zum Schutz vor Wind und Regen und einem Unterteil aus Strick, das hohe Elastizität gewähr-leistet. So lässt sich jede Bewegung des Golfschwungs ungehindert ausführen. Als Witterungsschutz kann zusätzlich das Shelter-Cover übergestreift werden.

Begründung der Jury
Die Konzeption der Golfjacke präsentiert ein zweilagiges Modell, das der nötigen Bewegungsfreiheit wie dem Schutz vor Wind und Regen gerecht wird.

Elevate Tincup
Functional Jacket
Funktionsjacke

Manufacturer
PF Concept, Roelofarendsveen, Netherlands
In-house design
Helina Guleria
Web
www.pfconcept.com

This high-quality functional sports jacket for outdoors provides light-weight, breathable and water repellent protection. Made of 100 per cent nylon diamond ripstop with cire coating, it features elasticated cuffs, an inner storm flap with chin guard and an adjustable-fit bottom hem. Ventilation via the back yoke allows the release of body heat. Reflective features on the front and back provide extra visibility.

Statement by the jury
Highly functional and designed in an urban outdoor style, the Elevate Tincup fulfils the high demands expected of sophisticated sportswear.

Die hochwertige Funktionsjacke für den Outdoor-Sport bietet leichten, atmungs-aktiven, wasserabweisenden Schutz. Sie wurde aus 100 Prozent Nylon-Dia-mond-Ripstop mit Cire-Beschichtung her-gestellt und besitzt elastische Ärmelbünd-chen, eine Abdeckleiste mit Kinnschutz und einen Bund mit einstellbarer Passform. Die Belüftung über die hintere Passe dient der Abgabe der Körperwärme. Reflektie-rende Elemente hinten und vorn sorgen für eine extra Sichtbarkeit.

Begründung der Jury
Hochfunktionell und im urbanen Outdoor-Stil entworfen, erfüllt die Elevate Tincup die Leistungsanforderungen, die an geho-bene Sportkleidung gestellt werden.

SLO'O Bike to Work Pants
Cycling Pants
Fahrradhose

Manufacturer
TMAX strategy & marketing company,
Taipei, Taiwan
In-house design
Mark Peng
Web
www.sloolife.com

The Slo'o Bike To Work Cycling Pants
have been designed to offer style and
functionality for the growing number of
people cycling to work. The combination
of ordinary looking work pants, which
include a cycling seat pad (made of elastic fabric) inside, spares cyclists the need
to change their pants at work. Thanks
to an innovative technique the pad fixes
easily onto the pants, can be swapped
between models for short and long rides,
and is just as easily removed.

Statement by the jury
These cycling pants fascinate as a syn-
thesis of sporting cycling pants and
pants suitable for work. A fixable and
easy-to-remove inner pad ensures the
desired comfort whilst cycling.

Die Slo'o Bike to Work Cycling Pants bieten
sich der wachsenden Anzahl Menschen, die
mit dem Rad zur Arbeit fahren, als stil- und
funktionsgerechte Lösung an. Die Verbindung aus herkömmlich aussehender Hose
mit Sitzpolster aus elastischem Material
am Gesäß erspart es Radfahrern, sich vor
Arbeitsbeginn umziehen zu müssen. Dank
einer innovativen Technik lassen sich die
Polster leicht anbringen, schnell für kurze
oder lange Fahrten durch ein anderes Modell austauschen und ebenso leicht wieder
entfernen.

Begründung der Jury
Diese Fahrradhose begeistert durch ihre
Synthese aus sportlicher Radlerhose und
einer bürotauglichen Hose. Ein innenlie-
gendes, einfach zu entnehmendes Pad
bietet darüber hinaus den gewünschten
Komfort auf dem Rad.

D-FORCE PROTECTOR
Body Protection
Körperschutz

Manufacturer
Korea OGK Co., Ltd.,
Seongnam, South Korea
In-house design
Kyung-Jun Bae, Seung-Hwan Yoo
Web
www.ogk.co.kr

The D-Force snow protector for winter sports uses D-Force foams to minimise the deviation in impact resistance due to temperature differences. This is achieved through the addition of additives and polymers that allow the adjustment of Tg (glass transition temperature). The protector thus maintains its hardness under all conditions. As protection against germs that easily grow in humid environments, it uses meshes that are anti-bacterially treated.

Statement by the jury
Made of specialist foams, the D-Force snow protector is an effective impact and shock absorber. In addition, the material also provides good ventilation, thus aiding hygiene.

Der D-Force Snow Protector für den Wintersport setzt D-Force-Schaumstoffe ein, um die temperaturbedingte Abweichung in der Stoßfestigkeit zu minimieren. Dazu werden Zusatzstoffe und Polymere eingesetzt, die eine Justierung der Glasübergangstemperatur ermöglichen. Der Protector hält so seine Härte bei Minusgraden im Toleranzbereich. Gegen Bakterien, die sich in feuchter Umgebung leicht vermehren, kommen antibakteriell behandelte Netze zum Einsatz.

Begründung der Jury
Dank spezialisierter Schaumstoffe schützt der D-Force Snow Protector wirksam etwa bei Stürzen. Durch seine gute Ventilation erfüllt das verwendete Material auch hohe Hygieneanforderungen.

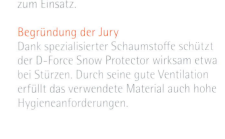

Friction
Cycling Clothing
Radsportkleidung

Manufacturer
Draggin Jeans Pty Ltd,
Port Melbourne, Australia
In-house design
Web
www.dragginjeans.net

This cycling clothing aims to protect cyclists' skin from painful road rash caused by a fall. Made of strong fibres specifically engineered for cyclists, the clothing is called Friction because it prevents painful gravel rash. It incorporates Roomoto, an innovative and highly protective fabric offering climate control, which takes heat and sweat away from the skin. An antibacterial treatment ensures comfort whilst cycling.

Statement by the jury
This cycling clothing fulfils the high demands of sports people, guaranteeing both breathability and skin protection in case of accidents.

Diese Radsportkleidung schützt die Haut von Radsportlern vor schmerzhaften Schürfwunden bei Stürzen. Hergestellt aus reißfesten, speziell für Radfahrer entwickelten Fasern, trägt sie den Namen Friction, denn sie verhindert schmerzhafte Hautabschürfungen. Die Verwendung des innovativen Roomoto-Stoffs sorgt für hohe Sicherheit und Klimakontrolle, indem Hitze und Schweiß von der Haut abgeleitet werden. Eine antibakterielle Ausrüstung gewährleistet hohen Komfort während der Fahrt.

Begründung der Jury
Die Radsportkleidung wird dem hohen Anspruch von Sportlern gerecht, nicht nur Atmungsaktivität zu gewährleisten, sondern auch die Haut bei Stürzen zu schützen.

POMICE JERSEY
Cycling Jersey
Fahrradtrikot

Manufacturer
Briko, Milan, Italy
In-house design
Paola Bertoncini
Web
www.briko.com

Pomice Jersey combines functional materials and an ergonomic fit. The front features a lightweight woven fabric of 73 g/sqm, structured with ultra-thin 20 denier yarn and 4-ways stretch. This fabric has an outstanding compactness that provides adequate muscle support and stability, and at the same time guarantees free movement. The shoulders and the back comprise a visually striking fabric that provides optimum ventilation.

Statement by the jury
This cycling jersey is impressive due to its ergonomic fit and lightweight materials that possess outstanding ventilation characteristics.

Pomice Jersey kombiniert funktionale Materialien mit ergonomischer Passform. Die Vorderseite besteht aus einem leichten Gewebe mit 73 g/qm und ist mit ultradünnem 20-Denier-Garn und 4-Wege-Stretch strukturiert. Der Stoff besitzt eine herausragende Dichtigkeit, die vollwertige Muskelunterstützung und -stabilität bietet, zugleich aber Bewegungsfreiheit lässt. Die Schultern und die Rückenpartie bestehen aus einem visuell auffälligen Gewebe, das für optimale Belüftung sorgt.

Begründung der Jury
Dieses Trikot beeindruckt durch einen ergonomischen Sitz und leichte Materialien, die für hervorragende Atmungsaktivität sorgen.

CEP Ortho+ Achilles Support Short Socks
Sportkompressionsstrümpfe

Manufacturer
medi GmbH & Co. KG, Bayreuth, Germany
In-house design
Bernd Hörath
Web
www.medi-corporate.com

The Ortho+ Achilles Support Short Socks are a sport brace that provide support for the achilles tendon. Highly adaptive pads made of smart memory foam take the pressure off the Achilles tendon and stimulate the microcirculation. The pads are placed to make sure they do not hurt or rub against the shoe. The integrated medi compression stimulates blood circulation in the foot and reduces any swelling. Furthermore, the knitted fabric of the socks is especially breathable and moisture conducting.

Statement by the jury
The sport brace merges scientific know-how with innovative materials to create a high-tech product that provides support for the achilles tendon.

Die Ortho+ Achilles Support Short Socks sind Sportkompressionsstrümpfe, die die Achillessehne stützen. Hochadaptive Pads aus Smartmemory Foam nehmen Druck von der Sehne und erhöhen die Mikrozirkulation. Die Pads sind so angebracht, dass sie trotz Schuh nicht unangenehm drücken oder scheuern. Die eingesetzte Medi-Compression-Technologie verbessert die Durchblutung im Fuß und reduziert so Schwellungen. Zudem ist das Gestrick der Strümpfe besonders atmungsaktiv und feuchtigkeitsleitend.

Begründung der Jury
Die Sportkompressionsstrümpfe verbinden wissenschaftliche Fakten und innovative Materialien zu einem Hightech-Produkt, das der Achillessehne Halt gibt.

Terraclaw 220
Trail Running Shoe
Trailschuh

Manufacturer
Inoveight Ltd, Staveley, Great Britain
In-house design
Matthew Head
Web
www.inov-8.com

The lightweight Terraclaw 220 trail running shoe facilitates racing at high speeds over all terrain. The Ray-Wrap lacing system, which aligns with the foot's first metatarsal, provides a close, secure and comfortable fit. The distinctive X-Lock overlay holds the foot in place, while the wider toe box allows the toes to splay, ensuring confidence and comfort in all underfoot conditions. The Dual-C outsole features two different rubber compounds, delivering superb grip.

Statement by the jury
The sophisticated fit of the running shoe convinces with its novel, off-centre positioned lacing system that lends it high comfort even when worn for a long time.

Der leichte Trailschuh Terraclaw 220 ist hohen Geschwindigkeiten in jedem Gelände gewachsen. Das Ray-Wrap-Schnürungssytem, das am Mittelfußknochen des Fußes ausgerichtet ist, vermittelt ein sicheres und komfortables Tragegefühl. Die weite Zehenbox behält auch bei längeren Läufen eine komfortable Passform, die durch das im Vorfußbereich angebrachte X-Lock-Fixierungssytem genügend Halt bietet. Die DUAL-C Sohle setzt sich aus zwei verschiedenen Gummimischungen zusammen, die hervorragende Griffigkeit gewährleisten.

Begründung der Jury
Die ausgereifte Passform des Sportschuhs überzeugt durch das neuartige, schräg versetzte Schnürsystem, das ihm hohen Komfort selbst bei langem Tragen verleiht.

KI xtrap neo
Shoes
Schuhe

Manufacturer
Innus Korea,
Busan, South Korea

In-house design
Gyudeog Kim

Web
www.innus.kr
www.kioriental.com

Optimum fit

During long runs both shoes and feet are subject to constantly changing load and stress, particularly when the terrain and surface conditions are rough. By fundamentally questioning the construction of running shoes, the design of the KI xtrap neo introduces an entirely new approach to the relationship between shoe and wearer. The innovative xtrap concept was developed to serve as a kind of safety belt for shoes by supporting both the sides and the rear of the outer leather part with a "cross line", manufactured as a single body and hence eliminating the otherwise conventional separation between insole and midsole. This approach ensures that the internal space of the shoe can adapt to provide the most optimum fit for the user. In addition, 22 ergonomic wave balance pads have been inserted to support the ankles and thus help maintain the balance of the body. The pads react independently to the terrain condition and apply targeted pressure to the foot to provide an overall load effect. In order to avoid adhesion of insole and outsole, an innovative press-moulding technique was employed in the assembly. Applied on the outer leather, it offers the benefit of reducing the need for environmentally unfriendly solvents by more than 50 per cent. Thanks to a remarkable concept, these shoes have emerged as a product that impresses with its lightness and ability to provide enhanced individual fit.

Bester Halt

Beim Lauftraining sind die Füße wie auch die Schuhe stetig wechselnden Belastungen ausgesetzt, wenn sich etwa die Bedingungen im Gelände ändern. Indem sie den Aufbau eines Trainingsschuhs grundsätzlich hinterfragt, ermöglicht die Gestaltung des KI xtrap neo ein völlig neues Verhältnis zwischen Schuh und Läufer. Das innovative xtrap-Konzept fungiert dabei als eine Art Sicherheitsgurt für Schuhe, wodurch sowohl die Seitenflächen als auch die Hinterseite des äußeren Lederparts mit einem „Fadenkreuz" unterstützt werden. Dieses ist aus einem Stück gefertigt, sodass keine Trennung mehr zwischen Einlage und Zwischensohle besteht. Auf diese Weise kann der vorhandene Platz innerhalb des Schuhs immer hin zur individuell besten Passform optimiert werden. Um das Fußgelenk zusätzlich in seiner Balance zu unterstützen, wurden bei diesem Trainingsschuh 22 ergonomische Wave Balance Pads eingearbeitet. Sie erzeugen beim Gehen einen entsprechenden Druck, durch den auf das Gelände reagiert wird und Unebenheiten in ihrer Wirkung abgeschwächt werden. Um das Verkleben von Einlege- und Laufsohle zu vermeiden, kommt bei dem Schuh eine innovative Pressformtechnik zum Einsatz. Auf das äußere Leder angewandt, hat sie den Vorteil, dass sich dadurch der Gebrauch umweltverschmutzender Lösemittel um die Hälfte reduziert. Dank eines bemerkenswerten Konzeptes entstand so ein Produkt, das durch seine Leichtigkeit und die Möglichkeit einer individuellen Passform beeindruckt.

Statement by the jury

Featuring an impressive construction, the KI xtrap neo delivers the best possible ergonomic fit for its wearer. Complemented by additionally supporting wave balance pads, this running shoe can adapt perfectly to any ground surface, and is even suitable for outdoor activities. The captivating unity of sole and insole was achieved through a special press-moulding technique. Overall, the shoe is extraordinarily light and durable, and manufactured in an environmentally friendly approach.

Begründung der Jury

Durch seinen eindrucksvollen Aufbau bietet der KI xtrap neo dem Träger den jeweils ergonomisch besten Halt. In Kombination mit den zusätzlich unterstützenden Wave Balance Pads kann sich dieser Trainingsschuh jeder Belastungssituation auch im Outdoorsektor perfekt anpassen. Bestechend ist die auf einer speziellen Pressformtechnik basierende Einheit aus Sohle und Einlage. Der Schuh ist dabei außerordentlich leicht sowie langlebig und umweltfreundlich gefertigt.

Designer portrait
See page 44
Siehe Seite 44

Heimathlet
Fitness Furniture
Fitnessmöbel

Manufacturer
heimathlet, Weimar, Germany
In-house design
Elisa Kirbst
Web
www.heimathlet.net

The Heimathlet (home athlete) is a piece of fitness furniture that integrates health-related sports equipment into everyday living areas, intelligently combining functionality with design. The forward leaning rungs, gym rings and a mat reveal the training potential of this piece of fitness equipment, which can even be used in confined spaces. It is suitable for doing exercises to build muscle mass, enhance muscle endurance, and as part of rehabilitation, yoga or Pilates exercises.

Statement by the jury
The Heimathlet has emerged as a highly elegant symbiosis of furniture and sports equipment for the home providing for a myriad of fitness needs.

Der Heimathlet ist ein Fitnessmöbel, das Gesundheitssport in den Wohnbereich integriert und Funktionalität intelligent mit Design kombiniert. Der nach vorn gelagerte Sprossenaufbau samt Turnringen und Matte offenbart das Trainingspotenzial dieses Fitnessstudios auf kleinem Raum, geeignet für Muskelaufbau, Kraftausdauer, Rehabilitation, Yoga oder Pilates. Zugleich zeigt er sich als eleganter Freischwinger, der sich für optimalen Sitzkomfort jeder Körpergröße intuitiv anpassen lässt.

Begründung der Jury
Der Heimathlet ist eine elegant gestaltete Symbiose aus Möbel und Sportgerät für zu Hause, die eine Vielzahl sportlicher Übungen erlaubt.

Bowflex Max Trainer®
Cardio Machine
Kardiogerät

Manufacturer
Nautilus, Inc., Vancouver, USA
In-house design
Web
www.nautilusinc.com

The Bowflex Max Trainer is a space-saving cardio machine for the home that is easy on the joints. It provides an entire body workout and helps burn more calories as it engages the upper body much more than elliptical machines. In order to achieve faster speed and higher calorie burn, Max Trainer combined high-intensity training with proprietary resistance technology. Its sleek design is reminiscent of a sports motorcycle.

Statement by the jury
High performance and unobtrusive design make this home trainer particularly outstanding as well as convincing in terms of the health benefits it provides.

Der Bowflex Max Trainer ist ein raumsparendes, gelenkschonendes Kardiogerät für zu Hause. Es trainiert den ganzen Körper und verbrennt eine erhöhte Anzahl von Kalorien, weil der Oberkörper wesentlich stärker gefordert wird als etwa bei Ellipsentrainern. Um eine höhere Geschwindigkeit und Kalorienverbrennung zu erzielen, verbindet der Max Trainer hochintensives Training mit urheberrechtlich geschützter Widerstandstechnik. Sein schlankes Design erinnert an ein Sportmotorrad.

Begründung der Jury
Hohe Leistung und eine unaufdringliche Gestaltung stechen bei diesem Hometrainer besonders hervor und überzeugen von seiner positiven gesundheitlichen Wirkung.

HomeRun
Treadmill
Laufband

Manufacturer
Skandika, Essen, Germany
Design
SEG Superweigh Enterprise Co., Ltd.,
Caotun Township, Nantou County, Taiwan
Web
www.skandika.com
www.superweighfitness.com.tw

The carefully designed HomeRun tread-mill offers the ability to work out both at home and in professional environ-ments. A non-slip surface area allows the workout to be combined with any activity on a notebook or tablet PC. The engine is constantly ventilated which guarantees the quality and durability of this sports equipment. With two handles, the treadmill coming with cup hold-ers transforms into a table or a modern sideboard. It folds up easily.

Statement by the jury
With the double benefit of being able to work out while reading or working; the HomeRun treadmill presents itself as a contemporary and premium lifestyle product.

Mit dem durchdacht gestalteten Home-Run-Laufband lässt sich zu Hause wie im Büro trainieren, und das bei jeglicher Tätigkeit am Notebook oder Tablet auf der rutschfesten Ablagefläche. Der Motor wird ständig gelüftet und sichert die Qualität und Haltbarkeit des Sportgerätes. Durch jeweils zwei Griffe verwandelt sich das mit Getränkehaltern ausgestattete Laufband in einen Tisch oder ein modernes Sideboard. Es lässt sich leicht zusammenklappen.

Begründung der Jury
Sport treiben und dabei lesen oder arbeiten – mit diesem Doppelnutzen präsentiert sich das Laufband HomeRun als zeitgemäß und zugleich edel gestaltetes Lifestyleprodukt.

Office Bike
Ergociser
Ergometer

Manufacturer
Skandika, Essen, Germany
Design
SEG Superweigh Enterprise Co., Ltd.,
Caotun Township, Nantou County, Taiwan
Web
www.skandika.com
www.superweighfitness.com.tw

The Office Bike offers up to 60 watts of power for training workouts to enhance endurance and burn calories in both private and professional environments. The elaborate design with a piano-black finish is eye-catching in modern offices or at home. The training device can also act as a recumbent bike. Based on a flywheel made of stainless steel, the magnetic drive system is not only maintenance-free, but also guarantees a smooth and silent rotation.

Statement by the jury
The Office Bike fascinates with its or-ganic design that turns it into an at-tractive as well as versatile piece of sports equipment for use both at home and in the work place.

Das Office Bike bietet 60 Watt Leistung und ermöglicht Konditionstraining und Kalorienverbrennung im privaten wie be-ruflichen Umfeld. Ein ausgefeiltes Design in Klavierlack-Optik macht es zum Blick-fang in modernen Umgebungen. Das auf Magnettechnologie basierende Antriebs-system über eine Schwungscheibe aus Edelstahl ist wartungsfrei und verspricht einen geschmeidigen wie geräuscharmen Rundlauf. Das Trainingsgerät kann zudem als Recumbent Bike genutzt werden.

Begründung der Jury
Das Office Bike beeindruckt mit seiner organischen Formgebung, die es zu Hause oder bei der Arbeit zu einem attraktiven wie vielseitig nutzbaren Sportgerät macht.

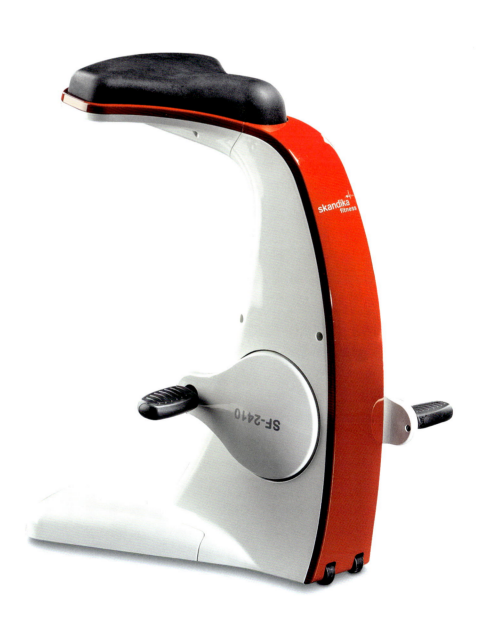

MYRUN TECHNOGYM
Fitness Equipment
Fitnessgerät

Manufacturer
Technogym SpA, Cesena, Italy
In-house design
Web
www.technogym.com

Myrun Technogym is designed to offer an outstanding running experience and feedback on one's running technique, as well as personalised training programmes. It integrates a treadmill and a native app that syncs to tablets and helps improve the user's running technique. An innovative running surface adapts to the way the user runs and thus reduces the risk of injury by absorbing impacts without taking momentum away from the run. The fitness device saves both space and energy.

Myrun Technogym wurde für eine nicht alltägliche Lauferfahrung mit Feedback zur Lauftechnik sowie mit personalisierten Trainingsprogrammen entworfen. Es umfasst ein Laufband und eine eigene App, die sich mit Tablets synchronisiert und dabei hilft, die Technik des Sportlers zu verbessern. Eine innovative Laufoberfläche passt sich der Laufart des Benutzers an und verringert so die Verletzungsgefahr, indem sie den Aufprall absorbiert, ohne das Laufmoment aus der Bewegung zu nehmen. Das Fitnessgerät ist sowohl platz- als auch energiesparend.

Statement by the jury
The fitness device distinguishes itself through its clear design, individual training programmes and a special running surface that ensures high safety and comfort.

Begründung der Jury
Das Fitnessgerät zeichnet sich durch eine klare Gestaltung, individuelle Trainingsprogramme und eine spezielle Lauffläche aus, die besondere Sicherheit und Komfort bietet.

Carving Balance
Fitness Device
Fitnessgerät

Manufacturer
Carver Fitness B.V., Eindhoven, Netherlands
Design
Remi van Oers, Eindhoven, Netherlands
Web
www.carvingfitness.com
www.remivanoers.com

The new Carving Balance provides balance training in its purest form of muscle activity and user-engagement. While performing the lateral movements, users turn away from the circular base causing their bodies to become out of balance. This imbalance generates maximum muscle activity and burns a high amount of calories. The elegant design of this fitness device highlights this movement; it is user-friendly and available in six different colours of high-quality synthetic leather.

Statement by the jury
The Carving Balance strikes a convincing balance between elegant design and functionality and thus enhances the pleasure of working out.

Der neue Carving Balance bietet Balance-Training in seiner reinsten Form der Muskelbeanspruchung und Aktivität. Beim Durchführen der Seitwärtsbewegung dreht sich der Benutzer aus der Kreisbasis heraus, wodurch der Körper aus der Balance gerät. Diese Verlagerung generiert eine maximale Muskelaktivität und verbrennt somit viele Kalorien. Die Bewegung wird durch die elegante Anmutung des Fitnessgeräts unterstrichen; anwenderfreundlich konstruiert, ist das Gerät in sechs verschiedenen Farben hochwertigen Synthetikleders erhältlich.

Begründung der Jury
Carving Balance findet gestalterisch die passende Ausgewogenheit zwischen stilvoller Anmutung und Funktionalität und erhöht so die Lust am Training.

FitQuest
Fitness Measurement Machine
Fitness-Messgerät

Manufacturer
MIE Medical Research Ltd.,
Leeds, Great Britain
Design
Neutronium Ltd (Richard Luxton),
Bristol, Great Britain
Darren Whiteman Multimedia
(Darren Whiteman), Bath, Great Britain
Web
www.mie-uk.com
www.neutroniumltd.co.uk
www.darrenwhiteman.com

FitQuest is a fitness measurement machine that takes less than four minutes for a fitness assessment, combining five tasks into four simple exercises, plus heart rate recovery and body mass composition analysis. It derives measurements from each task and compiles them into an overall Fitness Quotient, from which it provides an overall fitness assessment. The machine measures human performance for an individual with accurate results and compares this to a data set of over 50,000 users.

Statement by the jury
The reduced design of the measuring machine surprises with enormous technology and the sheer amount of data it compiles, and provides, of its user.

FitQuest ist ein Fitness-Messgerät, das in weniger als vier Minuten die Fitness auswerten kann, indem es fünf Aufgaben in vier einfachen Übungen verknüpft, plus Analyse der Herzfrequenzerholung und Körpermassenzusammensetzung. Aus all den hieraus gewonnenen Daten errechnet das Gerät einen allgemeinen Fitnessquotienten und liefert so eine Auswertung der Gesamtfitness. Das Gerät misst die individuelle Leistung eines Menschen akkurat und vergleicht das Ergebnis mit einem Datensatz von über 50.000 Anwendern.

Begründung der Jury
Das schlichte Design des Messgerätes überrascht mit einer beachtlichen Technologie und der Vielzahl der Werte, die es dem Benutzer zur Verfügung stellen kann.

SLX Fusion
Ski

Manufacturer
Elan, d.o.o., Begunje na Gorenjskem, Slovenia
Design
Gigodesign d.o.o., Ljubljana, Slovenia
Web
www.elanskis.com
www.gigodesign.com

SLX Fusion are race skis with distinct left and right construction. Optimal material distribution through the volume of the skis was achieved through analysis of force distribution. The combination of Power Spine technology with the Amphibio Profile, marked by the asymmetric tip and tail, enables effective transmission of force from a skier's legs to the snow. The "spine arch" construction is highly robust and reinforced with a carbon layer along its entire length.

Statement by the jury
Tailored for speed, the SLX Fusion delivers a fast and pleasurable skiing experience. In addition, the difference between the left and the right ski enhances manoeuvrability.

Der SLX Fusion ist ein Rennski mit echter Links- und Rechtskonstruktion. Die optimale Materialaufteilung auf das Produktvolumen wurde mithilfe einer Analyse der Kraftverteilung ermittelt. Durch die Kombination aus Power-Spine-Technologie und Amphibio-Profil mit asymmetrischem Tip und Tail lässt sich die Kraft des Skifahrers effektiv in den Schnee übertragen. Die Konstruktion mit „Rückgrat" ist sehr stabil und entlang der kompletten Skilänge mit einer Carboneinlage verstärkt.

Begründung der Jury
Der auf Geschwindigkeit ausgerichtete SLX Fusion verspricht sportliches Fahrvergnügen. Die Unterscheidung zwischen linkem und rechtem Ski erhöht zudem seine Beweglichkeit.

Elan Amphibio 16 Ti2 Fusion
Ski

Manufacturer
Elan, d.o.o., Begunje na Gorenjskem, Slovenia
Design
Gigodesign d.o.o., Ljubljana, Slovenia
Web
www.elanskis.com
www.gigodesign.com

Elan Amphibio 16 Ti2 Fusion with 4D technology is a versatile, fast and light all-mountain ski. Alongside the innovative combination of a rocker and camber profile, it stands out with differently shaped titanium inlays in front of, and behind, the binding, which create high torsion resistance. The front part of the ski features a convex shape for precise turning, while the concave back allows an easier exit from the turn.

Statement by the jury
This ski merges technology and know-how into a shape that promises a high degree of fun and manoeuvrability to professional skiers.

Der Amphibio 16 Ti2 Fusion mit 4-D-Technologie ist ein vielseitiger, schneller und gleichzeitig leichter Allmountain-Ski. Neben der innovativen Kombination von Rocker- und Camberprofil liegt seine Besonderheit in den unterschiedlich gewölbten Titaneinlagen vor und hinter der Bindung, die zu hoher Torsionssteifigkeit führen. Das konvex geformte Vorderteil erleichtert die präzise Schwungführung in der Kurve, während die konkave Rückseite die Schwungausfahrt vereinfacht.

Begründung der Jury
Technologie und Know-how verschmelzen bei diesem Ski zu einer Ausführung, die dem sportlichen Fahrer ein hohes Maß an Spaß und Wendigkeit verspricht.

SUPER AIRFLOW GOGGLES
Snow Goggles
Schneebrille

Manufacturer
Korea OGK Co., Ltd.,
Seongnam, South Korea
In-house design
Jeong-Min Kim, Sang-Ki Suh
Web
www.ogk.co.kr

Super Airflow Goggles, with their improved anti-fog coating treatment solve the problem of fogged-up goggle lenses, which causes poor visability. This is produced by warmth inside the lens coming together with cold air outside the goggles. When buckles either side of the goggles are pressed, fogged-up lenses are opened forward to let out the heat inside and clear the lenses. By pressing the button in the centre of the buckle, the lenses can also be removed and replaced. A set of three different lenses has been designed.

Statement by the jury
These snow goggles solve the problem of fogging through a sophisticated mechanism and, complemented by a set of different lenses, deliver a wide scope of performance features.

Die Schneebrille Super Airflow mit ihrer verbesserten Anti-Fog-Beschichtung löst das Problem, dass Schneebrillen beschlagen und so die Sicht behindern, wenn die Wärme im Innern der Brille auf die kalte Außentemperatur trifft. Beim Drücken der beidseitigen Schnallen öffnen sich die beschlagenen Gläser nach vorn, damit die Wärme im Innern entweichen und sich der Beschlag lösen kann. Wird die Mitte der Schnalle heruntergedrückt, können die Gläser abgenommen und durch eine von drei weiteren Glasvarianten ausgetauscht werden.

Begründung der Jury
Diese Schneebrille löst das Problem des Beschlagens mithilfe eines ausgereiften Mechanismus und bietet mit verschiedenen Gläsern darüber hinaus einen hohen Leistungsumfang.

FuseForm™ Brigandine 3L Jacket
Ski Apparel
Skibekleidung

Manufacturer
The North Face, Stabio, Switzerland
In-house design
Web
www.thenorthface.eu

The waterproof FuseForm Brigandine 3L jacket has been created for extreme mountain conditions; built specifically for big mountain skiing and snowboarding. The FuseForm technology allows to interweave the materials used for the jacket without any scams: the robust Cordura for areas that are subject to wear and tear with the two-way-stretch material for flexible zones such as arm and elbow area.

Statement by the jury
The high demands that a functional jacket for mountain activities has to meet are met by the FuseForm Brigandine 3L three-layer jacket – thanks to the use of sophisticatedly integrated high-quality materials.

Die wasserdichte 3-Lagen-Jacke FuseForm Brigandine 3L wurde für extreme Bedingungen am Berg entwickelt; hergestellt insbesondere für Alpinski oder Snowboarden. Mithilfe der FuseForm-Technologie ist es möglich, die für die Jacke verwendeten Materialien ohne Nähte miteinander zu verweben: das robuste Cordura für stark beanspruchte Bereiche mit dem Zwei-Wege-Stretchmaterial, das an flexiblen Stellen wie Arm- und Ellbogenbereich verwendet wird.

Begründung der Jury
Die hohen Ansprüche, die an eine Funktionsjacke für das Gebirge gestellt werden, erfüllt die FuseForm Brigandine 3L mithilfe hochwertiger und durchdacht kombinierter Materialien.

AMEO PowerBreather PB01
Training Device for Swimmers
Trainingshilfe für Schwimmer

Manufacturer
AMEO Sport GmbH, Starnberg, Germany
Design
Tom Ayton // aytonlab (Matthias Ocklenburg),
Darmstadt, Germany
Web
www.powerbreather.com
www.aytonlab.com

The AMEO PowerBreather PB01 is an innovative sporting equipment that allows swimmers to breathe fresh air under water. Unlike in conventional snorkels, a new valve technique, which was developed to incorporate ergonomics and design, reduces the risk of exhalation air (remaining in the tubes and containing CO_2) being re-inhaled into the lungs i.e. reciprocated breathing. Thanks to the sequential FreshAir system, the product ensures an optimal oxygen supply over an unlimited period of time.

Der AMEO PowerBreather PB01 ist ein innovatives Sportgerät, mit dem sich beim Schwimmen frei unter Wasser atmen lässt. Eine neue Ventiltechnik, die unter Ergonomie- und Designaspekten entwickelt wurde, verhindert – anders als beim herkömmlichen Schnorcheln –, dass CO_2-haltige Ausatemluft im Luftrohr verbleibt und in die Lunge zurückgeht (Pendelatmung). Durch das sequenzielle FreshAir-System sichert das Produkt die bestmögliche Sauerstoffversorgung über einen beliebig langen Zeitraum.

Statement by the jury
Swimming freely and being provided with sufficient fresh air even when being under water: this is the solution that the novel AMEO PowerBreather PB01 delivers while fitting snugly around the head.

Begründung der Jury
Ungehindert schwimmen und dabei selbst unter Wasser mit genügend Sauerstoff versorgt sein – dies erlaubt der neuartige AMEO PowerBreather PB01, der ganz bequem am Kopf sitzt.

Marine-B
Bodyboard

Manufacturer
T.O R&D Labs, Busan, South Korea
In-house design
Taeho Kim
Web
www.tornd.com

Marine-B is a neoprene-covered body board made of eco-friendly materials. The inside foam core is made of EPS (SE 2000). The cover can be removed, replaced, washed, and the board can be easily stored and carried. Furthermore, it is available in many contemporary colours. It can be used in open water or in a swimming pool.

Marine-B ist ein neoprenbeschichtetes Bodyboard aus umweltfreundlichen Materialien mit einem Kern aus EPS-Schaum (SE 2000). Seine Hülle lässt sich abnehmen, austauschen und waschen, und das Board sich einfach verstauen und tragen. Darüber hinaus ist es in mehreren zeitgemäßen Farbtönen erhältlich und eignet sich zum Wasserrutschen sowohl im offenen Wasser wie im Schwimmbad.

Statement by the jury
Robust eco-friendly materials and a simple design make the body board an uncomplicated and fun piece of sports equipment.

Begründung der Jury
Robuste, umweltfreundliche Materialien und eine schlichte Gestaltung machen das Bodyboard zum unkomplizierten Sport- und Fungerät.

Easybreath
Snorkelling Mask
Tauchermaske

Manufacturer
Tribord, Hendaye, France
In-house design
Cédric Caprice
Design
Fritsch-Durisotti, Conflans, France
Web
www.tribord.com
www.fritsch-durisotti.com

The Easybreath snorkelling mask was developed because breathing underwater with a snorkel is often considered uncomfortable and unhygienic. With this full-face mask, which allows breathing through the mouth and nose, breathing underwater becomes as easy and natural as on land. Thanks to its large size, it offers users an unobstructed 180 degree field of vision, while its double air-flow system prevents the mask from fogging.

Statement by the jury
Easybreath marks a new dimension in snorkelling masks. Breathing in a natural way makes snorkelling easier and enhances the fun and adventure of the experience.

Die Tauchermaske Easybreath wurde entwickelt, weil das Atmen unter Wasser mit einem Schnorchel häufig als störend, unbequem und unhygienisch empfunden wird. Mit dieser Integralmaske lässt sich unter Wasser genauso leicht und natürlich durch Mund und Nase atmen wie an Land. Dank ihrer Größe bietet sie ein Sichtfeld von 180° Grad. Aufgrund eines Doppelluftstrom-Systems besitzt sie zudem einen Antibeschlagschutz.

Begründung der Jury
Easybreath markiert eine neue Dimension der Tauchermaske. Natürliches Atmen erleichtert das Tauchen mit Schnorchel und erhöht den Spaß- und Erlebnisfaktor.

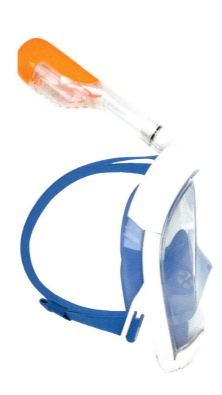

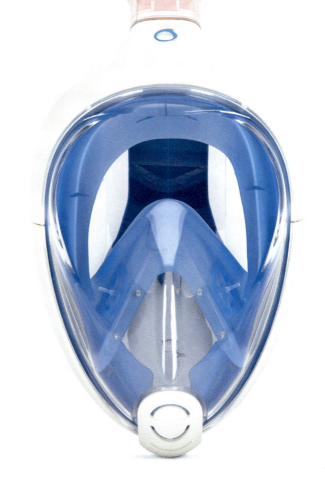

Sleipnir
Anemometer
Windmesser

Manufacturer
Vaavud, Copenhagen, Denmark
In-house design
Andreas Bruun Okholm
Design
Attention, Copenhagen, Denmark
Web
http://vaavud.com
www.attention-group.com

The Vaavud Sleipnir is a device that measures the wind speed and direction for outdoors and sports enthusiasts. The asymmetrical blades ensure that the rotor accelerates slightly when the largest blade is hit by the wind. A built-in high-frequency sensor measures the speed of the rotor and determines the wind direction by comparing the point of maximum velocity to the compass direction. The data is automatically shared with others via the Vaavud smartphone app.

Statement by the jury
The Vaavud Sleipnir has emerged as an essential tool for sport and leisure activities that turns a smart phone into a meteorological device.

Der Vaavud Sleipnir ist ein Messgerät zur Bestimmung der Windgeschwindigkeit und -richtung für Outdoor-Fans und -Sportler. Die asymmetrischen Klingen stellen sicher, dass der Rotor leicht beschleunigt, wenn die größte Klinge in Kontakt mit dem Wind kommt. Ein eingebauter Hochfrequenzsensor misst die Rotordrehzahl und ermittelt die Richtung durch Vergleichen des Punktes der maximalen Geschwindigkeit mit der Richtung des Kompasses. Die Daten werden über die Vaavud Smartphone-App automatisch mit anderen geteilt.

Begründung der Jury
Mit dem Vaavud Sleipnir wurde ein essenzielles Instrument für Sport und Freizeit entwickelt, das ein Smartphone zum meteorologischen Werkzeug macht.

SNOOZY
Temporary Sleeping Unit
Temporäre Schlafeinheit

Manufacturer
AirClad, Fast Architecture, Antwerp, Belgium
In-house design
Web
www.airclad.com

Snoozy is a flat-packed portable sleeping unit for two, which can be set up and used virtually anywhere. Capable of being assembled in just a few minutes, it is suitable for short-term events, permanent installations or an emergency situation. It has a bespoke lightweight frame system made out of extruded aluminium designed to allow easy integration of all structural elements to work together seamlessly. The main exterior is wrapped in a sealed air-inflated skin made from heavy-duty flexible fabric.

Snoozy ist eine Schlafeinheit für zwei Personen in Form eines mobilen „Inflatables", das sich fast überall aufstellen lässt. In nur wenigen Minuten einsatzbereit, eignet es sich für kurze und längere Aufenthalte sowie Notsituationen. Es besteht aus einem maßgefertigten leichten Rahmensystem aus extrudiertem Aluminium, was eine einfache Integration und das nahtlose Zusammenwirken aller Strukturelemente ermöglicht. Die Hauptaußenschicht ist in eine luftdichte Traglufthaut aus strapazierfähigem, flexiblem Gewebe gehüllt.

Statement by the jury
Snoozy merges sophisticated construction and high usability with a futuristic appearance into an eye-catching product.

Begründung der Jury
Bei Snoozy verschmelzen überlegte Konstruktion, hoher Nutzen und futuristische Anmutung zu einer Einheit, die die Blicke auf sich zieht.

Turbo
Tent
Zelt

Manufacturer
Sportsman Corporation, New Taipei, Taiwan
In-house design
Web
www.turbomotorhome.com

This roomy family tent features a frame that can be set up in less than a minute. Its strong aluminium poles are integrated into the tent body so they need not be assembled or disassembled. Near-vertical walls and a generous peak height of seven feet create a great deal of usable space. The front door and the two windows are equipped with mesh which ensures privacy, keeps bugs out and allows airflow. The tent body and fly are made from rugged, waterproof 150D ripstop polyester.

Dieses geräumige Familienzelt verfügt über eine Rahmengestaltung, die sich in weniger als einer Minute aufstellen lässt. Seine starken Aluminiumstangen sind in den Zeltkörper integriert, sodass sie weder auf- noch abgebaut werden müssen. Fast senkrechte Wände und ein großzügiges Höhenverhältnis von fast sieben Fuß bieten viel Nutzraum. Die Vordertür und die beiden Fenster weisen ein Netz auf, das Blicke von außen verhindert, Fliegen abhält und kühlende Luft hindurchlässt. Der Zeltkörper und das Außenzelt bestehen aus stabilem, wasserdichtem 150-Denier-Ripstop-Polyester.

Statement by the jury
The outstanding features of the spacious family tent Turbo are its ability to be put up easily and quickly, and its highly durable and waterproof materials.

Begründung der Jury
Der schnelle Aufbau zu einem großzügigen Zelt für die ganze Familie und besonders haltbares, wasserabweisendes Polyester sind die herausragenden Vorzüge des Turbo.

NIGOR® Pio Pio
Lightweight Tent
Leichtgewichtszelt

Manufacturer
Nigor Net BV, Utrecht, Netherlands
In-house design
Geor Nicolaas
Web
www.nigor.net

Pio Pio is an ultralight one-person tent of only 830 grams and with compact storage size. The inner height of 105cm enables sitting upright, while the inner tent length of 260 cm offers plenty of space for additional gear. The GorLyn 10 fly sheet is made of high-endurance ripstop nylon 6.6, which can be repaired e.g. in the event the fly is punctured by, for instance, a thorn, the hole can be closed by rubbing the fabric, which has undergone a three-step silicone coating process.

Pio Pio ist ein mit 830 Gramm ultraleichtes 1-Personen-Zelt mit kleinem Packmaß. Die Innenhöhe von 105 cm erlaubt aufrechtes Sitzen, und bei der Innenlänge von 260 cm bleibt genug Platz für die Ausrüstung. Das GorLyn-10-Außenzelt ist aus hochfestem Ripstop-Nylon 6.6 hergestellt. GorLyn hat „selbstheilende" Eigenschaften: Falls das Außenzelt durch einen Stachel punktiert wird, lässt sich das Loch durch Reiben des Materials, das in drei Schritten silikonbeschichtet wurde, schließen.

Statement by the jury
The spacious Pio Pio tent attracts attention above all others with its innovative fabric that is capable of having holes repaired by simply rubbing the surface.

Begründung der Jury
Das geräumige Zelt Pio Pio fällt in erster Linie durch sein innovatives Gewebe ins Auge: Löcher können durch Reiben einfach wieder geschlossen werden.

Oscar
Ultra Light Sleeping Bag
Ultraleichter Schlafsack

Manufacturer
Nordisk Company A/S, Silkeborg, Denmark
In-house design
Sarah Groth
Web
www.nordisk.eu

The Oscar three-season sleeping bag weighs less than 500 grams and is made of seven-denier ripstop nylon. Due to Thermo Dry Eco insulation made from recycled plastic it is highly durable. Cold spots are eliminated thanks to an optimised construction, while the foot box features a trapezoidal construction for increased insulation and comfort. Placed on the upper side for easy access; the short YKK two-way zipper allows for optimal ventilation.

Der Drei-Jahreszeiten-Schlafsack Oscar wiegt weniger als 500 Gramm und ist aus 7-Denier-Ripstop-Nylon gefertigt, wobei ihn eine Thermo-Dry-Eco-Füllung aus recyceltem Kunststoff leistungsfähig macht. Die spezielle Konstruktion eliminiert Kältebrücken und die Trapezform des Fußteils erhöht Isolation und Komfort. Ein kurzer Zweiwege-Reißverschluss, der auf der Oberseite angebracht ist, ermöglicht eine optimale Belüftung.

The Oscar sleeping bag distinguishes itself by being extremely light, having eco-friendly insulation and offering a high degree of comfort.

Begründung der Jury
Ein ausgesprochen geringes Gewicht, eine umweltfreundliche Füllung und eine auf hohen Komfort ausgerichtete Gestaltung sind die Pluspunkte, die den Schlafsack Oscar auszeichnen.

Lite Cot
Lightweight Foldable Cot
Leichte faltbare Liege

Manufacturer
Helinox Inc,
Incheon, South Korea

In-house design
Jake Lah

Web
www.helinox.com

reddot award 2015
best of the best

Comfortable lightweight

Camping beds and foldable cots are indispensable items. Although easy to set up to provide a place to sleep, they are often disregarded as hardly more than a purely functional, inconvenient solution. The Lite Cot by Helinox, however, reinterprets such foldable cots as a bed that is both functional and highly comfortable. The design of this cot is based on the use of an innovative duralumin alloy, chosen for its specific properties, and its effective implementation. The aluminium frame construction meticulously incorporates all components to ensure a perfectly balanced strength-to-weight ratio. The elastic polyethylene fibre used for the lying surface is custom-made to make sure that it does not stretch over time. This makes it highly durable and keeps it in shape for long-lasting relaxing and sleeping comfort. Overall, the cot is very light, weighing only 1.24 kg, and foldable into an exceptionally small storage size, allowing it to be easily stowed even in a car. An additional, well thought-out design aspect of this cot is that it can also be set up using only three legs and yet be fully functional, when the weight of the user is below 80 kg. This allows users to save an extra 120 grams of weight for even lighter transportation. A coherent design approach has thus led to the emergence of a cot with enticing novel characteristics. Merging functionality with comfort, the cot is suitable for use almost anywhere including at home as a stylish guest bed of aesthetic appearance.

Komfortables Leichtgewicht

Die Campingliege oder das Reisebett sind unverzichtbar. Rasch in eine Schlafgelegenheit verwandelt, werden sie jedoch oft nur als eine unbequeme Notlösung empfunden. Die Lite Cot von Helinox interpretiert eine solche faltbare Liege neu, hin zu einem funktionalen und überaus komfortablen Bett. Die Gestaltung basiert auf dem Einsatz einer innovativen Duralumin-Legierung, wobei deren besondere Eigenschaften effektiv umgesetzt wurden. Die Konstruktion des Aluminiumrahmens verwirklicht ein perfekt aufeinander abgestimmtes Stärke- und Gewichtsverhältnis aller Komponenten. Um die Bespannung aus einem elastischen Polyethylen-Fasermaterial dauerhaft in einem idealen Zustand zu halten, wird diese maßgefertigt. Die Liege ist deshalb sehr langlebig und sie hält ihre Form für einen gleichbleibend guten Schlafkomfort. Mit einem Gewicht von nur 1,24 kg ist sie zudem leicht und, da sie sich ausgesprochen klein zusammenfalten lässt, erlaubt sie ein einfaches Verstauen auch im PKW. Ein klug durchdachter Aspekt der Gestaltung dieser Liege ist darüber hinaus die Möglichkeit, sie auf nur drei Beinen aufzustellen, wenn die darauf Schlafenden weniger als 80 kg wiegen. Auf diese Weise lassen sich noch einmal 120 Gramm des Tragegewichtes einsparen. Ein schlüssiges Gestaltungskonzept bringt so eine Liege mit bestechend neuen Eigenschaften hervor. Durch ihre Funktionalität und ihren Komfort ist sie überall einsetzbar, durch ihre Ästhetik wird sie überdies zu einem formschönen Gästebett.

Statement by the jury

The use of an innovative duralumin alloy has resulted in the emergence of the Lite Cot by Helinox as a highly functional product. Its design perfectly implements the positive properties of the material to ensure the best strength-to-weight ratio. The foldable cot is very light and offers the highest sleeping comfort. Since it is intuitively assembled and disassembled, it can be used in almost any setting, delivering versatile usability. It thus suggests itself as a companion that easily adapts to any situation at any time.

Begründung der Jury

Der Einsatz einer innovativen Duralumin-Legierung führt bei der Liege Lite Cot von Helinox zu einem hochfunktionalen Produkt. Ihre Gestaltung nutzt auf perfekte Weise die positiven Eigenschaften des Materials, um das beste Stärke-Gewichts-Verhältnis zu erreichen. Diese faltbare Liege ist sehr leicht und bietet höchsten Schlafkomfort. Da sie zudem selbsterklärend auf- und wieder abzubauen ist, kann sie vielfach eingesetzt werden. Sie ist damit jederzeit ein sich gut der Situation anpassender Begleiter.

Designer portrait
See page 46
Siehe Seite 46

Swivel Chair
Lightweight Swivel Chair
Leichter Drehstuhl

Manufacturer
Helinox Inc, Incheon, South Korea
In-house design
Jake Lah
Web
www.helinox.com

This portable 360 degree swivel chair aims to combine function and comfort with portability, simplicity and durability. At the core of the innovation is a sophisticatedly engineered hub piece. The chair, which can pack down into a compact format, weighs only 1.2 kg but can support a weight of 145 kg. Thanks to the single-corded pole structure, it can be set up quickly and easily. The frame is made of high-strength aluminium which contributes to its low weight and provides a high degree of reliability and robustness.

Die Gestaltung des portablen, um 360 Grad drehbaren Swivel Chair stand vor der Herausforderung, Funktion und Komfort mit Tragbarkeit, Schlichtheit und Beständigkeit zu verbinden. Kern der Innovation ist das ausgezeichnet konstruierte Hubstück. Der Stuhl lässt sich kompakt zusammenpacken und wiegt 1,2 kg, kann aber selbst ein Gewicht von 145 kg tragen. Aufgrund des einzeln verschnürten Stangensystems ist er schnell und einfach aufzubauen. Das Gestänge aus hochfestem Aluminium trägt zu seinem geringen Gewicht bei und bietet eine hohe Zuverlässigkeit und Haltbarkeit.

Statement by the jury
The concept of the swivel chair is sophisticated down to the very last detail, focusing not only on providing versatile usability but also comfort, and based on the use of durable materials.

Begründung der Jury
Der Swivel Chair ist bis ins Kleinste durchdacht und legt ebenso Wert auf eine hohe Benutzerfreundlichkeit wie auf Komfort und widerstandsfähige Materialien.

Camp Chair
Campingstuhl

Manufacturer
Helinox Inc, Incheon, South Korea
In-house design
Jake Lah
Web
www.helinox.com

The Camp Chair sets a new standard in terms of its functionality and use of sophisticated materials. It is not only compact, robust and very light, weighing just 1,188 grams, but it also offers a compact storage size and ensures high seating comfort. Features such as the optimised DAC frame tubing made from special alloy, the tailor-made seat and the seat top spacer pole are seamlessly integrated to form a product of high functionality.

Mit seinem überzeugenden Design und ausgesuchten Materialien setzt der Camp Chair Maßstäbe. Er ist nicht nur kompakt, robust und mit nur 1.188 Gramm Gewicht sehr leicht, sondern besitzt zudem ein kleines Packmaß und ist sehr komfortabel. So integrieren sich Designmerkmale wie der optimierte Rohrdurchmesser des aus einer besonderen Legierung hergestellten DAC-Gestänges, eine maßgeschneiderte Sitzform und die obere Sitzabstandsstange nahtlos in die hohe Funktionalität.

Statement by the jury
For both day trips and longer stays, the simple and robust design of the camping chair presents itself as a reliable solution for outdoor activities.

Begründung der Jury
Ob für Tages- oder längere Ausflüge – der schlicht und stabil gestaltete Campingstuhl präsentiert sich als zuverlässige Lösung für unterwegs.

Cot Chair
Chair
Stuhl

Manufacturer
Osung Duralumin Co., Ltd., Incheon, South Korea
In-house design
Web
www.duralumin.co.kr

Cot Chair can be characterised by three words: compact, light and comfortable. With a weight of only 1.38 kg and small packing size, it is very light and easy to carry anywhere. The shape of the chair was inspired by a hammock, which makes it very comfortable and ensures users to be relaxed when seated. It can be used on the beach as well as camping field, mountains or even at the office during a break.

Der Cot Chair lässt sich mit drei Wörtern charakterisieren: kompakt, leicht und bequem. Mit einem Gewicht von nur 1,38 kg und geringer Verpackungsgröße ist er sehr leicht und handlich überall hin mitzunehmen. Die Gestaltung basiert auf einer Hängematte, was den Sitz sehr komfortabel macht, sodass die Benutzer sich sofort entspannen können. Er eignet sich für den Einsatz am Strand, beim Campen, in den Bergen und sogar für die Pause im Büro.

Statement by the jury
The experience of sitting in a chair similar to a hammock and its simple construction make the Cot Chair an outstandingly comfortable seat for outdoor or indoor use.

Begründung der Jury
Das Gefühl, wie in einer Hängematte zu sitzen, und seine einfache Konstruktion machen den Cot Chair zur auffallenden Entspannungsgelegenheit für drinnen und draußen.

Chammock Original
Hanging Hammock Chair
Hängemattenstuhl

Manufacturer
Chammock Co., Taichung, Taiwan
In-house design
Web
www.chammockco.com

Chammock Original is a hanging hammock chair that features a compact storage size and a weight of only 270 grams. For maximum safety, it is constructed from high quality nylon and lightweight aluminium carabiners. Easy to assemble in any space, it can be hung between trees, posts or boat masts. A strap between the carabiners allows for simple width adjustment for optimal comfort.

Statement by the jury
This compact and easy to stow hammock offers not only easy assembly but also scores highly with its construction based on durable, high-quality materials.

Der Chammock Original ist ein Hängemattenstuhl mit einer kompakten Verpackungsgröße und wiegt nur 270 Gramm. Für größtmögliche Sicherheit wurde er aus hochwertigem Nylon und leichten Aluminiumkarabinern konstruiert. Er ist überall mühelos aufzubauen und kann zwischen Bäumen, Pfosten oder Bootsmasten aufgehängt werden. Eine Lasche zwischen den Karabinern ermöglicht eine rasche Weitenjustierung für den optimalen Komfort.

Begründung der Jury
Der kompakt verstaubare Hängemattenstuhl verspricht nicht nur eine einfache Handhabung, sondern überzeugt auch mit hochwertigen und widerstandsfähigen Materialien.

Gigseat
Ergonomic Outdoor Seat
Ergonomische Sitzschale

Manufacturer
Gigtrigger as, Bodø, Norway
Design
EGGS Design as (Carl André Nørstebø, Carl-Gustaf Lundholm), Trondheim, Norway
Minoko Design as (Audun Sneve), Levanger, Norway
Web
www.gigseat.com
www.gigseat.de
www.eggsdesign.no
www.minoko.no

You can also find this product in
Dieses Produkt finden Sie auch in
Living
Page 445
Seite 445

Gigseat is a seat for outdoor use, specifically designed for use in uneven and hilly terrain. It can be used on both sides, according to the terrain gradient. A triangular and removable plug in the back prevents the seat from sliding and keeps the user's bottom dry. On the adjacent side, a triangular hole in the seat provides a drink holder. Made of 100 per cent recyclable thermoplastic, Gigseat is suitable for most outdoor surfaces including grass, stone, sand and snow, and is connectable to another seat, for example, for use at events.

Statement by the jury
Easy transportability and robust execution lend the Gigseat flexibility of use, not least outside while on the go.

Der Gigseat ist eine Sitzschale für draußen und speziell für unebenes und hügeliges Terrain konstruiert. Er kann, je nach Gefälle, beidseitig verwendet werden. Der entfernbare Stöpsel im Rücken verhindert, dass er rutscht, und hält den Po trocken; die dreikantige Aussparung am vorderen Rand kann als Trinkgefäßhalter genutzt werden. Der Gigseat aus recycelbarem Thermoplast ist für die meisten Untergründe wie Gras, Stein, Sand und Schnee geeignet und lässt sich mit weiteren Sitzschalen, etwa bei Großveranstaltungen, zusammenfügen.

Begründung der Jury
Die einfache Transportierbarkeit und die robuste Ausführung machen den Gigseat nicht nur unterwegs in der Natur flexibel in der Anwendung.

GoBites Trio
Cutlery
Besteck

Manufacturer
humangear, inc., San Francisco, USA
In-house design
Design
LUNAR California, San Francisco, USA
Web
www.humangear.com
www.lunar.com

The GoBites Trio is a three-piece utensil set and convenient storage case in one. The innovatively arranged design of the box offsets the fork and spoon so that the knife can be positioned exactly in the centre. This efficient packing arrangement turns the product into an easy-to-transport set of tools, including a toothpick, which are all presented and removable separately. In addition, the knife also features a useful bottle-opener.

GoBites Trio besteht aus einem dreiteiligen Besteckset samt Aufbewahrungsbox. Die innovativ gestaltete Box wurde so konstruiert, dass Gabel und Löffel nicht exakt übereinanderliegen, sodass dazwischen genau das Messer passt. Diese geschickte Anordnung macht das Produkt zu einem einfach zu transportierenden Set – sogar mit Zahnstocher –, dem die Besteckteile separat zu entnehmen sind. Das Messer ist praktischerweise mit einem Flaschenöffner ausgestattet.

Statement by the jury
The utensils and storage case score highly with their sophisticated design, which ensures this product to be a handy resource for everyday and leisure purposes.

Begründung der Jury
Das Set aus Besteck und Box beeindruckt mit seiner ausgeklügelten Gestaltung. Sie prädestiniert es als praktischen Begleiter in Alltag und Freizeit.

GoBites Duo
Cutlery
Besteck

Manufacturer
humangear, inc., San Francisco, USA
In-house design
Design
LUNAR California, San Francisco, USA
Web
www.humangear.com
www.lunar.com

GoBites Duo is a smart combination of a fork and a spoon, featuring a simple interlocking mechanism. This mechanism not only allows the fork and spoon to nest together compactly for transport and storage, but it also allows the handles to be connected end-to-end, creating a more comfortable and easy-to-use tool of significantly longer length. Both utensils can also be used separately and are easy to clean.

Statement by the jury
Outdoor enthusiasts require light and easy-to-use utensils and the GoBites Duo meets this demand with a clear and sophisticated concept.

Die intelligente Kombination aus Gabel und Löffel GoBites Duo verfügt über einen einfachen Verbindungsmechanismus. Dieser ermöglicht nicht nur die kompakte Verschachtelung von Gabel und Löffel für Transport und Lagerung, sondern bewirkt auch, dass sie so miteinander verbunden werden können, dass ein besser handhabbares, deutlich längeres Utensil entsteht. Die Besteckteile können auch einzeln benutzt werden und sind leicht zu reinigen.

Begründung der Jury
Unterwegs sind leichte und einfach zu benutzende Utensilien gefragt – ein Anspruch, dem das GoBites Duo durch seine klare, durchdachte Gestaltung gerecht wird.

GoBites Uno
Cutlery
Besteck

Manufacturer
humangear, inc., San Francisco, USA
In-house design
Design
LUNAR California, San Francisco, USA
Web
www.humangear.com
www.lunar.com

GoBites Uno is a fork and spoon combination utensil following the principle of a two-in-one tool. It is the result of combining ergonomic materials with user testing. The "U" shape at either end allows natural and comfortable grip in either fork or spoon mode. The tines on the sides of the fork feature tapered edges that help users easily cut food without a knife. A diagonal ridge visually highlights the blending of the spoon and the fork.

Statement by the jury
High functionality and a well thought-through design idea form a successful symbiosis in the GoBites Uno. The compact combination tool also convinces ergonomically.

GoBites Uno ist ein konsequent entwickeltes Kombinationsbesteck aus Gabel und Löffel auf Basis des 2-in-1-Prinzips und Ergebnis ergonomischer Material- sowie Benutzertests. Die U-Form seiner Enden ermöglicht eine natürliche und bequeme Handhabung in Gabel- wie Löffelstellung. Mithilfe der Schrägkanten an den Zinken der Gabel lässt sich das Essen ohne Messer teilen. Eine diagonale Furche zeigt visuell an, wo Gabel und Löffel ineinander übergehen.

Begründung der Jury
Hohe Funktionalität und eine schlüssige Gestaltungsidee gehen bei GoBites Uno eine gelungene Symbiose ein. Auch ergonomisch überzeugt das handliche Kombinationsbesteck.

RangerWood 55
Swiss Army Knife
Schweizer Taschenmesser

Manufacturer
Victorinox AG, Ibach (Schwyz), Switzerland
In-house design
Web
www.victorinox.com

Its 130 mm length and wide range of functions characterise the RangerWood 55 as an ideal outdoor tool. It features a locking blade with a mechanism that easily unlocks by pressing the Victorinox cross & shield button. The elegant walnut wood scales are designed ergonomically to ensure a perfect grip, allowing users to work with it pleasant and safely. Within the ten available functions, it also integrates a wood saw, can opener, corkscrew and a bottle opener with integrated lockable screwdriver.

Statement by the jury
With its many different features the pocket knife distinguishes itself as a highly versatile tool that is ideally suited to a wide range of outdoor activities.

Eine Länge von 130 mm und eine große Auswahl an Werkzeugen zeichnen den RangerWood 55 als ideales Outdoor-Tool aus. Er verfügt über eine Feststellklinge, die einfach per Druck auf das Victorinox Cross & Shield entriegelt werden kann. Die ergonomisch geformten Nussbaumholz-Schalen ermöglichen einen angenehmen Griff für sicheres Arbeiten. In den zehn Grundfunktionen sind auch eine Holzsäge, ein Kapselheber mit integriertem feststellbarem Schraubendreher, ein Dosenöffner und ein Korkenzieher integriert.

Begründung der Jury
Das Taschenmesser erweist sich dank seiner zahlreichen Funktionen als sehr vielseitig und damit hervorragend geeignet für unterschiedlichste Aktivitäten.

Indestructible Powerpack 6000
Battery Charger
Ladegerät

Manufacturer
VARTA Consumer Batteries GmbH & Co. KGaA, Ellwangen, Germany
Design
GNOSIS Product Development,
Shenzhen, China
Web
www.varta-consumer.com
www.gnosispd.com

The Indestructible Powerpack 6000 provides extra power for all common smartphones, tablets and similar devices and allows recharging through two USB power connections. Its shell is dust-, water- and shock-proof and can be used in almost any environment thanks to its robust ABS plastic and aluminium construction. The 6000 mAh version has two USB ports, one with 1.0 A, and the other 2.4 A, and is compatible with almost all widely used USB cables and electronic devices.

Das Indestructible Powerpack 6000 ermöglicht die mobile Stromversorgung von Smartphones, Tablets und Ähnlichem sowie das Laden USB-ladefähiger Geräte. Es ist staubdicht, wasserfest und stoßfest und lässt sich dank seiner stabilen Bauweise in ABS-Kunststoff und Aluminium in jeder Umgebung benutzen. Die 6000 mAh-Version verfügt über zwei USB-Ports mit 1,0 A bzw. 2,4 A und ist mit den gängigen USB-Kabeln und -Endgeräten kompatibel.

Statement by the jury
The compact construction of the Powerpack together with its high functionality offer versatile use for people on the go.

Begründung der Jury
Geschützt in einer kompakten Konstruktion bietet die hohe Funktionalität des Powerpacks unterwegs einen vielseitigen Nutzwert.

Jumpr Quad
Battery Charger
Ladegerät

Manufacturer
Ascent Solar Technologies, Inc., Thornton, USA
Design
Nexiom Company Limited (Lei Zheng, Vincent Lau), Hong Kong
Web
www.ascentsolar.com
www.nexiom.cc

The recharger Jumpr Quad enables Ni-MH batteries to be used as a main power storage device. It transforms the output of ordinary batteries and Ni-MH batteries to USB 5 volt, and is therefore suitable as a backup power supply to recharge handheld devices and as a charger for Ni-MH batteries via a USB port. With the integration of a torch function for emergency use this power bank is designed for outdoor use and can be attached by a top hook to a bag or belt.

Mit dem Ladegerät Jumpr Quad lassen sich Nickel-Metallhydrid-Akkus als Haupt-Energiespeichergerät verwenden. Es wandelt die Leistung von herkömmlichen Batterien und Nickel-Metallhydrid-Akkus auf USB 5 Volt um und eignet sich dadurch als Back-up-Stromversorgung zum Aufladen von Mobilgeräten und als Ladegerät für die genannten Akkus über USB-Anschluss. Der Jumpr Quad mit integrierter Taschenlampenfunktion für Notfälle wurde für Outdoor-Aktivitäten entwickelt und lässt sich mit seinem Bügel an einer Tasche oder dem Gürtel befestigen.

Statement by the jury
This charging device offers versatile use for outdoor activities and stands out in terms of its usability and the durability of the product materials.

Begründung der Jury
Dieses Ladegerät bietet vielseitige mobile Nutzungsmöglichkeiten und besticht durch widerstandsfähige Materialien und praktische Handhabung.

Generatr S100
Battery Charger
Ladegerät

Manufacturer
Ascent Solar Technologies, Inc., Thornton, USA
Design
Nexiom Company Limited (Lei Zheng, Vincent Lau), Hong Kong
Web
www.ascentsolar.com
www.nexiom.cc

Generatr S100 is a lightweight and handheld device designed to provide AC and DC power for outdoor activities. In addition to a 120 watt AC output, it also features 19 volt, 12 volt, and USB outputs. With its high capacity of 100 watt-hours it supports all recharging demands including notebooks and handheld devices. Its core component for energy storage is a late-model high-density lithium polymer battery with an efficient energy management system.

Generatr S100 ist ein leichtes mobiles Ladegerät, das bei Outdoor-Aktivitäten Wechsel- (AC) und Gleichstrom (DC) liefert. Neben einem AC-Ausgang mit 120 Watt bietet das Gerät je einen 19-Volt-, einen 12-Volt- sowie einen USB-Ausgang. Mit seiner hohen Kapazität von 100 Watt-stunden unterstützt es jegliche Auflade-anforderungen etwa von Notebooks und Mobilgeräten. Die Hauptkomponente zur Stromspeicherung ist ein moderner Lithium-Polymer-Akku mit hoher Energie-dichte und einem effizienten Energiema-nagementsystem.

Generatr Y1200
Battery Charger
Ladegerät

Manufacturer
Ascent Solar Technologies, Inc., Thornton, USA
Design
Nexiom Company Limited (Lei Zheng, Vincent Lau), Hong Kong
Web
www.ascentsolar.com
www.nexiom.cc

Generatr Y1200 provides energy for outdoor activities and work, featuring a 1000 watt high power AC output, as well as 19 volt, 12 volt and USB outputs. The 1200 watt-hour capacity enables it to fully support high-power demand items such as televisions, refrigerators, computers and handheld devices. Designed with a metal skeleton structure, the product can also be used with life support equipment thanks to its long-lasting power storage technology.

Generatr Y1200 liefert mit einem 1.000-Watt-Wechselstrom-Hochleistungs-ausgang sowie 19-Volt-, 12-Volt- und USB-Ausgängen Strom für Outdoor-Aktivitäten und zum Arbeiten. Mit einer Kapazität von 1.200 Wattstunden unterstützt er die hohen Energieanforderungen von Fernseh-geräten, Kühlschränken, Computern und Mobilgeräten uneingeschränkt. Das in Metallskelettbauweise gestaltete Produkt eignet sich dank langlebiger Energiespei-chertechnologien auch für den Einsatz mit lebenserhaltenden Geräten.

Statement by the jury
Supported by a durable external structure, this high-capacity charger is perfectly suitable for diverse outdoor activities.

Begründung der Jury
Unterstützt durch ein robustes Gehäuse ist das leistungsfähige Ladegerät hervor-ragend für unterschiedliche Aktivitäten im Freien geeignet.

ABScond Flow
Avalanche Backpack
Lawinenrucksack

Manufacturer
VAUDE Sport GmbH & Co. KG, Tettnang, Germany
In-house design
Philipp Ziegler
Web
www.vaude.com

The spacious ABScond Flow avalanche backpack is made of tough Cordura with ABS Twinbag technology and is perfect for freeriding and ski day tours. Its fully adjustable shoulder straps allow the backpack to adapt comfortably to the length of the back, which in combination with the moulded contact back and an adjustable hip belt, ensure a stable fit. It comes with a number of additional features, such as a ski and snowboard attachment and a zip, which can expand the volume by four litres.

Der geräumige Lawinenrucksack ABScond Flow besteht aus robustem Cordura mit ABS-Twinbag-Technologie und eignet sich sehr gut zum Freeriden und für Tagesski-touren. Seine verstellbaren Schulterträger lassen sich bequem an den Rücken anpas-sen, was zusammen mit dem geprägten Kontaktrücken und verstellbaren Hüftgurt für Stabilität sorgt. Er besitzt zahlreiche Extras wie etwa eine Skifixierung, Snow-boardbefestigung und einen Reißverschluss für weiteres Volumen von vier Litern.

Statement by the jury
Apart from providing a higher degree of safety, the design of the avalanche backpack also places great importance on versatile and functional usability.

Begründung der Jury
Neben der hohen Sicherheit als zentraler Komponente legt die Gestaltung des Lawinenrucksacks ebenso Wert auf einen vielseitigen praktischen Nutzen.

Filo Messenger Bag
Fahrradtasche

Manufacturer
Filo LLC, Taipei, Taiwan
In-house design
Web
www.facebook.com/filodesigner

The aim of the Filo messenger bag, made of water-repellent polyester, was to enhance safety for the cyclist. The cover flap features a reflective fabric that makes the cyclist more visible in the dark. The spacious interior can even be increased when pulling the two zippers on the sides. The bag can be worn diagonally across only one shoulder or – for better weight distribution – like a rucksack on both shoulders.

Ziel war es, mit der Filo Messenger Bag aus wasserabweisendem Polyester eine verkehrssichere Tasche zu entwerfen. Der Überschlag besteht aus reflektierendem Material, mit dem der Radfahrer in der Dunkelheit gut gesehen wird. Der geräumige Innenraum lässt sich durch zwei Reißverschlüsse sogar noch vergrößern. Die Tasche kann sowohl schräg über nur eine Schulter als auch – um das Gewicht besser zu verteilen – wie ein Rucksack von beiden Schultern getragen werden.

Aeolus
Shoulder Bag/Windbreaker
Umhängetasche/Windjacke

Manufacturer
Taiwan Textile Research Institute,
New Taipei, Taiwan
In-house design
Wei-Hung Chen , James Hsu
Web
www.ttri.org.tw

Aeolus is a shoulder bag which can be transformed into a windbreaker. As a bag it can carry a mobile phone, wallet, cosmetic bag, keys, cards etc., or it can easily be transformed into a windbreaker. Made of plant-based nylon fabric, it keeps the wearer warm and is comfortable to wear during exercise thanks to its permeability. When it is used as a windbreaker, there is a pocket at the back to hold personal belongings. Moreover, Aeolus can also be fixed onto bike grips and thus is easily adaptable to an individual user's needs.

Statement by the jury
Adaptable, practical and original; these are the characteristics with which the versatile Aeolus emerges as a true all-rounder for everyday use.

Aeolus ist zugleich Umhängetasche und Windjacke. Als Umhängetasche nimmt sie Mobiltelefon, Portemonnaie, Schlüssel und anderes auf und kann mit wenigen Handgriffen in eine Windjacke verwandelt werden. Aus pflanzenbasiertem Nylon gefertigt, hält sie warm und lässt sich aufgrund ihrer guten Atmungsaktivität zum Sport zu tragen. In einer Tasche am Rücken können persönliche Dinge verstaut werden. Außerdem kann Aeolus am Fahrradlenker befestigt werden und so ganz nach Bedarf zum Einsatz kommen.

Begründung der Jury
Wandlungsfähig, praktisch und originell – mit diesen Charakteristika präsentiert sich die vielseitige Aeolus als echter Allrounder im Alltag.

Strato
Men's Ultra Light Down Jacket
Ultraleichte Männer-Daunenjacke

Manufacturer
Yeti GmbH, Görlitz, Germany
In-house design
Oliver Reetz
Web
www.yetiworld.com

Combining modern technical components with innovative workmanship, the Strato is a truly lightweight jacket weighing just 165 grams thanks to the high quality Crystal Down filling. It provides two fully functional zipper pockets, one inner pocket, one integrated packsack, as well as an elasticated hem and cuffs. It is made from an ultra light and highly durable fabric that is also breathable. The jacket can be packed into its own left-hand pocket and used as a pillow.

Moderne technische Komponenten, eine innovative Verarbeitung und die hochwertige Crystal Down Füllung sorgen dafür, dass Strato mit 165 Gramm ein absolutes Leichtgewicht ist. Die Daunenjacke bietet zwei vollwertige Reißverschlusstaschen, eine Innentasche, eine integrierte Packtasche, einen elastischen Saum plus Bündchen und besteht aus einem ultraleichten und strapazierfähigen Gewebe, das zudem atmungsaktiv ist. Sie kann in ihre eigene Seitentasche gepackt und als Kissen benutzt werden.

Thimble
Fingerhut
Needlework Utensil
Handarbeitswerkzeug

Manufacturer
Prym Consumer Europe GmbH,
Stolberg, Germany
Design
Atelier Papenfuss, Weimar, Germany
Web
www.prym-consumer.com
www.atelierpapenfuss.de

The ergonomic thimble ensures secure positioning on the needle's eye and an ideal transmission of force thanks to its structured surface. The needlework utensil consists of two different components. The hard component protects the finger from the needle. The flexible component is perforated for optimal ventilation, comfortable use and a secure fit of the thimble.

Statement by the jury
The material of the thimble has a soft, velvety look. The high-contrast colour scheme provides a visually appealing indication of the areas' functions.

Mit seiner strukturierten Oberfläche ermöglicht der ergonomische Fingerhut eine sichere Positionierung auf dem Nadelöhr und eine optimale Kraftübertragung. Das Handarbeitswerkzeug besteht aus zwei unterschiedlichen Komponenten. Der harte Bestandteil schützt den Finger vor der Nadel. Die flexible Komponente ist durchlässig und sorgt dadurch für eine optimale Belüftung, hohen Tragekomfort und einen sicheren Sitz des Fingerhuts.

Begründung der Jury
Samtig weich erscheint das Material des Fingerhuts. Die kontrastreiche Farbgebung gibt einen optisch ansprechenden Hinweis auf die Funktion der Bereiche.

Tripod
Dreifuß
Needlework Utensil
Handarbeitswerkzeug

Manufacturer
Prym Consumer Europe GmbH,
Stolberg, Germany
Design
Atelier Papenfuss, Weimar, Germany
Web
www.prym-consumer.com
www.atelierpapenfuss.de

This Tripod is a device for manual punching and riveting using a hammer. The needlework utensil can be positioned anywhere on a textile surface. Its three feet make it very stable and provide optimal guiding and centring of the tools, thus achieving neat results. Up to 300 different punching and riveting inserts can be used together with the needlework utensil.

Statement by the jury
The clear design promises an intuitive understanding of how to use the Tripod. The rounded proportions look appealing.

Der Dreifuß ist eine Vorrichtung zum manuellen Stanzen und Vernieten unter Zuhilfenahme eines Hammers. Das Handarbeitswerkzeug kann beliebig auf einer textilen Fläche positioniert werden. Die drei Füße sorgen für einen stabilen Stand und eine optimale Führung und Zentrierung der Elemente, somit wird ein sauberes Ergebnis erzielt. Es können bis zu 300 verschiedene Stanz- und Nieteinsätze verwendet werden.

Begründung der Jury
Die klare Formensprache verspricht ein intuitives Verstehen, wie der Dreifuß benutzt wird. Die gerundeten Proportionen wirken charmant und freundlich.

Mastodonts
Wooden Toy
Holzspielzeug

Manufacturer
Wodibow, Segovia, Spain
In-house design
Pablo Saracho
Web
www.wodibow.com

Mastodonts is a family of toys made of wood and magnets. 22 individual pieces, including one single body, can be put together in order to create four different animals – just by changing the head, mouth and ears. It only becomes apparent once the last piece is in place which animal has been built. It is made of beech wood and magnets, coated with an olive oil and beeswax finish and free from plastic, paint, varnish and glue.

Statement by the jury
This family of toys merges a design suitable for children with the use of natural materials into an elegantly crafted toy based on a well thought through concept.

Die Mastodonts sind eine Tierfamilie aus Holz und Magneten. Aus insgesamt 22 Bestandteilen, darunter ein einziger Körper, lassen sich durch Austausch von Kopf, Mund und Augen vier verschiedene Tiere gestalten. Welches Lebewesen entstanden ist, sieht man erst, wenn das letzte Teil an seinem Platz ist. Durch die Verwendung von Buchenholz und Magneten sowie einer Oberflächenbehandlung aus echtem Bienenwachs und Olivenöl wurde auf Kunst- oder Klebstoffe, Farben und Lacke verzichtet.

Begründung der Jury
Eine kindgerechte Gestaltung und natürliche Materialien verbinden sich in dieser Tierfamilie zu einem durchdacht konzipierten wie formschönen Spielzeug.

ARCKIT
Architectural Model Kit
Architekturmodell-Baukasten

Manufacturer
Arckit, Dublin, Ireland
In-house design
Damien Murtagh
Web
www.arckit.com

Arckit is a scaled architectural model building design tool, allowing the user to physically design and bring ideas to life without the need to measure, cut, glue and stick materials together. Available in three kit sizes, it is based on modern building techniques and consists of a series of interconnecting modular components that click together to create multiple building types.

Statement by the jury
Arckit allows architectural ideas and models to be created, and later modified, playfully and easily, thus turning the art of architecture into a direct hands-on experience.

Arckit ist ein mehrteiliges Gestaltungswerkzeug für den Bau von Architekturmodellen. Ideen lassen sich physisch zum Leben erwecken, ohne dabei Materialien vermessen, zuschneiden und verkleben zu müssen. Das Set ist in drei Bausatzgrößen erhältlich. Es basiert auf moderner Bautechnik und besteht aus einer Reihe modular miteinander verbindbarer Komponenten zur Erstellung unterschiedlichster Gebäudetypen.

Begründung der Jury
Mit Arckit lassen sich Architekturentwürfe und -ideen spielend einfach umsetzen und jederzeit auch wieder modifizieren. Damit wird Baukunst unmittelbar plastisch.

NTR
Microphone
Mikrofon

Manufacturer
RØDE Microphones,
Sydney, Australia

In-house design
RØDE Microphones

Web
www.rode.com

reddot award 2015
best of the best

Sculptural elegance

In the field of audio engineering, ribbon microphones are often considered the purest of studio microphones. Their high sound quality is ranked above all due to their special reproduction quality and fast transient response, such as with the plucking sound of a guitar string. By placing the ribbon motor separate to the microphone body, the NTR microphone adopts a distinctive and self-evident design vocabulary. This new construction approach has led to a structural geometry of minimalist appearance in the ribbon motor's rejection zone, which in turn delivers maximum transparency around the ribbon and minimises resonance. The motor itself is coupled to the body by a special suspension so effective that the microphone requires no external shock mounting in studio conditions. In addition, this new ribbon microphone pairs the ribbon to an ultra-low noise transformer that features low impedance and delivers a high output. The NTR is built with premium-quality active electronic components. It can be used with a wide range of different preamps and, unlike other ribbon microphones, eliminates the need for any additional gain. Sonically, it offers excellent sound accuracy as well as precise reproduction of delicate high-frequency detail. A fascinating design that also captivates the eyes with an impressive appearance of classic elegance.

Skulpturale Eleganz

Im Bereich der Tontechnik gelten Bändchenmikrofone als die klangtechnisch reinsten Studiomikrofone. Ihre hohe Klangqualität basiert vor allem auf ihrer besonderen Fähigkeit der Schallabbildung und Transientenansprache bei Klangereignissen wie beispielsweise dem Anreißen einer Gitarrensaite. Bei dem Mikrofon NTR führt die Gestaltung mit einem außerhalb des Gehäuses platzierten Bändchenmotor zu einer eindrucksvollen und eigenständigen Formensprache. Diese neue Anordnung erlaubt eine minimalistisch anmutende Strukturgeometrie im Dämpfungsbereich des Motors, was wiederum zu einer maximalen Transparenz in der Nähe des Bändchens führt und Resonanzprobleme verhindert. Der Motor selbst ist in einer speziellen Lagerung verankert, weshalb im Studioumfeld flexibel ohne die für die Stabilisierung üblichen Spinnen gearbeitet werden kann. Bei diesem neuen Bändchenmikrofon ist zudem das Bändchen mit einem rauscharmen Wandler verbunden, der über einen hohen Ausgangspegel und eine extrem niedrige Impedanz verfügt. Das NTR ist mit hochwertigen elektronischen Komponenten ausgestattet und es kann mit vielen unterschiedlichen Mikrofonvorverstärkern arbeiten. Im Gegensatz zu anderen Bändchenmikrofonen muss dabei keine zusätzliche Pegelanhebung erfolgen. Auf diese Wiese bietet es eine exzellente Klangverarbeitung sowie eine filigrane und präzise Abbildung des Höhenbereiches – und fasziniert dabei zugleich mit der eindrucksvollen Anmutung klassischer Eleganz.

Statement by the jury

Lending the NTR ribbon microphone its distinctive elegant lines was facilitated through the innovation of placing the ribbon motor outside the microphone case. Its impressive sculptural appearance is complemented by sonically outstanding reproduction accuracy that captures high frequencies precisely and authentically. Thanks to the use of state-of-the-art electronic components, the NTR is suitable for use in many recording setups as it does without the need for additional gain requirements.

Begründung der Jury

Bei dem Bändchenmikrofon NTR ermöglicht die innovative Anordnung des Bändchenmotors außerhalb des Mikrofongehäuses eine elegante Linienführung. Seine beeindruckend skulpturale Anmutung geht einher mit einer außergewöhnlich guten klanglichen Wiedergabequalität, die auch die Höhenbereiche sorgfältig und präzise abbildet. Durch den Einsatz hochwertiger elektronischer Komponenten ist das NTR sehr vielseitig verwendbar, da es überdies keine zusätzliche Pegelanhebung benötigt.

Designer portrait
See page 48
Siehe Seite 48

DUBS Acoustic Filters
Earplugs
Ohrstöpsel

Manufacturer
Doppler Labs, Inc., New York, USA
In-house design
Dan Wiggins, Jacob Palmborg
Web
www.getdubs.com

Dubs Acoustic Filters are advanced tech earplugs acoustically engineered to reduce the volume while preserving the clarity of sound at music venues and in other loud environments. Developed as a solution to the growing problem of noise-induced hearing loss in young adults, they redefine how the earplug should look, feel and sound. The special design sits flush within the ear and masks the complexity of the Dubs proprietary audio filters. The product is durable, reusable and comes with a convenient carrying case.

Die akustischen Filter Dubs sind High-Tech-Ohrstöpsel zur Lautstärkedämpfung mit gleichzeitig klarem Klang, etwa für Konzerte oder andere laute Umgebungen. Aufgrund zunehmender lärmbedingter Hörverluste bei jungen Erwachsenen interpretieren sie Klang, Taktilität und das Aussehen eines Ohrenstöpsels neu. Die spezielle Gestaltung sitzt bündig am Ohr und verdeckt die Komplexität des geschützten Dub-Audiofilters. Das Produkt ist langlebig, wiederverwertbar und mit handlichem Etui erhältlich.

Statement by the jury
The Dubs Acoustic Filters captivate with their organic design that sits flush within the ear, as well as with sophisticated technology that offers good sound quality.

Begründung der Jury
Die Dubs Acoustic Filters bestechen sowohl durch ihre organische Gestaltung, die bündig mit dem Gehörgang abschließt, als auch durch ihre ausgereifte Technik, die gute Klangqualität bietet.

Maschine Studio
Production System

Manufacturer
Native Instruments GmbH,
Berlin, Germany
In-house design
Web
www.native-instruments.com

Maschine Studio is a groove production system for tactile beatmaking. Two high-resolution colour displays visualise the features of the Maschine software. This allows fluid content browsing, control of effects and mixing directly on the hardware device, so that users can focus exclusively on creativity. Eight touch-sensitive knobs reveal key parameters when triggered. Its minimalist design supports a productive creative process.

Statement by the jury
Suitable for producers and performers alike, the Maschine Studio lends itself as an ideal tool; an efficient, versatile, high-performance design that ensures intuitive operation.

Maschine Studio ist ein Groove Production System für taktiles Beatmaking. Zwei hochauflösende Farbdisplays visualisieren die Features der Maschine-Software: Flüssiges Browsing, Effekt-Steuerung oder Mixing sind direkt auf der Hardware möglich, sodass sich der Anwender auf den kreativen Prozess konzentrieren kann. Acht berührungsempfindliche Drehregler bieten Zugriff auf die wichtigsten Einstellungen. Ein minimalistisches Design unterstützt präzises Performen.

Begründung der Jury
Für Produzenten und Performer bietet sich das Maschine Studio als ideales Tool an: Leistungsstark, vielseitig und effizient gestaltet, ermöglicht es eine intuitive Steuerung.

Traktor Kontrol S8
DJ Controller

Manufacturer
Native Instruments GmbH, Berlin, Germany
In-house design
Web
www.native-instruments.com

Traktor Kontrol S8 offers a new DJ mixing and live performance dynamic. The controller features enhanced performance controls for a multisensory DJ experience. Track information, waveforms and library browsing are revealed on two high-resolution displays so that users can focus on the performance. Touch-sensitive knobs, faders and LED-guided touch strips trigger the views and panels on the displays. The robust exterior of the device features anodised aluminium top-panels.

Statement by the jury
The DJ controller convinces with an enormous range of features in a compact design and presents itself as an ideal all-round solution for DJs.

Traktor Kontrol S8 bietet eine neue Mixing- und Performance-Möglichkeiten für DJs. Der Controller ermöglicht einen fortschrittlichen Touch-and-see-Workflow für multisensorisches DJing. Browsing, Track-Infos und Wellenformen werden auf zwei hochauflösenden Displays angezeigt, womit der Fokus auf der Performance liegt. Berührungsempfindliche Drehregler, Fader und Touch-Strips mit LED-Guide steuern die Anzeigen und Panels auf den Displays. Das robuste Gehäuse verfügt über eloxierte Aluminium-Top-Panels.

Begründung der Jury
Der DJ-Controller überzeugt durch seinen enormen Leistungsumfang auf kompaktem Raum und präsentiert sich so als optimale Komplettlösung für DJs.

Alpine Music Earplugs
Hearing Protection
Gehörschutz

Manufacturer
Alpine Hearing Protection,
Soesterberg, Netherlands
In-house design
Web
www.alpine.eu

These earplugs are equipped with exchangeable sets of special AlpineAcousticFilters with low, medium and high protection rates. Alongside optimal protection, these filters guarantee listening pleasure without loss of clarity. Stored in a solid travel box, the hearing protection plugs are made of thermoplastic AlpineThermoShape material, which by means of body heat shapes itself inside the ear thus ensuring a highly comfortable fit.

Statement by the jury
A comfortable fit, very good hearing protection and uncompromising sound quality – this triad characterises the high quality of the MusicSafe Classic and MusicSafe Pro earplugs.

Diese Ohrstöpsel sind mit speziellen austauschbaren AlpineAcousticFiltersets ausgerüstet, die niedrige, mittlere und hohe Dämpfungswerte bieten. Neben optimalem Schutz gewährleisten die Filter Musikgenuss ohne Verlust der Klangqualität. Der in einer soliden Travelbox erhältliche Gehörschutz wird aus thermoplastischem AlpineThermoShape-Material hergestellt, das sich mittels Körperwärme dem Ohr anpasst und dadurch bequemes Tragen ermöglicht.

Begründung der Jury
Komfortable Passform, sehr guter Gehörschutz, uneingeschränktes Klangniveau – dieser Dreiklang charakterisiert die hohe Qualität der Ohrstöpsel MusicSafe Classic und MusicSafe Pro.

GEWA Germania
E-Violin

Manufacturer
GEWA music GmbH, Adorf, Germany
In-house design
Heinrich Drechsler, Constantin Savulescu
Web
www.gewamusic.com

The Germania E-Violin deviates from the standard skeletonised outline of most electrical stringed instruments. Instead, it reflects the traditional violin shape with a minimalist interpretation. Retaining the essential key reference points of a traditional instrument including feel and playability, it offers high ergonomic comfort, and replaces the ebony with flexwood, which offers comparable feel and sonic properties.

Statement by the jury
The distinctive shape of the E-Violin turns it into a true eye-catcher. Thanks to the preservation of the original proportions – complemented by optimised ergonomics – it retains its classic playable feel.

Die E-Violine Germania verlässt die ausgetretenen Pfade skelettierter Umrisse bisheriger elektrischer Streichinstrumente und zitiert die traditionelle Violinform in einer minimalistischen Interpretation. Dabei bleibt die traditionelle Spielbarkeit erhalten: Die Parameter entsprechen den klassischen Maßen und gewähren hohen ergonomischen Komfort. Statt Ebenholz wird Flexwood mit ebenbürtigem Spielgefühl für das Griffbrett verwendet.

Begründung der Jury
Ihre markante Formgebung macht die E-Violine zu einem echten Blickfang. Dank Wahrung der Proportionen bleibt – bei zudem optimierter Ergonomie – das klassische Spielgefühl erhalten.

CLP-585
Digital Piano

Manufacturer
Yamaha Corporation, Hamamatsu, Japan
In-house design
Sunao Okamura, Mami Sato,
Takenori Omachi
Web
www.yamaha.com

The CLP-585 was designed to be a modern standard piano, which is respectful of the instrument's 300 year history. It includes the sounds of both the Yamaha CFX and the Bösendorfer, providing players an authentic experience of two of the most famous concert grand pianos. The compact control panel is located on the left side of the keyboard. Whilst offering the same generous space for musical performances as an acoustic piano, the CLP-585 delivers a natural performance feel in a compact body.

Statement by the jury
The harmoniously well-balanced design of the CLP-585 combines the sound and classic appearance of famous concert grand pianos with the electronic control elements that clearly identify it as a digital piano.

Das CLP-585 wurde als modernes Standardpiano entwickelt, das die über 300-jährige Geschichte des Klaviers ehrt. Es ist mit dem Klang des Yamaha CFX sowie des Bösendorfers ausgestattet und bietet somit das authentische Spielerlebnis zweier weltberühmter Konzertflügel. Das kompakte Bedienfeld ist links neben der Tastatur angebracht. Mit ebensoviel Raum für die musikalische Darbietung wie bei einem akustischen Klavier vermittelt das CLP-585 in seiner kompakten Gesamtgröße ein natürliches Spielgefühl.

Begründung der Jury
Die Gestaltung des CLP-585 verbindet den Klang sowie die klassische Formgebung berühmter Konzertpianos auf sehr harmonische Weise mit den Steuerelementen, die es deutlich als E-Piano kennzeichnen.

Crossway
E-Piano

Manufacturer
Shanghai LKK Integrated Design Co., Ltd.,
Shanghai, China
In-house design
Chunming Ma, Qingwei Zhao
Web
www.lkkdesign.com

This electronic piano was designed for music students as well as beginners of the piano. The large 22" touch screen is perfectly matched to the software, videos and teaching system. In appearance, it follows the classical aesthetics of a piano and integrates modern metal and transparent acrylic materials. The transparent cover offers a revealing view of the workings of the piano, including the keys and the screen.

Statement by the jury
The concept of this e-piano focuses strictly on its function as an educational tool and for practising piano playing. In addition, it also convinces with its elegant lines.

Das elektronische Klavier wurde für die Musikerziehung sowie für Anfänger entworfen. Die Ausstattung mit einem 22"-Sensorbildschirm ist auf die Software, Videos und das Lernsystem abgestimmt. Dabei folgt seine Anmutung der klassischen Ästhetik eines Klaviers und integriert moderne Metalle sowie transparentes Acryl. Die durchsichtige Abdeckung gibt den Blick auf die Mechanik, die Tastatur und den Bildschirm frei.

Begründung der Jury
Die Konzeption dieses E-Pianos ist ganz auf seine Funktion hin, den Unterricht und das Üben des Klavierspiels, ausgerichtet. Auch visuell überzeugt seine elegante Linienführung.

Komplete Kontrol S-Series
Keyboard Controller

Manufacturer
Native Instruments GmbH,
Berlin, Germany
In-house design
Web
www.native-instruments.com

This keyboard series, of minimalist design, debuts several technological innovations such as a Light Guide function, which provides colour-coded visual feedback from the software, integrated arpeggiator, and scale and chord mappings – all of which allow users to focus on the keyboard. Two touch strips replace traditional pitch and mod wheels, providing precision control over the advanced touch-interface for bending, warping, and sound automation.

Statement by the jury
The compact Kontrol S series stands out with its clear design arrangement. In addition, thanks to a smart colour coding system, users can fully concentrate on the keyboard and their own creativity.

Die minimalistisch gestaltete Keyboard-Serie führt mehrere technologische Neuerungen ein wie die Light-Guide-Funktion, die eine farbcodierte Visualisierung der Software, des integrierten Arpeggiators sowie des Scale- und Chord-Mappings liefert. So bleibt immer das Keyboard im Fokus. Zwei Touch-Strips ersetzen traditionelle Pitch- und Mod-Wheels, die das Touch-Interface präzise steuern und intuitives Bending, Warping oder Sound-Automationen ermöglichen.

Begründung der Jury
Die Komplete Kontrol S-Series fällt durch ihre übersichtliche Gestaltung ins Auge. Dank intelligenter Farbcodierung kann sich der Anwender zudem ganz der kreativen Nutzung des Keyboards widmen.

Letto dayBed
Dog Bed
Hundebett

Manufacturer
MiaCara, Herzogenaurach, Germany
Design
Bhoom, Creative Design Unit
(Gerd Couckhuyt), Kortrijk, Belgium
Web
www.miacara.com
www.bhoom.eu

Letto dayBed is a dog bed with an eye-catching design displaying striking and timeless lines. The sturdy body of the bed is made from anodised aluminium with a real wood veneer. The all-round raised side panels give the dog a sense of security, while the fitted mattress is pleasantly soft, reversible and fully washable. The soft, yet robust fabric has a distinct structure and, together with the aluminium frame, is available in different colours. The wooden feet were designed in a style reminiscent of the 1950s.

Statement by the jury
The design of the Letto dayBed for dogs presents a stylish aesthetic and is paired with materials that promote premium and durable quality.

Letto dayBed ist ein Hundebett, das durch eine zeitlose und ausdrucksstarke Linienführung auf sich aufmerksam macht. Sein Korpus aus mit Echtholzfurnier beschichtetem, eloxiertem Aluminium ist sehr widerstandsfähig. Die rundum erhöhten Seitenteile geben dem Vierbeiner Geborgenheit, die eingepasste Polsterauflage ist angenehm weich, wendbar und komplett waschbar. Der softe, aber robuste Stoff ist lebendig strukturiert und wie das beschichtete Aluminium in verschiedenen Farben erhältlich. Die Holzfüße wurden im Stil der 1950er Jahre entworfen.

Begründung der Jury
Die Gestaltung des Letto dayBed präsentiert sich in einer stilvollen Ästhetik und setzt auch bei seinen Materialien auf eine hochwertige und langlebige Qualität.

PARAMOUR Pleasure Partners Set
Erotic Toy

Manufacturer
L'amourose, Paris, France
In-house design
DC Du
Web
www.lamourose.com

Paramour is a set of erotic toys which are unobtrusive in design and allows partners to partake in each other's pleasure. The three interchangeable silicon toys are built around a single vibration bullet and are discreet in appearance. The remote control, which is shaped as a tactile pad, can be used intuitively in order to control a partners' pleasure. It offers five vibration modes, has two touch settings and also allows control via a smartphone.

Statement by the jury
The Paramour partner set, with its organically designed lines, rests pleasantly in the hand, conveying a strong sense of intuitive operation.

Paramour ist ein Erotikset in dezenter Anmutung, mit dem beide Partner an der Lust des anderen teilhaben können. Eine Vibrationskugel lässt sich in drei verschiedene Silikon-Spielzeuge integrieren. Die Fernbedienung in Form eines taktilen Kissens kann intuitiv gesteuert und mit ihr die Lust des Partners beeinflusst werden. Sie hat fünf Schwingungsmodi, zwei Berührungseinstellungen und ist auch über das Smartphone bedienbar.

Begründung der Jury
Durch seine organisch gestaltete Formgebung liegt das Partnerset Paramour gut in der Hand und vermittelt eine intuitive Bedienbarkeit.

Pleasure Object 2
Erotic Toy

Manufacturer
JOYDIVISION international AG,
Hannover, Germany
In-house design
Ignacio A. Nolasco, Oliver Redschlag
Web
www.joydivision-international-ag.de

This product line is characterised by a body that varies in size. To achieve the desired size, these toys are pressed together; they then automatically expand back to their original size. Thanks to the rounded shape of the ends, the toys adapt to the user's anatomy while the slender shaft offers an ergonomic and comfortable grip, as well as different stimulation possibilities. The toys are manufactured from medically approved and dermatologically tested Silikomed.

Statement by the jury
Alongside flexibility, thanks to which the Pleasure Object 2 adapts itself snugly to the user's anatomy, this product distinguishes itself with its remarkable design language.

Diese Produktlinie ist durch einen variablen Körper gekennzeichnet. Er lässt sich auf die gewünschte Nutzungsgröße zusammendrücken und nimmt anschließend automatisch wieder sein ursprüngliches Volumen ein. Durch die geschwungene Gestaltung passen sich die Produkte der menschlichen Anatomie an und bieten durch den schlanken Schaft eine ergonomische Handhabung sowie verschiedene Stimulationsvarianten. Sie sind aus hautverträglichem Silikomed hergestellt.

Begründung der Jury
Neben der Flexibilität, mit der sich das Pleasure Object 2 an den individuellen Körper anschmiegt, zeichnet sich das Produkt durch eine bemerkenswerte Formensprache aus.

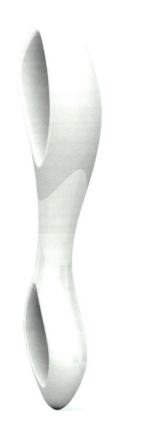  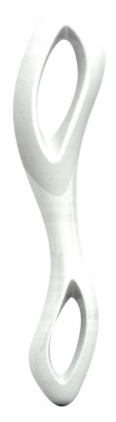

Vesper
Vibrator

Manufacturer
Crave, San Francisco, USA
In-house design
Ti Chang
Web
www.lovecrave.com

Vesper is a vibrator that can be worn like a pendant on a necklace. The multi-speed vibrator, contained in a solid slim device, can be adjusted manually to varying speeds and charged via USB. The pendant and matching chain are both made from highly polished 316 stainless steel, but are separable, allowing the user to decide whether to use as jewellery or as a vibrator. The device is water-resistant and available in three colours.

Statement by the jury
Vesper fascinates through its inventive combination of an erotic toy and a piece of jewellery. The minimalist design and the use of stainless steel further enhance its high-quality appeal.

Vesper ist ein Vibrator, der sich als Schmuck an einer Halskette tragen lässt. Die Geschwindigkeit des schlanken, starken Geräts ist individuell einstellbar, es kann via USB aufgeladen werden. Der Anhänger und die passende Halskette sind aus hochpoliertem 316 Edelstahl gefertigt und können separiert werden, sodass das Produkt wahlweise als Schmuck oder als Vibrator dient. Das Gerät ist in drei Farben verfügbar und wasserfest.

Begründung der Jury
Vesper fasziniert durch seine originelle Verbindung aus Erotic Toy und Schmuckstück. Die minimalistische Formgebung und das Material Edelstahl akzentuieren den wertigen Charakter zusätzlich.

RUYI
Erotic Toy

Manufacturer
Tianailu, Shanghai, China
In-house design
Li Lun
Web
www.tianailu.com

Ruyi is an erotic toy, with a name derived from a gift in ancient China and meaning "best wishes to others". This product is wholly water-resistant thanks to a full coating of silicon gel and comes in a specially shaped case that doubles as a docking station base. The case therefore provides wireless charging for Ruyi and privacy, as its function is indistinguishable from the outside. This erotic toy from the Far East is also a stylish lifestyle product.

Statement by the jury
The design of Ruyi merges a shapely erotic toy with an elegantly matched case into a harmonious unity. As such, the product is easy to handle, and promotes itself as a true lifestyle accessory.

Ruyi ist ein Liebesspielzeug, dessen Name aus dem chinesischen Altertum stammt und „deinem Wunsch entsprechend" bedeutet. Das komplett mit Silikagel ummantelte Produkt ist vollständig wasserdicht und wird mit einem elegant geformten Etui geliefert, das zugleich als Ladesockel dient. Darin kann Ruyi kabellos geladen werden und ist damit vor den Augen Dritter geschützt. Das Intimspielzeug aus dem Fernen Osten ist zugleich stilvolles Lifestyleprodukt.

Begründung der Jury
In der Gestaltung von Ruyi verschmelzen ein formschönes Erotic Toy und ein dazu passend elegantes Etui zu einer harmonischen Einheit. Das handliche Produkt avanciert so zum Lifestyleaccessoire.

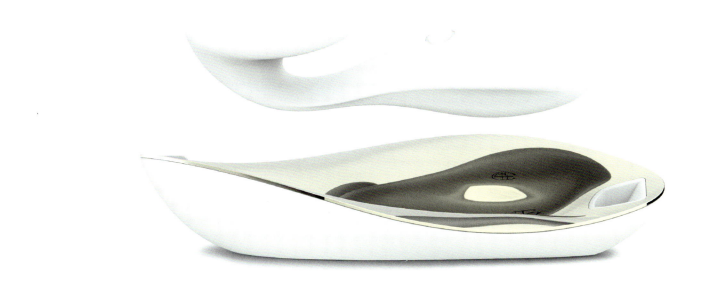

EMMA
Massage Device
Massagegerät

Manufacturer
SVAKOM Design USA Limited,
Newark, Delaware, USA
Design
Shenzhen acme life industrial design Co., Ltd.,
Shenzhen, China
Web
www.svakom.net
www.acmelife.asia

The ball-shaped design, made from safe and soft silica gel provides close contact and, as it can be operated hands-free, has a maximised range of contact. Special mechanisms like the double springs increase the intensity of the vibration and ensure a perfect combination of vibration efficiency and shock absorption. The novel temperature control technique improves the osmotic force of vibration to ensure muscles relax quickly. With the additional massage heads, Cat and Rabbit, Emma has six vibration modes in order to provide a pleasant massage experience.

Die ballförmige Gestaltung aus sicherem, weichem Silikagel ermöglicht innigen Kontakt und verbesserten Berührungsumfang und befreit von der Einschränkung handbetriebener Geräte. Spezielle Mechanismen wie die Doppelfederung erhöhen die Vibrationsintensität in einer perfekten Kombination aus Effizienz und Abdämpfung. Für eine schnelle Muskelentspannung verbessert die neuartige Temperaturkontrolltechnik den osmotischen Vibrationsdruck. Mit den Massagezusatzköpfen Cat und Rabbit bietet Emma eine angenehme Massageerfahrung in sechs Vibrationsstufen.

Rabbit

Cat

iroha
Vibrator

Manufacturer
TENGA Co., Ltd., Tokyo, Japan
In-house design
Ayumi Mochizuki, Tomoko Nakajima
Web
http://iroha-tenga.com

iroha was created in order to be able to respond to a bodies' natural need to experience pleasure. The three items showcase an organic design and, by means of their differing shapes, help to indicate the various erogenous zones they are intended for. The subtle white pink and green pastel colours help to distinguish each item. They rest pleasantly in the hand and offer intuitive operation.

Statement by the jury
All three iroha products catch the eye with a compact and easy to handle design that conveys a sense of intuitive and self-explanatory operation.

iroha wurde als Antwort auf das natürliche Bedürfnis nach Vergnügen mit sich selbst entworfen. Die drei Gegenstände zeigen eine organische Gestaltung und gehen in ihrer jeweils anderen Formgebung auf unterschiedliche Intimbereiche ein. Durch ihre dezenten Pastellfarben Weiß, Rosa und Lindgrün werden sie einfach voneinander unterschieden. Sie liegen angenehm in der Hand und sind intuitiv bedienbar.

Begründung der Jury
Alle drei iroha-Produkte springen durch ihre kompakte handliche Formgebung ins Auge und vermitteln dabei eine einfache selbsterklärende Steuerung.

iroha FIT
Vibrator

Manufacturer
TENGA Co., Ltd., Tokyo, Japan
In-house design
Ayumi Mochizuki, Tomoko Nakajima
Web
http://iroha-tenga.com

FIT is an addition to the iroha family. With its sensually soft and pliable design, it is made to gently fit the contours of the female body. The product is available in two models each with its own shape and subtle colour: one with a sleek straight shaft and the other with a gently rippled shaft. They are intuitive to use via two easy-to-operate buttons.

Statement by the jury
FIT impresses with its sleek, slightly upward tapering shaft. It thus embraces the female body while offering optimum handling.

FIT ist ein Zusatz zur iroha-Familie. Mit seiner sinnlich weichen und biegsamen Gestaltung passt sich das Produkt sanft den Konturen des weiblichen Körpers an. Es ist in dezenten Farbtönen und zudem in zwei formal unterschiedenen Ausführungen erhältlich: einmal mit schlichtem geraden Schaft und einmal mit einem harmonisch gewellten. Die Steuerung erfolgt intuitiv über zwei einfach zu bedienende Knöpfe.

Begründung der Jury
FIT beeindruckt durch seinen schlanken, sich nach oben leicht verjüngenden Körper. Damit passt er sich den weiblichen Formen an und bietet zugleich eine optimale Handhabung.

219

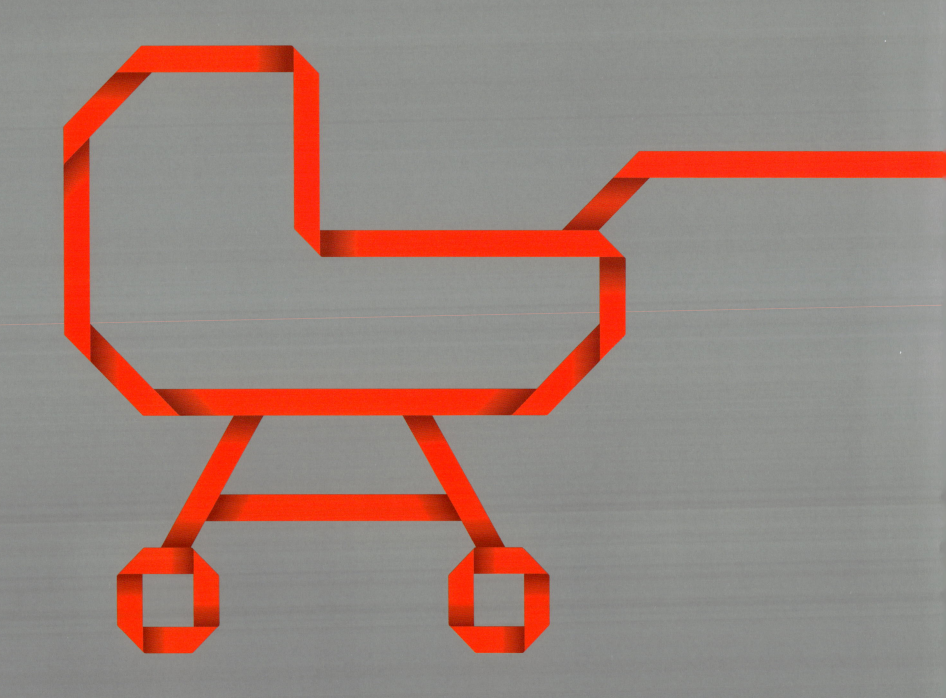

Babies and children
Baby und Kind

Joolz Geo
Stroller
Kinderwagen

Manufacturer
Milk Design B.V.,
Amsterdam, Netherlands

In-house design
Milk Design B.V.

Web
www.my-joolz.com

reddot award 2015
best of the best

Elegant multi-talent

The stroller is a key component for the mobility of parents with children, as it allows users to master the many different scenarios they meet in everyday life. In order to be prepared for these challenges, the design of the Joolz Geo stroller puts the maxim of flexibility centre stage. The resulting design concept is based on three different configurations that are compatible with one another, offering parents many solutions for everyday life with children. In addition, the assortment comprises interchangeable elements including a spacious and robust shopping basket. The Mono configuration can easily be expanded with the Duo extension for a second child by adding a second cot or seat, to conveniently transport a sibling. In the Twin configuration, this stroller is also suitable for newborn twins. The sophisticated multifunctional concept of the Joolz Geo merges with an appearance of classical elegance, as well as a clear line design that highlights its competence. The stroller delivers parents highly advanced ergonomics and robustness, as well as real-life practicability. Whichever type of configuration parents choose, the Joolz Geo remains extremely manoeuvrable and easy to handle. It thus offers extremely high comfort to children while parents can enjoy the benefits of form and function.

Elegantes Multitalent

Der Kinderwagen steht im Mittelpunkt der Mobilität mit Kindern, bei der täglich viele unterschiedliche Szenarien gemeistert werden müssen. Um diesen Herausforderungen gewachsen zu sein, stand bei der Gestaltung des Kinderwagens Joolz Geo vor allem die Maxime der Flexibilität im Vordergrund. Das daraus resultierende Konzept basiert auf drei verschiedenen Konfigurationen, die untereinander kompatibel sind. Mit diesen eröffnen sich viele Lösungen für den Alltag mit Kindern. Das Sortiment umfasst zudem auswechselbare Elemente wie beispielsweise einen stabilen, großen Einkaufskorb. Das Mono-Modell lässt sich leicht erweitern für ein zweites Kind zur Duo-Konfiguration, wobei der Joolz Geo mit einer zweiten Wanne oder einem zweiten Sitz ausgestattet wird, um ein Geschwisterkind einfach mitzutransportieren. In der Twin-Konfiguration ist dieser Kinderwagen außerdem für neugeborene Zwillinge geeignet. Das ausgereifte multifunktionale Konzept des Joolz Geo verbindet sich mit der Anmutung klassischer Eleganz sowie einer klaren Linienführung, die seine Kompetenz unterstreicht. Dieser Kinderwagen bietet den Eltern neben seiner lebensnahen Praktikabilität eine hochentwickelte Ergonomie und Solidität. Welche Konfiguration die Eltern auch nutzen, der Joolz Geo bleibt extrem wendig und einfach zu handhaben. Die Kinder haben es dadurch ausgesprochen bequem, während die Eltern die Vorteile von Form und Funktion genießen.

Statement by the jury

The Joolz Geo stroller merges elegant line design with a perfectly thought-through modular concept. Parents can choose between three different configurations to easily adapt the stroller to any given real-life requirement. Thanks to its sophisticated functionality, each step of changing the stroller's configuration is easy and self-explanatory. In addition, the stroller features ergonomics that enables parents to master everyday challenges with children smoothly and stress-free.

Begründung der Jury

Der Kinderwagen Joolz Geo vereint eine elegante Linienführung mit einem perfekt durchdachten Konzept der Modularität. Die Eltern können aus drei verschiedenen Konfigurationen wählen und diesen Kinderwagen deshalb leicht jeder Situation anpassen. Auf der Grundlage einer ausgereiften Funktionalität lässt sich jeder Handgriff leicht und selbsterklärend bewerkstelligen. Auch die Ergonomie dieses Kinderwagens ist dabei so entwickelt, dass sich der Alltag mit Kindern entspannt bewältigen lässt.

Designer portrait
See page 50
Siehe Seite 50

Mountain Buggy Urban Jungle – Luxury Collection
Stroller
Kinderwagen

Manufacturer
Phil&Teds Most Excellent Buggy Company,
Wellington, New Zealand
In-house design
Web
www.philandteds.com
www.mountainbuggy.com

The luxury collection of this buggy series is characterised outwardly by high quality leather features and classical patterns such as maritime stripes or houndstooth. With regard to functional aspects, the stroller offers, for example, a lockable hand brake, a large storage capacity and a new folding technology, due to which it can be folded using one hand and then locks automatically. In the carrycot with good air circulation, the child can adopt three positions for transport: lying flat, inclined (for reflux) or parent facing seated.

Die Luxuskollektion dieser Kinderwagen-serie ist äußerlich geprägt von hochwertigen Lederdetails und klassischen Mustern wie maritimen Streifen oder Hahnentritt. Unter funktionalen Aspekten bietet der Wagen z. B. eine feststellbare Handbremse, viel Stauraum sowie eine neue Falttechnik, dank derer er sich mit einer Hand zusammenfalten lässt und dann automatisch arretiert. In der Babytragetasche mit guter Luftzirkulation kann das Kind in drei Positionen transportiert werden: liegend, halb aufgerichtet oder den Eltern zugewandt sitzend.

Statement by the jury
The use of classical materials and patterns provides the pushchair with a timelessly elegant appearance. Furthermore, its many practical detail solutions are convincing.

Begründung der Jury
Die Verwendung klassischer Materialien und Muster verleiht dem Kinderwagen ein zeitlos elegantes Erscheinungsbild. Zudem überzeugt er mit vielen praktischen Detaillösungen.

Mountain Buggy nano
Travel Stroller
Reisebuggy

Manufacturer
Phil&Teds Most Excellent Buggy Company,
Wellington, New Zealand
In-house design
Web
www.philandteds.com
www.mountainbuggy.com

The Mountain Buggy nano is a very light buggy, which weighs only 5.9 kg and folds with just a few, quick actions. In a specially fitted bag, it can be carried slung simply over the shoulder, stowed away or taken on a flight as hand luggage. By means of an integrated adaptor, baby seats can be fitted to the nano, meaning the buggy can be used from the birth onwards.

Der Mountain Buggy nano ist mit einem Gewicht von 5,9 kg ein sehr leichter Buggy, der sich mit wenigen Handgriffen kompakt zusammenfalten lässt. In einer maßgeschneiderten Tasche lässt er sich unkompliziert über der Schulter tragen, verstauen oder als Handgepäck auf Flugreisen mitnehmen. Mithilfe eines integrierten Adapters lassen sich auch Babyschalen auf dem nano befestigen, sodass der Buggy von Geburt an genutzt werden kann.

Statement by the jury
This high quality buggy impresses due to its elegant form and, because of its small pack size, it is an ideal travel companion.

Begründung der Jury
Dieser qualitativ hochwertige Buggy besticht durch eine elegante Formgebung und wird durch sein geringes Packmaß zum idealen Reisebegleiter.

Phil&Teds Verve
Inline Double Stroller
Doppel-Buggy

Manufacturer
Phil&Teds Most Excellent Buggy Company,
Wellington, New Zealand
In-house design
Web
www.philandteds.com
www.mountainbuggy.com

This double stroller with slimline aluminium frame offers a great deal of room
and 23 different options for the transportation of one or two children. The
flexible stroller can be folded to half its
size, even with the second seat attached,
so can be stowed even in smaller cars.
With its swivel-mounted front wheels,
Verve is effortless to push and can be
manoeuvred with one hand or parked
with Verve's one-touch hand brake.

Dieser Doppelbuggy mit schlankem Alumi-
niumgestell bietet viel Platz und 23 verschiedene Möglichkeiten, um ein oder zwei
Kinder zu transportieren. Der flexible Buggy
lässt sich auch mit dem zweiten Sitz auf die
Hälfte seiner Größe zusammenklappen und
findet damit selbst in kleineren Autos Platz.
Mit seinen zwei schwenkbaren Vorder-
rädern lässt sich Verve mühelos schieben,
mit einer Hand manövrieren oder mit der
One-Touch-Handbremse parken.

Solstice
Jogging Stroller
Sportkinderwagen

Manufacturer
Burley Design LLC, Eugene, USA
In-house design
Web
www.burley.com

Solstice is a jogging stroller which folds and unfolds effortlessly and, due to an automatic locking device, stands on its wheels also when folded up. When the safety belt is opened, it retracts thanks to the innovative spring integrated technology, allowing the child to get in and out easily. The stroller has a foot parking brake and a height adjustable handlebar; it is well sprung and the front wheel tracking is adjustable.

Statement by the jury
This dynamically designed jogging stroller surprises with many practical functions which are emphasised by the prominent yellow setting knobs.

Der Solstice ist ein Sportkinderwagen, der sich schnell auf- und zusammenklappen lässt und mithilfe einer automatischen Arretierung auch zusammengeklappt auf seinen Rädern stehenbleibt. Beim Öffnen des Anschnallgurtes zieht sich dieser dank des innovativen Federmechanismus zurück und ermöglicht leichtes Ein- und Aussteigen. Der Wagen mit Fußparkbremse und höhenverstellbarem Schiebegriff ist gut gefedert, das vordere Schwenkrad lässt sich feststellen.

Begründung der Jury
Dieser dynamisch gestaltete Sportkinderwagen überrascht mit vielen praktischen Funktionen, die durch die gelb hervorgehobenen Einstellknöpfe betont werden.

Baby Jogger city mini® ZIP
Stroller
Kinderwagen

Manufacturer
Newell Rubbermaid, Atlanta, USA
In-house design
Web
www.newellrubbermaid.com

The eye-catching feature of the city mini ZIP is its sophisticated one-hand folding system. By this means, the stroller folds three dimensionally, so that when folded it only takes up an eighth of its original volume. The stroller seat reclines completely and is equipped with a windproof modular cover. A large sun canopy protects from wind and weather and offers UV protection of 50+.

Statement by the jury
This lightweight stroller can be folded simply like an umbrella and when unfolded offers every necessary comfort.

Hervorstechendes Funktionsmerkmal des city mini ZIP ist sein ausgefeiltes Einhand-schließsystem. Mit seiner Hilfe wird der Wagen aus drei Richtungen zusammenge-klappt, sodass er gefaltet nur noch ein Achtel des ursprünglichen Volumens ein-nimmt. Der Sitz des Kinderwagens ist komplett zurücklehnbar und mit einer windabweisenden modularen Abdeckung ausgestattet. Ein großes Sonnenverdeck schützt bei Wind und Wetter und bietet einen UV-Schutz von 50+.

Begründung der Jury
Dieser leichte Kinderwagen lässt sich einfach wie ein Regenschirm zusammen-klappen und bietet auseinandergefaltet jeden nötigen Komfort.

CYBEX PRIAM
Stroller Travel System
Kinderwagen-Reisesystem

Manufacturer
CYBEX GmbH, Bayreuth, Germany
In-house design
Web
www.cybex-online.com
Honourable Mention

The Priam pays homage to the works of Ray and Charles Eames with design influence from their furniture classics. The result is a versatile travel system: a carry cot, an infant car seat and a seat unit can be used on the same frame. Therefore, the Priam firstly serves as pram, then, as the child gets older, as a forward- or rear-facing stroller. The travel system is sturdy, light and suitable for any terrain thanks to various wheel and ski attachments.

Der Priam versteht sich als Hommage an Ray und Charles Eames mit gestalterischen Anlehnungen an deren Möbelklassiker. Das Ergebnis ist ein vielseitiges Reisesystem, bei dem auf einem Grundgestell Kinderwagenaufsatz, Babyschale und Sitzeinheit verwendet werden können. So dient der Priam erst als Kinderwagen, später als ein wahlweise vor- oder rückwärtsgerichteter Sportwagen. Das Reisesystem ist robust, leicht und kann dank verschiedener Rad- und Skiaufsätze in jedem Terrain genutzt werden.

Statement by the jury
The Priam stroller makes a timeless impression in spite of a modern formal language, and it convinces with regard to its functionality due to its high degree of flexibility.

Begründung der Jury
Der Priam Kinderwagen wirkt trotz einer modernen Formensprache zeitlos und überzeugt in funktionaler Hinsicht durch seine hohe Flexibilität.

231

i-Gemm™
Infant Car Seat
Babyschale

Manufacturer
Joie Children's Products Co., Limited, United Kingdom
In-house design
Web
www.joiebaby.com

The i-Gemm baby seat is designed in such a way that it protects particularly the infant's head, neck and chest. The so-called Tri-Protect headrest reduces the risk of injury by means of a multi-layer protection principle and special memory foam. An infant insert, soft harness pads, a cover extension which provides additional shade, as well as the easily adjustable headrest with settings for seven different positions provide for the newborn's comfort.

Die Babyschale i-Gemm ist so gestaltet, dass sie Kopf, Nacken und Brust des Babys besonders schützt. Die sogenannte Tri-Protect-Kopfstütze reduziert mithilfe eines Mehrschicht-Schutzprinzips und eines speziellen Memory-Schaums das Verletzungsrisiko. Für den Komfort des Babys sorgen eine Neugeboreneneinlage, weiche Gurtpolster, eine Verdeckerweiterung, die zusätzlichen Schatten spendet, sowie die siebenfach verstellbare Kopfstütze, die sich mit einer Hand justieren lässt.

Statement by the jury
The maxim of greatest safety, which applies to child seats, defines the functional design of the infant seat which at the same time offers a high degree of comfort for the child.

Begründung der Jury
Die bei Kindersitzen geltende Maxime höchster Sicherheit prägt die funktionale Gestaltung der Babyschale, die gleichzeitig hohen Komfort fürs Kind bietet.

REBL™
Child Car Seat
Kinderautositz

Manufacturer
Nuna International, Erp, Netherlands
In-house design
Web
www.nuna.eu
Honourable Mention

The Rebl child seat has a seating unit which rotates 360 degrees, easing the actions of getting in and out. An insert for newborns, an adjustable headrest and seven recline positions provide pleasant rearward transport of children from birth up to a body size of 105 cm (i-Size standard). A high degree of side impact protection is provided by special memory foam and the energy absorbent seat shell; whereas the nine-position Isofix anchor points assure optimal attachment to the vehicle.

Der Kindersitz Rebl hat eine um 360 Grad drehbare Sitzeinheit, die das Ein- und Aussteigen erleichtert. Ein Neugeboreneneinsatz, eine verstellbare Kopfstütze und sieben Liegepositionen ermöglichen eine angenehme rückwärtsgerichtete Beförderung von Kindern von der Geburt an bis zu einer Körpergröße von 105 cm (i-Size-Standard). Für einen hohen Seitenaufprallschutz sorgen ein spezieller Memory-Schaum und die energieabsorbierende Sitzschale, während neunfach verstellbare Isofix-Konnektoren eine optimale Verbindung zum Fahrzeug sicherstellen.

Statement by the jury
With its clear formal language the Rebl indicates its high degree of functionality and gives the impression of safety and reliability so that it can be entrusted with the child.

Begründung der Jury
Mit seiner klaren Formensprache transportiert der Rebl seine hohe Funktionalität und wirkt sicher und zuverlässig, sodass man ihm den Nachwuchs gern anvertraut.

AxissFix
Child Car Seat
Kinderautositz

Manufacturer
Maxi-Cosi, Dorel Juvenile Europe, Cholet, France
In-house design
Yann Naslain
Design
Estech, Buc (Yvelines), France
Web
www.maxi-cosi.com
www.estech-design.com

The child car seat AxissFix can be pivoted by 360 degrees. This means a child up to a body size of 84 cm travels in the car facing rearward, whereby the head and neck are well protected. When the child grows, the seat can be simply rotated to face forward until the child is about four years old or 105 cm tall. The seat is easily installed by means of the Isofix system and fulfils the i-Size standard.

Der Kinderautositz AxissFix ist um 360 Grad drehbar. So kann ein Kind bis zu einer Körpergröße von 84 cm rückwärtsgerichtet im Auto mitreisen, wodurch Kopf und Nacken gut geschützt sind. Wird das Kind größer, kann der Sitz einfach gedreht und in Fahrtrichtung verwendet werden, bis das Kind etwa vier Jahre alt ist oder 105 cm misst. Der Sitz lässt sich mithilfe des Isofix-Systems installieren und erfüllt den i-Size-Standard.

BeSafe iZi Modular
Child Seat System
Kindersitzsystem

Manufacturer
HTS BeSafe AS, Krøderen, Norway
Design
HTS BeSafe NPD, Utrecht, Netherlands
Web
www.besafe.com

BeSafe iZi Modular is a modular child seat system, consisting of one Isofix compatible base and two child seats which conform to the EU standard (i-Size). When the child grows out of the baby seat, the next seat is used on the same base. The base module can be extended to allow the child more legroom, which means children up to the age of about four years (105 cm tall) can be transported facing the rear.

BeSafe iZi Modular ist ein modulares Kindersitzsystem, bestehend aus einer Isofix-kompatiblen Basis und zwei Kindersitzen, die dem EU-Standard (i-Size) entsprechen. Ist das Kind dem Babysitz entwachsen, wird der Folgesitz auf derselben Basis genutzt. Das Basismodul kann zur Vergrößerung der Beinfreiheit ausgezogen werden, sodass auch Kinder bis zu einem Alter von ca. vier Jahren (105 cm Körpergröße) entgegen der Fahrtrichtung transportiert werden können.

Statement by the jury
This child seat system convinces by its well-considered functionality, enhanced safety factor and a high degree of seating comfort for the child.

Begründung der Jury
Dieses Kindersitzsystem überzeugt durch eine durchdachte Funktionalität, verbesserte Sicherheit und einen hohen Sitzkomfort für den Nachwuchs.

235

Stokke® Steps™
High Chair
Hochstuhl

Manufacturer
Stokke AS,
Ålesund, Norway

In-house design
Anders August Kittilsen,
Hilde Angelfoss

Design
Permafrost AS
(Andreas Murray,
Tore Vinje Brustad),
Oslo, Norway

Web
www.stokke.com
www.permafrost.no

reddot award 2015
best of the best

Step by step

In the world of toddlers, the high chair plays a central role as it allows them to partake directly in the world of grown-ups, until they can finally sit on a normal chair by themselves at the table. Designed to closely meet the expectations of all parties involved, the Stokke Steps emerged as a high chair solution that focuses on the newborn and has the ability to be used up until school age. Particularly innovative is the formal and functionally convincing combination of bouncer and high chair. Stokke Steps is based on a modular system that covers and accommodates a child's needs as it grows. While the youngest children need more support, older children can mostly sit unaided usually from the age of six months. In addition, the bouncer can also be used separately as a higher baby seat that allows interaction between parent and child. The high chair features a clear design language that lends the individual modules an age-appropriate appearance. The baby set highlights a sense of playfulness, while the junior chair with its precise lines blends well into almost any interior. The functional concept of this high chair allows easy and self-explanatory adjustment of all modules, which are made of high-quality materials including wood and leather. The Stokke Steps high chair thus ensures high longevity and provides the ability to also be later used for siblings or even by future generations.

Schritt für Schritt

In der Welt der Kleinkinder spielt der Hochstuhl eine zentrale Rolle, denn durch diesen können sie direkt am Leben der Erwachsenen teilhaben, bis sie schließlich auf normalen Stühlen am Tisch sitzen können. Eine nah an den Erwartungen aller Beteiligten entwickelte Gestaltung führt mit dem Stokke Steps zu einem Hochstuhl, der von Beginn an eingesetzt werden kann und das Kind bis hin zum Schulalter begleitet. Innovativ ist insbesondere die formal wie funktional schlüssige Einheit aus Hochstuhl und Babywippe. Stokke Steps basiert auf einem modularen System, welches den Bedürfnissen des wachsenden Kindes entspricht. So benötigen die jüngsten Kinder mehr Unterstützung, während Kinder ab sechs Monaten meist schon ohne fremde Hilfe sitzen können. Zusätzlich kann die Wippe separat auch als erhöhter Babysitz verwendet werden, der die Interaktion zwischen Eltern und Kind ermöglicht. Gestaltet ist der Hochstuhl mit einer klaren Formensprache, wobei die einzelnen Module eine altersentsprechende Anmutung besitzen. Das Baby-Set wirkt verspielt, während sich der Junior-Stuhl mit seinen präzisen Linien gut der bestehenden Inneneinrichtung anpasst. Das funktionale Konzept dieses Hochstuhls erlaubt ein leichtes und selbsterklärendes Anpassen der Module, die aus hochwertigen Materialien wie Holz und Leder gefertigt sind. Der Hochstuhl Stokke Steps ist deshalb sehr langlebig und kann auch gut für Geschwister oder sogar von zukünftigen Generationen genutzt werden.

Statement by the jury

The Stokke Steps is a high chair that grows with the child, featuring an adaptability that closely accommodates the changing needs of both children and parents. The innovative concept of age-appropriate, multifunctional modules even comprises a soft bouncer so that the system can be used from the very beginning. This high chair is manufactured from high-quality materials and appeals with its finely crafted details. Showcasing a clear language of form, it blends into any interior.

Begründung der Jury

Der Stokke Steps ist ein mitwachsender Hochstuhl, der durch seine Anpassungsfähigkeit exakt den Bedürfnissen von Kindern und Eltern gerecht wird. Das innovative Konzept altersentsprechender, multifunktionaler Elemente integriert auch eine weiche Babywippe, sodass er von Anfang an eingesetzt werden kann. Dieser Hochstuhl ist aus sehr hochwertigen Materialien gefertigt, wobei auch die sorgfältig ausgearbeiteten Details begeistern. Durch seine klare Formensprache fügt er sich in jedes Interieur ein.

Designer portrait
See page 52
Siehe Seite 52

SipSnap
Drinking Lids
Trinkdeckel

Manufacturer
Double Double, San Francisco, USA
In-house design
Michelle Ivankovic, Sativa Turner
Web
www.sipsnap.com

SipSnap is a universal silicone drinking lid that stretches to fit over any cup. There are two versions: SipSnap TOT incorporates a chew-resistant spout to create a spill-proof sippy cup, while SipSnap KID accommodates any straw to provide a splash-proof option for older children and adults. Offering control over cup materials as well as the flexibility to use what is at hand, SipSnaps can be transported in their case for use on the go.

Statement by the jury
These colourful, pleasantly feeling drink covers are easy to use, hygienic and flexible and ease the everyday life with children.

SipSnap ist ein universeller, dehnbarer Trinkdeckel aus Silikon, der auf jedes Trinkgefäß passt. Es gibt ihn in zwei Varianten: SipSnap TOT mit kaufester Trinktülle für einen auslaufsicheren Trinkbecher und SipSnap KID, der in Verbindung mit einem Strohhalm älteren Kindern und Erwachsenen spritzsicheres Trinken erlaubt. Die Deckel sind für Gefäße aus jedem Material geeignet und lassen sich in ihrem Behälter für den Einsatz unterwegs transportieren.

Begründung der Jury
Diese farbenfrohen, haptisch angenehmen Trinkdeckel sind leicht zu nutzen, hygienisch und flexibel einsetzbar und erleichtern so den Alltag mit Kindern.

Avent My Penguin Sippy Cup
Becher

Manufacturer
Royal Philips, Eindhoven, Netherlands
In-house design
Web
www.philips.com

The Avent My Penguin Sippy Cup, with its ergonomic design resembling a penguin, encourages toddlers to drink independently. Its soft learning handles are slip-free and easy to hold. Thanks to the curved, soft drinking spout, toddlers do not need to tilt their heads too far back to drink and can adopt a natural drinking position. The cup is easily assembled and easily cleaned.

Statement by the jury
These cheerful cups combine ergonomic aspects with playful elements and thus help toddlers to learn the skills of drinking.

Mit seiner ergonomischen, an einen Pinguin erinnernden Gestaltung ermuntert der Avent My Penguin Sippy Cup Kleinkinder zum selbständigen Trinken. Seine weichen Lerngriffe sind rutschfest und gut zu halten. Dank des abgewinkelten, weichen Trinkschnabels müssen Kleinkinder ihren Kopf beim Trinken nicht zu stark nach hinten neigen und können eine natürliche Trinkposition einnehmen. Der Becher ist leicht zusammenzusetzen und zu reinigen.

Begründung der Jury
Diese fröhlichen Becher kombinieren ergonomische Aspekte mit spielerischen Elementen und erleichtern Kleinkindern so das Trinkenlernen.

Avent Easy Sip Cup
Becher

Manufacturer
Royal Philips, Eindhoven, Netherlands
In-house design
Web
www.philips.com

Due to its soft silicone drinking spout, the Avent Easy Sip Cup allows toddlers a smooth transition from the breast to independent drinking. The ergonomic design and the rippled form provide a steady grip, even without handles, and promote development of motor skills. The drinking cup is easy to clean and spill-free. There is a version for babies from six months and a version for toddlers from twelve months.

Statement by the jury
This drinking cup, due to its soft silicone spout, is particularly gentle on the gums and ergonomically suited exactly for small hands.

Der Avent Easy Sip Cup ermöglicht Kleinkindern mit seinem weichen Silikon-Trinkschnabel einen sanften Übergang von der Brust zum eigenständigen Trinken. Die ergonomische Gestaltung und die gerippte Form bieten auch ohne Griffe festen Halt und fördern die Entwicklung der motorischen Fähigkeiten. Der leicht zu reinigende Trinkbecher ist auslaufsicher. Es gibt ihn in einer Ausführung für Babys ab sechs Monate und einer für Kleinkinder ab zwölf Monate.

Begründung der Jury
Dieser Trinklernbecher ist mit seinem weichen Silikonschnabel besonders zahnfleischfreundlich und ergonomisch genau auf kleine Hände abgestimmt.

Minbie
Baby's Teat
Trinksauger

Manufacturer
Shanghai Baron Technologies Co., Ltd., China
In-house design
Design
I-Sip Trading Pty Ltd (Julia Wilson,
John Nielsen, Jon Seddon), Minbie,
Melbourne, Australia
Web
www.minbie.com.au

Minbie is a soft silicon baby's teat which by means of its special form transfers the natural breastfeeding technique of newborns to that of drinking from the bottle. The nipple tapers slightly upwards and becomes flatter so that it lies tight on the palate. The nipple is slightly off-centre and has a small bulge on the tongue side, helping the baby to orientate and to stabilise the nipple when drinking.

Statement by the jury
With its form based on that of the mother's breast, the teat eases the baby's adjustment from breast to bottle feeding.

Minbie ist ein weicher Silikon-Trinksauger, der mit seiner besonderen Form die natürliche Trinktechnik von Neugeborenen an der Brust auf das Trinken aus der Flasche überträgt. Der Sauger verjüngt sich nach oben hin leicht und wird flacher, sodass er eng am Gaumen anliegt. Das Mundstück ist leicht dezentral platziert und hat eine kleine Wölbung an der Zungenseite, was dem Baby dabei hilft, sich zu orientieren und den Sauger beim Trinken zu stabilisieren.

Begründung der Jury
Mit einer der Mutterbrust nachempfundenen Formgebung erleichtert dieser Trinksauger Babys die Umstellung von Brust- auf Flaschenfütterung.

Avent Baby Food and Bottle Warmer
Babyflaschenwärmer

Manufacturer
Royal Philips, Eindhoven, Netherlands
In-house design
Web
www.philips.com

The baby bottle warmer heats breast milk gently and evenly. During warming, the milk circulates so that heat accumulation is avoided. A sensor detects and controls the milk temperature in order to prevent overheating. The operating panel design is clear, and an indicator light displays the warming progress. The milk is kept warm for 20 minutes, after which the device switches off automatically.

Statement by the jury
Thanks to its high design quality, this baby bottle warmer fits seamlessly into a modern kitchen environment and furthermore convinces with its intelligent functionality.

Der Babyflaschenwärmer wärmt Muttermilch behutsam und gleichmäßig auf. Die Milch zirkuliert während der Erwärmung, sodass es nicht zum Hitzestau kommen kann. Ein Sensor erfasst und kontrolliert die Milchtemperatur, um Überhitzung zu vermeiden. Das Bedienfeld ist übersichtlich gestaltet, und eine Fortschrittsanzeige zeigt mit Licht den Erwärmungsverlauf an. Die Milch wird 20 Minuten lang warm gehalten, bevor sich das Gerät automatisch abschaltet.

Begründung der Jury
Dank seiner hochwertigen Gestaltung fügt sich dieser Babyflaschenwärmer nahtlos in moderne Kücheneinrichtungen ein und überzeugt darüber hinaus mit einer intelligenten Funktionalität.

Avent Baby Monitor DECT SCD501
Babyfon

Manufacturer
Royal Philips, Eindhoven, Netherlands
In-house design
Web
www.philips.com

The Avent DECT Baby Monitor SCD501 is a compact device for monitoring babies. With a range of up to 300 metres, a reliable connection – based on DECT technology, which guarantees zero interference – and a high sound quality it offers a great level of security. Five LED lights on the parents' unit signal noises so that parents can monitor the baby's activity even without hearing the sound. The baby monitor can also be used as a night light.

Statement by the jury
With soft lines and a reduced design the baby monitor appears reliable and meets this expectation also from a technological viewpoint.

Der Avent DECT-Baby-Monitor SCD501 ist ein kompaktes Gerät zur Überwachung von Babys. Mit einer Reichweite von bis zu 300 Metern, einer zuverlässigen Vebindung – die zugrunde liegende DECT-Technologie garantiert null Interferenzen – und seiner hohen Tonqualität bietet er große Sicherheit. Fünf LED-Leuchten an der Elterneinheit zeigen Geräusche an, sodass Eltern die Aktivität ihres Babys auch ohne Ton überwachen können. Das Babyfon lässt sich auch als Nachtlicht nutzen.

Begründung der Jury
Mit sanften Linien und einer reduzierten Gestaltung wirkt das Babyfon zuverlässig und löst diese Erwartung auch unter technologischen Aspekten ein.

Avent uGrow Smart Baby Monitor
Babyfon

Manufacturer
Royal Philips, Eindhoven, Netherlands
In-house design
Web
www.philips.com

This handy baby monitor transmits video and audio live streaming from the child's room to the smartphone or tablet. The device uses so-called SafeConnect technology to automatically optimise the bandwidth between networks for a reliable connection. All settings can be controlled via an app. The baby monitor supports up to ten cameras and also offers additional functions such as multicolour night light as well as humidity and temperature monitoring.

Statement by the jury
The uGrow delights with its modern technology and a form, which is reduced to the essentials and integrates unnoticeably in every nursery.

Dieses handliche Babyfon sendet einen Video- und Audio-Livestream aus dem Kinderzimmer auf Smartphone oder Tablet-PC. Das Gerät nutzt die sogenannte SafeConnect-Technologie, um die Bandbreite zwischen Netzwerken für eine zuverlässige Verbindung automatisch zu optimieren. Alle Einstellungen können über eine App gesteuert werden. Das Babyfon unterstützt bis zu zehn Kameras und bietet außerdem Funktionen wie Nachtlicht oder Feuchtigkeits- und Temperaturüberwachung.

Begründung der Jury
Das uGrow begeistert durch moderne Technik und eine auf das Wesentliche reduzierte Form, durch die es sich unauffällig in jedes Kinderzimmer integriert.

JooN Phone
Kids' Phone
Kindertelefon

Manufacturer
Infomark, Seongnam, South Korea
Design
Purplerain product (Dong-Suk Kim), Anyang, South Korea
Web
www.infomark.co.kr
www.purplerain-product.com
Honourable Mention

JooN is a wearable telephone for children, which provides not only a telephone function, but also location information as well as safety functions such as an SOS button. Via GPS receiver, the JooN transmits precise position data to the parents' smartphone. A safety zone can be pre-configured so that upon leaving it an SMS alarm is sent. The phone is hard to break and has a strap of environmentally friendly TPSiV plastic.

Statement by the jury
The sturdy, child-friendly design and its simple operation make this children's telephone an uncomplicated companion.

Das JooN ist ein tragbares Telefon für Kinder, das neben einer Telefonfunktion sowohl Standortinformationen als auch Sicherheitsfunktionen wie eine SOS-Taste bietet. Per GPS-Empfänger meldet das JooN präzise Positionsdaten an das Smartphone der Eltern. Eine Sicherheitszone lässt sich vorkonfigurieren, bei deren Verlassen ein SMS-Alarm gesendet wird. Das Telefon ist schwer zerbrechlich, das Armband aus dem umweltfreundlichen Kunststoff TPSiV gefertigt.

Begründung der Jury
Die robuste, kindgerechte Gestaltung und seine einfache Handhabung machen dieses Kindertelefon zu einem unkomplizierten Begleiter.

Bobux Xplorer
Children's Footwear
Kinderschuhe

Manufacturer
Bobux International, Auckland, New Zealand
In-house design
Andrew Johnston, Parveen Chand
Web
www.bobux.com

The Xplorer has been developed through a user-centred design process, especially for children who are just learning to walk and move alternately between crawling, pulling themselves up and taking their first steps. The materials used, such as a soft, hard-wearing EVA sole, leather upper and special protective toe cap, make the Xplorer light and flexible and at the same time offer the protection and durability of a walking shoe.

Der Xplorer wurde in einem benutzer-zentrierten Gestaltungsprozess speziell auf Kinder zugeschnitten, die gerade das Laufen lernen und dabei noch zwischen Krabbeln, Aufrichten und ersten Schritten hin- und herwechseln. Die verwendeten Materialien wie eine weiche, strapazier-fähige EVA-Sohle, Leder-Obermaterial und spezielle Zehenschutzkappen machen den Xplorer leicht und flexibel, bieten aber gleichzeitig den Schutz und die Haltbarkeit eines Laufschuhs.

Statement by the jury
The successful symbiosis of a soft first shoe and a stable walking shoe is brought into additional focus by the distinctive design of the sole.

Begründung der Jury
Die gelungene Symbiose aus weichem Erstschuh und stabilem Laufschuh wird durch die markante Gestaltung der Sohle zusätzlich in den Fokus gerückt.

VATOOL
Educational Toy
Lernspielzeug

Manufacturer
BPABLE Corp., Seoul, South Korea
In-house design
Jaeho Lee, Daehwan In
Web
www.vatool.com
Honourable Mention

Vatool (VariousTool) is an educational toy for teaching children worldwide to practice the idea of upcycling. The various connection elements make it possible to screw PET bottles of every shape and size together in a simple way. The game can be enlarged by additional angle pieces, so that children can combine the bottles in many different ways. In this manner, PET bottles can be used to produce imaginative objects and figures.

Vatool (VariousTool) ist ein Lernspielzeug, mit dem Kindern weltweit die Idee des Upcyclings vermittelt wird. Die unterschiedlichen Verbindungselemente ermöglichen es, PET-Flaschen jeder Form und Größe auf einfache Weise miteinander zu verschrauben. Das Spiel lässt sich durch zusätzliche Winkelstücke erweitern, sodass Kinder die Flaschen in unterschiedlichster Weise kombinieren können. So entstehen aus PET-Flaschen phantasievolle Objekte und Figuren.

Statement by the jury
Vatool impresses with its simple technical solution by means of which children can create something new from PET bottles.

Begründung der Jury
Vatool besticht durch eine einfache technische Lösung, mit deren Hilfe Kinder aus PET-Flaschen etwas Neues erschaffen können.

Green Science
Educational Toys
Lernspielzeug

Manufacturer
4M Industrial Development Limited,
Hong Kong
In-house design
Web
www.4m-ind.com

Aim of the educational toy series Green Science is to stimulate children's interest in environment and sustainability issues in a playful way. A range of experiment sets, among them a weather station, a potato clock and a tin can calculator, stimulate the natural curiosity of children and thereby teach them knowledge on matters concerning global warming, recycling and alternative energy resources.

Statement by the jury
The sets of the Green Science series delight due to their playful-experimental overall approach to environmental issues, from which also the parents could learn something.

Ziel der Lernspielzeugserie Green Science ist es, spielerisch das Interesse von Kindern an Umwelt- und Nachhaltigkeitsthemen zu wecken. Verschiedene Experimentiersets, darunter eine Wetterstation, eine Kartoffeluhr und ein Blechdosen-Taschenrechner, machen sich die natürliche Neugier von Kindern zunutze und vermitteln dabei Wissen rund um die Themenbereiche Erderwärmung, Recycling und alternative Energiequellen.

Begründung der Jury
Die Sets der Green-Science-Serie begeistern durch ihre spielerisch-experimentelle Herangehensweise an Umweltthemen, durch die auch Eltern noch einiges dazulernen können.

Albert
Smart Education Robot
Intelligenter Lernroboter

Manufacturer
SK telecom, Seoul, South Korea
In-house design
Web
www.sktelecom.com
Honourable Mention

Albert is a small education robot with which children can communicate by means of a smartphone and the associated "Smart Robot Application". The smartphone is installed onto the robot's stomach and connected to it via Bluetooth. Albert can thus be controlled via the smartphone, read books out loud or play with the child. His friendly outward appearance hides a range of sensors which allow Albert to react to his environment.

Statement by the jury
With the child-friendly design of his appearance, Albert seems to be a nice playmate who, thanks to the smartphone-based solution, is always ready for a new game.

Albert ist ein kleiner Lernroboter, der mithilfe eines Smartphones und der dazugehörigen „Smart Robot Application" mit Kindern interagiert. Das Smartphone wird auf den Bauch des Roboters gesteckt und per Bluetooth mit ihm verbunden. So kann Albert über das Smartphone gelenkt werden, Bücher vorlesen oder Spiele mit dem Kind spielen. Hinter dem sympathischen Äußeren verbergen sich diverse Sensoren, die Albert auf seine Umwelt reagieren lassen.

Begründung der Jury
Albert erscheint mit seinem kindgerecht gestalteten Äußeren als netter Spielkamerad, der dank der Smartphone-basierten Lösung mit immer wieder neuen Spielen aufwarten kann.

Mister Crocodile on a stick
Herr Krokodil am Stiel
Wooden Toy
Holzspielzeug

Manufacturer
Wilsonic Design, Trzin, Slovenia
In-house design
Peter Rojc, Metod Burgar, Nina Mihovec
Web
www.wilsonicdesign.com

"Mister Crocodile on a stick" is a contemporary interpretation of the hobbyhorse made of solid wood. The form of the crocodile head design is friendly and soft, at the same time it has minimalist lines. The head can also be used as a blackboard. "Mister Crocodile on a stick" thus becomes a toy, which encourages the creativity and fantasy of children and inspires them to role-play. The toy is also available with other animal heads.

Statement by the jury
This wooden toy is a charming new interpretation of the classical hobbyhorse and as "Mister Crocodile on a stick" provides variation for children's rooms.

„Herr Krokodil am Stiel" ist eine zeitgemäße Interpretation des Steckenpferds aus Massivholz. Die Form des Krokodilkopfs ist freundlich und weich gestaltet, zugleich aber minimalistisch in der Linienführung. Der Kopf kann auch als Schreibtafel genutzt werden. So wird „Herr Krokodil am Stiel" zu einem Spielzeug, das Bewegung, Kreativität und Phantasie fördert und zu neuen Rollenspielen anregt. Das Spielgerät ist auch mit anderen Tierköpfen erhältlich.

Begründung der Jury
Dieses Holzspielzeug ist eine charmante Neuinterpretation des klassischen Steckenpferds und bringt als „Herr Krokodil am Stiel" Abwechslung ins Kinderzimmer.

My First Dream House
Mein erstes Traumhaus
Play Tent
Spielzelt

Manufacturer
afun®, Yangpyeong, South Korea
In-house design
Kangwook Shin, Mihyung Kim
Web
www.afun.co.kr
Honourable Mention

This play tent offers children a refuge where they can play without being seen. The four frame members are made of birch wood. Their edges are rounded in order to minimise risk of injury. They are simply fitted together and form a stable base. The fabric of high quality cotton is available in different patterns and colours and opens on all four sides.

Statement by the jury
My First Dream House is convincing due to its simple design; it offers a feeling of security even though it is easily accessible from all sides.

Dieses Spielzelt bietet Kindern einen Rückzugsort zum unbeobachteten Spielen. Die vier Rahmenelemente sind aus Birkenholz gefertigt. Sie haben abgerundete Kanten, um die Verletzungsgefahr zu minimieren, lassen sich einfach zusammenstecken und bilden einen stabilen Überbau. Der Stoffbezug besteht aus hochwertiger Baumwolle, ist in unterschiedlichen Mustern und Farben erhältlich und lässt sich an allen vier Seiten öffnen.

Begründung der Jury
My First Dream House besticht durch seine schlichte Gestaltung, vermittelt Geborgenheit und ist dennoch von allen Seiten gut zugänglich.

PlanHome™ Table & Chair
Children's Furniture
Kindermöbel

Manufacturer
Plan Creations Co., Ltd., Bangkok, Thailand
In-house design
Web
www.plantoys.com

Table and chair of the PlanHome collection feature an elegant bent wood design. The multifunctional table has a tabletop with chalkboard coating and two storage compartments; the chair is ergonomically designed according to the children's physique. The durable furniture is made from the sustainable material PlanWood which, among other materials, is composed of sawdust resulting from the manufacture of the company's toy production.

Statement by the jury
This children's furniture is both sturdy and yet charming and gives style to the children's room due to its reduced, organic formal language.

Tisch und Stuhl der PlanHome-Kollektion sind durch ein elegantes Design aus gebogenem Holz gekennzeichnet. Der Multifunktionstisch hat eine Tischplatte mit Tafelbeschichtung und zwei Ablagefächer; der Stuhl ist ergonomisch auf den Körperbau von Kindern abgestimmt. Die haltbaren Möbelstücke sind aus dem nachhaltigen Material PlanWood gefertigt, das unter anderem aus Sägespänen hergestellt wird, die bei der Spielzeugproduktion des Herstellers abfallen.

Begründung der Jury
Diese Kindermöbel sind zugleich stabil und anmutig und bringen mit ihrer reduzierten, organischen Formensprache Stil ins Kinderzimmer.

Ultralight Airplanes
Ultraleicht-Gleiter

Manufacturer
Franckh-Kosmos Verlags-GmbH & Co. KG,
Stuttgart, Germany
Design
Genius Toy Co., Ltd. (Ole Vestergaard
Poulsen), Taichung, Taiwan
Web
www.kosmos.de
www.gigo.com.tw

Ultralight Airplanes is an experiment kit
for constructing different types of flying
gliders with up to five different wing
configurations. The individual parts and
bamboo rods were designed to enable
a creative and safe assembly to kids from
eight years of age. Without needing any
extra building materials, tools or even
glue, all bamboo rods can be connected
easily with the joints and stimulate an
individual development of new planes.

Statement by the jury
The experimentation kit set stands out
with an easy and intuitive construction
method for building toy planes, and impresses with its use of natural materials.

Ultraleicht-Gleiter ist ein Experimentierkasten, mit dem sich Gleitflugzeuge zu fünf
verschiedenen Flugzeugmodellen konstru-
ieren lassen. Die einzelnen Bauteile und
Bambusstäbe wurden entworfen, um krea-
tives und sicheres Bauen für Kinder ab
acht Jahren zu ermöglichen. Ohne weiteres
Baumaterial, Werkzeug oder Klebstoff
lassen sich die Bambusstäbe über die Ver-
bindungsstücke zusammenstecken und
fördern das individuelle Weiterentwickeln
neuer Flugzeugmodelle.

Begründung der Jury
Der Experimentierkasten macht durch die
einfache, sofort verständliche Bauweise der
Flugzeuge auf sich aufmerksam und über-
zeugt durch die Verwendung natürlicher
Materialien.

LittleBig Bike
Children's Bicycle
Kinderfahrrad

Manufacturer
LittleBig Bikes, Greystones, Ireland
In-house design
Simon Evans
Web
www.littlebigbikes.com
Honourable Mention

The LittleBig is designed in such a way
that it is first used as a small balance
bike with which children can develop
their sense of balance and their motor
skills. The innovative frame design makes
it easily possible to convert the small
balance bike into a bigger one with a
higher saddle. When the child reaches
the stage that it can learn to cycle, the
crank and pedals can be added so that
the LittleBig becomes a proper pedal
bicycle.

Statement by the jury
The idea of a 3-in-1 bicycle which
children can use for up to five years is
convincing, also from the viewpoint of
sustainability.

Das LittleBig ist so konzipiert, dass es
zunächst als kleines Laufrad genutzt wird,
mit dem Kinder ihren Gleichgewichtssinn
und ihre motorischen Fähigkeiten schulen.
Die innovative Gestaltung des Rahmens
ermöglicht ein einfaches Transformieren
des kleinen in ein größeres Laufrad mit
höherem Sattel. Ist das Kind soweit, dass es
Radfahren lernen kann, lassen sich Tret-
kurbel und Pedale hinzufügen, sodass aus
dem LittleBig ein vollwertiges Fahrrad wird.

Begründung der Jury
Die Idee eines 3-in-1-Fahrrads, das Kinder
bis zu fünf Jahre lang nutzen können,
überzeugt auch unter dem Gesichtspunkt
der Nachhaltigkeit.

Fashion, lifestyle and accessories
Mode, Lifestyle und Accessoires

Lite-Biz
Cabin-Size Luggage
Kabinengepäck

Manufacturer
Samsonite Europe NV,
Oudenaarde, Belgium

In-house design
Samsonite Design &
Development Team
(Diego Recchia)

Web
www.samsonite.com

reddot award 2015
best of the best

A reliable partner

In a globalised world, air travel has become part of the everyday life of many people. The Lite-Biz luggage collection is primarily targeted at modern business travellers. Its design was heavily influenced by architecture and the automotive industry. This cabin bag combines a functionality that suits the world in which it will be used with an appearance of strength and robustness. Rectilinear ribs on the front of the case give it the necessary structure and, in conjunction with the contrasting angled ribs on the sides, ensure a high level of resilience. Of particular interest to frequent travellers is the integrated external pocket on the front, which provides easy access to travel documents and other personal belongings. The design of this area has been maximised to offer as much space as possible by making good use of the height axis and the axis towards the front. The cabin bag is also equipped with an innovative single-point TSA-approved locking system, making it possible to open and close the front and main compartment with just one touch. In order to withstand the daily stresses to which it is exposed, the suitcase is extremely hard-wearing and its large, smooth-running double wheels also make it very easy to manoeuvre. The design of Lite-Biz has focused on the requirements of airport use and has resulted in a luggage collection that seamlessly fits in with the mobility of today's everyday life and enhances it with its aesthetic appeal.

Verlässlicher Partner

In einer globalisierten Welt ist das Reisen mit dem Flugzeug für viele Menschen gelebter Alltag. Die Gepäckkollektion Lite-Biz richtet sich vor allem an die Zielgruppe des modernen Geschäftsreisenden, wobei sie in ihrer Gestaltung maßgeblich von der Architektur und Automobilindustrie inspiriert wurde. Dieses Kabinengepäck verbindet eine situationsnahe Funktionalität mit dem Ausdruck von Stärke und Stabilität. Geradlinig gestaltete Rippen im Frontbereich geben ihm die nötige Struktur und sorgen, in Verbindung mit dazu kontrastreich geneigten Rippen auf der Seite, für ein hohes Maß an Belastbarkeit. Für Vielreisende überaus durchdacht ist insbesondere die Ausstattung mit einer in den Frontbereich integrierten Außentasche. Man hat damit einen schnellen Zugriff auf Reisedokumente oder andere Utensilien, wobei dieser Bereich gestalterisch so maximiert wurde, dass er größtmöglichen Platz bietet, indem er sowohl die gesamte Achse in der Höhe wie auch nach vorne ausnutzt. Das Kabinengepäck verfügt ferner über ein innovatives Ein-Punkt-TSA-Verschlusssystem, wodurch sich das Front- und Hauptfach mit nur einem Handgriff schließen und öffnen lassen. Um den tagtäglichen Belastungen gewachsen zu sein, ist es extrem widerstandsfähig und mit leichtgängigen großen Doppelrollen sehr wendig. Dank einer sehr nahe an den Bedingungen auf Flughäfen orientierten Gestaltung entstand mit Lite-Biz eine Gepäckkollektion, die den Alltag heutiger Mobilität nahtlos mitvollzieht und durch ihre Ästhetik bereichert.

Statement by the jury

Lite-Biz is an elegant item of cabin-size luggage that has followed a consistent design concept in every respect. It is extremely light, hard-wearing and convenient for the long stretches most travellers have to cover in the airport. Its smooth-running double wheels ensure it can easily be manoeuvred into any direction. One of its winning functional details is its perfectly constructed outer pocket. This gives travellers immediate access to their travel documents or laptop without having to open the suitcase.

Begründung der Jury

Lite-Biz ist ein in jeder Hinsicht konsequent gestaltetes, elegant anmutendes Kabinengepäck. Es ist außerordentlich leicht, widerstandsfähig und handlich für die langen Wege im Flughafen, wo es sich mittels leichtgängiger Rollen gut in alle Richtungen bewegen lässt. Ein bestechend funktionales Detail ist seine perfekt konstruierte Außentasche. Ohne den Koffer ganz öffnen zu müssen, hat man so die Reisedokumente oder den Laptop stets sofort griffbereit.

Designer portrait
See page 54
Siehe Seite 54

Lite-Shock
Luggage Line
Gepäck

Manufacturer
Samsonite Europe NV, Oudenaarde, Belgium
In-house design
Samsonite Design & Developement Team
(Erik Sijmons)
Web
www.samsonite.com

Taking nature as their source of inspiration, the suitcases of this collection stand out from the competition. Their ripple-effect shell design is based on the pattern that a stone creates when dropped into water. The product design mirrors its function, which focuses primarily on durability and resilience. Weighing between 1.7 and 2.5 kg, depending on size, the Lite-Shock suitcase can take an even larger amount of luggage without exceeding the weight limits imposed by airlines companies.

Dank der Natur als Inspirationsquelle stechen die Koffer dieser Serie unter ihren Mitstreitern hervor: Ihre wellenartige Oberfläche basiert auf dem Muster, das beim Fall eines Steines ins Wasser entsteht. Somit verläuft die Gestaltung des Produktes parallel zu seiner Funktion, die vorrangig auf Widerstandsfähigkeit und Belastbarkeit zielt. Bei einem Gewicht, das je nach Größe 1,7 bis 2,5 kg beträgt, können viele Utensilien im Inneren des Lite-Shock verstaut werden, ohne, dass die Grenze des von den Fluglinien jeweils erlaubten Check-in-Gewichtes überschritten wird.

Statement by the jury
The distinctive design feature of Lite-Shock is the ripple motif that defines the suitcase not only with regard to its aesthetic appeal, but also in functional terms thanks to its shock-absorbing capability.

Begründung der Jury
Markantes Designmerkmal des Lite-Shock ist das Wellenmotiv, das den Koffer nicht nur in ästhetischer, sondern dank seiner stoßabsorbierenden Wirkung auch in funktionaler Hinsicht prägt.

Lock'n'Roll
Suitcase
Koffer

Manufacturer
Samsonite NV, Oudenaarde, Belgium
Design
Therefore Ltd (Richard Miles),
London, Great Britain
Web
www.americantourister.eu

Curved lines, paired with a unique mix of texture and sporty colour accents give Lock'n'Roll its dynamic appearance. The cubic shape provides maximum packing space, while the 3-point lock enables fast and secure locking. Thanks to injection molding and innovative engineering, this Made-in-Europe collection represents the perfect synergy of strength and lightness: the biggest size (Spinner 75cm) only weighs 4.6 kg.

Statement by the jury
This suitcase not only offers an innovative locking systems, but is also appealing thanks to its sporty design.

Gestalterische Elemente wie eine gerundete Linienführung, eine einzigartige Kombination von Oberflächenstrukturen und sportliche Farbakzente geben Lock'n'Roll sein dynamisches Aussehen. Die kubische Form bietet maximalen Packraum, während das 3-Punkt-Verschlusssystem ein schnelles und sicheres Verschließen ermöglicht. Dank des verwendeten Spritzgussverfahrens und der innovativen Konstruktion bietet diese Made-in-Europe-Serie die perfekte Synergie aus Robustheit und Leichtigkeit: der Größte Koffer (Spinner 75 cm) wiegt gerade einmal 4,6 kg.

Begründung der Jury
Dieser Koffer bietet nicht nur ein innovatives Verschlusssystem, sondern gefällt auch wegen seines sportlichen Designs.

Lightway
Suitcase
Koffer

Manufacturer
Samsonite NV, Oudenaarde, Belgium
In-house design
American Tourister Design &
Development Team
Web
www.americantourister.eu
Honourable Mention

The Lightway collection is one of the lightest on the market: the smallest size only weighs 1.2 kg. Lightway offers remarkable lightness and strength thanks to an innovative, lightweight construction, using fiber glass pultrusions and polypropylene injected reinforcements. On top of being among the lightest cases, Lightway offers a large packing capacity.

Statement by the jury
The suitcases of this collection are very light but, at the same time, robust thanks to a new, innovative manufacturing process.

Die Lightway-Serie zählt zu den leichtesten ihrer Art: Der kleinste Koffer wiegt nur 1,2 kg. Die innovative, leichte Konstruktion eines Gestänges aus Fiberglas mit Verstärkungen aus Polypropylen sorgt für hohe Widerstandsfähigkeit und ein auffallend geringes Gewicht. Trotz seiner enormen Leichtigkeit besitzt Lightway ein großes Packvolumen.

Begründung der Jury
Diese Kofferserie ist dank einer neuen, innovativen Art der Verarbeitung sehr leicht und zugleich widerstandsfähig.

R-21
Suitcase
Koffer

Manufacturer
Samsonite Asia, Hong Kong
In-house design
Nobuo Maeda
Design
Moto Design Inc. (Song Min Hoon)
Web
www.samsonite.com
www.motodesign.com

What makes these suitcases stand out is their eye-catching construction and bright colours. The distinctive design and practical functions are geared towards the mobility of the user. Whether used for city breaks or longer trips, R-21 can hold a lot of luggage, before it is folded up like a book and closed. This makes handling significantly easier, as its contents are always readily available when travelling.

Statement by the jury
The striking design and new functional concept behind the R-21 collection result in a contemporary interpretation of the classic suitcase.

Die Besonderheit des Koffers liegt in seiner auffälligen Struktur und der leuchtenden Farbe. Die individuelle Gestaltung und die praktischen Funktionen orientieren sich an der Mobilität des Nutzers. Ob für Städtetrips oder längere Reisen – im R-21 kann viel Gepäck verstaut werden, bevor er sich schließlich wie ein Buch zuklappen und verschließen lässt. Dies erleichtert die Handhabung maßgeblich, da der Inhalt während der Reise stets griffbereit bleibt.

Begründung der Jury
Mit ihrer auffälligen Gestaltung und einem neuen Funktionskonzept zeigt die Serie R-21 eine zeitgemäße Interpretation des klassischen Koffers.

TRAVEL METER No. 6703
Suitcase
Koffer

Manufacturer
T & S Co., Ltd., Suitcase Designing Company,
Koshigaya, Japan
In-house design
Ruicheng Qi
Web
www.tands-luggage.jp

This hard-case suitcase Travel Meter No. 6703 is equipped with built-in digital LCD scales, which show the gross weight when the suitcase is stood on end and thus keep preparation before a journey to a minimum, as well as reduce the risk of having to pay for excess luggage. The No. 6703 suitcase can weigh contents up to 50 kg which find ample storage space inside an additional protective inner frame. A metal logo, attached to the top of the suitcase just above the integrated scales, also serves to protect it.

In den Hartschalenkoffer Travel Meter No. 6703 ist eine digitale LCD-Waage eingearbeitet, die nach dem Aufstellen das Gewicht des Inhalts anzeigt und so den Aufwand vor dem Reiseantritt möglichst gering hält. Das Volumen des No. 6703 umfasst bis zu 50 kg Gepäck, das im Schutz eines zusätzlichen Innenrahmens ausreichend Platz findet. Über der integrierten Waage auf der Oberseite des Koffers ist ein Logo aus Metall angebracht, das gleichzeitig auch als Schutz für die Waage dient.

Statement by the jury
The built-in scales are a practical addition to the features of this suitcase, which appeals thanks to its timeless, high-quality appearance.

Begründung der Jury
Mit der integrierten Waage bietet dieser Koffer einen praktischen Zusatznutzen. Zudem gefällt er mit einem zeitlosen, hochwertigen Äußeren.

ANCHOR + No. 6701
Suitcase
Koffer

Manufacturer
T & S Co., Ltd., Suitcase Designing Company,
Koshigaya, Japan
In-house design
Ruicheng Qi
Web
www.tands-luggage.jp

Anchor + No. 6701 is equipped with a system that makes transporting luggage when travelling easier. Pressing down the carrying handle blocks the wheels of the case immediately. This brake system was inspired by the image of a ship casting anchor. Just like an anchor, the brake stops the suitcase firmly. If the handle is pulled out again, the wheels are automatically released. Whether travelling on a bus, train or by air, this suitcase gives its user enormous flexibility and increased security.

Anchor + No. 6701 ist mit einem System ausgestattet, das den Transport von Gepäck auf Reisen erleichtert. Beim Hineindrücken des Tragegriffs werden die Räder des Koffers direkt blockiert. Diese Idee der Bremse wurde vom Werfen eines Schiffsankers inspiriert, da sie den Koffer ebenso unmittelbar stoppt. Zieht man den Griff heraus, werden die Rollen automatisch wieder in Betrieb gesetzt. Unterwegs mit dem Bus, der Bahn oder dem Flugzeug unterstützt der Koffer seinen Benutzer durch große Flexibilität und hohe Sicherheit.

Statement by the jury
This suitcase offers increased convenience by integrating the button for the locking brake into the case handle, which makes for intuitive use.

Begründung der Jury
Dieser Koffer erhöht den Komfort dadurch, dass der Knopf für die Feststellbremse direkt in den Koffergriff integriert ist und so eine intuitive Bedienung erlaubt.

Princess Power Box
Suitcase
Koffer

Manufacturer
Princess Traveller, Breda, Netherlands
In-house design
Web
www.princesstraveller.com

This suitcase contains a built-in charging station to supply mobile phones or tablets with power any time, anywhere. As soon as increased use on long journeys has drained the battery power, devices can be connected individually or simultaneously. Once the so-called Power Box in the 100 per cent recyclable suitcase has provided sufficient charge, the connected devices have enough power to run for up to seven hours.

Statement by the jury
Thanks to an integrated power block, this suitcase makes it possible to charge electronic devices while travelling which makes it well-suited to a modern, mobile lifestyle.

In diesen Koffer wurde eine Station integriert, die das Aufladen des Mobiltelefons oder Tablets überall und zu jeder Zeit möglich macht. Wenn deren Akku infolge des erhöhten Gebrauchs auf längeren Reisen nicht mehr ausreicht, können sie einzeln oder auch simultan angeschlossen werden. Sobald die sogenannte Power Box des zu 100 Prozent recycelbaren Koffers ausreichend geladen ist, gibt sie technischen Geräten Strom für bis zu sieben Stunden.

Begründung der Jury
Dank eines integrierten Powerblocks ermöglicht dieser Koffer das Laden elektronischer Endgeräte auf Reisen und passt deshalb gut zu einem modernen, mobilen Lebensstil.

Porsche Design Roadster 3.0 Trolley 550
Luggage
Reisegepäck

Manufacturer
Müller & Meirer Lederwarenfabrik GmbH, Kirn, Germany
In-house design
Porsche Design Studio, Zell am See, Austria
Web
www.mueller-und-meirer.com
www.porsche-design.com

Launched more than 15 years ago, the Roadster 3.0 luggage collection now presents itself to the world from a new angle. Weighing up to 30 per cent less than its predecessor, the new collection consistently adheres to a purist design concept. The brushed handles are made from a type of aluminium that is also employed in aircraft construction. The wheels of the trolley run smoothly and quietly and are shielded by aluminium protectors from damage caused by steps and curbs.

Statement by the jury
The Roadster 3.0 luggage collection combines timeless puristic design with a high level of functionality and is set to become a long-lasting companion.

Sie wurde zum ersten Mal vor mehr als 15 Jahren präsentiert und zeigt sich nun von einer neuen Seite: die Gepäckserie Roadster 3.0. Mit bis zu 30 Prozent weniger Gewicht als ihr Vorgänger verfolgt sie heute ein durchgehend puristisches Design. Für die matt schimmernden Griffe wurde ein Aluminium verwendet, das auch im Flugzeugbau gängig ist. Die Rollen der Trolleys laufen leicht und leise und werden durch Aluminiumprotektoren vor Schäden durch Treppen oder Bordsteinkanten geschützt.

Begründung der Jury
Die Gepäckstücke der Kollektion Roadster 3.0 kombinieren eine klare, puristische Gestaltung mit hoher Funktionalität und werden so zu langlebigen Begleitern.

Pivotal Soft Case
Travel Luggage
Reisegepäck

Manufacturer
Pivotal, Richmond, USA
In-house design
Leighton Klevana
Web
www.pivotalgear.com

The user-friendliness of this suitcase is maximised by its ergonomic patented Pivot-Grip handle which rotates through 360 degrees. This type of construction not only makes transport easier, but greatly reduces twisting and straining of the arm and wrist. A divider system in the main compartment makes it possible to neatly organise the contents. If desired, this can be removed. When not in use, the suitcase can be folded flat for storage to a size of 15 x 30 x 91 cm.

Statement by the jury
An innovative handle and a flexible division of the interior are the prominent features of this cleanly yet distinctively designed travel bag.

Die Benutzerfreundlichkeit des Koffers erhöht sich durch einen ergonomisch patentierten Pivot-Griff, der um 360 Grad rotiert. Neben einem erleichterten Transport werden durch eine derartige Konstruktion auch ein Verdrehen und die Belastung des Arm- und Hüftgelenks entscheidend verringert. Ein Unterteilungssystem im Hauptfach des Koffers hält das Gepäck übersichtlich und kann auf Wunsch herausgenommen werden. Der Koffer selbst lässt sich bei Nichtgebrauch ebenfalls auf ein geringes Packmaß von 15 x 30 x 91 cm zusammenfalten.

Begründung der Jury
Ein innovativer Handgriff und eine flexible Innenraumaufteilung sind die hervorstechenden Merkmale dieser klar und zugleich markant gestalteten Reisetaschen.

Mendoza TAGA
Luggage
Gepäck

Manufacturer
Thomas Mendoza International Co., Ltd., Hong Kong
In-house design
James Lau, Paul Chong
Web
www.mendoza-bag.com
Honourable Mention

The concept of the Mendoza Taga was guided by its functional properties and its ambitious design aims. Besides long durability and increased ease of use, the focus of this suitcase is its aesthetic appearance, which distinguishes it greatly from other luggage items. All of its functions have been carefully thought through, so that the design of this handy suitcase has resulted in a reliable accessory.

Der Entwurf des Mendoza Taga orientiert sich an funktionalen Eigenschaften und einem hohen gestalterischen Anspruch. Neben einer langen Haltbarkeit und bequemer Handhabung liegt der Schwerpunkt des Koffers auf seinem ästhetischen Erscheinungsbild und einem Aussehen, das ihn stark von anderen Gepäckstücken unterscheidet. Durchdacht in jeder seiner Funktionen ist mit dem Entwurf dieses handlichen Koffers ein verlässlicher Begleiter entstanden.

Statement by the jury
With its distinctive, technology-based design, the Mendoza Taga creates a robust, functional impression.

Begründung der Jury
Mit seiner markant-technischen Gestaltung hinterlässt der Mendoza Taga einen robusten, funktionalen Eindruck.

Nimbus All Weather Suitcase

Suitcase
Reisegepäck

Manufacturer
Lojel, Kowloon, Hong Kong
In-house design
Web
www.lojel.com
Honourable Mention

Nimbus is a series of suitcases that can withstand both sudden rain showers and up to six hours of continual rain. This resistance has won its certification according to the international IPX3 standard for water resistance. After a thorough analysis of all of its components, the Nimbus all-weather suitcase has proved to be a product that protects the belongings of its owner from harm and keeps itself dry.

Zertifiziert nach dem internationalen Wasserbeständigkeitsstandard IPX3 halten die Koffer dieser Serie das Gepäck bei plötzlichen Regenschauern und andauerndem Niederschlag bis zu sechs Stunden trocken. Nach genauer Analyse der einzelnen Bestandteile ist mit dem Allwetter-Reisegepäck Nimbus ein Produkt entstanden, das Schaden vom Hab und Gut seines Besitzers fernhält und dabei selbst trocken bleibt.

Statement by the jury
This suitcase is compelling for its water-resistant properties, while its colourful details act as eye-catchers.

Begründung der Jury
Dieser Koffer überzeugt durch wasserabweisende Eigenschaften und macht mit farbenfrohen Details auf sich aufmerksam.

Pháin
Backpack Bag
Rucksacktasche

Manufacturer
Ideoso Design Inc., Taipei, Taiwan

Design
Chun-Chieh Wang

Web
www.ideoso.com.tw

reddot award 2015
best of the best

A punchy message

Bags or backpacks made of all manner of recycled materials are currently very trendy, because they allow users to demonstrate their individuality and environmental awareness. In this category, the Pháin backpack bag is a delightful new interpretation that is bound to attract many envious glances. It is made of unprinted cement bags in their original colour – a material that has to be tough to fulfil its primary function. These bags consist of thick kraft paper on the outside and, on the inside, are made of durable and tear-resistant, glued PP-woven fabric. The particular aesthetic appeal of these backpack bags lies in the combination of a seemingly rough material with the expressive design of the straps. Heavy-duty slings commonly used by cranes are used as surprisingly functional adjustable straps on the back. The brass buckles used to adjust the straps have been arranged to reflect the nature of a construction site on which this idea was based. At the same time, the pockets give the backpack the appearance of an accessory. Another aspect of the originality of the backpack bag's design is its particular shape. Inspired by the functionality of diving bags, it is open at the top and can be opened and closed by folding over the top edge. In this way the Pháin backpack bag keeps thieves at bay, but nonetheless ensures that everything inside can quickly be found.

Ausdrucksstarke Botschaft

Taschen oder Rucksäcke aus Recyclingmaterialien unterschiedlichster Art liegen im Trend, denn ihre Besitzer kommunizieren durch sie Individualität und Umweltbewusstsein. Die Rucksacktasche Pháin findet für diese Produktgattung eine überaus reizvolle neue Form, mit der diese Tasche in der Menge die Blicke auf sich zieht. Gestaltet ist sie aus unbedruckten Zementsäcken in Originalfarbe, einem Material, das alleine schon durch seine Bestimmung sehr fest sein muss. Auf der Außenseite bestehen diese Säcke aus dickem Kraftpapier und auf ihrer Innenseite aus einem langlebigen und reißfesten, verklebten PP-Gewebestoff. Ihre besondere Ästhetik beziehen diese Rucksacktaschen aus der Kombination des rau wirkenden Materials mit einer ausdrucksstarken Gestaltung der Tragegurte. Als verblüffend funktionale Verstellriemen dienen hier Schwerlast-Tragegurte auf der Rückseite, wie sie gewöhnlich bei Baustellenkränen eingesetzt werden. Die über Messingschnallen längenverstellbaren Gurte sind so angeordnet, dass sie den Baustellencharakter gut widerspiegeln, aber zugleich den Taschen die Anmutung von Accessoires verleihen. Ein weiterer Aspekt der originellen Gestaltung der Rucksacktasche liegt in ihrer spezifischen Form. Angelehnt an die Funktionalität von Tauchtaschen ist sie oben offen und lässt sich öffnen und schließen, indem der obere Rand umgeknickt wird. Die Rucksacktasche Pháin ist dadurch diebstahlsicher und alle Dinge in ihrem Inneren können rasch gefunden werden.

Statement by the jury

In an extraordinary way, Pháin turns cement bags into desirable fashion objects. These backpack bags are functional, durable and comfortable to wear. The colour and detailing are well matched and ensure Pháin is an eye-catching accessory. Thanks to a committed design concept, the idea of sustainability is conveyed in an attention-grabbing manner.

Begründung der Jury

Auf unvergleichliche Weise werden durch Pháin Zementsäcke in begehrenswerte Fashionprodukte verwandelt. Diese Rucksacktasche ist funktional, langlebig und sie lässt sich komfortabel tragen. Gut abgestimmt sind die Farbgebung wie auch die Details, mit denen Pháin zu einem Blickfänger wird. Mittels eines engagierten gestalterischen Ansatzes wird hier der Gedanke der Nachhaltigkeit aufmerksamkeitsstark transportiert.

Designer portrait
See page 56
Siehe Seite 56

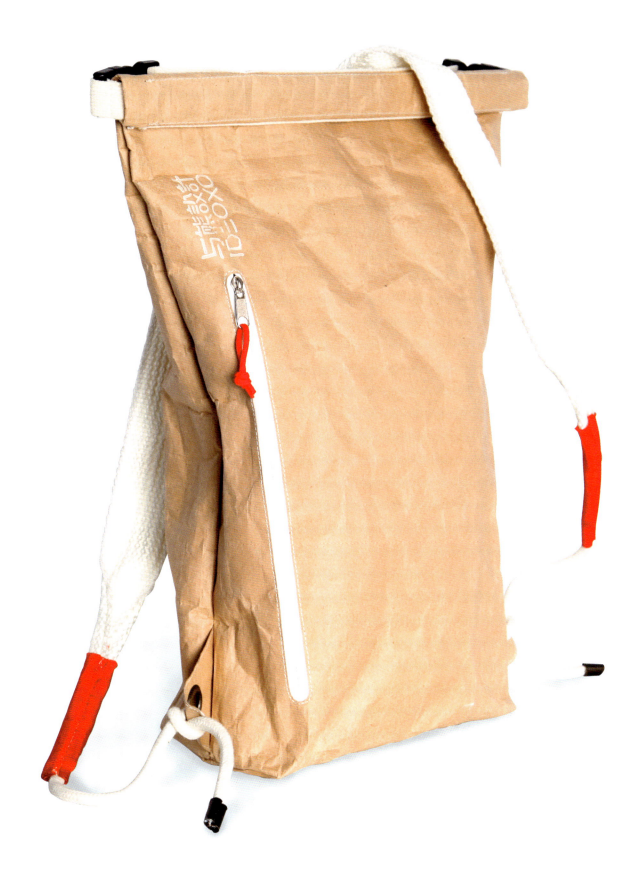

Manufacturer
nobrainer GmbH, Cologne, Germany
In-house design
Crumpler Design Team
Web
www.crumpler.eu

This intricately crafted travel collection encompasses both smaller and larger bags and can be used in a wide range of situations. As the suitcases are made of 1000 Denier nylon fabric and 330 Denier Ripstop nylon, they are both water and tear resistant. The interior is made of ultra-light 210 T Nylon ensuring that lightness and durability remain key characteristics of the resulting products.

Die sorgfältig verarbeitete Gepäckkollektion umfasst sowohl kleinere als auch größere Modelle und kommt bei unterschiedlichsten Anlässen zum Einsatz. Aufgrund einer Verarbeitung aus 1.000-Denier-Nylongewebe und 330-Denier-Ripstop-Nylon sind die Reisetaschen sowohl reiß- als auch wasserfest. Das Innere ist mit einem ultraleichten 210 T-Nylongewebe ausgekleidet, sodass die Produkte schließlich Leichtigkeit und Strapazierfähigkeit zu ihren charakteristischen Merkmalen zählen können.

Statement by the jury
The Track Jack collection stands out for its particularly light weight and a functionality that has been thought through down to the smallest detail.

Begründung der Jury
Die Track Jack Kollektion zeichnet sich durch besondere Leichtigkeit und eine bis ins Detail durchdachte Funktionalität aus.

Leitz Complete Smart Traveller
Business Travelling Bags
Business-Reisetaschen

Manufacturer
Esselte Leitz GmbH & Co KG,
Stuttgart, Germany
In-house design
Maja Tsubaki
Web
www.leitz.com

Whilst the interior of this range of travel bags is very roomy, they are nonetheless notable for their particular lightness; a feature which is worth its weight in gold, particularly on long business trips. Passport and business cards fit neatly into the front pocket and are easily accessible if needed. Transparent mesh pockets enable the user to see what is inside them and an integrated cleaning cloth helps to keep screens of smartphones and laptops clean.

Während das Innere der Reisetaschen dieser Serie sehr geräumig ist, zeichnen diese sich trotz ihrer Größe durch eine besondere Leichtigkeit aus. Das bewährt sich vor allem auch auf längeren Geschäftsreisen. Im Vorderfach der Taschen können Ausweis und Visitenkarten untergebracht werden und sind dort bei Bedarf rasch griffbereit. Durchsichtige Mesh-Pockets gewähren Einblick in ihren Inhalt, ein integriertes Reinigungstuch hält die Bildschirme von Smartphones und Laptops sauber.

Statement by the jury
Perfect in its functional alignment with the needs of business travellers, these travel bags are easy to recognise thanks to their bright green colour.

Begründung der Jury
Funktional optimal auf die Ansprüche Geschäftsreisender abgestimmt, haben die Reisetaschen dank leuchtend grüner Farbakzente zudem einen hohen Wiedererkennungswert.

Mobiler's Backpack
Recycle Backpack
Recycling-Rucksack

Manufacturer
Mondo Design Co., Ltd., Tokyo, Japan
In-house design
Ryota Nakamura
Web
www.seal-brand.com

This rucksack is an environmentally friendly product made of tyre inner tubes. The use of this robust material makes it highly durable as well as resistant to water and other harmful environmental influences. Each individual design differs in shape. Individuality is therefore one of the rucksack's key characteristics. Mobiler's Backpack also offers numerous storage options inside. It is able to safely carry everything from a notebook to a tablet.

Statement by the jury
This rucksack, made of recycled inner tubes, combines functional design with sustainability and results in a product with a high recognition value.

Bei diesem Rucksack handelt es sich um ein umweltfreundliches Produkt, für das Reifenschläuche wiederverwertet wurden. Die Verwendung von robustem Material führt zu einer langen Lebensdauer und macht ihn resistent gegen Wasser und andere schädliche Umwelteinflüsse. Jeder einzelne Entwurf ist unterschiedlich in seiner Gestalt, somit zählt Individualität zu seinen stärksten Merkmalen. Es befinden sich zahlreiche Möglichkeiten zur Unterbringung im Inneren des Mobiler's Backpack, auch ein Notebook oder Tablet sind in ihm sicher verwahrt.

Begründung der Jury
Dieser Rucksack aus recycelten Schläuchen vereint eine funktionale Gestaltung mit Nachhaltigkeit zu einem Produkt mit hohem Wiedererkennungswert.

HP Core Conservative Backpack
Backpack
Rucksack

Manufacturer
Hewlett-Packard, Palo Alto, USA
In-house design
Youjin Nam
Web
www.hp.com

Whilst the exterior of this rucksack is mottled grey with black accents, the interior is neutral navy-blue in colour. The shoulder straps are padded and breathable and the back is equipped with integrated air channels and an adjustable sternum strap for stability and comfort. Aside from the built-in audio charging port, there is a separate pocket on the front for storing mobile devices, thus making them easily accessible.

Statement by the jury
The Core Conservative Rucksack combines a high-quality, sober appearance with a well-organised interior, suitable for use in business environments.

Während äußerlich ein meliertes Grau und schwarze Akzente dominieren, gestaltet sich das Innere des Rucksacks in einem neutralen Marineton. Die Schultergurte sind gepolstert und atmungsaktiv, eine Rückseite mit integrierten Luftkanälen und ein verstellbarer Brustgurt sorgen für Stabilität und schonenden Tragekomfort. Besonderes Merkmal ist neben einem eingebauten Audio-Ladeanschluss ein Extrafach an der Vorderseite, in dem mobile Endgeräte mit schnellem Zugriff untergebracht werden können.

Begründung der Jury
Der Core-Conservative-Rucksack kombiniert ein hochwertiges, sachliches Erscheinungsbild mit einem gut organisierten Innenleben für den Einsatz im Business-Umfeld.

HP Odyssey Backpack
Sport Backpack
Sport-Rucksack

Manufacturer
Hewlett-Packard, Palo Alto, USA
In-house design
Youjin Nam
Web
www.hp.com
Honourable Mention

With its padding, the HP Odyssey protects the back of its wearer on outdoor trips or on campus. The rucksack consists of a durable air-mesh material and offers ample storage space inside. Alongside a number of pockets for equipment and belongings, a dedicated compartment ensures safe storage of notebooks, tablets or e-books. Strategically placed straps make the backpack comfortable to wear on the shoulders.

Statement by the jury
The HP Odyssey scores high for its well-thought-out functionality, which focuses particularly on the safe and comfortable transport of sensitive electronic equipment.

Mit seiner Polsterung schützt der HP Odyssey auf dem Outdoor-Trip oder dem Campus den Rücken seines Trägers. Der Rucksack besteht aus stabilem Air-Mesh-Material und bietet in seinem Inneren vielseitigen Stauraum. Neben diversen Taschen für Zubehör können in einem passenden Fach Notebook, Tablet oder E-Book sicher untergebracht werden, während die strategisch platzierten Riemen bequem auf den Schultern sitzen.

Begründung der Jury
Der HP Odyssey überzeugt mit einer durchdachten Funktionalität, die insbesondere auf den sicheren und komfortablen Transport empfindlicher elektronischer Geräte ausgerichtet ist.

HP Trend Backpack
Backpack
Rucksack

Manufacturer
Hewlett-Packard, Palo Alto, USA
In-house design
Youjin Nam
Web
www.hp.com

Good functionality was the guiding idea behind the design of this rucksack. One of the key characteristics the development focused on was its light weight. The bright red interior contrasts with the unobtrusive grey of the exterior fabric, which includes easily accessible pockets. The front of the rucksack is equipped with an integrated port to allow charging of a mobile or MP3 player.

Statement by the jury
The design of this rucksack paid particular attention to the transport of mobile devices. Red highlights create eye-catching accents both inside and out.

Mit guter Funktionalität als leitendem Gedanken zählte bei der Entwicklung dieses Rucksacks auch seine auffallende Leichtigkeit zu den Hauptmerkmalen. Das leuchtende Rot im Inneren kontrastiert mit der dezenten grauen Farbwahl für das Außenmaterial, das außerdem mit schnell zugänglichen Taschen versehen ist. Der vordere Teil des Rucksacks beinhaltet einen integrierten Anschluss zum mobilen Aufladen von Mobiltelefon oder MP3-Player.

Begründung der Jury
Dieser Rucksack ist speziell im Hinblick auf den Transport mobiler Geräte durchdacht gestaltet. Rote Details setzen außen wie innen auffällige Akzente.

HP Trend Messenger
Messenger Bag
Tasche

Manufacturer
Hewlett-Packard, Palo Alto, USA
In-house design
Youjin Nam
Web
www.hp.com

The many different storage options offered by this bag make it a very attractive accessory. Notebooks measuring up to 15.6 inches can be fitted into the padded inner compartment and there is additional room for transporting important belongings in the numerous pockets closed with zip fasteners. On top of this, the HP Trend Messenger bag's outer pockets also offer a lot of extra space. The flap edges are covered with a reflective material. It is secured with multiple hooks, loop fasteners and side buckles.

Statement by the jury
This messenger bag with its rectilinear design not only offers considerable space for devices and belongings, but also transports them safely.

Vielseitige Möglichkeiten zum Verstauen und ein modisches Äußeres machen die Umhängetasche zu einem attraktiven Begleiter. Ein Notebook mit bis zu 15,6" ist im gepolsterten Innenfach sicher untergebracht, wichtiges Zubehör kann indes in den zahlreichen Fächern mit Reißverschluss transportiert werden. Nicht zuletzt bieten die Außentaschen der HP Trend Messenger Bag eine Menge Platz. Die Verschlussklappe ist mit einer Kante aus reflektierendem Material versehen und zudem durch mehrere Haken, Schlaufenverschlüsse und Seitenklammern gesichert.

Begründung der Jury
Diese geradlinig gestaltete Messenger Bag bietet nicht nur viel Platz für Geräte und Utensilien, sondern transportiert diese auch gut gesichert.

Porsche Design Agnodice
Women's Handbag
Damen-Handtasche

Manufacturer
Porsche Design Group,
Ludwigsburg, Germany
In-house design
Porsche Design Studio,
Zell am See, Austria
Web
www.porsche-design.com

The puristic design of the Agnodice handbag is reminiscent of a traditional doctor's bag and was inspired by the Greek physician Agnodice, after whom the handbag is named. In addition to the elegant black model, the bag is available in ten different leathers. Four different options for the metal fittings and a choice of two different colours for the lining allow even greater individualisation of this design classic.

Statement by the jury
Taking the timeless puristic doctor's bag as its inspiration, Agnodice makes a strong impression with its many customisable options.

Die puristisch gestaltete Agnodice-Handtasche erinnert an eine traditionelle Arzttasche. Die Inspiration dafür lieferte die griechischen Ärztin Agnodice, der das Modell auch seinen Namen verdankt. Neben dem edlen in Schwarz gehaltenem Modell ist die Tasche in zehn verschiedenen Leder-Varianten erhältlich. Für noch mehr Individualität kann zwischen vier verschiedenen Optionen für die Metallbeschläge sowie zwei Farben für das Innenfutter gewählt werden.

Begründung der Jury
Auf der Basis einer puristisch-zeitlosen Arzttasche als Vorbild beeindruckt die Agnodice mit vielen Personalisierungsmöglichkeiten.

Oberwerth – Freiburg
Photo Bag
Fototasche

Manufacturer
net SE/Globell-Deutschland,
Koblenz, Germany
In-house design
Regina Immes
Web
www.oberwerth.de

The principal components of this handmade camera bag – Cordura fabric and tanned cow-hide – make it light and give it a durable shape. A separate insert can be taken out to change the bag's use while an additional rivet on the strap reduces tugging and, prevents it from fraying. As fasteners the bag uses a special secure fastening solution developed by the automotive industry in the 1960s, thereby preventing unintentional opening of the bag.

Statement by the jury
This versatile bag is appealing due to its high-quality workmanship, lovingly crafted details and an attractive retro look.

Die wesentlichen Bestandteile der handgefertigten Kameratasche sind die Materialien Cordura und gegerbtes Rindsleder. So erhält sie eine leichte und dennoch stabile Form und kann durch ein separat herausnehmbares Innenteil auch zweckentfremdet werden. Eine zusätzliche Niete am Gurt mindert den Zug und verhindert so dessen Ausreißen. Bei den Verschlüssen der Tasche handelt es sich um eine Entwicklung für die Autoindustrie der 1960er Jahre, deren spezielle Befestigungs- und Sicherungslösung ein versehentliches Öffnen verhindert.

Begründung der Jury
Diese vielseitige Tasche besticht durch hochwertige Verarbeitung, liebevoll gestaltete Details und einen charmanten Retro-Look.

Leather Wallet with GPS & Bluetooth Technology
Small Leather Goods
Kleinlederwaren

Manufacturer
Torero Corporation Pvt Ltd, Kolkata, India
In-house design
Web
www.torerocorp.com
Honourable Mention

This handcrafted wallet made of soft Spanish cowhide stands out due to its elegance and durability. With four card pockets, two compartments for every type of currency and two multi-use pockets, your valuables will be stored securely. Despite the ample interior space, this wallet is extremely slim and therefore fits neatly into both handbags and back pockets. Thanks to an integrated GPS system that works via Bluetooth in synchronization with a free Cross Tracker app on the smartphone and tablet, the wallet's owner will always be able to find his or her way around.

Das spanische Rindsleder fühlt sich weich an und wurde in Handarbeit zu einer Geldbörse verarbeitet, die sich durch ihre Eleganz und Langlebigkeit auszeichnet. Mit vier Kartenfächern, zwei Fächern für jegliche Währungseinheiten und zwei Mehrzwecktaschen sind die Wertsachen in ihr sicher verstaut. Trotz großem Stauraum im Inneren ist die Brieftasche sehr flach gehalten und somit leicht in der Hand- oder Gesäßtasche untergebracht. Dank eines integrierten GPS-Systems, das über Bluetooth in Synchronisation mit kostenloser Cross Tracker App auf dem Smartphone oder Tablet funktioniert, findet der Besitzer dieses Portemonnaies immer und überall den Weg.

Statement by the jury
This real leather wallet is a successful combination of classic and modern attributes, and will make everyday transactions easier and more enjoyable.

Begründung der Jury
Diese Echtlederbrieftasche ist eine gelungene Kombination klassischer und moderner Attribute, die den Alltag verschönern und zugleich erleichtern.

Travel Slippers
Shoes
Schuhe

Manufacturer
Studio shoedesign, Arnhem, Netherlands
In-house design
Drs. Renate Volleberg
Web
www.shoedesign.nl

These Dutch slippers take patterns as their starting point. A 2D design concept ensures they accurately adapt to the feet and enclose them in soft felt. The fabric is cut with a laser just like a piece of paper and then the feet will given its final shape. The travel slippers keep feet cosy at home, but are also faithful companions on short or long journeys.

Statement by the jury
The two-dimensional, paper-cut-type patterns are very original design features of these felt slippers that become three-dimensional once worn.

Diese Hausschuhe aus den Niederlanden wurden auf der Basis von Mustern konstruiert. Nach einem 2-D-Konzept entwickelt, passen sie sich genau den Füßen an und umschließen diese mit weichem Filz. Wie ein Stück Papier wird das Material mittels eines Lasers zurechtgeschnitten und schließlich durch den Fuß in seine finale Form gebracht. Die Travel Slippers wärmen die Füße daheim und zeigen sich auch auf kurzen oder langen Reisen als treue Begleiter.

Begründung der Jury
Originelles Designmerkmal dieser Filzpantoffeln sind ihre zweidimensionalen, scherenschnittartigen Muster, die am Fuß dreidimensional werden.

Birkenstock EVA Sandale
Shoes
Schuhe

Manufacturer
Birkenstock GmbH & Co. KG,
Neustadt (Wied), Germany
In-house design
Web
www.birkenstock.com

The newly conceived and developed EVA sandals are reproductions of the most popular Birkenstock models. They reinterpret the originals in a new, minimalist design language, focusing on the key characteristics of Birkenstock: the anatomically moulded deep footbed with toe grip and the bone profile on the sole. EVA (ethylene vinyl acetate) is an ultralight, sturdy and odour-neutral plastic which is pleasantly soft, extremely flexible and yet highly durable. This makes the EVA sandals particularly suitable for wear in Wellness and outdoor areas such as in the sauna, on the beach or by the seaside.

Statement by the jury
With a shape reminiscent of the Birkenstock classics, these foot-friendly plastic shoes are appealing thanks to their new features such as extreme light weight and washability.

Die neu konzipierten und entwickelten EVA-Sandalen sind Reproduktionen der beliebtesten Birkenstock-Modelle. Sie interpretieren ihre jeweiligen Original-Vorbilder in einer neuen minimalistischen Formensprache, konzentriert auf die für Birkenstock charakteristischen Kernmerkmale: das anatomisch geformte Tieffußbett mit Zehengreifer und das Knochenprofil in der Laufsohle. EVA (Ethylen-Vinylacetat) ist ein ultraleichter, robuster und dabei völlig geruchsfreier Kunststoff. Das Material ist angenehm weich, extrem elastisch und dennoch sehr widerstandsfähig. Deshalb eignen sich die EVA-Sandalen besonders für Wellness- und Outdoor-Bereiche wie Sauna, Strand oder Meer.

Begründung der Jury
Der Form nach eine Reminiszenz an Birkenstock-Klassiker überzeugen diese fußfreundlichen Kunststoffschuhe durch neue Eigenschaften wie extreme Leichtigkeit und Abwaschbarkeit.

Wedge Sandal 1 240 053
Shoes
Schuhe

Manufacturer
Deichmann SE, Essen, Germany
In-house design
Web
www.deichmann.com

These wedge sandals have a purist form which allows them to be worn in combination with a wide range of clothes. The use of a variety of materials gives them a distinctive style that can be interpreted as either sporty or elegant depending on taste. The sandals go with both skirts and dresses and are feminine and playful in style even when worn with trousers. The wedge heel stretches the leg and its shape means it is comfortable to wear when walking.

Statement by the jury
This straightforward wedge sandal is so versatile it can be combined with a large range of fashion, but still offer a high degree of comfort.

Die Sandale mit Keilabsatz verfolgt eine puristische Linie, die vielfältige Kombinationsmöglichkeiten zulässt. Die Verwendung verschiedener Materialien verleiht ihr einen prägnanten Stil, der je nach Geschmack sowohl sportlich als auch elegant interpretiert werden kann. Die Sandale passt somit zu Rock oder Kleid und wirkt auch zur Hose weiblich und verspielt. Der Absatz streckt das Bein und schont mit seiner komfortablen Passform die Füße beim Gehen.

Begründung der Jury
Diese unkomplizierte Keilsandale ist modisch vielseitig kombinierbar und bietet gleichzeitig hohen Komfort.

Porsche Design Beverly Hills
Shoes
Schuhe

Manufacturer
Porsche Design Group,
Ludwigsburg, Germany
In-house design
Porsche Design Studio,
Zell am See, Austria
Web
www.porsche-design.com

This men's moccasin of Porsche Design's spring/summer collection 2015 is based on the brand's first shoe from 1977. Updated with several small innovations, the outer sole drawn upwards at the heel is characteristic of this slip-on shoe. The so-called pull-tab allows to slip in and out easily, while a specially developed insole ensures the desired sturdiness and comfort of the shoe.

Statement by the jury
Inspired by a Porsche Design classic, this men's shoe stands out thanks to the new ergonomic details it offers.

Der Mokassin aus der Frühjahrs-/Sommer-Kollektion 2015 von Porsche Design ist angelehnt an den ersten Schuh der Marke aus dem Jahre 1977. Mit kleinen Neuerungen überarbeitet, bleibt die nach oben gezogene Sohle an der Ferse auch für diese Ausführung des Mokassins charakteristisch. Das sogenannte Pull-Tab erleichtert das An- und Ausziehen der Schuhe maßgeblich, während ein eigens entwickeltes Fußbett den Schuh stabil und komfortabel macht.

Begründung der Jury
Angelehnt an einen Porsche Design-Klassiker, beeindruckt dieser Herrenschuh mit neuen, ergonomischen Details.

Triumph Magic Wire
Bra
BH

Manufacturer
Triumph Intertrade AG,
Bad Zurzach, Switzerland

Design
Triumph International AG
(Klemens Moeslinger),
Heubach, Germany

Web
www.triumph.com

reddot award 2015
best of the best

Natural silhouette

When it comes to lingerie, comfort and fit are key, as these articles of clothing are worn close to the skin. The concept for the Triumph Magic Wire bra is the result of worldwide studies which concluded that the supporting metal or plastic wires, frequently used in bras, become uncomfortable if worn over longer periods of time. The innovative use of silicone in combination with a special moulding process made it possible to develop an item of lingerie that offers completely new properties. The Magic Wire is a bra that is very comfortable to wear thanks to the embedding of the silicone wire in between padded layers. This ensures that only the velvet-soft micro material of the bra comes into direct contact with the wearer's skin. A built-in stabiliser offers the additional benefit of a good push-in effect. For larger models its support function is reinforced by small rods that are integrated into the side-seams. The bra is given its shape by an innovative adhesive technology, which makes the seam-edges virtually invisible under clothing and ensures there are no pressure points. The Triumph Magic Wire is made with a gently shimmering jacquard weave in appealing, contemporary fashion colours. A feminine appearance is combined with advanced functionality to give the female bust a beautiful silhouette.

Natürliche Silhouette

Im Bereich der Lingerie sind der Komfort und die Passform sehr wichtig, denn diese Kleidungsstücke werden direkt am Körper getragen. Die Gestaltung des Triumph Magic Wire basiert auf weltweiten Studien, die zu dem Ergebnis führten, dass insbesondere die bei BHs häufig verwendeten stützenden Bügel aus Metall oder Kunststoff bei längerem Tragen als unangenehm empfunden werden. Der innovative Einsatz des Materials Silikon in Verbindung mit speziellen Mould-Verfahren ermöglicht ein Lingerieprodukt mit neuen Eigenschaften. Der Magic Wire ist ein BH mit sehr viel Komfort dank eines aus Silikon gefertigten, zwischen gepolsterte Lagen eingesetzten Bügels. Deshalb hat nur das samtweiche Mikro-Material des BHs direkten Kontakt zur Haut der Trägerin. Durch einen eingearbeiteten Stabilisator bietet der Magic Wire zusätzlich sehr gute Push-in-Effekte, wobei die stützende Funktion bis in große Größen mithilfe in die Seitennaht eingelegter Stäbchen noch verstärkt wird. In Form gebracht wird dieser BH durch eine innovative Klebetechnologie. Die Abschlüsse der Nähte sind daher unter der Kleidung nahezu unsichtbar und es gibt keine Druckstellen. Der Triumph Magic Wire besteht aus einem sanft schimmernden Jacquardgewebe in ansprechenden, aktuellen Modefarben. Eine feminine Anmutung verbindet sich so mit einer innovativen Funktionalität, die der Büste der Frau eine schöne Silhouette verleiht.

Statement by the jury

The Magic Wire bra offers the wearer new features as it fits very close to the body and thereby creates the ideal silhouette. Its perfect support function is impressive, particularly as it is achieved without the conventional wires traditionally used in lingerie. Here, a new type of bra has been created to suit the needs of the wearer by noticeably improving its comfort. In doing so, silicone, as a material, has gained in relevance and is likely to find new applications in design.

Begründung der Jury

Der Magic Wire-BH bietet neue Trageeigenschaften, indem er sich eng an den Körper anschmiegt und dabei eine ideale Silhouette zeichnet. Beeindruckend ist seine perfekt stützende Funktion, die er ohne die im Bereich der Lingerie gängigen Bügel erreicht. Entsprechend den Bedürfnissen der Trägerinnen wurde hier eine neue Art von BH mit einem spürbar besseren Komfort gestaltet. Das Material Silikon erlangt dadurch eine neue Bedeutung, die auch dem Design neue Wege eröffnet.

Designer portrait
See page 58
Siehe Seite 58

ITEM m6 Tights Invisible
Compression Tights –
looks like 15 denier
Kompressive Strumpfhose in
15-DEN-Optik

Manufacturer
medi GmbH & Co. KG, Bayreuth, Germany
In-house design
Web
www.medi-corporate.com
www.item-m6.com

These compression tights are available
in a variety of colours to suit different
skin tones. Despite their strength, the
thickness of the material used visually
only corresponds to 15 denier making
them barely noticeable when worn. In an
effect akin to make-up, they hide little
irregularities of the skin, accentuate
the slim line of the legs thanks to their
shape-defining compression behaviour,
and a supporting insert does the same
for the bottom.

Statement by the jury
The clearly defined compression effect
of these tights offers support and shapes
the legs, while at the same time being
barely visible.

Die Kompressionsstrumpfhose ist in Farb-
nuancen erhältlich, die auf verschiedene
Hauttöne abgestimmt sind. Entgegen ihrer
Dicke entspricht die Stärke des verwendeten Materials optisch nur 15 Denier
und fällt somit beim Tragen kaum auf. Mit
einem Make-up-ähnlichen Effekt kaschiert
sie kleine Unebenheiten der Haut, model-
liert durch einen definierten Kompressi-
onsverlauf die schlanke Linie der Beine und
mittels einer stützenden Einlage auch den
Po.

Begründung der Jury
Dank eines exakt definierten Kompressi-
onsverlaufs bietet diese Strumpfhose Halt
und formt die Beine, zugleich ist sie kaum
sichtbar.

Excella® Blade
Fastening Element
Verbindungselement

Manufacturer
YKK Corporation, Tokyo, Japan
In-house design
Tsutomu Saito
Design
Nanahiko Mio, Milan, Italy
Web
www.ykk.com

The discrete design of this zip fastener contributes to its adaptability to almost any style of clothing. Its teeth are like the herringbone zip fastener and receive their unique form from a special manufacturing method. Called Excella Blade, this modern zip fastener can be widely used: from baggage and handbags to clothing and even purses – wherever functional strength and excellent taste are important.

Das zurückhaltende Äußere des Reißverschlusses trägt dazu bei, dass er sich so gut wie jedem Modestil anpasst. Seine Zähne stehen im Fischgrätenverband und verdanken ihre Form einer besonderen Montagetechnik. Die Anwendungsbereiche der modernen Verschlussvariante namens Excella Blade reichen von Reisegepäck und Taschen über Kleidung bis hin zu Geldbörsen, wo sie für funktionale Stärke und guten Geschmack steht.

Statement by the jury
This elegant, minimalist zip fastener with its re-designed, blade-shaped teeth demonstrates flexibility of use and its appearance impresses with pure elegance.

Begründung der Jury
Dieser elegante, minimalistische Reißverschluss mit neu gestalteten, klingenförmigen Zähnen harmoniert mit verschiedensten Modestilen und lässt sich flexibel einsetzen.

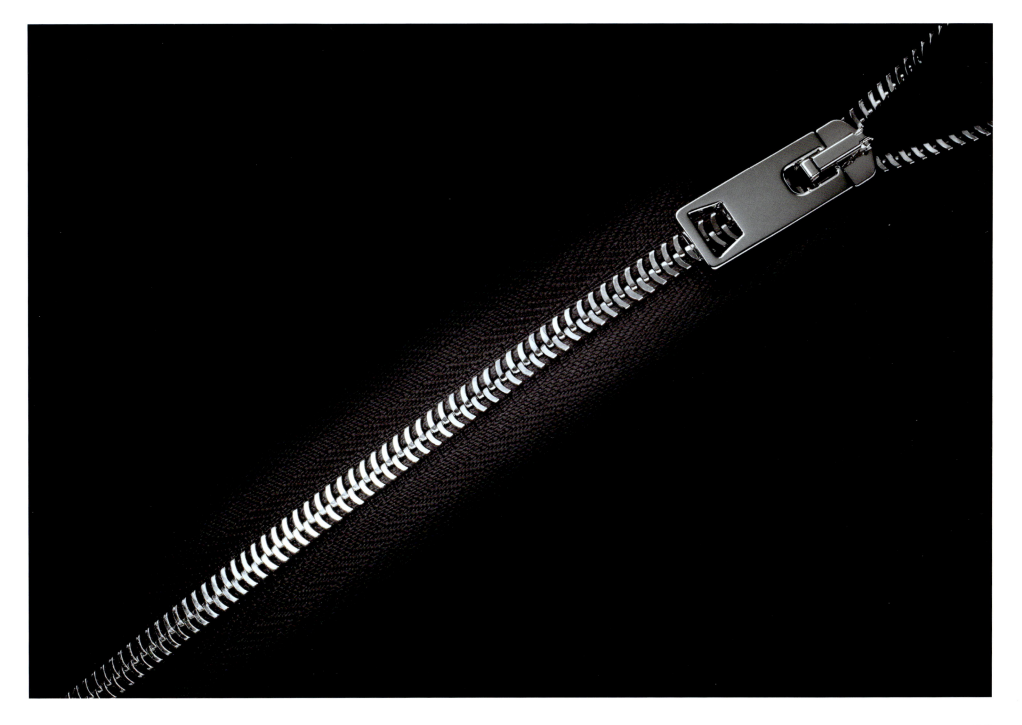

Bamboo Eyewear
Brillen aus Bambus

Manufacturer
Raymond-L
International Co., Ltd.,
Taipei, Taiwan

In-house design
Raymond Lao

Web
www.raymond-l.com

reddot award 2015
best of the best

A fascinating material

Bamboo is a material of which there is an almost unlimited supply in nature. The plant grows very quickly
and can be used in many different areas. However, to
date its use in eyewear design has been tricky as
reading glasses require a high degree of stability and
flexibility. This bamboo spectacle frame from the
Naturalism.T series is the result of a two-year development process. One of its sources of inspiration was
the advantage offered by a knotting technique traditionally used in China. The outcome was a thin frame
measuring only 2.3 mm, which makes it very comfortable to wear. The very natural appearance associated
with bamboo is combined with a high level of stability. The design uses an innovative connection and pin
technique so that the frame is very light (11 grams)
and very flexible. Both the hinges and bow springs are
made of titanium to strengthen the join between the
frame and the temples. As a result, the glasses are adjustable and easily fit different head shapes and sizes.
The bamboo grain ensures that every frame is unique.
All frames are hand-polished to give them a smooth
surface. In this way, an extremely elegant appearance
is attractively combined with the natural quality of
this sustainable raw material.

Die Faszination des Materials

Bambus ist ein in der Natur schier unendlich zur Verfügung stehendes Material. Diese Pflanze wächst sehr
schnell und sie kann in vielen Bereichen verwendet
werden. Im Brillendesign ist ihr Einsatz jedoch bislang
schwierig, da hier eine hohe Stabilität und Flexibilität
erreicht werden muss. Die Brillengestelle aus Bambus
der Serie Naturalism.T basieren auf einem zwei Jahre
dauernden Entwicklungsprozess, der sich unter anderem mit den Vorteilen einer in China traditionell verwendeten Verknüpfungstechnik befasste. Das Ergebnis
sind nur 2,3 mm dünne Gestelle, die einen sehr guten
Tragekomfort bieten. Die mit dem Bambus verbundene
Anmutung von Natürlichkeit geht dabei einher mit
einer hohen Stabilität. Durch die Gestaltung mit einer
innovativen Verbindungs- und Stecktechnik sind diese
nur elf Gramm wiegenden Brillengestelle leicht sowie
sehr flexibel. Sowohl die Gelenke als auch die Bogen-
federn sind aus Titan gefertigt, um die Verbindung zwischen der Fassung und den Bügeln zu verstärken. Die
Brillen sind dadurch justierbar und lassen sich problemlos an verschiedene Kopfformen anpassen. Jedes dieser,
durch das Material Bambus in seiner Maserung einzigartige Brillengestell wird von Hand poliert und erhält
auf diese Weise eine weiche Oberfläche. Eine überaus
elegante Anmutung verbindet sich so reizvoll mit der
Natürlichkeit dieses nachhaltigen Materials.

Statement by the jury

The use of bamboo gives these reading glasses an extremely elegant appearance and texture. Their particular shape is an expression of the ground-breaking
use of a particular knotting and pin technique. As a
result, the glasses are comfortable and very durable.
The great attention to detail, the titanium hinges and
the plastic elements are very eye-catching. These
glasses make an important contribution to the topic
of sustainability.

Begründung der Jury

Das Material Bambus verleiht dieser Brille eine außerordentlich elegante Anmutung und Textur. Ihre besondere Form ist Ausdruck einer für dieses Material
innovativ genutzten Verknüpfungs- und Stecktechnik,
wodurch sie komfortabel und langlebig ist. Bei dieser
Brille begeistert zudem die sorgfältige Ausarbeitung
der Details wie des Titanscharniers und der Kunststoff-
elemente. Sie setzt damit wichtige Signale in Fragen
des Umweltschutzes.

Designer portrait
See page 60
Siehe Seite 60

LINDBERG New 1800 Horn
Eyewear
Brillen

Manufacturer
LINDBERG, Åbyhøj, Denmark
In-house design
Web
www.lindberg.com

Working with horn, whose natural grey nuances appear on the surface of these glasses, gives them their individual colour and underlines the sublime expression of Lindberg's design philosophy. The ultra-thin front is attached to the titanium temples by screwless hinges. The extreme flexibility of the temples makes wearing the New 1800 Horn more comfortable, while the consistently slender lines of the product give it a very discrete and elegant look.

Der Werkstoff Horn, dessen natürliche graue Nuancen in der Oberfläche der Brille hervortreten und ihr dadurch eine individuelle Farbe verleihen, unterstreicht die Design-Philosophie von Lindberg. Die dünne Front ist über schraubenlose Scharniere mit Bügeln aus Titan verbunden, die sich durch hohe Flexibilität auszeichnen und zum Tragekomfort der New 1800 Horn beitragen. In ihrer Form folgt die Konstruktion einer durchgehend schlanken Linie, wodurch sie eine dezente und elegante Wirkung erhält.

Statement by the jury
An extremely narrow horn front and flexible titanium temples define these minimalist glasses. They make them light and determine their timeless appearance.

Begründung der Jury
Eine extrem dünne Hornfront und flexible Titanbügel definieren diese minimalistische Brille. Sie verleihen ihr Leichtigkeit und prägen ihr zeitloses Erscheinungsbild.

seeoo
Reading Glasses
Lesebrille

Manufacturer
Lasnik OG, seeoo, Rosental an der Kainach, Austria
In-house design
Gerald Lasnik
Web
www.seeoo.eu
Honourable Mention

The new type of stainlesssteel bridge system used for these reading glasses makes it possible to give them an exceptionally flat frame. When closed, the palladium-coated spectacles are only 4 mm wide. To keep the glasses compact throughout, the temples were made as short as possible. The silicone parts can be changed according to taste, giving the glasses a different look with each of the different colours used.

Das neuartige Stegsystem der Lesebrille aus Edelstahl erlaubt eine besonders flache Anfertigung. Im geschlossenen Zustand misst das mit Palladium beschichtete Modell in der Höhe nur 4 mm. Um die Kompaktheit der Brille durchgängig beizubehalten, wurden die Bügel auf die nötige Länge gekürzt. Die Silikonteile können je nach Geschmack ausgetauscht werden und verleihen der Brille in unterschiedlichen Farben immer wieder einen neuen Look.

Statement by the jury
The seeoo reading glasses are impressive particularly because they can be folded down extremely flat.

Begründung der Jury
Die seeoo-Lesebrille beeindruckt vor allem dadurch, dass sie sich extrem flach zusammenklappen lässt.

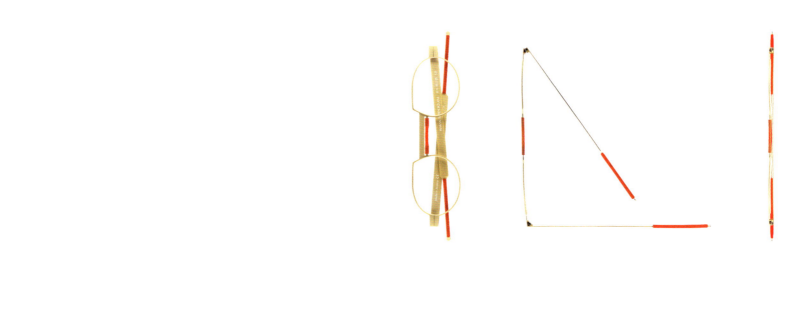

DESIGN/D3
Spectacle Frame
Brillenfassung

Manufacturer
MARKUS T, Markus Temming GmbH,
Gütersloh, Germany
In-house design
Web
www.markus-t.com

These glasses are manufactured in Germany by a family-owned company. Minimalist in appearance, they are delicately made with discreet technical detailing. A titanium wire with a diameter of 1 millimetre gives each of the different styles of glasses – Full Rim, Top Rim or Double Bridge – their individual character. The temples do not have a metal core so that opticians can easily shorten them and adapt the glasses to the wearer. A screwless hinge in the temple and TMi, a specially developed synthetic material, also ensure great stability.

Statement by the jury
The light, purist design of the D3 collection glasses has produced captivatingly delicate frames and innovative solutions for small details.

Diese Brille wird in Deutschland in eigener Manufaktur hergestellt. Ihr Äußeres reduziert sich auf eine filigrane Ausarbeitung mit dezenten technischen Details. Ein Draht aus Titan von 1 mm Durchmesser verleiht der Brille ihren jeweiligen Style: Full Rim, Top Rim oder Double Bridge. Die Bügel des Modells kommen ohne Metallkern aus und können vom Optiker gekürzt und so an den Träger angepasst werden. Neben einem schraubenlosen Scharnier im Bügel gewährleistet der eigens entwickelte Kunststoff TMi eine hohe Formbeständigkeit.

Begründung der Jury
Die leichten, puristisch gestalteten Brillen der D3-Serie begeistern mit ihren überaus filigranen Gestellen und innovativen Detaillösungen.

Titan One
Ophthalmic Glasses
Optische Brille

Manufacturer
Silhouette International, Linz, Austria
In-house design
Web
www.silhouette.com

With is natural lines, the limited edition Signature Collection creates a harmonious unity between wearer and spectacles. Both the temples and the bridge are made of a single piece of high-tech titanium and provide the spectacles with harmonious contours. The design of the open temples unostentatiously rounds off the elegant concept of Titan One.

Statement by the jury
The innovative frame construction made from a single piece of titanium wire gives Titan One a particular lightness and a minimalist appearance.

Die limitierte Signature Collection bringt durch ihre natürliche Linienführung Träger und Brille in eine harmonische Einheit. Die Bügel und der Steg sind aus einem einzigen Stück Hightech-Titan gefertigt und bilden eine harmonische Kontur für die Gläser. Die offen gestalteten Bügelenden formen einen dezenten Abschluss, der das elegante Konzept der Titan One passend abrundet.

Begründung der Jury
Die innovative Rahmenkonstruktion aus einem einzigen Stück Titandraht verleiht der Titan One besondere Leichtigkeit und eine minimalistische Anmutung.

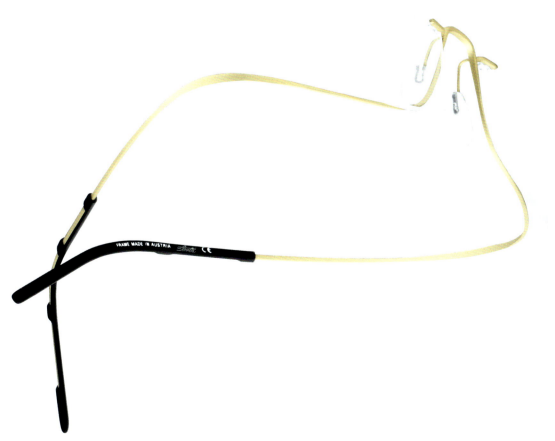

Geometric Collection Diamond Shades
Sunglasses
Sonnenbrille

Manufacturer
Eyewear Solutions GmbH, Hartberg, Austria
Design
13&9 Design, Graz, Austria
Martin Lasnik, Rosental, Austria
Web
www.13and9design.com

These sunglasses for men and women take their inspiration from architecture, strictly using converging lines into an unified concept. The precise manufacture and the exact proportions between angles and edges give the sunglasses a high degree of symmetry. The cellulose acetate frames undergo a complex manufacturing process in which they are first milled, then sanded and sandblasted, and lastly polished to achieve the final surface finish.

Die Sonnenbrille für Damen und Herren ist inspiriert von der Architektur und verbindet ausschließlich in Fluchtpunkten zusammenlaufende Linien zu einem einheitlichen Konzept. In ihrer präzisen Anfertigung und den ausgewogenen Proportionen zwischen den Winkeln und Kanten erzielt das Modell vor allem eine genaue Symmetrie. Der Rahmen aus Celluloseazetat wurde in einem aufwendigen Prozess von einer Fräsmaschine bearbeitet und erhält nach anschließendem Schleifen und Sandstrahlen durch eine Politur seine finale Oberflächenanmutung.

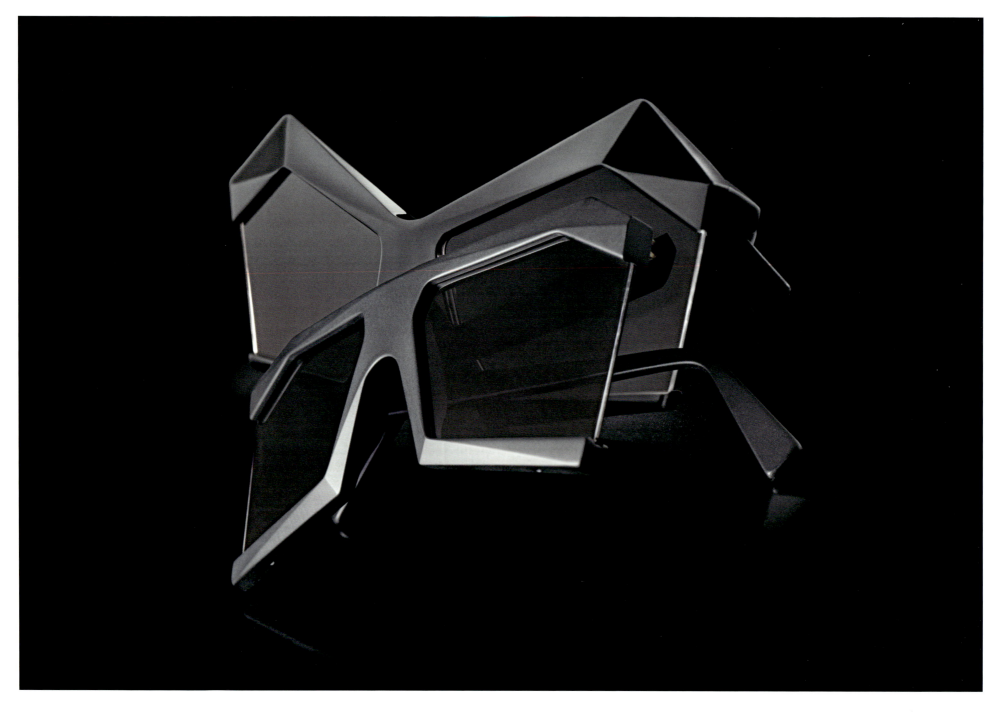

Glass & Glam

Sunglasses
Sonnenbrille

Manufacturer
Glass & Glam, Brussels, Belgium
Design
Rusak (Serge Rusak), Liège, Belgium
Web
www.glassandglam.com
www.rusak.be

By using silicone these sunglasses
adjust to fit the wearer's face and are
extremely durable in use. Glass & Glam
is both robust and flexible and can be
used as sunglasses and prescription
spectacles. The hinges and bridge fuse
with the frame so that they seem to
come from one and the same mould. An
accompanying protective case fits into
a vehicle's cup holder and also has an
extra feature, which enables it to stand
on flat surfaces.

Statement by the jury
With a flexible silicone frame, Glass &
Glam presents Mod. 52, a reinterpretation of classic sunglasses that is bound
to be an eye-catching accessory.

Durch die Verwendung von Silikon passt
sich die Sonnenbrille beim Tragen dem
Gesicht an und erweist sich im Gebrauch
als außerordentlich belastbar. Glass & Glam
ist robust und flexibel zugleich und kann
als Sonnen- oder Korrekturbrille eingesetzt
werden. Scharniere und Bügel verschmelzen mit dem Rahmen, sodass sie wie aus
einem Guss erscheint. Eine ergänzende
Schutzhülle ermöglicht das Befestigen im
Auto und enthält außerdem eine Halterung zum sicheren Abstellen auf geraden
Flächen.

Begründung der Jury
Die Glass & Glam interpretiert mit Mod. 52
klassische Sonnenbrillen mit ihrem Gestell
aus flexiblem Silikon neu und wird dadurch
zum Blickfang.

KO–211
Eyewear
Brillen

Manufacturer
Sunreeve Co., Ltd., Fukui, Japan
Design
Ken Okuyama Design Co., Ltd., Ken Okuyama,
Tokyo, Japan
Web
www.kenokuyamaeyes.com
www.kenokuyamadesign.com

KO-211 is a half-rim frame model with a truss structure shape which reflects the image of buildings and car parts for Okuyama's taste. The stylish glasses make a totally different impression from the front view and the top view. The thick part of the front is fabricated by high-tech CNC milling instead of metal pressing.

KO-211 ist ein Halbrand-Brillenmodell mit einer Fachwerk-Konstruktion, in der sich auch Okuyamas Stilgefühl für Gebäudearchitektur und Autoteile widerspiegelt. Von oben betrachtet wirkt die elegante Brille völlig anders als von der Vorderseite. Die präzise und komplexe Struktur des Mittelteils konnte durch den Einsatz von Hightech-CNC-Fräsmaschinen anstelle von Metallpressen erreicht werden.

Statement by the jury
Ultra-modern technologies come together to achieve this design and create eyewear of great style appeal.

Begründung der Jury
Modernste Technologien fließen in diesem Entwurf zusammen und überzeugen mit viel Stil.

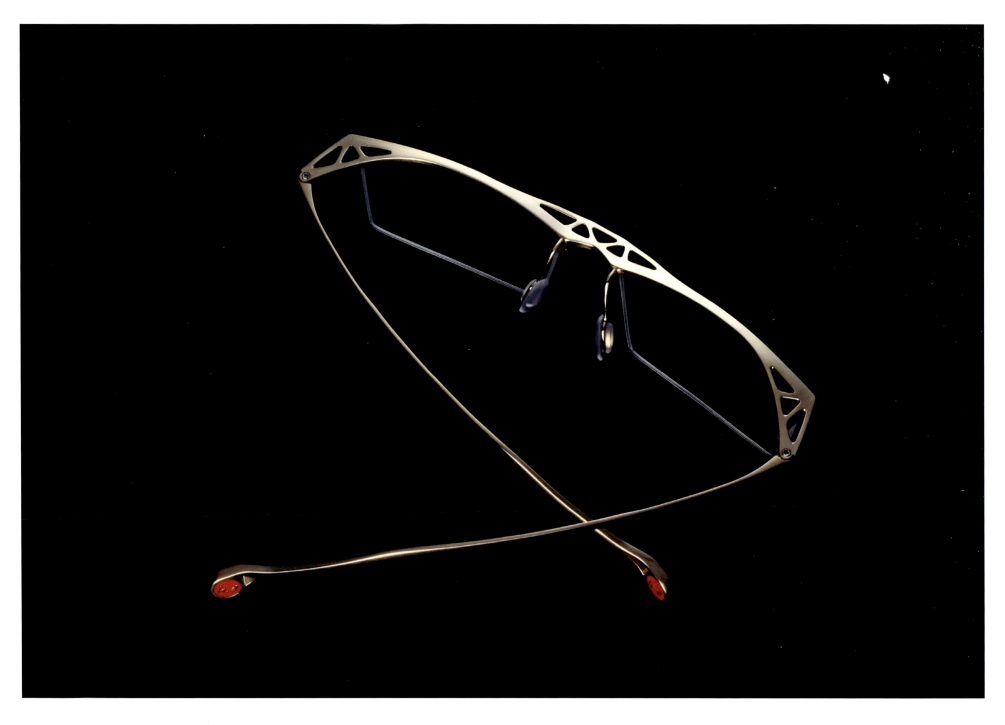

KO-212
Eyewear
Brillen

Manufacturer
Sunreeve Co., Ltd., Fukui, Japan
Design
Ken Okuyama Design Co., Ltd., Ken Okuyama, Tokyo, Japan
Web
www.kenokuyamaeyes.com
www.kenokuyamadesign.com

KO-212 is a stylish full-rim frame model with a truss structure shape which reflects the image of buildings and car parts for Okuyama's taste. It's in a neat look from the front view, thanks to the diagonally cut side line, and provides a totally different impression from the top view. As for colour variations, matt finish colours have been chosen for the chic look.

KO-212 ist ein modisches Vollrand-Brillen-modell mit einer Fachwerk-Konstruktion, die Okuyamas ästhetisches Empfinden für Gebäude und Autoteile aufgreift. Dank der diagonal geschnittenen Seitenlinie macht das Modell von vorne einen eleganten Eindruck. Von oben betrachtet wirkt das stilvolle Modell völlig anders als von der Vorderseite. Erhältlich ist die Fassung in raffinierten Mattfarben.

Statement by the jury
Precise geometric milling grooves on the upper edge of the glasses give these otherwise purist spectacles a slightly futuristic appearance.

Begründung der Jury
Präzise geometrische Ausfräsungen am oberen Brillenrand verleihen dieser ansons-ten puristisch gestalteten Brille eine leicht futuristische Anmutung.

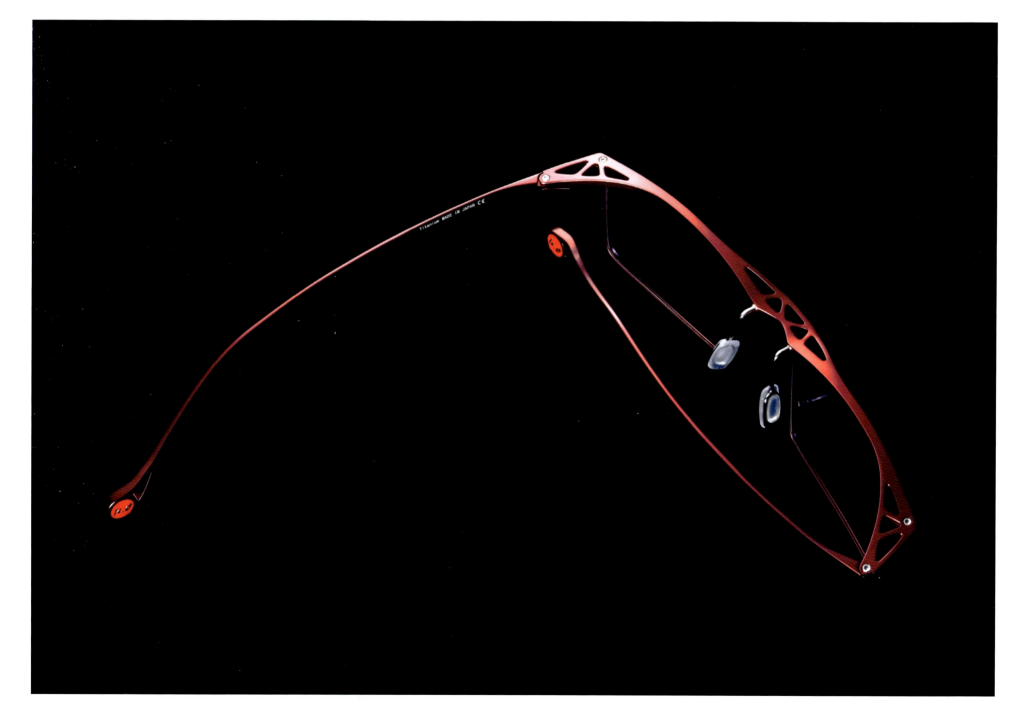

Marksman Twist

27" Automatic Umbrella
27"-Automatikschirm

Manufacturer
PF Concept, Roelofarendsveen, Netherlands
In-house design
Marijn Molenaar
Web
www.pfconcept.com

This umbrella gives an everyday product a new twist. Instead of pushing a button as usual, it is opened by twisting the handle, thus revealing a nifty invention. The shaft and ribs are in fibreglass, which makes the umbrella light and robust in use. The spokes are tipped with protective ends to minimise risk of injury to others in crowded places.

Mit diesem Produkt wurde ein klassischer Alltagsgegenstand neu interpretiert: Statt wie üblich auf Knopfdruck entfaltet sich der Schirm beim Drehen des Griffes und entpuppt sich so als raffinierte Erfindung. Sowohl Schaft als auch Gestänge sind aus Glasfaser angefertigt und trotz ihrer Robustheit leicht. Die Speichen haben Schutzenden, um die Gefahr zu minimieren, andere an belebten Plätzen zu verletzen.

Statement by the jury
The Marksman Twist scores highly with its original opening solution that turns an everyday movement into a new experience.

Begründung der Jury
Der Marksman Twist punktet mit seiner originellen Aufspannlösung, die dazu beiträgt, dass ein alltäglicher Handgriff neu erlebt wird.

SwissCard Nailcare
Swiss Army Knife
Schweizer Taschenmesser

Manufacturer
Victorinox AG, Ibach (Schwyz), Switzerland
Design
Painsith Consult (Hermann K. Painsith), Klagenfurt, Austria
Web
www.victorinox.com
www.paicon.at

The SwissCard Nailcare boasts the usual attributes of a Swiss Army Knife as well as an additional glass nail file. Its grains have not been applied to the nail file's body in the normal way, but have instead been incorporated into the glass as fine dust pores and therefore the nail file never loses effectiveness. It is easily cleaned with warm water and fits neatly into the purse, handbag or jacket pocket.

Die SwissCard Nailcare bietet neben den gängigen Funktionen eines Schweizer Taschenmessers auch eine Glasnagelfeile. Ihre Schleifkörner sind nicht wie gewohnt mechanisch auf dem Feilenkörper aufgebracht, sondern wurden als staubfeine Poren in das Glas eingearbeitet, sodass die Feile weder an Kraft noch Schliff verliert. Sie lässt sich unter warmem Wasser einfach reinigen und anschließend gut im Geldbeutel, der Hand- oder Jackentasche verstauen.

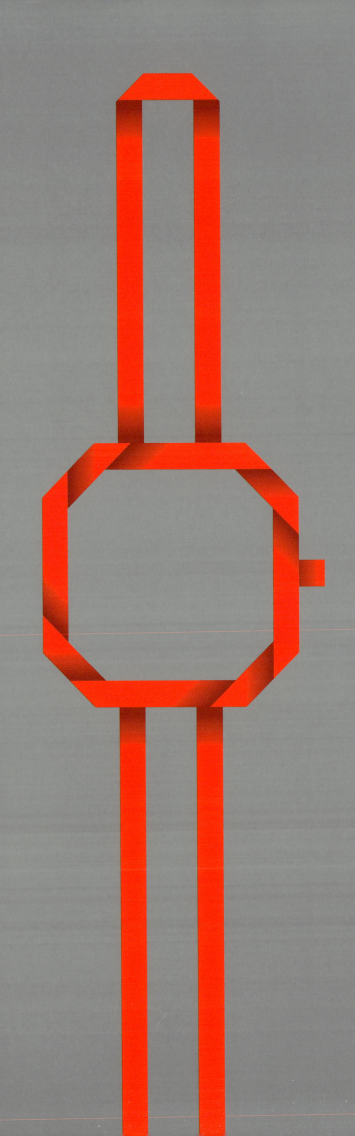

Watches and jewellery
Uhren und Schmuck

Apple Watch
Smartwatch

Manufacturer
Apple, Cupertino, USA

In-house design
Apple

Web
www.apple.com

reddot award 2015
best of the best

Experiencing time

The invention of pocket watches and later wrist-watches gave everybody the possibility to check on the time anywhere and thus plan activities accordingly. The Apple Watch now adds a further interpretation to the measuring of time. Coming with a design language that is inspired by traditional watchmaking, it features an innovative interface that enables users to customise the watch by allowing them to choose between different, more personal watch faces. Users can thus configure the interface in the style of a classic chronograph, for instance, or opt for a more playful display such as a fluttering butterfly. Both the interface and the housing of the Apple Watch have been created and perfectly matched specifically for such a small device. A digital crown makes navigating the watch highly intuitive and amazes users with its versatility and many possibilities. The innovative Force Touch display is extremely sensitive and recognises even the lightest touch. Another exciting experience is how it senses the difference between a tap and a press, as the Taptic Engine enables a new vocabulary of both acoustic and sensory notifications and responses. Offering a myriad of new interactive possibilities, the Apple Watch also allows extensive customising such as the material used for the wristband or the enclosure, ranging from stainless steel to a model in gold – for a new way of experiencing time.

Erlebnis Zeit

Seit der Erfindung der Taschenuhren und später auch der Armbanduhren hatte der Mensch die Möglichkeit, die Uhrzeit selbst zu ermitteln und danach zu handeln. Die Apple Watch ist nun Ausdruck einer weitergehenden Interpretation der Zeitmessung. In ihrer Formensprache angelehnt an die traditionelle Uhrengestaltung, bietet sie dem Nutzer mit einem innovativen Interface die Möglichkeit einer individuellen Darstellung, denn ihr Zifferblatt lässt sich in vielfältigster Weise variieren. So kann der Nutzer das Interface im Stil eines klassischen Chronographen konfigurieren oder er entscheidet sich für eine verspielte Oberfläche, über die ein Schmetterling flattert. Das Interface wie auch die äußere Gestaltung der Apple Watch wurden den speziellen Anforderungen eines solch kleinen Gerätes perfekt angepasst. Die Navigation über eine digitale Krone geschieht intuitiv, wobei der Nutzer verblüfft ist von ihren vielseitigen Möglichkeiten. Überaus sensibel reagiert auch das innovative Force Touch-Display auf leichte Berührungen. Ein spannendes Erlebnis ist es zudem, wie dieses sofort den Unterschied zwischen den Aktionen Tippen oder Drücken erkennt, wobei die Taptic Engine ein neues Vokabular an akustischen und sensorischen Mitteilungen und Reaktionen bietet. Erlaubt die Apple Watch auf diese Weise viele neue interaktive Möglichkeiten, lässt sich auch ihre Ausstattung wie etwa das Material des Armbandes oder des Gehäuses von Edelstahl bis hin zu einer Version in Gold vielseitig dem jeweiligen Stilempfinden anpassen – die Zeit wird damit neu erfahren.

Statement by the jury

The Apple Watch convinces with a new concept of a timepiece that distinguishes itself fundamentally from other smartwatches. In a fascinating manner it bridges the gap between the worlds of digital computer and traditional watch design. With its perfect design down to the last detail, the watch delivers a high degree of precision and the experience of an innovative interface that offers users a myriad of different customisable configurations.

Begründung der Jury

Die Apple Watch begeistert mit einem neuen Konzept für eine Uhr, durch das sie sich grundlegend von anderen Smartwatches unterscheidet. Auf faszinierende Art und Weise schlägt sie eine Brücke zwischen der digitalen Computer- und der traditionell verhafteten Uhrenwelt. Dem Nutzer bietet diese in all ihren Details perfekt gestaltete Uhr ein Höchstmaß an Präzision und das Erleben eines innovativen Interfaces, das er individuell äußerst vielfältig selbst konfigurieren kann.

SmartBandTalk, SmartWatch 3

Manufacturer
Sony Mobile Communications Inc.,
Tokyo, Japan
In-house design
Web
www.sonymobile.com

The SmartBandTalk is equipped with an E-paper display, which makes it easy to read by daylight and ensures it is not so bright that it will disturb you at night. The case is available in black or white and can be changed within seconds without requiring any tools. SmartWatch 3 not only has a finely textured exterior, but offers users a colour choice between black, white, pink or lime and also comes in metal or leather to suit the wearer's taste.

Statement by the jury
The large choice of different combinations turns the "bracelet and watch" duo into an extremely versatile product with its own original identity.

Das SmartBandTalk hat ein E-Paper-Display, das bei Sonnenlicht gut benutzbar ist und auch in der Nacht nicht mit Helligkeit stört. Das Gehäuse ist sowohl in Schwarz als auch in Weiß erhältlich und kann innerhalb von Sekunden und ohne Werkzeug gewechselt werden. In Ergänzung zur fein strukturierten Außenseite passt sich Smart-Watch 3 in den zur Verfügung stehenden Farben Schwarz, Weiß, Pink oder Limette, nach Wunsch auch in Metall oder Leder, an den Geschmack ihres Trägers an.

Begründung der Jury
Die große Auswahl an Kombinationsmöglichkeiten macht das Duett „Armband und Uhr" zu einem sehr vielseitigen Produkt mit origineller Persönlichkeit.

Horological Machine No. 6 "Space Pirate"
Watch
Uhr

Manufacturer
MB&F, Geneva, Switzerland

In-house design
MB&F

Design
Through the Looking Glass Sàrl
(Eric Giroud),
Geneva, Switzerland

Web
www.mbandf.com

reddot award 2015
best of the best

Mysterious worlds

The particular fascination of science fiction resides in the stories of unknown worlds that it stages. The design of the Space Pirate projects just such a scenario. Its organically shaped case and the four indicated "eyes" make it appear almost like a friendly alien from outer space. Having a case made of shimmering titanium alloy that is also used in aircraft design, the watch can withstand temperatures of up to 400 degrees Celsius. Despite this, the case is outstandingly light and offers good wearing comfort thanks to its rounded edge design. The watch's exoskeleton is a fascinating element that signifies a bold reinterpretation of familiar watch design. Two of the four semi-spherical indicators in the corners display the time in hours and minutes – the other two are related to the automatic winding mechanism of the watch to reduce wear and tear. Another exciting detail is the 60-second flying tourbillon located underneath the central dome. The system is so highly sensitive that two semi-spherical, titanium protective covers are used to block harmful UV light. Furthermore, the watch showcases a sophisticated and incredibly spectacular clockwork: the real glass panel reveals how the delicate automatic winding system moves with the help of its axe-shaped winding rotor. It is details such as these that time and again transport owners of the Space Pirate watch into new, mysterious worlds.

Geheimnisvolle Welten

Science-Fiction hat eine besondere Faszination darin, dass diese Geschichten unbekannte Welten inszenieren. Die Gestaltung der Space Pirate zeichnet solch ein Szenario nach. Mit einem organisch geformten Gehäuse und vier angedeuteten gewölbten „Augen" erinnert sie an ein freundliches Wesen aus dem Weltall. Die schimmernde Legierung ihres Titanpanzers wird auch in der Luftfahrt verwendet, weshalb diese Uhr Temperaturen bis 400 Grad Celsius aushalten kann. Dennoch ist ihr Gehäuse ungewöhnlich leicht und bietet mit abgerundet gestalteten Ecken einen guten Tragekomfort. Faszinierend ist an dieser Uhr ihr Aufbau und die damit einhergehende Neudefinition bekannter Uhrenelemente. So zeigen an den Ecken des Gehäuses zwei der vier runden Displays die Zeit in Stunden und Minuten an – zwei weitere sind mit dem automatischen Aufzug verbunden, um den Verschleiß zu verhindern. Ein spannendes Detail ist ein unter der zentralen Kuppel angeordnetes fliegendes Sekundentourbillon. Das System ist so empfindlich, dass es durch zwei halbkugelförmige Titanabdeckungen vor schädlicher Sonneneinstrahlung geschützt wird. Geradezu spektakulär ist auch das hochentwickelte Uhrwerk: Durch einen Glasboden hindurch sieht man, wie sich das filigrane automatische Aufzugssystem mittels eines Aufzugsrotors in Form einer Streitaxt bewegt. Mit solchen Details entführt die Space Pirate ihren Besitzer so immer wieder in neue, geheimnisvolle Welten.

Statement by the jury

Inspired by the world of science fiction, this wristwatch stages its own cosmos. It dissolves the boundaries of classical wristwatch design and advances it through an exciting futuristic approach. The distinctive shape of the clockwork simply fascinates and its mechanism mesmerises. Featuring the appearance of a friendly being from another world, this wristwatch seems to tell its own tale – a world of adventure for the beholder.

Begründung der Jury

Angelehnt an die Welt der Science-Fiction inszeniert diese Armbanduhr ihren eigenen Kosmos. Klassische Zeitanzeiger werden in ihrer Bedeutung aufgelöst und das Uhrendesign auf spannende Weise futuristisch weiterentwickelt. Der Betrachter ist fasziniert von der besonderen Form des Uhrwerkes, welches auch in seiner Mechanik verblüfft. Mit der Anmutung eines freundlichen Wesens aus einer anderen Sphäre scheint diese Uhr ihre eigene Geschichte zu erzählen – ein Abenteuer für den Betrachter.

Designer portrait
See page 62
Siehe Seite 62

Horological Machine No. 5 CarbonMacrolon
Watch
Uhr

Manufacturer
MB&F, Geneva, Switzerland
In-house design
Maximilian Büsser
Design
Through the Looking Glass Sàrl (Eric Giroud), Geneva, Switzerland
Web
www.mbandf.com
Honourable Mention

The time display of this watch consists of two dials that rotate in opposite directions with jumping hours and comes equipped with a prism that has an integrated magnifier to improve legibility. This futuristic model also has an in-built mechanical automatic movement protected by a waterproof inner case. To improve its resistance to scratches, the watch is made of a dense polycarbonate resin reinforced by carbon nanotubes.

Die Zeitanzeige der Uhr besteht aus zwei gegenläufig drehenden Ziffernscheiben mit springender Stunde und verfügt über ein Prisma mit integrierter Lupe. Neben einer dadurch begünstigten Ablesbarkeit trägt das futuristische Modell ein mechanisches Automatikwerk in sich, das durch ein wasserdichtes Innengehäuse geschützt wird. Um Kratzer zu vermeiden, wurde dichtes Polycarbonatharz verwendet, das durch Carbon-Nanoröhren zusätzlich verstärkt wird.

Porsche Design Timepiece No. 1
Chronograph

Manufacturer
Porsche Design Group, Ludwigsburg, Germany
In-house design
Porsche Design Studio, Zell am See, Austria
Web
www.porsche-design.com

Timepiece No. 1 is the first to be developed completely in-house. Its design is based on the same principles as Professor Ferdinand Alexander Porsche's legendary black Chronograph I. The aesthetics of the dial further reflect the purism of Porsche Design. The hands and bar indices are filled with white luminescent material so the large dial is perfectly readable at all times. The model is equipped with a mechanical ETA Valjoux 7750 chronograph movement that has a power reserve of 48 hours. The blackened sapphire crystal case back shows the inner workings of the energy-optimised Porsche Design rotor.

Timepiece No. 1 ist die erste komplett eigens entwickelte Uhr nach denselben Ansprüchen, die Professor Ferdinand Alexander Porsche bereits an den legendären mattschwarzen Chronograph I stellte. Die Ästhetik des Zifferblatts der Timepiece No. 1 spiegelt den Purismus von Porsche Design wider. Die Leuchtzeiger und Stabindizes sind mit weißer Leuchtmasse gefüllt, sodass das große Zifferblatt jederzeit bestmöglich ablesbar ist. Technisch ist das Modell mit einem mechanischen Chronographenwerk ETA Valjoux 7750 ausgestattet, das eine Gangreserve von 48 Stunden aufweist. Der geschwärzte Saphirglasboden ermöglicht einen freien Blick auf den energieoptimierten Porsche Design-Rotor.

Statement by the jury
With a focus on a high functionality, this timepiece is nonetheless memorable for its purist design.

Begründung der Jury
Mit einem Fokus auf hoher Gebrauchstauglichkeit bleibt der Zeitmesser dank seiner puristischen Gestaltung in Erinnerung.

Pontos Date
Watch
Uhr

Manufacturer
Maurice Lacroix SA, Saignelégier, Switzerland
In-house design
Sandro Reginelli
Web
www.mauricelacroix.com

The sandblasted black case of this watch has been coated with PVD and has a semi-matt finish. The facets of the 40 mm case break the light in an unusual way, while the short case lugs sharply taper downwards. This makes the watch significantly more comfortable to wear as it gently adapts to the wrist. The anti-reflective coating on both sides of the sapphire crystal glass of the watch ensures improved legibility.

Statement by the jury
The impressive look of the watch is due to ingenious little details that are so well coordinated that they produce a perfect appearance.

Das sandgestrahlte Gehäuse in Schwarz wurde mit PVD beschichtet und seidenmatt ausgeführt. Die Facetten des Gehäuses von 40 mm Durchmesser brechen das Licht auf besondere Art, während die kurzen Bandanstöße sich nach unten verjüngen. Dies erhöht den Komfort beim Tragen maßgeblich, da sie sich schonend an das Handgelenk anschmiegen. Das beidseitig entspiegelte Saphirglas der Uhr gewährleistet ein komfortables Ablesen der Zeit.

Begründung der Jury
Der beeindruckende Anblick der Uhr wird durch ausgeklügelte Details möglich, die gut aufeinander abgestimmt ihre makellose Erscheinung bedingen.

Pontos S Supercharged
Watch
Uhr

Manufacturer
Maurice Lacroix SA, Saignelégier, Switzerland
In-house design
Sandro Reginelli
Web
www.mauricelacroix.com

The brushed, polished case of this timepiece is made of stainless steel and measures 48 mm overall. Red and white elements contrast with the black dial and therefore make it very easy to read. Both the hour and minute hands are diamond cut, luminous and colour-coordinated with the patented, rotational inner bezel. The chronograph function includes a central second hand and a 30-minute and 12-hour counter.

Statement by the jury
The very detailed design of this watch is appealing, not least because of its smart, sporty character.

Das gebürstete und polierte Gehäuse ist aus Edelstahl, sein Durchmesser beträgt 48 mm. Rote und weiße Elemente stehen im Kontrast zum schwarzen Zifferblatt, das dadurch sehr leicht abgelesen werden kann. Sowohl der Stunden- als auch der Minutenzeiger ist diamantgeschliffen, mit Leuchtstoff beschichtet und farblich auf die patentierte, drehbare Lünette im Inneren abgestimmt. In der Chronographenfunktion sind ein zentraler Sekundenanzeiger sowie ein 30-Minuten- und 12-Stunden-Zähler enthalten.

Begründung der Jury
Die detailliert durchdachte Gestaltung der Uhr überzeugt nicht zuletzt auch mit ihrem sportlich-schicken Charakter.

The Bradley: A Tactile Timepiece
The Bradley: Ein fühlbarer Zeitmesser
Wristwatch
Armbanduhr

Manufacturer
Eone Timepieces Inc, Washington, D.C., USA
In-house design
Web
www.eone-time.com

The Bradley was designed so that people with visual impairment can also use it. The time can be told by touching two ball bearings: one gives the minutes, the other the hours. The movement of the ball bearings on the watch face is controlled by magnets. A gentle shaking of the wrist will restore the ball bearings to the position corresponding to the correct time.

The Bradley wurde so konstruiert, dass auch Menschen mit Sehbehinderung sie nutzen können: Die Zeit kann durch die Berührung zweier Kugeln abgelesen werden, von denen die eine die Minuten und die andere die Stunden wiedergibt. Die Kugeln sind mit Magneten gekoppelt, die deren Bewegung auf dem Zifferblatt ermöglichen. Sie bringen sie außerdem durch sanftes Schütteln des Handgelenks immer wieder zur korrekten Zeit zurück.

Metro Datum Gangreserve
Wristwatch
Armbanduhr

Manufacturer
NOMOS Glashütte/SA, Roland Schwertner KG,
Glashütte, Germany
Design
Berlinerblau GmbH (Mark Braun),
Berlin, Germany
Web
www.nomos-glashuette.com

This understated, hand-wound watch by NOMOS Glashütte does entirely without flourishes and indicates the time with elegantly tapered hands. The watch face not only displays the date, but also the power reserve and minute indexes. The slightly curved sapphire crystal glass of the watch protects the DUW 4401 manually wound movement ticking away inside. This caliber is equipped with the in-house NOMOS swing system.

Statement by the jury
The precise remodelling of this watch has given it a cosmopolitan air, presenting excellent features in a refreshing new way.

Die dezent gestaltete Handaufzugsuhr von NOMOS Glashütte kommt ganz ohne Schnörkel aus und zeigt die Zeit mit ausgesprochen feinen Zeigern an. Auf ihrem Zifferblatt befindet sich neben der Minuterie das Datum sowie die Gangreserveanzeige. In ihrem Inneren tickt das Handaufzugswerk DUW 4401 mit hauseigenem NOMOS-Swing-System, zu sehen durch den Boden aus Saphirglas.

Begründung der Jury
Die präzise Ausarbeitung in Verbindung mit einem urbanen Touch macht diese Uhr zu einem aparten Modell mit attraktiven Eigenschaften.

Compass wristwatch
Kompass-Uhr
Wristwatch
Armbanduhr

Manufacturer
Shenzhen Ciga Design Co., Ltd.,
Shenzhen, China
In-house design
Jianmin Zhang, Xin Jiang
Web
www.ciga.com.cn
Honourable Mention

The Compass wristwatch is particularly suitable for an outdoor life as it is shatterproof and watertight. The red second hand has two functions: as a compass needle it points north and as a second hand it shows the precise time in seconds. The colour accents of the hour and minute hands make them stand out from the black face and leather strap so that they are easy to read. A silver-coloured case completes the appearance of the practical sports accessory.

Statement by the jury
The usefulness of the Compass wristwatch is due to the variety of functions it offers; something that will not only appeal to sport fans.

Diese Kompass-Uhr eignet sich aufgrund ihrer bruchsicheren und wasserfesten Konstruktion ganz besonders für den Outdoor-Bereich. Der rote Zeiger weist mit seiner Doppelfunktion in Richtung Norden und zeigt ebenso die Sekunden an. Auch die farblich akzentuierten Stundenzeiger heben sich vom in Schwarz gehaltenen Zifferblatt und Lederband ab und sind somit gut erkennbar. Eine silberfarbene Fassung rundet das Gesamtbild des praktischen Sport-Accessoires ab.

Begründung der Jury
Die Kompass-Uhr überzeugt mit ihrer Vielfalt an Gebrauchsmöglichkeiten und bereitet damit nicht nur Sportbegeisterten viel Freude.

Jacob Jensen Strata series
Watch
Uhr

Manufacturer
S. Weisz Uurwerken BV, Amstelveen, Netherlands
Design
Jacob Jensen Design, Jacob Jensen Brand Products,
Højslev, Denmark
Web
www.weiszwatches.com
www.jacobjensen.com

Designer Timothy Jacob Jensen was inspired by Nature when creating this watch. Its elegant lines recall the rock and earth strata that can be found in certain parts of Denmark. The circular case is made of stainless steel and is framed by a clearly visible leather band. Protected by mineral crystal glass, the time can be told from a simple silvery watchface. The size of the case is 36 mm for the ladies' watch and 41 mm for the men's version.

Designer Timothy Jacob Jensen ließ sich beim Entwurf dieser Uhr von der Natur inspirieren. Die eleganten Linien erinnern an Lagen von Stein und Erde, wie sie in Teilen Dänemarks zu finden sind. Das kreisförmige Gehäuse besteht aus Edelstahl und wird von einem deutlich sichtbaren Band aus Leder gehalten. Unter einem Schutz aus Mineralglas ist die Zeit auf einem schlichten silberfarbenen Zifferblatt zu erkennen. Der Durchmesser des Gehäuses beträgt bei der Herrenuhr 41 und bei der Damenuhr 36 mm.

Statement by the jury
The design of this watch, inspired by Nature, leaves a lasting impression due to its distinctive style.

Begründung der Jury
Die sich an Formen der Natur orientierende Gestaltung dieser Uhr hinterlässt durch ihren prägnanten Auftritt einen bleibenden Eindruck.

Skeleton Pure Water
Wristwatch
Armbanduhr

Manufacturer
Armin Strom AG, Biel, Switzerland
In-house design
Claude Greisler
Web
www.arminstrom.com

ARMIN STROM's Skeleton Pure celebrates the mechanical architecture of a movement by highlighting the layered construction of the individual components. The skeletonised hand winding calibre ARM09-S is from the outset conceived both as mechanical construction and as a design entity that demonstrates a perfect balance between the aesthetic and the functional. The blue 3D PVD coating enhances the dimensions and lends a new definition to the movement construction. When wound, the crown wheels, innovatively applied on the dial side, exhibit impressive animation.

Die Skeleton Pure von ARMIN STROM zelebriert die mechanische Architektur eines Uhrwerks durch die Betonung der geschichteten Konstruktion der einzelnen Werksteile. Das skelettierte Handaufzugskaliber ARM09-S wurde von Anfang an sowohl als mechanische Konstruktion wie auch als Designeinheit entwickelt und veranschaulicht so ein ideales Gleichgewicht von Ästhetik und Funktion. Die blaue 3D-PVD-Beschichtung betont das Format und definiert die gesamte Werkskonstruktion neu. Innovativ zeigen sich die zifferblattseitig angelegten Aufzugsräder, welche dem Träger beim Aufziehen eine eindrucksvolle Darbietung bescheren.

MSH01
Wristwatch Movement
Uhrwerk

Manufacturer
MeisterSinger GmbH & Co. KG,
Münster, Germany
In-house design
Manfred Brassler
Web
www.meistersinger.de
www.meistersinger.net

The new MeisterSinger MSH01 manual wind mechanical movement is the basis for a new generation of single-hand watches crafted by the Münster-based manufacturer. Within three years, the team at MeisterSinger's long-standing cooperation partner in Switzerland designed a calibre from scratch that is aesthetically unmistakable. The MSH01 is very robust, and its design focuses on the essentials. A long power reserve, a uniquely designed movement and its design potential for subsequent watch concepts were at the forefront of the development.

Statement by the jury
The skilled interaction of the individual components of this watch in both form and function give it a distinctive, unified appearance.

Das neue Handaufzugswerk MSH01 von MeisterSinger ist die Basis für eine neue Produktgeneration des Münsteraner Herstellers von Einzeigeruhren. Innerhalb von drei Jahren konstruierte das Team bei MeisterSingers langjährigem Kooperationspartner in der Schweiz von Grund auf ein Kaliber mit ästhetischer Eigenständigkeit. Das MSH01 ist sehr robust, die Konstruktion auf das Wesentliche konzentriert. Im Vordergrund der Entwicklung standen eine große Kraftreserve, einzigartige Werkgestaltung und das Potential der Werkkonstruktion für weitere Uhrenkonzepte.

Begründung der Jury
Das in Form und Funktion gekonnte Zusammenspiel der einzelnen Bestandteile macht das markante, einheitliche Erscheinungsbild dieser Uhr aus.

HIRSCH Performance Collection
Bracelets / Straps for Watches
Armbänder für Uhren

Manufacturer
HIRSCH Armbänder GmbH,
Klagenfurt, Austria
In-house design
Web
www.hirschag.com

This bracelet combines the materials it uses – leather and caoutchouc – to create a harmonious entity. Available in seven different models, the bracelets of this collection are all flexible, but at the same time also very robust and resistant to harmful external influences. In the HIRSCH Performance Collection, the marriage of the internal and external functions with the appearance of the bracelet not only make it useful, but also an extremely chic accessory for daily use.

Statement by the jury
Two independent materials have here been combined to create a product of outstanding quality through their complementary characteristics.

Die verwendeten Materialien Leder und Kautschuk wurden zusammengeführt und bilden in diesem Armband eine Einheit. Jedes für sich ist als Komponente einer 7-teiligen Produktpalette flexibel und zugleich sehr widerstandsfähig gegen schädliche Einflüsse. Die HIRSCH Performance Collection verkörpert funktionell wie optisch eine Symbiose und ist nicht nur ein nützliches, sondern auch ein schönes Accessoire für den täglichen Gebrauch.

Begründung der Jury
Zwei eigenständige Materialien wurden hier so miteinander kombiniert, dass ihre sich ergänzenden Eigenschaften zu einem ausgesprochen qualitätvollen Produkt führen.

Tira
Bangle
Armreif

Manufacturer
Ulla + Martin Kaufmann,
Hildesheim, Germany

In-house design
Ulla + Martin Kaufmann

Web
www.ulla-martin-kaufmann.de

Opulent virtuosity

Gold, with its unique properties, has been used in jewellery design for centuries. It is very easy to shape and process, possesses a certain density and has a distinctive brilliance. The Tira bangle is made of 750 gold, a gold alloy that is commonly used in jewellery due to its malleable yet solid properties. The fascination of this bangle, with its highly sensuous appeal, is found in the novel design approach it follows by exploiting the properties of the material. Through a delicate interplay it merges the contrasting aspects of utmost material hardness with the greatest possible flexibility. This bangle gently embraces the arm of its wearer, showcasing the full beauty of the material. In order not to obstruct the unity of material and form, the simple yet functional push button was transformed coherently and unobtrusively into the gold. Poignantly logical in functionality, this allows the piece of jewellery to be open and closed safely and easily. In addition, it lends the golden coils of the bangle a seamlessly organic appearance that draws attention with a perfect fit. Thus Tira is the successful expression of a design that aims at placing the bangle in a light yet opulent context.

Virtuose Opulenz

Wegen seiner besonderen Eigenschaften wird Gold seit Jahrtausenden für die Schmuckgestaltung genutzt. Es lässt sich sehr gut in Form bringen, besitzt dennoch Schwere und hat einen spezifischen Glanz. Der Armreif Tira besteht aus 750er Gold und damit einer Legierung, die im Schmuckbereich oft eingesetzt wird, da sie sich hervorragend für die Bearbeitung eignet. Die Faszination dieses überaus sinnlich anmutenden Armreifs liegt vor allem darin, wie es seiner Gestaltung gelingt, die Eigenschaften des Materials neu auszuloten. In einem feinsinnigen Zusammenspiel vereinen sich die Gegensätze von höchster Materialhärte und gleichzeitig größtmöglicher Flexibilität. Dieser Armreif schmiegt sich weich an den Arm seiner Trägerin an und zeigt dabei die ganze Schönheit des Materials. Um die Einheit von Stoff und Form nicht zu durchbrechen, wurde das Prinzip des Druckknopfes schlüssig in das Material Gold übersetzt. In seiner Funktionalität bestechend logisch, lässt sich dieses Schmuckstück so leicht und sicher verschließen. Die goldenen Windungen des Armreifes wirken deshalb auch nahtlos organisch und setzen sich am Arm perfekt in Szene. Damit ist Tira der gelungene Ausdruck einer Gestaltung, deren Ziel es war, diesen Armreif in einen leichten, aber dennoch opulenten Kontext zu setzen.

Statement by the jury

The Tira bangle made of 750 gold fascinates with its approach to exploring the possibilities of the material. In a virtuoso manner it rearranges the properties of gold into an object of the highest elegance and grace. Visual lightness merges with hardness and flexibility. The perfect craftsmanship and functionality of this bangle project an exquisite quality.

Begründung der Jury

Bei dem Armreif Tira aus 750er Gold fasziniert die Art und Weise des Umgangs mit den Möglichkeiten des Materials. Virtuos werden die Eigenschaften des Goldes neu komponiert zu einem Objekt höchster Eleganz und Grazie. Visuelle Leichtigkeit verbindet sich mit Flexibilität und Härte. Durch seine handwerkliche Perfektion und Funktionalität erreicht dieser Armreif eine exquisite Qualität.

Designer portrait
See page 64
Siehe Seite 64

Manufacturer
Bernhard Elsässer, Halle/Saale, Germany
In-house design
Web
www.bernhard-elsaesser.com

This delicate necklace made of staples is available in lengths of 420 mm and 780 mm. It also comes in three different versions: silver, gold-plated silver as well as rhodium-plated silver anthracite. A closure in the form of a hook makes it possible to adjust the length of the necklace to individual taste so that it can be flexibly adapted to the wearer's preferences.

Die filigrane Heftklammerkette ist in den Längen 420 mm und 780 mm erhältlich. Außerdem kann unter den drei Ausführungen Silber, Silber-Gold plattiert sowie Silber-Anthrazit rhodiniert gewählt werden. Ein Verschluss in der Form eines Hakens ermöglicht es, dass die Kette individuell in der Länge verändert werden kann und sich so flexibel den Bedürfnissen ihrer Trägerin anpasst.

Statement by the jury
The original design of this necklace is based on the idea of putting staples into a new context. The result is an exceptionally filigreed, charming piece of jewellery.

Begründung der Jury
Die originelle Gestaltung dieser Kette basiert auf der Idee, Heftklammern in einen neuen Kontext zu stellen. Das Ergebnis ist ein äußerst graziles und charmantes Schmuckstück.

squeezed
Necklace
Halskette

Manufacturer
Heike Walk, Cologne, Germany
In-house design
Web
www.heikewalk.de

The shape of this necklace was inspired by a simple candy cone. The plasticity of the matt, gleaming segments made of stainless steel is accentuated by a flat, corded ribbon. This acts as the connecting element for the luxurious hand-made necklace, which is available in different lengths as well as in a very varied range of colours.

Ihre Formgebung erhält die Kette dank der Inspiration durch ein simples Zuckertütchen. Die Plastizität der matt glänzenden Segmente aus Edelstahl wird durch das flache Ripsband in ihrer Wirkung unterstrichen. Dieses fungiert als verbindendes Element dieses Halsschmucks, der in Handarbeit gefertigt wird. Die Kette ist in unterschiedlichen Längen erhältlich und bietet außerdem ein abwechslungsreiches Spektrum an Farben.

Statement by the jury
The cheerful appearance of this necklace is the result of a harmonious interplay of contrasting shapes, colours and materials.

Begründung der Jury
Die fröhliche Anmutung dieser Kette entsteht durch ein harmonisches Zusammenspiel von kontrastierenden Formen, Farben und Materialien.

Niessing Kristallit
Jewellery
Schmuck

Manufacturer
Niessing Manufaktur GmbH & Co. KG, Vreden, Germany
In-house design
Nina Georgia Friesleben
Web
www.niessing.com

A special in-house laser process gives this distinctive pendant its fine lines which also indicate the folding edges and in doing so clearly define its geo-metrical shape. This piece of jewellery is further enhanced by a Pavé frame con-sisting of fine, brilliant-cut diamonds. The pendant, whose form is reminiscent of a crystal, is available in both gold and platinum.

Statement by the jury
The manufacture of these crystalline pendants combines traditional crafts-manship with modern technology. The result are elegant pieces of jewellery that radiate both clarity and sensuality.

Durch ein hauseigenes Laserverfahren erhalten diese ausdrucksstarken Anhänger ihre feinen Linien, die auch die Falzkan-ten vorgeben und so deren geometrische Körper genau definieren. Die Schmuckstü-cke werden außerdem verziert von einer Pavé-Fassung, die aus feinen Brillanten besteht. Die Anhänger, deren Formen an Kristalle erinnern, sind sowohl in Gold als auch Platin erhältlich.

Begründung der Jury
Für die Fertigung dieser kristallinen Anhän-ger wird Handwerkskunst mit moderner Technik kombiniert. Das Ergebnis sind edle Schmuckstücke, die zugleich Klarheit und Sinnlichkeit ausstrahlen.

Mater
Pendant
Anhänger

Manufacturer
Vanessa Robert + Alice Bodanzky, Rio de Janeiro, Brazil
In-house design
Design
Vanessa Robert, Rio de Janeiro, Brazil
Alice Bodanzky, Rio de Janeiro, Brazil
Peter Curet, Leiden, Netherlands
Web
www.materjoias.com.br
Honourable Mention

The jewelry relies on computational design processes to eternalize for expecting mothers the extraordinary moment that characterizes the genesis of a human being. The jewelry's shape is determined by a dedicated software that combines two types of pregnancy-related data: the image of the fetus, as it appears on a printed ultrasound, and a recorded sound, which can be the heartbeat, a message or even a song. The outcome alludes to the position of the future child in the womb and is always unique, just like every newborn. The jewelry is cast in silver or gold, after being fabricated in resin by a 3D printer. This project received financial support from the Research Foundation of the State of Rio de Janeiro (FAPERJ).

Das Schmuckstück entsteht durch computer-gestützte Design-Prozesse, um für werdende Mütter den besonderen Moment zu verewigen, der mit der Entstehung eines menschlichen Wesens einhergeht. Die Gestaltung des Schmucks geschieht dabei durch eine eigens entwickelte Software, die zwei schwangerschaftsbezogene Daten-quellen kombiniert: das Ultraschallbild des Kindes und eine Audio-Aufnahme. Diese kann der Herzschlag, eine gesprochene Nachricht, oder sogar ein Song sein. Das Ergebnis spielt auf die Haltung des Kindes im Mutterleib an und ist individuell, so wie jedes Neugeborene selbst. Das fertige Juwel wird in Silber oder Gold gegossen, nachdem mittels 3D-Drucker eine Harz-form erstellt wurde. Dieses Projekt wurde von der Stiftung zur Förderung der Forschung im Bundesstaat Rio de Janeiro (FAPERJ) unterstützt.

Statement by the jury
Computer-aided design has here made it possible to create a very personal and emotional item of jewellery.

Begründung der Jury
Mithilfe eines computergestützten Design-prozesses entsteht hier ein sehr persönli-ches und emotionales Schmuckstück.

Dreh dich...im Kreis
Rings
Fingerringe

Manufacturer
Jutta Ulland, Ahaus, Germany
In-house design
Web
www.jutta-ulland.de

The smaller bands of this finger ring can be twisted out of the frame and, in doing so, move into the centre of the circle. This creates a ring band which rapidly wraps itself around the finger and eventually runs back into the frame. Ring band and ring head thus achieve a fluid transition to the frame in all available versions. These pieces of jewellery come in yellow gold, white gold, silver gold-plated or silver.

Die kleineren Bänder des Fingerrings drehen sich aus dem Rahmen heraus und bewegen sich so in das Innere des Kreises. Von dort aus bildet sich die Ringschiene, welche sich mit Schwung um den Finger wickelt und schließlich wieder zurück in den Rahmen läuft. Ringschiene und Ringkopf bilden so bei allen Varianten einen fließenden Übergang zum Rahmen. Die Schmuckstücke werden in Gelbgold, Weißgold, Silber goldplattiert oder Silber gefertigt.

Statement by the jury
The delicate, entwined design of these rings is captivating. Depending on the angle from which they are viewed, they always appear new and different.

Begründung der Jury
Diese Ringe bezaubern mit ihrer zarten, verschlungenen Gestaltung, die je nach Perspektive immer wieder anders und neu wirkt.

AFTER WORK PARTY
Rings
Fingerringe

Manufacturer
Hans D. Krieger KG, Idar-Oberstein, Germany
Design
BIRTHE BEERBOOM BERLIN, Berlin, Germany
Web
www.birthe-beerboom.com
www.kriegernet.com

The ring is part of a collection that depicts scenes of modern cosmopolitan life on highest-quality materials. The clear, linear shape of the ring, with its slightly tensed surface, accentuates the scenes on it. The rotating mechanism inside makes it possible to display the various images in turn, while the inner ring band stays in contact with the finger. Thanks to its flat construction, the ring is comfortable to wear. It is available in 18-carat white and rose gold with optional diamonds.

Der Ring ist Teil einer Kollektion, die Szenen moderner kosmopolitischer Lebensart auf feinsten Materialien abbildet. Die klare, geradlinige Form des Rings mit leicht gespannter Oberfläche hebt die Zeichnungen hervor. Die Drehmechanik im Inneren ermöglicht es, durch die Szene zu wandern, während die innere Ringschiene am Finger anliegt. Durch die flach gearbeitete Konstruktion ist der Ring angenehm zu tragen. Er ist erhältlich in 18-karätigem Weiß- und Roségold, wahlweise mit feinen Brillanten.

Accessories
Aeroplane interior fittings
Automobile technology
Automobiles
Buses
Caravan equipment
Caravans and motor homes
Car entertainment
Construction, utility and
transport vehicles
Fork-lift trucks
Jet skis
Motorbikes and mopeds
Off-road vehicles
Railway vehicles
Trailers
Tyres
Vehicle lighting systems
Wheel rims

Anhänger
Automobile
Automobiltechnik
Bau-, Nutz- und
Transportfahrzeuge
Busse
Car-Entertainment
Caravan-Ausstattung
Caravans und Wohnmobile
Fahrzeugbeleuchtungen
Felgen
Flugzeug-Inneneinrichtung
Gabelstapler
Geländefahrzeuge
Jet-Skis
Motorräder und -roller
Reifen
Schienenfahrzeuge
Zubehör

Vehicles
Fahrzeuge

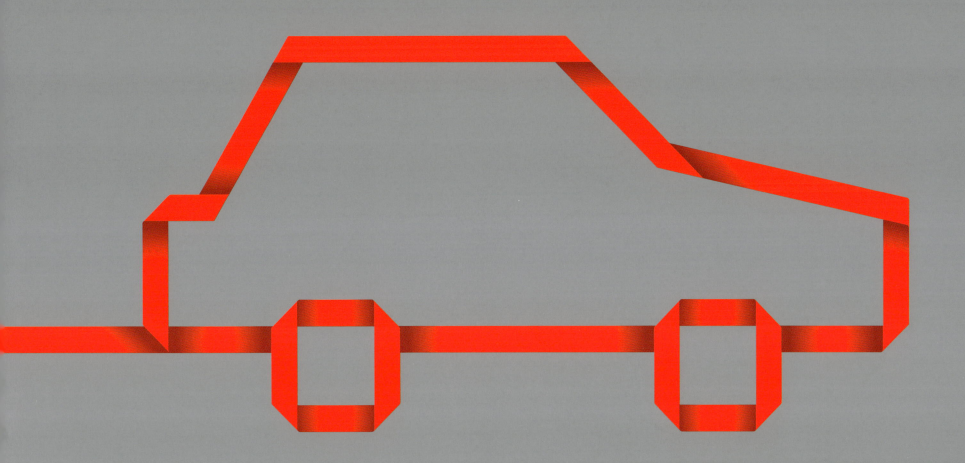

Mercedes-AMG GT
Sports Car
Sportwagen

Manufacturer
Daimler AG, Stuttgart, Germany

In-house design
Daimler AG

Web
www.daimler.com

reddot award 2015
best of the best

Of beauty and intelligence

A special challenge in the design of automobiles is to continuously find an appropriate form for innovative concepts of motorisation. The Mercedes-AMG GT is designed with proportions that give the car a strong, expressive character and skilfully reflect its technical potential. A visual unity of positively arched surfaces, a dome-shaped arched roof line and frameless doors lend this sports car a sculptural appeal. Its broad shoulders and narrow tail light graphics create an emotional overall appearance. This impression is continued in the interior in a highly consistent way, embodying an equally high degree of emotion and sensual purity. The purist design, which clearly visualises the advanced technology of the instruments, perfectly matches the high-grade materials. The sport seats and steering wheel also create an elegant, appealing impression. The driver experiences precision and a surrounding where every detail has been carefully considered. A maxim that is also echoed in the AMG Drive Unit controls: arranged like eight cylinders in a V-layout, they emphasise the powerful and high-tech impression of the centre console. The combination of boldly sculpted surfaces and flowing lines turn the Mercedes into a sports car that successfully realises its objective of creating a fusion of beauty and intelligence.

Von Schönheit und Intelligenz

Eine besondere Herausforderung im Design von Fahrzeugen liegt immer auch darin, eine angemessene Form für innovative Konzepte der Motorisierung zu finden. Der Mercedes-AMG GT ist gestaltet mit Proportionen, die ihm eine starke Ausdruckskraft geben und gekonnt seine fahrtechnischen Möglichkeiten widerspiegeln. Eine gestalterische Einheit aus bewusst überspannten Flächen, einer kuppelförmig gespannten Dachlinie sowie rahmenlosen Türen verleiht diesem Sportwagen eine skulpturale Anmutung. Emotionalisierend wirken seine breiten Schultern und eine schmal gehaltene Heckleuchtengrafik. Das Interieur führt diesen Eindruck auf stimmige Weise fort, wobei es ein ebenso hohes Maß an Emotion und sinnlicher Klarheit verkörpert. Die puristische, die hochentwickelte Technik gut visualisierende Gestaltung der Anzeigen passt perfekt zu den verwendeten hochwertigen Materialien. Eine edel anmutende Wirkung haben auch die Sportsitze und das Lenkrad. Der Fahrer erlebt Präzision und eine Umgebung, bei der sorgfältig auf jedes Detail geachtet wurde. Eine Maxime, die sich auch an den Bedienelementen der AMG Drive Unit zeigt: Im Stil von acht Zylindern in V-Anordnung platziert, unterstreichen diese den kräftigen und technischen Ausdruck der Mittelkonsole. In der Verknüpfung kraftvoll modellierter Flächen und fließender Linien entstand damit ein Sportwagen, der überaus gelungen seine Zielsetzung, eine Verbindung von Schönheit und Intelligenz, realisiert.

Statement by the jury

The proportions of the Mercedes-AMG GT perfectly unite to form a statement of purity and precision. Its sculptural appeal, which is created by positively arched surfaces and a dome-shaped arched roof line, is fascinating. Each detail is assigned a corresponding functionality in the design, and nothing appears superfluous or artificial. Thus, the Mercedes-AMG GT embodies a very good example of a sophisticated sports car design.

Begründung der Jury

Auf perfekte Weise vereinen sich die Proportionen des Mercedes-AMG GT zu einer Aussage von Klarheit und Präzision. Faszinierend ist seine skulpturale Anmutung, die durch überspannte Flächen sowie die Gestaltung mit einer kuppelförmig gespannten Dachlinie entsteht. Jedem Detail ist gestalterisch eine entsprechende Funktionalität zugeordnet, nichts wirkt überflüssig oder aufgesetzt. Der Mercedes-AMG GT verkörpert damit ein sehr gutes Beispiel für ein ausgereiftes Sportwagen-Design.

Designer portrait
See page 66
Siehe Seite 66

Mazda MX-5
Roadster

Manufacturer
Mazda Motor Corporation,
Hiroshima, Japan

In-house design
Masashi Nakayama

Web
www.mazda.com

reddot award 2015
best of the best

Developed with passion

Introduced in 1989, the Mazda MX-5 represents the renaissance of the roadster. Now in its fourth generation, the look is sportier than ever. With a design language that consciously accentuates the contrast between stillness and motion, the new model radiates a magnified degree of agility. The car weighs 100 kg less than its predecessor, yet is exceptionally comfortable. To convey the desired sportiness, its designers moved the cabin further toward the rear and lowered the hip point of the seats. One remarkable effect achieved was the special way light reflects off the new MX-5's body surface. These reflections change with the angle of light, making the car seem alive. And by merging the body surface into the door trim, the boundary between the interior and exterior virtually dissolves. The interior itself has been optimised for driving with the top down. Equipped with an easy-to-use soft top that looks and feels good, innovative wind control, and headrests with integrated speakers, the sophisticated cabin offers drivers outstanding comfort. As a highly inspiring reinterpretation, the new Mazda MX-5 delivers an exciting, contemporary roadster experience.

Lebendig weiterentwickelt

Der Mazda MX-5 steht seit 1989 für die Renaissance des Roadsters. Die nun vierte Generation zeigt eine eindringlich sportive Anmutung. Mit einer Formensprache, die bewusst den Gegensatz zwischen Ruhe und Bewegung auf die Spitze treibt, strahlt dieser Roadster ein hohes Maß an Agilität aus. Er ist um 100 kg leichter als sein Vorgänger und dabei überaus komfortabel. Um den gewünschten Eindruck von Sportlichkeit zu erzielen, wurde die Kabine weiter nach hinten gerückt und der Hüftpunkt niedriger gesetzt. Ein bemerkenswerter Effekt wird im Exterieur hervorgerufen durch besondere Lichtreflexe auf der Lackierung. Sie verändern sich mit dem Einfallswinkel des Lichts und lassen den neuen MX-5 außerordentlich lebendig erscheinen. Die Karosserieoberfläche geht zudem visuell in die Türverkleidung über, sodass die Grenzen zwischen Innen und Außen verschmelzen. Das Interieur des Mazda MX-5 ist auf das Fahren insbesondere mit heruntergelassenem Verdeck ausgerichtet. Ausgestattet mit einem bedienungsfreundlichen und ästhetisch wie haptisch ansprechenden Softtop, einer innovativen Luftflusskontrolle sowie in die Kopfstützen eingelassenen Lautsprechern, bietet es dem Fahrer hochentwickelten Komfort. Als eine sehr inspirierte Neuinterpretation ermöglicht der neue Mazda MX-5 damit ein spannendes, zeitgemäßes Roadster-Erlebnis.

Statement by the jury

The design of the fourth and latest generation Mazda MX-5 has managed without fault to pick up where its predecessor left off. This new roadster possesses a highly expressive and rather masculine identity. In an impressive manner, the car has also been developed into an all-rounder that should appeal to many different target groups. It is lighter, more agile and offers outstanding ride comfort as well as a perfect driving performance.

Begründung der Jury

In der jetzt vierten Version gelingt es der Gestaltung des Mazda MX-5 nahtlos an die Vorgängermodelle anzuknüpfen. Dieser neue Roadster besitzt eine sehr ausdrucksstarke Identität, die mit einer eher maskulinen Anmutung einhergeht. Auf beeindruckende Weise wurde er dabei weiterentwickelt hin zu einem Allround-Auto, welches nunmehr viele Zielgruppen anspricht. Er ist leichter, agiler und bietet einen ausgezeichneten Fahrkomfort sowie eine perfekte Performance.

Designer portrait
See page 68
Siehe Seite 68

Ferrari FXX K
Sports Car
Sportwagen

Manufacturer
Ferrari SpA,
Maranello (Modena), Italy

Design
Ferrari Design
(Flavio Manzoni, Werner Gruber),
Maranello (Modena), Italy

Web
www.ferrari.com

Dynamic power

Ferrari has been synonymous with motor racing since 1929, when the famous Scuderia was founded. Ever since becoming a fully-fledged manufacturer in 1947, the company has always bestowed upon its road-going models the technology developed by the racing division. Derived from the LaFerrari supercar, the FXX K is the latest in a programme of technological laboratory prototypes created for a highly select group of clients and the fruit of both GT Research and Development as well as experimentation in Formula 1. It is a track-only edition built in an extremely limited series. The car's staggering performance is attested by two significant figures: a total power output of 1050 cv, delivered from the naturally aspirated V12 engine combined with an electric motor, and a maximum torque in excess of 900 Nm. It magnifies the LaFerrari's styling concept and unleashes its untapped potential without being subject to homologation requirements or regulatory restrictions.
A plethora of aerodynamic innovations, meticulously developed in collaboration with the designers, serve to enhance not only its efficiency but also its purposeful aesthetics. All body panels have been modified to various extents, with most of the alterations aimed at generating more downforce. A lower suspension set up, wider front and rear track and increased camber combine to reinforce the FXX K's aggressive stance.

Dynamische Kraft

Der Name Ferrari ist seit der Gründung der berühmten Scuderia im Jahre 1929 ein Synonym für den Motorsport. Dieser versierte Hersteller stattet bereits seit 1947 seine straßentauglichen Modelle mit Technologien aus, die in seiner Rennabteilung entwickelt wurden. Abgeleitet vom Supercar LaFerrari stellt sich der FXX K nun als das neueste Modell in einer Reihe technologischer Laborprototypen dar. Er ist das Ergebnis der GT-Forschung und -Entwicklung sowie von Experimenten in der Formel 1 und wurde als Rennstrecken-Version nur in sehr geringer Stückzahl für einen ausgewählten Kundenkreis kreiert. Die beeindruckende Performance dieses Ferrari zeigt sich in zwei wesentlichen Eckdaten: einer Gesamtleistung von 1050 PS dank eines V12-Saugmotors in Verbindung mit einem elektrischen Motor sowie einem maximalen Drehmoment von über 900 Nm. Gestalterisch wird das Styling-Konzept des LaFerrari aufgegriffen und in seinen Möglichkeiten ausgeschöpft, ohne Zulassungsvorgaben und anderen regulatorischen Beschränkungen zu unterliegen. Vielfältige aerodynamische Innovationen, die gemeinsam mit den Designern entwickelt wurden, erhöhen nicht nur die Leistungsfähigkeit des Fahrzeugs, sondern begründen auch seine eindringliche Ästhetik. Alle Karosserieteile wurden dabei in unterschiedlichem Maße modifiziert, wobei die meisten Änderungen auf eine höhere Anpresskraft abzielten. Tiefergelegt mit einer verbreiterten Spur sowie gestaltet mit einem erhöhten Sturz, präsentiert sich der FXX K mit einer starken dynamischen Ausstrahlung.

Statement by the jury

As a further development derived from the LaFerrari high-performance racing car, complemented by a further refinement of all elements, the Ferrari FXX K showcases an appearance of extreme dynamics that embodies true novelty. An engaging design has developed a car whose tremendous engine power is striking at first sight. Impressive is also the visually well-defined interplay between interior and exterior. The Ferrari FXX K is perfectly balanced and delivers a sensational driving experience.

Begründung der Jury

Als Weiterentwicklung des Hochleistungssportwagens LaFerrari und einer damit einhergehenden Verfeinerung aller Elemente zeigt der Ferrari FXX K einen Ausdruck von extremer Dynamik, der in dieser Form neu ist. Eine engagierte Gestaltung ließ hier ein Auto entstehen, dessen enorme Motorkraft auf den ersten Blick deutlich wird. Beeindruckend ist die gut ausgearbeitete visuelle Verbindung zwischen Exterieur und Interieur. Dieser Ferrari ist perfekt ausbalanciert und bietet ein sensationelles Fahrerlebnis.

Designer portrait
See page 70
Siehe Seite 70

Ferrari California T
Sports Car
Sportwagen

Manufacturer
Ferrari SpA, Maranello (Modena), Italy
Design
Ferrari Design (Flavio Manzoni, Andrea Militello),
Maranello (Modena), Italy
Pininfarina SpA, Cambiano, Italy
Web
www.ferrari.com
www.pininfarina.it

The design of the Ferrari California T displays elegance, sportiness and exclusivity and thus reflects the long tradition of the California model. Its retractable hard top and the 2+2 configuration make it versatile. A balanced relationship of convex and concave surfaces characterise its modern appearance, whereas the flanks at the same time pay homage to the pontoon fender design of the 250 Testa Rossa. The front section is characterised by a wide radiator grille; the rear section is particularly aerodynamic with an innovative, triple-fence diffuser.

Die Gestaltung des Ferrari California T vermittelt Eleganz, Sportlichkeit und Exklusivität und steht damit in der langen Tradition des Modells California. Sein faltbares Hardtop und die 2+2-Konfiguration machen ihn vielseitig. Ein ausgewogenes Verhältnis von konvexen und konkaven Flächen prägt sein modernes Erscheinungsbild, während die Seitenpartie gleichzeitig Anleihen an das Ponton-Kotflügeldesign des 250 Testa Rossa zeigt. Die Front ist geprägt von einem breiten Kühlergrill, das Heck ist mit einem neuen Dreifachdiffusor besonders aerodynamisch gestaltet.

Ferrari LaFerrari
Sports Car
Sportwagen

Manufacturer
Ferrari SpA, Maranello (Modena), Italy
Design
Ferrari Design (Flavio Manzoni, Werner Gruber),
Maranello (Modena), Italy
Web
www.ferrari.com

The slender silhouette and striking proportions of the LaFerrari result from the close interaction between its advanced architecture, its integrated hybrid drivetrain and rigorous aerodynamic language. It remains true to the archetype of the classical V12 sports car with a mid-mounted engine. The cockpit and the engine are located within the wheelbase in order to achieve the best possible weight distribution, while mounting the KERS and gearbox at the rear led to a better visual balance between the front and rear overhangs. Both body and chassis are constructed in hand laid-up carbon fibre.

Die schlanke Silhouette und die markanten Proportionen des LaFerrari resultieren aus dem engen Zusammenspiel zwischen einem durchdachten Fahrzeugaufbau, dem integrierten Hybridantrieb und einer sehr aerodynamischen Formensprache. Er bleibt dem Prototyp des klassischen V12-Sportwagens mit Mittelmotor treu. Cockpit und Motor sind innerhalb des Radstands platziert, um die bestmögliche Gewichtsverteilung zu erreichen, während KERS und Getriebe im hinteren Bereich positioniert sind, um die optische Ausgewogenheit zwischen Front- und Decküberhang zu verbessern. Karosserie und Chassis sind aus von Hand aufgelegter Kohlefaser gefertigt.

Statement by the jury
The aerodynamic design of the LaFerrari, evocative of Formula 1 cars, expresses at first sight the performance and power of this super sports car.

Begründung der Jury
Die aerodynamische Gestaltung des LaFerrari mit Anleihen an Formel-1-Wagen kommuniziert die Leistung und Kraft dieses Supersportwagens auf den ersten Blick.

BMW 2 Series Convertible
BMW 2er Cabrio
Passenger Car
Personenkraftwagen

Manufacturer
BMW Group, Munich, Germany
In-house design
Web
www.bmwgroup.com

Dynamic lines and balanced proportions define the appearance of the BMW 2 Series Convertible. The front appears wide and sporty. A rising crease line adds a dynamic wedge shape to the vehicle's side. The beltline embraces the passenger compartment with one flowing movement, creating the boat deck style that is typical of BMW convertibles. The designers attached major importance to the enclosed appearance of the interior: The borders of the front door inserts continue in the borders of the rear door inserts, thus visually connecting the driver's and the rear seat passengers' area.

Dynamische Linien und ausgewogene Proportionen bestimmen das Erscheinungsbild des BMW 2er Cabrios. Die Front wirkt breit und sportlich. Die hinter dem Vorderrad entspringende Sickelinie prägt die Keilform des Fahrzeugs. Die Brüstungslinie umschließt den Fahrgastraum in einer fließenden Bewegung und formt so den für BMW Cabrios typischen Bootsdeck-Charakter. Im Interieur laufen die Kanten der vorderen Türspiegel dynamisch-fließend in den Kanten der hinteren Türspiegel fort, wodurch Fahrer- und Fondbereich harmonisch miteinander verbunden werden.

Statement by the jury
With an exterior and interior equally harmonious in design, the BMW 2 Series Convertible convinces as a classic convertible with fabric top.

Begründung der Jury
Das außen wie innen gleichermaßen harmonisch und wertig gestaltete BMW 2er Cabrio überzeugt als klassisches Cabriolet mit Stoffverdeck.

Audi TT Family

Passenger Car
Personenkraftwagen

Manufacturer
AUDI AG, Ingolstadt, Germany
In-house design
Web
www.audi.com

For the new generation of the TT family, the exterior has been newly interpreted in order to combine elegance with a masculine, sporty impression. Smooth surfaces and a clear, purist design characterise the appearance. The Singleframe grill is wide and flat, the four rings are located on the bonnet. In the interior, the instrument cluster and infotainment system are combined in a central virtual cockpit. The innovative air conditioning control is integrated directly in the air vents. All operating elements are ergonomically grouped around the driver.

Bei der neuen Generation der TT-Familie wurde das Exterieur neu interpretiert, um Eleganz mit einer maskulinen, sportlichen Anmutung zu verbinden. Straffe Flächen und eine klare, puristische Gestaltung prägen das Erscheinungsbild. Der Singleframe-Grill ist breit und flach, die vier Ringe sitzen auf der Motorhaube. Im Innenraum wurden Kombiinstrument und Infotainmentsystem in einem zentralen virtuellen Cockpit zusammengefasst. Die innovative Klimaanlagenbedienung ist direkt in die runden Luftausströmer integriert, alle Bedienelemente sind ergonomisch um den Fahrer gruppiert.

Statement by the jury
The new TT family continues with the purist design of the first TT, but with more accentuation. The interior enthrals with the new virtual cockpit.

Begründung der Jury
Die neue TT-Familie führt die puristische Gestaltung des ersten TT stärker akzentuiert fort, das Interieur begeistert mit dem neuen virtuellen Cockpit.

Volkswagen Passat Variant
Passenger Car
Personenkraftwagen

Manufacturer
Volkswagen AG, Wolfsburg, Germany
In-house design
Web
www.volkswagen.de

A high side line, the flat roof and a lengthened wheelbase characterise the elegant, sporty appearance of the new Passat Variant. Due to its long wheelbase, the Passat gains not only more legroom but also features luggage space with a volume of 650 litres. The interior design is clear and of high quality; elements such as a concise horizontal band of air vents and ambient lighting emphasise the features. The driver is aided by digital instruments such as the Active Info Display and a head-up display.

Eine hohe Seitenlinie, das abgeflachte Dach und ein verlängerter Radstand prägen das elegante, sportliche Erscheinungsbild des neuen Passat Variant. Durch den langen Radstand gewinnt der Passat an Raum, sodass er nicht nur mehr Beinfreiheit bietet, sondern auch ein Gepäckraumvolumen von 650 Litern aufweist. Die Gestaltung des Innenraums ist klar und wertig, Elemente wie ein prägnantes Horizontalband mit durchgehenden Luftausströmern oder eine Ambientebeleuchtung setzen Akzente. Der Fahrer wird durch digitale Instrumente wie das Active-Info-Display und ein Head-up-Display unterstützt.

Statement by the jury
The Passat Variant is appealing with its elongated silhouette and clear lines, giving it elegance and discreet sportiness.

Begründung der Jury
Der Passat Variant gefällt mit seiner langgestreckten Silhouette und einer klaren Linienführung, die ihm Eleganz und eine dezente Sportlichkeit verleihen.

Volkswagen Passat
Passenger Car
Personenkraftwagen

Manufacturer
Volkswagen AG, Wolfsburg, Germany
In-house design
Web
www.volkswagen.de

The new Passat has been designed in order to build a bridge into the next higher class of car. It has arisen from a new platform, the new modular traverse matrix, which forms the basis for its new proportions. A longer wheelbase, shorter overhangs, a wider front panel and a flatter silhouette give the Passat a dynamic, contemporary appearance. The interior is of high quality and provides ample space, whereas technologies such as an Active Info Display and a head-up display provide a high level of convenience.

Der neue Passat wurde mit dem Anspruch gestaltet, eine Brücke in die nächsthöhere Fahrzeugklasse zu schlagen. Er entstand auf einer neuen Plattform, dem modularen Querbaukasten, der auch die Grundlage für seine neuen Proportionen legt. Ein längerer Radstand, kürzere Überhänge, eine breitere Front und eine flachere Silhouette verleihen dem Passat ein dynamisches, zeitgemäßes Erscheinungsbild. Der hochwertig ausgestattete Innenraum bietet viel Platz, während Technologien wie ein Active-Info-Display und ein Head-up-Display für hohen Bedienkomfort sorgen.

Statement by the jury
Elements such as the emphasised wheel arches and the more strongly sculptured bonnet confer a sporty appearance on the new Passat.

Begründung der Jury
Elemente wie die betonten Radkästen und die stärker skulpturierte Motorhaube verleihen dem neuen Passat ein sportliches Aussehen.

Volkswagen Golf Sportsvan

Passenger Car
Personenkraftwagen

Manufacturer
Volkswagen AG, Wolfsburg, Germany
In-house design
Web
www.volkswagen.de

The superior, independent appearance of the Golf Sportsvan is defined by extended outer dimensions in comparison to the Golf Plus in combination with a new, sharply contoured design. Thanks to a longer wheelbase, the van makes a more sporty impression in spite of its height. A pronounced character line on the exterior combines objectivity with elegant lines. The interior is spacious, of high-quality design and provides a clear and fresh impression. A dominating element is the newly designed dashboard with a central touchscreen in the centre console.

Das souveräne, eigenständige Erscheinungsbild des Golf Sportsvan wird bestimmt durch gestrecktere Außenmaße im Vergleich zum Golf Plus in Verbindung mit einem neuen, scharf konturierten Design. Dank eines längeren Radstands wirkt der Van trotz seiner Höhe sportlicher. Eine ausgeprägte Charakterlinie verbindet beim Exterieur Sachlichkeit mit einer eleganten Linienführung. Der Innenraum ist geräumig, hochwertig gestaltet und vermittelt einen klaren und frischen Eindruck. Dominierendes Element ist die neu gestaltete Schalttafel mit einem zentralen Touchscreen in der Mittelkonsole.

BMW 2 Series Active Tourer
BMW 2er Active Tourer
Passenger Car
Personenkraftwagen

Manufacturer
BMW Group, Munich, Germany
In-house design
Web
www.bmwgroup.com

The BMW 2 Series Active Tourer combines compact dimensions and functionality with a powerful form language. Rising lines in the side and a steeply sloping bonnet give the car a dynamic, wedge-shaped appearance. The long wheelbase allows generous space in the interior. A sweeping instrument panel, a large windscreen and the optional panorama roof contribute to a generous interior atmosphere. The elevated sitting position provides an overview in urban traffic. All instruments are positioned conveniently for the driver.

Der BMW 2er Active Tourer vereint kompakte Abmessungen und Funktionalität mit einer kraftvollen Formensprache. Aufwärts ziehende Linien in der Seite und eine stark geneigte Motorhaube verleihen dem Fahrzeug ein dynamisches, keilförmiges Erscheinungsbild. Der lange Radstand erlaubt ein geräumiges Interieur. Eine weitläufige Instrumententafel, eine große Frontscheibe und das optionale Panoramadach tragen zu einer großzügigen Innenraumatmosphäre bei. Die erhöhte Sitzposition bietet Übersicht im Stadtverkehr, sämtliche Bedienelemente sind auf den Fahrer ausgerichtet.

Statement by the jury
The dynamic-looking BMW 2 Series Active Tourer combines compact dimensions with a feel of generous spaciousness.

Begründung der Jury
Der dynamisch wirkende BMW 2er Active Tourer verbindet kompakte Dimensionen mit einem großzügigen Raumgefühl.

Hyundai i20
Passenger Car
Personenkraftwagen

Manufacturer
Hyundai Motor Company, Seoul, South Korea
Design
Hyundai Motor Group
(Peter Schreyer)
Hyundai Design Centre Europe
(Thomas Bürkle)
Web
www.hyundai.com

The new i20 has a clear and elegant design. A long bonnet, more vertically aligned A-pillars and a distinctive character line along the side of the vehicle characterise its autonomous appearance. The car's bold impression is emphasised at the front by the hexagonal grille and the LED headlamps. At the rear, black glossy C-pillars give the impression of a floating roofline.

Statement by the jury
This compact car convinces with balanced proportions and a form which indicates power and lightness at the same time.

Der neue i20 zeigt eine klare und elegante Linienführung. Eine lange Motorhaube, senkrechter stehende A-Säulen und eine markante Charakterlinie über die gesamte Fahrzeuglänge prägen sein eigenständiges Erscheinungsbild. Die kraftvolle Anmutung des Fahrzeugs wird im Frontbereich durch den Hexagonalgrill und die LED-Scheinwerfer unterstrichen. Am Heck vermitteln die schwarz glänzenden C-Säulen den Eindruck einer schwebenden Dachlinie.

Begründung der Jury
Dieser Kleinwagen überzeugt mit ausgewogenen Proportionen und einer Formgebung, die zugleich Kraft und Leichtigkeit vermittelt.

Hyundai i20 Coupe
Passenger Car
Personenkraftwagen

Manufacturer
Hyundai Motor Company, Seoul, South Korea
Design
Hyundai Motor Group
(Peter Schreyer)
Hyundai Design Centre Europe
(Thomas Bürkle)
Web
www.hyundai.com

The unique silhouette of the i20 Coupe is defined by the raked A-, B- and C-pillars. The dynamic profile is emphasised by the sloping roofline, the slim black glossy C-pillar applications and the sporty rear spoiler. The front is characterised by the reversed hexagonal grille, which harmonises with the newly designed front bumper. The interior convinces with high-quality materials and an ergonomically designed driver environment.

Statement by the jury
Form language and stylish lines of the i20 Coupe express dynamic and sportiness, whereas the interior gains merit with its level of comfort.

Die kraftvolle Silhouette des i20 Coupe ist bestimmt von stark geneigten A-, B- und C-Säulen. Das dynamische Profil wird durch die geneigte Dachlinie, die schwarzen Applikationen an der C-Säule und den sportlichen Heckspoiler hervorgehoben. Die Frontpartie ist geprägt von einer eigenständigen Version des Hexagonalgrills, der mit dem neu gestalteten Stoßfänger harmoniert. Der Innenraum überzeugt mit hochwertigen Materialien und einer ergonomisch gestalteten Fahrerumgebung.

Begründung der Jury
Formensprache und Linienführung des i20 Coupe drücken Dynamik und Sportlichkeit aus, während der Innenraum mit Komfort punktet.

Hyundai Sonata
Passenger Car
Personenkraftwagen

Manufacturer
Hyundai Motor Company, Seoul, South Korea
In-house design
Peter Schreyer
Web
www.hyundai.com

The new Sonata combines enhanced safety functions with an elegant form language. Characteristic design elements such as the distinctive character line at the sides have been updated so that they blend harmoniously with the overall appearance. The coupé-like roofline accentuates the sporty appearance of the car. The interior applies high quality materials. The driver-oriented cockpit and intuitive HMI assure high driving comfort.

Statement by the jury
The Sonata offers a convincing combination of elegant appearance and safety as well as generous and upmarket interior fittings.

Der neue Sonata verbindet eine umfangreiche Sicherheitsausstattung mit einer eleganten Formensprache. Prägende Designelemente wie die markante Charakterlinie an den Seiten fügen sich harmonisch in das Gesamtbild ein. Die Coupé-artige Dachlinie betont den sportlichen Auftritt des Fahrzeugs. Das Interieur ist mit hochwertigen Materialien ausgestattet und das fahrerorientierte Cockpit und die intuitive Benutzerführung sorgen für hohen Komfort.

Begründung der Jury
Der Sonata bietet eine überzeugende Kombination aus elegantem Erscheinungsbild und Sicherheit sowie einer großzügigen und gehobenen Innenraumausstattung.

ŠKODA Fabia
Passenger Car
Personenkraftwagen

Manufacturer
ŠKODA Auto a.s., Mladá Boleslav, Czech Republic
In-house design
Web
www.skoda-auto.com

Concise horizontal lines define the outward appearance of the new ŠKODA Fabia. In the interplay with the reduced height of the car and the roof lightly sloping to the rear, the precise lines give the silhouette a powerful appearance. The sides are clearly and dynamically conceived, and the surfaces below the tailgate's edge are slanted slightly inwards. The front embodies the new form language of the manufacturer with a clear geometry of the cooler grill and crystalline lamp lines. Due to the emphasised horizontal, the Fabia's interior appears generously dimensioned and clearly arranged.

Eine prägnante horizontale Linienführung bestimmt das Erscheinungsbild des neuen ŠKODA Fabia. Im Zusammenspiel mit der reduzierten Fahrzeughöhe und dem leicht nach hinten abfallenden Dach lassen die präzisen Linien die Silhouette kraftvoll aussehen. Die Seitenpartien sind klar und dynamisch gestaltet, die Flächen unterhalb der Heckklappenkante leicht nach innen gestellt. Die Front greift die neue Formensprache des Herstellers mit einer klaren Kühlergeometrie und kristalliner Leuchtengrafik auf. Durch die betonte Horizontale wirkt der Fabia auch innen großzügig und aufgeräumt.

Statement by the jury
The ŠKODA Fabia with its streamlined exterior design presents a modern and sporty impression. The compact car offers a surprising amount of room inside.

Begründung der Jury
Der ŠKODA Fabia vermittelt mit seiner straffen Exterieurgestaltung einen modernen und sportlichen Eindruck. Im Innenraum überrascht der Kleinwagen mit viel Platz.

Mazda2
Passenger Car
Personenkraftwagen

Manufacturer
Mazda Motor Corporation, Hiroshima, Japan
In-house design
Ryo Yanagisawa
Web
www.mazda.com
Honourable Mention

The new Mazda2 is designed according to the manufacturer's own "KODO – Soul of Motion" philosophy, whereby the natural motions of animals are to be reflected in the vehicle. The headlamps suggest the eyes of a predator and confer a dynamic appearance to the front. The distinctive wing motif of the cooler grill continues through the headlamps in a contour line at the side of the car. Increasing the length and extending the wheelbase creates for the Mazda2 a generous interior, which is ergonomically and comfortably designed.

Der neue Mazda2 ist nach der hersteller-eigenen Designphilosophie „KODO – Soul of Motion" gestaltet, wonach sich die natürliche Bewegung von Tieren im Design des Fahrzeugs widerspiegeln soll. Scheinwerfer, die an die Augen eines Raubtieres erinnern, verleihen der Front ein dynamisches Aussehen. Das markante Flügelmotiv des Kühlergrills setzt sich durch die Scheinwerfer in einer Konturlinie an der Fahrzeugseite fort. Durch Zuwachs an Länge und einen größeren Radstand hat der Mazda2 einen großzügigen Innenraum, der ergonomisch und komfortabel gestaltet ist.

Statement by the jury
The Mazda2 is appealing with a dynamic and simultaneously elegant exterior design and imparts a sense of generous spaciousness inside.

Begründung der Jury
Der Mazda2 gefällt mit einer dynamischen und zugleich eleganten Exterieurgestaltung und vermittelt im Inneren ein großzügiges Raumgefühl.

smart fortwo
Passenger Car
Personenkraftwagen

Manufacturer
Daimler AG, Stuttgart, Germany
In-house design
Web
www.daimler.com

The design of the new smart fortwo is purist and progressive and reflects the philosophy of "FUN.ctional Design". Attributes which are typical for smart such as the silhouette with short overhangs, the tridion cell and the attractive face also characterise the new generation in a changed form. The interior offers more room at the sides; a characteristic design element is the "Loop", which features, for example, on the dashboard and doors. The sculptural two-section dashboard consists of a powerful and sensuous exterior section and a large, concave decorative section.

Die Gestaltung des neuen smart fortwo ist puristisch und progressiv und spiegelt die Philosophie des „FUN.ctional Design" wider. smart-typische Attribute wie die Silhouette mit den kurzen Überhängen, die tridion-Zelle und das sympathische Gesicht prägen in veränderter Form auch die neue Generation. Der Innenraum bietet mehr Platz an den Seiten, prägendes Designelement ist der „Loop", der sich z. B. an Instrumententafel und Tür findet. Die skulptural zweigeteilte Instrumententafel besteht aus einem kraftvoll-sinnlichen Außenteil sowie einem großen, konkaven Zierteil.

Statement by the jury
Thanks to its subtle changes, the smart fortwo remains true to itself also in the new generation. The interior gains merit with more room at the side.

Begründung der Jury
Der smart fortwo bleibt sich dank sanfter Veränderungen auch in der neuen Generation treu. Im Innenraum punktet er mit mehr Platz zur Seite hin.

Kia Soul EV
Passenger Car
Personenkraftwagen

Manufacturer
Kia Motors Europe GmbH, Frankfurt/Main, Germany
In-house design
Kia Design Center Europe
Web
www.kia.com

The electrically driven version of the new Kia Soul differs from the Soul variants with a conventional drive above all in the front design. The lower air inlet has a flatter shape, and the cooler grill is replaced by a cover plate, behind which the electrical connections are located. The cabin captivates by a bright, driver-friendly interior concept. An 8-inch navigation system provides information on charging stations nearby or the remaining available mileage. Innovative driver zone air conditioning saves electricity when the driver is alone in the car.

Die Elektro-Version des neuen Kia Soul unterscheidet sich vor allem in der Frontgestaltung von den Soul-Varianten mit konventionellem Antrieb. Der untere Lufteinlass ist flacher dimensioniert und der Kühlergrill durch eine Blende ersetzt, hinter der sich die Ladeanschlüsse befinden. Das Interieur besticht durch ein helles, fahrerorientiertes Innenraumkonzept. Ein 8-Zoll-Navigationssystem informiert über Ladestationen im Umkreis oder die verbleibende Reichweite. Eine innovative Fahrerzonenklimatisierung spart Strom, wenn der Fahrer allein unterwegs ist.

Statement by the jury
With a long range capacity, a simple control concept and high-quality interior design, the Kia Soul EV contributes to the acceptance of the electric car.

Begründung der Jury
Mit einer hohen Reichweite, einem einfachen Bedienkonzept und einer hochwertigen Innenraumgestaltung trägt der Kia Soul EV zur Akzeptanz des Elektroautos bei.

Volvo XC90
SUV

Manufacturer
Volvo Car Group,
Gothenburg, Sweden

In-house design
Thomas Ingenlath

Web
www.volvocars.com

reddot award 2015
best of the best

Pioneer of form

The cars of the Swedish company Volvo stand for a
high level of safety and a distinctive design. Conceived as a premium quality SUV with seven seats,
the Volvo XC90 marks the beginning of a new
Volvo product architecture. A striking feature is the
expressively designed face with the revised brand
emblem. In combination with the T-shaped daytime
running lights, this concept establishes a highly
recognisable brand identity. The interior design of the
car impresses with a tablet-like touchscreen control
console. Forming the heart of the entire in-car control
system, it allows the intuitive control of all functions
and also integrates a smartphone via the supported
Apple CarPlay and Android Auto operation systems.
The Volvo XC90 features an advanced standard safety
package with an innovative auto brake assistant for
intersections as well as special run-off road protection, which protects passengers in case of accidents
caused by the car driving off the road. The City Safety
auto brake system can furthermore automatically
detect vehicles, cyclists and pedestrians at day and
night. Consistent and convincing in its details, the
SUV can be equipped with a wide range of powerful
four-cylinder Drive-E engines. Also in this respect,
the essence of the brand is perpetuated in an extremely contemporary way.

Vorreiter der Form

Die Autos des schwedischen Unternehmens Volvo
stehen für ein hohes Maß an Sicherheit und eine
eigenständige Formensprache. Der als siebensitziges
Premium-SUV konzipierte Volvo XC90 ist Vorreiter
der neuen Volvo Produktarchitektur. Auffällig ist zunächst sein ausdrucksvoll gestaltetes Gesicht mit
dem überarbeiteten Markenemblem. In Verbindung mit
elegant anmutenden T-förmigen Tagfahrleuchten begründet dieses Konzept eine gut wiedererkennbare
Markenidentität. Im Interieur begeistert das Fahrzeug
mit einer wie ein Tablet bedienbaren Touchscreen-
Steuerkonsole. Als Herzstück des gesamten Bordsys-
tems ermöglicht diese die intuitive Steuerung aller
Funktionen, wobei auch das Smartphone mittels der
Betriebssysteme Apple CarPlay oder Android Auto
eingebunden werden kann. Der Volvo XC90 verfügt
serienmäßig über eine hochentwickelte Sicherheitsausstattung mit einem innovativen Notbremsassisten-
ten für Kreuzungen sowie einem speziellen Insassen-
schutz bei Unfällen durch Abkommen von der Fahrbahn
(Run-off Road Protection). Das Notbremssystem
City Safety ist zudem in der Lage, automatisch Fahrzeuge, Fahrradfahrer und Fußgänger bei Tag und
bei Nacht zu erkennen. Das schlüssig und in seinen
Details überzeugend gestaltete SUV kann mit einer
breiten Motorenpalette aus leistungsstarken Drive-E
Vierzylindern ausgestattet werden. Überaus zeitgemäß
wird auch in dieser Hinsicht die Essenz der Marke
weitergeführt.

Statement by the jury

At first glance the Volvo XC90 communicates driving
comfort and follows the demand for unconditional
safety. Impressive features are its innovative safety
package, including an auto brake assistant for intersections and the special passenger protection in case
the car runs off the road. The design successfully
balances an appearance of motorised power and dynamic, elegant lines. A design that is also entirely
reflected in the interior.

Begründung der Jury

Auf den ersten Blick kommuniziert der Volvo XC90
Fahrkomfort und den Anspruch bedingungsloser Sicherheit. Seine innovative Sicherheitsausstattung wie
ein Notbremsassistent für Kreuzungen und der besondere Schutz der Insassen beim Abkommen von der
Fahrbahn ist beeindruckend. Der Gestaltung gelingt die
Balance zwischen der Anmutung motorisierter Stärke
und einer dynamisch-eleganten Linienführung. Eine
Formensprache, die sich auch im Interieur kontinuierlich widerspiegelt.

Designer portrait
See page 72
Siehe Seite 72

Kia Sorento
Passenger Car
Personenkraftwagen

Manufacturer
Kia Motors Europe GmbH, Frankfurt/Main, Germany
In-house design
Kia Design Center Europe
Web
www.kia.com
Honourable Mention

The idea behind the design of the third generation Kia Sorento was to be perceived as street wolf. The result is an SUV which is characterised by a front with long headlamps drawn to the rear, prominent fog lights and an almost vertical cooler grill. Distinctive elements of the rear design are the powerful "shoulders" and the sharply contoured registration plate recess. The modern and spacious interior is designed along horizontal lines. The high proportion of soft-touch materials and leather create an upmarket ambience.

Die Idee hinter der Gestaltung der dritten Generation des Kia Sorento war es, ihn als Straßenwolf zu verstehen. Das Resultat ist ein SUV, dessen Front durch lange, weit nach hinten gezogene Scheinwerfer, auffällige Nebelscheinwerfer und einen fast aufrecht stehenden Kühlergrill geprägt ist. Markante Elemente der Heckansicht sind die kraftvollen „Schultern" und die scharf konturierte, tiefe Kennzeichenmulde. Das moderne und großzügige Interieur ist an horizontalen Linien ausgerichtet. Der hohe Anteil an Soft-Touch-Materialien und Leder schafft ein hochwertiges Ambiente.

Statement by the jury
With its concise front, the Kia Sorento appears powerful. The interior design of the SUV presents an impression of high quality.

Begründung der Jury
Mit seiner prägnanten Front wirkt der Kia Sorento kraftvoll. Die Innenraumgestaltung des SUV hinterlässt einen hochwertigen Eindruck.

Mazda CX-3
Passenger Car
Personenkraftwagen

Manufacturer
Mazda Motor Corporation, Hiroshima, Japan
In-house design
Youichi Matsuda
Web
www.mazda.com

The new Mazda CX-3 is a small SUV, which is characterised by powerful lines and striking proportions. The high cooler grill in combination with the concise LED headlamps emanates confidence. Distinctive features of the cabin are the black D-pillars and a wrap-around window effect, which generate the impression of motion and simultaneously enhance visibility. The new interior offers a comfortable and open atmosphere. The intuitive cockpit concept with head-up display and connectivity technology assures a high level of convenience and driving comfort.

Der neue Mazda CX-3 ist ein kleines SUV, das durch kraftvolle Linien und auffallende Proportionen gekennzeichnet ist. Der hohe Kühlergrill strahlt in Kombination mit den prägnanten LED-Scheinwerfern Selbstbewusstsein aus. Markante Merkmale der Kabine sind die schwarzen D-Säulen und scheinbar umlaufende Fenster, die einen Eindruck von Bewegung erzeugen und zugleich die Sicht verbessern. Das neue Interieur bietet eine komfortable und offene Atmosphäre. Das intuitive Cockpit-Konzept mit Head-up-Display und Konnektivitätstechnologie sorgt für hohen Fahrkomfort.

Statement by the jury
This small SUV with its distinctive exterior indicates presence and gains merit in the interior with ample room and a user-friendly design.

Begründung der Jury
Dieses kleine SUV zeigt mit seinem markanten Exterieur Präsenz und punktet im Innenraum mit viel Platz und einem nutzerorientierten Design.

Niesmann+Bischoff Flair 2015
Motor Home
Reisemobil

Manufacturer
Niesmann+Bischoff GmbH, Polch, Germany
In-house design
Hubert Brandl, Tobias Weiß
Design
Studio SYN (Lars Bergmann), Rüsselsheim, Germany
Web
www.niesmann-bischoff.com
www.studio-syn.de

The Flair 2015 convinces with an enclosed unit of front, sides and rear sections, reminiscent of a car. Horizontal light lines lend structure to the sides of the vehicle, the convex-concave lines continue in the front and rear sections. The rear view camera is located behind the brand logo and thus protected from dirt. Fully insulated chassis hatches are integrated in the sides and can be operated with one hand.

Der Flair 2015 überzeugt mit einer geschlossenen Einheit aus Bug, Seiten und Heck, die das Reisemobil gestalterisch stärker in die Nähe des Automobils rückt. Horizontale Lichtkanten strukturieren die Fahrzeugflanken, die konvex-konkave Linienführung setzt sich in Bug und Heck fort. Die Rückfahrkamera ist geschützt vor Schmutz hinter dem Markenlogo platziert. In die Seiten sind vollisolierte Busklappen integriert, die sich mit einer Hand bedienen lassen.

New City
Urban Bus
Stadtbus

Manufacturer
Carrocera Castrosua S.A., Santiago de Compostela, Spain
In-house design
Sergio Tarrio Sieira
Web
www.castrosua.com

Safety, functionality and comfort are the highlights of design for this urban bus. Accentuating lines determine its avant-garde appearance and add character. Daytime driving lights with LED technology in the form of a "C", standing for Castrosua, indicate the brand identity. The driver's cab is designed ergonomically, and its wrap-around windscreen provides good visibility.

Sicherheit, Funktionalität und Komfort standen bei der Gestaltung dieses Stadtbusses im Mittelpunkt. Akzentuierende Linien bestimmen sein avantgardistisches Erscheinungsbild und verleihen ihm Charakter. Das Tagfahrlicht mit LED-Technologie bildet ein „C", das für Castrosua steht, und prägt die Markenidentität. Der Fahrerplatz ist ergonomisch gestaltet und bietet mit umlaufender Windschutzscheibe eine gute Sicht.

Statement by the jury
The distinctive design of the New City confers this urban bus with a high degree of recognition, while functional details enhance the safety.

Begründung der Jury
Die markante Gestaltung des New City verleiht diesem Stadtbus einen hohen Wiedererkennungswert, während funktionale Details die Sicherheit verbessern.

Velaro Platform
Velaro Plattform
High-Speed Train
Hochgeschwindigkeitszug

Manufacturer
Siemens AG, Krefeld-Uerdingen, Germany
In-house design
Benno Schiefer, Ralf Staub, Nadine Hohlstein, Sander van Lieshout
Web
www.siemens.com

The Velaro is a high-speed train which is certificated in conformity with TSI standard and combines innovative technology with contemporary design. At speeds up to 400 km/h, complex crash absorbers increase the safety of the passengers. Furthermore, special attention was paid to environmental impact, so that the Velaro has an optimised cd value, consumes 20 per cent less energy and is 97 per cent recyclable. Thanks to the platform concept, the interior can be flexibly designed. For this conception, the needs of passengers with reduced mobility have been taken into consideration.

Der Velaro ist ein Hochgeschwindigkeitszug, zugelassen nach TSI-Norm, der innovative Technologie mit zeitgemäßem Design vereint. Bei Geschwindigkeiten von bis zu 400 km/h erhöhen komplexe Crash-Absorber die Sicherheit der Fahrgäste. Zudem wurde Wert auf Umweltverträglichkeit gelegt, sodass der Velaro einen optimierten cw-Wert hat, 20 Prozent weniger Energie verbraucht und zu 97 Prozent recyclebar ist. Dank des Plattformgedankens kann das Interieur flexibel gestaltet werden. Bei der Konzeption wurden Belange von Fahrgästen mit reduzierter Mobilität berücksichtigt.

Statement by the jury
With the platform concept, the Velaro offers great flexibility of interior design. Furthermore, the aerodynamic train scores points with superior safety and environmental compatibility.

Begründung der Jury
Mit dem Plattformkonzept bietet der Velaro große Flexibilität bei der Innenraumgestaltung, außerdem punktet der aerodynamische Zug mit hoher Sicherheit und Umweltverträglichkeit.

High-Speed Train ICx
High-Speed Triebzug ICx

Manufacturer
Bombardier Transportation,
Hennigsdorf, Germany
In-house design
Marco Caspari, Stefan Steilen,
Gero Garske, Michael Brogan
Web
www.bombardier.com

The ICx was developed by Bombardier and Siemens for Deutsche Bahn as successor to the ICE1 and IC trains, whereby Bombardier was responsible for the exterior design. A considerably improved aerodynamic form characterises the new exterior design of the 250 km/h train and imparts a high degree of formal aesthetic design quality. At the same time it displays typical design features of its predecessors and thus integrates harmoniously in the existing vehicle family of the DB long-distance rail vehicles.

Statement by the jury
The exterior design of the ICx successfully combines a clear, aerodynamic form with typical features of the train family.

Der ICx wurde als Nachfolger der ICE1- und IC-Züge von Bombardier und Siemens für die Deutsche Bahn entwickelt, wobei Bombardier für das Außendesign verantwortlich war. Eine deutlich verbesserte aerodynamische Form kennzeichnet das neue Exterieurdesign des 250 km/h schnellen Zuges und verleiht ihm seine hohe formal-ästhetische Gestaltungsqualität. Gleichzeitig zeigt er typische Designmerkmale seiner Vorgänger und integriert sich auf diese Weise harmonisch in die bestehende Fahrzeugfamilie des DB-Fernverkehrs.

Begründung der Jury
Das Exterieurdesign des ICx verbindet auf gelungene Weise eine klare, aerodynamische Formgebung mit typischen Merkmalen der Fahrzeugfamilie.

Solaris Tramino Braunschweig
Tram
Stadtbahn

Manufacturer
Solaris Bus & Coach S.A., Owińska, Poland
Design
studioFT (Jens Timmich), Berlin, Germany
Web
www.solarisbus.de
www.studioft.com

For the Tramino Braunschweig, integrated design solutions have been created which combine the operative demands on a modern public transport system with efficient technical solutions. The design of the vehicle unites high functionality and emotional content in order to achieve the greatest possible acceptance for the new tram. A distinctive exterior design and a spacious, modern and barrier-free interior characterise the outward appearance.

Statement by the jury
Due to a convincing overall concept, the Tramino Braunschweig creates acceptance by the passengers and upgrades the urban public transport system.

Für den Tramino Braunschweig wurden integrierte Designlösungen geschaffen, die operative Anforderungen an ein modernes öffentliches Verkehrssystem und leistungsfähige technische Lösungen kombinieren. Das Fahrzeugdesign vereint hohe Funktionalität und emotionale Komponenten, um eine möglichst große Akzeptanz für die neue Bahn zu erreichen. Ein markantes Exterieurdesign und ein großzügiges, modernes und barrierefreies Interieur prägen das Erscheinungsbild.

Begründung der Jury
Durch ein überzeugendes Gesamtkonzept schafft der Tramino Braunschweig Akzeptanz bei den Fahrgästen und wertet den öffentlichen Personennahverkehr auf.

S200 Calgary
Light Rail Vehicle
Stadtbahn

Manufacturer
Siemens Industry, Inc, Mobility Division, Sacramento, USA
Design
Tricon Design AG, Kirchentellinsfurt, Germany
Web
www.siemens.com/mobility
www.tricon-design.de

The S200 Calgary is a modern high-floor urban tram, the front end design of which is inspired by the goalie mask of the NHL's Calgary Flames, symbolising safety and protection. The large windscreen wraps around the A-pillars and provides good visibility. The impression of a visor is created, partly due to the dynamic shape of the surface. This theme is emphasised by a red line running around the entire vehicle. The concave surface of the roof panels which diminishes towards the front is similarly emphasised in red. The intelligent lighting concept assures high visibility.

Der S200 Calgary ist eine moderne Hochflur-Stadtbahn, deren Frontgestaltung von der Eishockeymaske der Calgary Flames inspiriert ist und Sicherheit und Schutz symbolisiert. Die große Frontscheibe läuft um die A-Säulen und bietet gute Sicht. Der Eindruck eines Visiers entsteht unter anderem durch die dynamische Flächengestaltung. Dieses Thema wird durch eine rote Linie verstärkt, die sich um das ganze Fahrzeug zieht. Die konkave Fläche der Dachschürzen, die zur Front hin ausläuft, wird ebenfalls durch rote Farbe betont. Ein intelligentes Lichtkonzept sorgt für gute Erkennbarkeit.

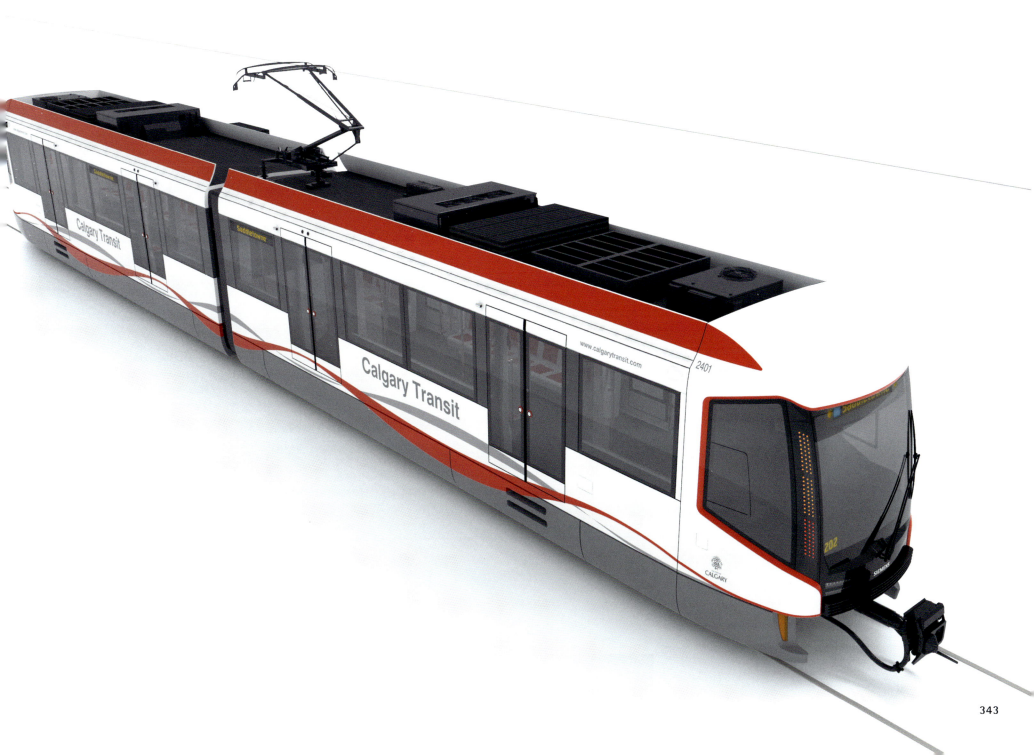

MT-07
Motorcycle
Motorrad

Manufacturer
Yamaha Motor Co., Ltd.,
Shizuoka, Japan

In-house design
Design Center
(Toshiyuki Yasunaga)

Design
GK Dynamics Incorporated
(Kazumasa Sasanami),
Tokyo, Japan

Web
www.yamaha-motor.co.jp
www.gk-design.co.jp/dynamics

reddot award 2015
best of the best

Pure riding pleasure

It is not just since the famous road movie "Easy Rider" about late 1960's America that riding a motorcycle has been connected with a lifestyle and a feeling of absolute freedom on the road. It is a special experience that lets one quickly forget everyday life. The design of the MT-07 motorcycle aims at visualising a lightness of motorcycling that, above all, places the fun of riding centre stage. Riding this motorcycle should feel like wearing one's favourite casual outfit and be an expression of their personal lifestyle. This idea is implemented through a fascinatingly purist design, lending the motorcycle a highly concise appearance. It impresses with a lightweight, slim and compact chassis, which provides a highly agile appeal. An in-line two-cylinder, 689-cc engine has been conclusively integrated into the overall well-proportioned form, offering riders a powerful torque. This combination perfectly matches the overall concept of the MT-07, since it adds a pleasant feeling of acceleration to the riding comfort. With these features, this motorcycle fascinates both, experienced riders, who do not insist on terms like "performance" or "top speed", and entry-level riders likewise. Thanks to its entirely successful design, it conveys a good sense of the pleasure of riding a motorcycle to both target groups.

Purer Fahrspaß

Nicht erst seit dem berühmten Roadmovie „Easy Rider" über das Amerika der späten 1960er Jahre verbindet sich mit dem Motorradfahren ein Lebensgefühl von unbegrenzter Freiheit auf der Straße. Es ist ein besonderes Erlebnis, das den Alltag rasch vergessen lässt. Die Gestaltung des Motorrads MT-07 will eine Leichtigkeit des Motorradfahrens visualisieren, bei der vor allem der Fahrspaß im Vordergrund steht. Das Fahren mit diesem Motorrad soll sich anfühlen wie das Tragen des Lieblingsoutfits in der Freizeit und dabei Ausdruck des persönlichen Lebensstils sein. Umgesetzt wird dies mittels einer faszinierend puristischen Gestaltung, durch die das Motorrad sehr prägnant wirkt. Es beeindruckt mit einem leichten, schlank und kompakt gebauten Chassis, welches ihm eine sehr agile Anmutung verleiht. Schlüssig in die insgesamt wohl proportionierte Form integriert wurde ein 689-ccm-Reihenmotor mit zwei Zylindern, der beim Fahren ein starkes Drehmoment bietet. Diese Kombination passt ausgezeichnet zum Gesamtkonzept der MT-07, da sie dem Fahrer zum Fahrkomfort auch ein angenehmes Beschleunigungsgefühl gibt. Derart ausgerüstet, begeistert dieses Motorrad sowohl versierte Fahrer, die nicht auf den Aspekten „Leistung" oder „Höchstgeschwindigkeit" beharren, wie auch Einsteiger gleichermaßen. Durch seine rundum gelungene Gestaltung vermittelt es beiden Zielgruppen ein gutes Gefühl für die Freuden des Motorradfahrens.

Statement by the jury

With its well-balanced design, the MT-07 motorcycle communicates power and agility. Purist and skilfully reduced to the essential elements, each detail appears to be exactly where it belongs and is directed at highest efficiency. The rear view is a particular visual highlight, while this motorcycle demonstrates a perfect relationship of design and riding experience. For the rider, this means pure riding pleasure.

Begründung der Jury

Mit seiner gut ausbalancierten Gestaltung kommuniziert das Motorrad MT-07 Kraft und Agilität. Puristisch und gekonnt auf die wesentlichen Elemente reduziert, wirkt jedes Detail an seinem Platz und auf höchste Effizienz ausgerichtet. Die Rückansicht ist ein besonderes visuelles Highlight, wobei diesem Motorrad ein perfektes Verhältnis zwischen Formensprache und Fahrerlebnis gelingt. Für den Fahrer bedeutet dies Fahrspaß pur.

Designer portrait
See page 74
Siehe Seite 74

BMW S 1000 RR
Motorcycle
Motorrad

Manufacturer
BMW Group, Munich, Germany
In-house design
Web
www.bmwgroup.com

The BMW S 1000 RR in its third generation meets high demands of performance and handling. The lines and proportions of the powerful superbike impart maximum speed and precision. The high rear and low front sections create in a side view a dynamic wedge shape. At the same time, the BMW S 1000 RR continues with the familiar RR design features such as asymmetrical side panels with their characteristic grills and the so-called split face with the asymmetrical headlights in a more dynamic form.

Die BMW S 1000 RR in ihrer dritten Generation erfüllt hohe Ansprüche an Leistung und Fahrverhalten. Linienführung und Proportionen des kraftvollen Superbikes vermitteln Höchstgeschwindigkeit und Präzision. Das hohe Heck und die tiefe Front formen in der Seitenansicht eine dynamische Keilform. Gleichzeitig führt die BMW S 1000 RR die bekannten RR-Gestaltungsmerkmale wie die asymmetrischen Seitenverkleidungsteile mit ihren charakteristischen Kiemen und das sogenannte Split-Face mit den asymmetrischen Scheinwerfern in einer noch dynamischeren Form fort.

BMW S 1000 XR
Motorcycle
Motorrad

Manufacturer
BMW Group, Munich, Germany
In-house design
Web
www.bmwgroup.com

The BMW S 1000 XR is a motorcycle from the segment adventure sport. It unites elements from the GS segment such as the flyline, the stylised beak and the relatively long suspension with elements from the touring segment such as the aerodynamically sophisticated front fairing and the intelligent storage concept. Elements such as the split face and the asymmetrical side fairing components, on the other hand, stem from the supersports segment. This novel combination signifies the versatile capabilities of the machine and merges emotion and function.

Die BMW S 1000 XR ist ein Motorrad aus dem Segment Adventure Sport. In ihr vereinen sich Elemente aus dem GS-Bereich wie die Flyline, der angedeutete Schnabel und die relativ langen Federwege mit Elementen aus dem Tourensegment, wie die aerodynamisch ausgefeilte Frontverkleidung und das intelligente Stauraumkonzept. Elemente wie das Split-Face und die asymmetrischen Seitenverkleidungsteile wiederum stammen aus dem Supersport-Segment. Diese neuartige Kombination kommuniziert die vielfältigen Einsatzmöglichkeiten der Maschine und verbindet Emotion und Funktion.

347

BMW R 1200 R
Motorcycle
Motorrad

Manufacturer
BMW Group, Munich, Germany
In-house design
Web
www.bmwgroup.com

The BMW R 1200 R is a versatile roadster. Dynamic proportions and an elevated, filigree rear provide a slight wedge shape and indicate the sportiness of the motorcycle. The headlight is compactly integrated in the fork head and thus provides an enclosed silhouette. The fairing is reduced to essentials, typical for a naked bike. Slots and intentionally designed openings give the motorcycle an impression of lightness. The ergonomic seat facilitates an optimal union of rider and bike and fatigue-free riding, even on long journeys.

Die BMW R 1200 R ist ein vielseitiger Roadster. Dynamische Proportionen und ein erhöhtes, filigranes Heck zeichnen eine leichte Keilform und kommunizieren ihre Sportlichkeit. Der Scheinwerfer ist kompakt in den Gabelkopf integriert und sorgt so für eine geschlossene Silhouette. Die Verkleidung ist auf das Nötigste reduziert, wie es für ein Naked-Bike typisch ist. Durchbrüche und bewusst gestaltete Öffnungen lassen das Motorrad leicht wirken. Die ergonomische Sitzbank ermöglicht eine optimale Verbindung von Fahrer und Maschine und eine ermüdungsfreie Fahrt auch auf langen Strecken.

Statement by the jury
Comfort, sportiness and a streamlined, muscular design characterise this versatile roadster.

Begründung der Jury
Komfort, Sportlichkeit und eine straffe, muskulöse Gestaltung kennzeichnen diesen vielseitigen Roadster.

BMW C evolution
Electrically Powered Maxi Scooter
Elektro-Maxi-Scooter

Manufacturer
BMW Group, Munich, Germany
In-house design
Web
www.bmwgroup.com

The BMW C evolution is the first all-electric powered scooter from BMW Motorrad, conceived as a modern, agile vehicle for the urban environment. Its distinctive form language appears futuristic and at the same time displays the characteristic elements of a BMW motorcycle, such as the so-called split face of the front. The clear and reduced design, the LED lamps and the TFT display impart modernity and innovative capacity. The bright white and electric green colours indicate on the one hand dynamics, on the other hand environmental compatibility due to emission-free driving.

Der BMW C evolution ist der erste vollelektrisch angetriebene Scooter von BMW Motorrad, der als modernes, agiles Fahrzeug für den urbanen Raum konzipiert ist. Seine markante Formensprache wirkt futuristisch und zeigt zugleich charakteristische Elemente eines BMW-Motorrads wie das sogenannte Split-Face der Front. Das klare und reduzierte Design, die LED-Leuchten und das TFT-Display vermitteln Modernität und Innovationskraft. Die Farben Hellweiß und Electricgreen stehen für Dynamik einerseits und Umweltverträglichkeit durch emissionsfreies Fahren andererseits.

KTM 1290 SUPER ADVENTURE
Motorcycle
Motorrad

Manufacturer
KTM Sportmotorcycle GmbH, Mattighofen, Austria
Design
KISKA GmbH, Anif (Salzburg), Austria
Web
www.ktm.com
www.kiska.com

The design of the KTM 1290 Super Adventure was based on the premise of creating a motorcycle which is very suitable for touring and simultaneously remains sporty and dynamic. The result is a comfortable travel enduro with simple, powerful and dynamic form language and a tank with greater capacity. Furthermore, wind protection has been enhanced by the development of an innovative, particularly aerodynamic double windscreen.

Die Gestaltung der KTM 1290 Super Adventure stand unter der Prämisse, ein Motorrad mit hoher Reisetauglichkeit unter gleichzeitiger Beibehaltung von Sportlichkeit und Dynamik zu schaffen. Das Ergebnis ist eine komfortable Reiseenduro mit einer einfachen, kräftigen und dynamischen Formensprache und einem größeren Tankvolumen. Zudem wurde der Windschutz durch die Entwicklung eines innovativen, besonders aerodynamischen doppelten Windschildes verbessert.

Statement by the jury
As a travel enduro, the KTM 1290 Super Adventure combines a powerful, aggressive appearance with unusually high comfort for long distance travel.

Begründung der Jury
Als Reiseenduro vereint die KTM 1290 Super Adventure ein kraftvolles, aggressives Auftreten mit ungewohnt hohem Komfort auf längeren Strecken.

TRICITY
3-Wheel Scooter
Dreirad-Motorroller

Manufacturer
Yamaha Motor Co., Ltd., Shizuoka, Japan
In-house design
Design Center (Hirotoshi Noguchi)
Web
www.yamaha-motor.co.jp

The front wheels of this 3-wheel 125 cc city scooter employ a special mechanism (Leaning Multi Wheel), which allows them to tilt with the chassis when cornering. The independently suspended front wheels and their connecting mechanism have an optimised weight distribution to make driving more fun. Furthermore, a flat running board for mounting and dismounting and a spacious storage compartment under the seat increase the convenience of the commuter vehicle.

Die Vorderräder dieses dreirädrigen 125er-Stadtrollers verwenden einen speziellen Mechanismus (Leaning Multi Wheel), der es ihnen erlaubt, sich bei Kurvenfahrten mit dem Chassis zu neigen. Die unabhängig aufgehängten Vorderräder und ihr Verbindungsmechanismus sorgen für eine optimierte Gewichtsverteilung und machen das Fahren angenehm. Zudem erhöhen ein flaches Trittbrett zum Auf- und Absteigen und ein geräumiges Staufach unter dem Sitz den Komfort des Pendlerfahrzeugs.

Statement by the jury
The tilting mechanism of the front wheels increases the fun of driving; emphasis on the wheels provides Tricity with a singular and yet familiar appearance.

Begründung der Jury
Der Neigemechanismus der Vorderräder erhöht den Fahrspaß, die Betonung der Räder verleiht Tricity ein eigenständiges und zugleich vertrautes Erscheinungsbild.

BRP Can-Am Spyder F3
Roadster

Manufacturer
BRP Inc., Valcourt, Canada
In-house design
Web
www.brp.com

The three-wheeled roadster Can-Am Spyder F3 was conceived as a cruiser; at the same time, great emphasis in the design was given to an autonomous appearance in order to set it apart from other cruisers. Its construction facilitates high performance and ergonomic comfort at a low driving position. For overall safety, the Spyder F3 provides stability when stationary and when being driven. Well-balanced proportions, smooth lines and a distinctive form language characterise the cool and at the same time dynamic overall impression of this roadster with borrowings from cruiser style.

Der dreirädrige Roadster Can-Am Spyder F3 ist als Cruiser konzipiert. Gleichzeitig wurde bei der Gestaltung Wert auf ein eigenständiges Erscheinungsbild gelegt, um ihn von anderen Cruisern zu differenzieren. Seine Bauweise ermöglicht eine hohe Leistung und ergonomischen Komfort bei einer tiefen Fahrposition. Für umfassende Sicherheit bietet der Spyder F3 Stabilität im Stand und beim Fahren. Stimmige Proportionen, fließende Linien und eine markante Formensprache prägen den lässigen und zugleich dynamischen Gesamteindruck des Roadsters mit Cruiser-Anleihen.

BRP Can-Am Outlander L
All-Terrain Vehicle
Geländewagen

Manufacturer
BRP Inc., Valcourt, Canada
In-house design
Web
www.brp.com

The Can-Am Outlander L was designed on the premise of developing an inexpensive and efficient vehicle and thereby retaining features such as precision-engineered steering and driver-oriented design. Powered by two economical engine options, the light-weight utility vehicle provides reliable performance and torque. Its nimble chassis and an ergonomic design assure good manoeuvrability even in difficult terrain.

Die Gestaltung des Can-Am Outlander L stand unter der Prämisse, ein preisgünstiges und effizientes Fahrzeug zu entwickeln und dabei Merkmale wie eine präzisionsgefertigte Lenkung und ein fahrerorientiertes Design beizubehalten. Angetrieben von zwei sparsamen Motorvarianten, liefert das leichte Nutzfahrzeug zuverlässige Leistung und Drehkraft. Sein wendiges Chassis und eine ergonomische Gestaltung sorgen auch in schwierigem Gelände für gute Manövrierbarkeit.

Statement by the jury
The ergonomic, driver-friendly design of the Can-Am Outlander L contributes to good controllability of this sporty off-road vehicle.

Begründung der Jury
Die ergonomische, fahrerorientierte Gestaltung des Can-Am Outlander L trägt zu einer guten Beherrschbarkeit dieses sportlichen Geländewagens bei.

BRP Sea-Doo SaR Search and Rescue
Jet Ski

Manufacturer
BRP Inc., Valcourt, Canada
In-house design
Web
www.brp.com

This jet ski was specially developed for rescue missions and offers high performance in the surf, white water, along rocky coasts and in floods. The rescue jet boat is manoeuvrable and robust; dual, rigid floaters and running boards provide the necessary stability and buoyancy in rough waters. Equipment for shallow water facilitates targeted assistance in case of natural disasters. Overall size and weight assure speedy deployment readiness.

Dieser Jet-Ski wurde speziell für Rettungseinsätze entwickelt und bietet hohe Leistung in der Brandung, in Wildwassern, an Felsküsten oder bei Fluten. Das Rettungsjetboot ist wendig und robust; duale, steife seitliche Schwimmkörper und Trittflächen verleihen die nötige Stabilität und Schwimmkraft in wilden Gewässern. Eine Ausstattung für flache Gewässer ermöglicht gezielte Hilfe bei Naturkatastrophen. Gesamtgröße und -gewicht gewährleisten eine rasche Einsatzbereitschaft.

Statement by the jury
In this jet boat, all details are optimised for rescue missions, and this is indicated by a powerful form language.

Begründung der Jury
Bei diesem Jetboot sind alle Details für den Rettungseinsatz optimiert, was durch eine kraftvolle Formensprache kommuniziert wird.

C-Explorer 3
Submersible
Tauchboot

Manufacturer
U-Boat Worx, Breda, Netherlands
In-house design
Web
www.uboatworx.com

The C-Explorer 3 is a submersible which allows three people to stay for several hours at up to 300 metres depth. A distinctive highlight is the spherical pressure body which is made completely of acrylic and facilitates a 360-degree panoramic view of the underwater world. The C-Explorer 3 can be equipped with cameras, spotlights or robotic arms, making it also suitable as exploratory submarine or for maintenance work.

Der C-Explorer 3 ist ein Tauchboot, mit dem sich drei Personen mehrere Stunden in bis zu 300 Metern Tiefe aufhalten können. Markantes Merkmal ist der komplett aus Acryl gefertigte, kugelförmige Druckkörper, der einen 360-Grad-Panoramablick auf die Unterwasserwelt ermöglicht. Der C-Explorer 3 kann mit Kameras, Strahlern oder Roboterarmen ausgestattet werden, sodass er sich auch als Forschungs-U-Boot oder für Wartungsarbeiten eignet.

Diverto QS 100
Tractor, Excavator, Loader, Mower
Traktor, Bagger, Radlader, Mäher

Manufacturer
Diverto Technologies BV,
Wemeldinge, Netherlands

In-house design
Leonard Huissoon &
Team Diverto 2014

Design
DB Industrial Design,
Sheffield, Great Britain

Web
www.diverto.com
www.dbindustrialdesign.com

reddot award 2015
best of the best

Multifunctional platform

Work tasks in urban areas, such as municipalities, in landscape architecture or forestry, are manifold and highly varied. Verges have to be mowed, trenches excavated and in winter snow ploughed. Based on a well thought-through concept, the Diverto QS 100 is aimed at presenting the possibility of fulfilling all these tasks with one machine. In this sense, it is a multifunctional machine platform that combines tractor, excavator, loader and mower. As a tractor, it is equipped with an on-road approved towing trailer and incorporates state-of-the art emission technology. The vehicle boasts rotating operation of up to 360 degrees allowing all tasks to be performed easily and effectively. Furthermore, two functions can be used simultaneously, in addition to being remote controlled, and a total of 200 tools and attachments can be connected, turning it into a highly versatile tool carrier. Showcasing an ergonomically designed operator's cabin, this innovative vehicle not only offers high comfort to the operator and also a second person, but also ensures a high degree of all-round visibility as well as safety for an effective work performance. The multifunctional concept of the Diverto QS 100 delivers a new kind of flexibility that also allows the working time to be planned effectively, thus setting new standards in its field.

Multifunktionale Plattform

Die Arbeiten im urbanen Bereich wie etwa in Kommunen, in der Landschaftsarchitektur oder der Forstwirtschaft sind sehr differenziert und vielfältig. Es müssen Böschungen gemäht, Gräben ausgebaggert oder im Winter Schnee geräumt werden. Auf der Grundlage eines gut durchdachten Konzepts ermöglicht es der Diverto QS 100, diese unterschiedlichen Aufgaben mit nur einem Fahrzeug auszuführen. Im Sinne einer multifunktionalen Plattform ist er Traktor, Bagger, Radlader und Mäher zugleich. Als Traktor ist er mit einem für die Straße zugelassenen Anhängerzug ausgestattet und verfügt über eine hochentwickelte Emissionstechnologie. Das Fahrzeug hat einen bis zu 360 Grad reichenden Arbeitsradius, dank dessen alle Arbeiten einfach und effektiv erledigt werden können. Zwei verschiedene Funktionen können dabei gleichzeitig, auch per Fernsteuerung, ausgeführt werden. Für die diversen Tätigkeiten lassen sich zudem etwa 200 unterschiedliche Werkzeuge anschließen. Dem Fahrer wie auch einer zweiten Person bietet dieses innovative Fahrzeug viel Komfort. Die ergonomisch gestaltete Fahrerkabine, die eine sehr gute Rundumsicht erlaubt, gewährleistet ein effektives und sicheres Arbeiten. Das multifunktionale Konzept des Diverto QS 100 ermöglicht eine neue Flexibilität, durch die auch die Arbeitszeit sehr sinnvoll geplant werden kann und dieses Fahrzeug damit neue Standards setzt.

Statement by the jury

Following an innovative approach, the Diverto QS 100 integrates a tractor, excavator, loader and mower into one single vehicle and thus separates itself from the classical form. It embodies a very well thought-through multifunctional platform for the most varied tasks, allowing them to be performed in a more coordinated and effective way. All details have been carefully honed, with an ergonomically designed operator cabin offering great comfort, safety and intuitive handling.

Begründung der Jury

Auf innovative Weise integriert der Diverto QS 100 einen Traktor, Bagger, Radlader und Mäher in nur einem Fahrzeug und löst sich damit von der klassischen Form. Er ist eine überaus gut durchdachte multifunktionale Plattform für die unterschiedlichsten Aufgaben, die dadurch besser koordiniert und effektiver ausgeführt werden können. Alle Details sind sorgfältig entwickelt, wobei die ergonomisch gestaltete Fahrerkabine dem Fahrer viel Komfort, Sicherheit und eine intuitive Bedienung bietet.

Designer portrait
See page 76
Siehe Seite 76

Ponsse Scorpion
Forest Machine
Forstmaschine

Manufacturer
Ponsse Oyj, Vieremä, Finland

In-house design
Ponsse Oyj

Design
LINK Design and
Development Oy,
Espoo, Finland

Web
www.ponsse.com
www.linkdesign.fi

reddot award 2015
best of the best

Perfect visibility

In forestry work, which has to comply with many strict regulations and is measured against the principle of effectivity, the cost effectiveness of forest machines also plays a central role. The Ponsse Scorpion was conceived with the aim of equally satisfying the demands of user-friendliness, productivity and safety, all at the same time. As a flagship of a new type of forest machines, it impresses with a design that unites all elements required for this type of machine into a harmoniously balanced form language. The product graphics of this harvester have also been well implemented as they not only achieve a strong and distinctive signalling effect when working in the forest, but also possess high recognition value. Another innovation in the area of forest machines is presented by the placement of the operator. The newly interpreted operator's cabin delivers unrestricted all-round visibility of the entire working environment. This augments work safety and allows highly comfortable controllability and intuitive operation of the machine. Since the Scorpion also possesses exceptional driving and standing stability as well as minimum surface pressure, it enables maximised harvesting productivity in terms of both environmental awareness and driver ergonomics. Its design thus impressively manages to lend this type of machine both a new design vocabulary and enhanced functionality, paired with a user-friendliness that also defines a new standard for the future.

Perfekte Rundumsicht

In der Forstwirtschaft, die sich an enge Vorgaben halten muss und an der Maxime der Effektivität gemessen wird, spielt auch die Wirtschaftlichkeit der Forstmaschinen eine zentrale Rolle. Der Ponsse Scorpion wurde konzipiert mit der Zielsetzung, Anforderungen wie Benutzerfreundlichkeit, Produktivität und Sicherheit gleichermaßen zu erfüllen. Als Flaggschiff einer neuen Art von Forstmaschinen beeindruckt er durch eine Gestaltung, die alle für eine solche Maschine nötigen Elemente in einer harmonischen Formensprache vereint. Gut gelöst ist auch die Produktgrafik dieses Harvesters, da sie eine markante Signalwirkung im Wald erreicht und zugleich einen hohen Wiedererkennungswert schafft. Eine Innovation im Bereich der Forstmaschinen ist darüber hinaus die Positionierung des Fahrers. Die neu interpretierte Fahrerkabine erlaubt ihm eine uneingeschränkte Rundumsicht auf sein gesamtes Arbeitsumfeld. Dies erhöht die Sicherheit, wobei er überaus komfortabel agieren und die Bedienelemente intuitiv steuern kann. Da der Scorpion zudem über eine hervorragende Fahr- und Standstabilität sowie einen nur minimalen Bodendruck verfügt, ermöglicht er eine maximierte Produktivität mit Rücksicht auf die Umwelt und die ergonomischen Bedürfnisse des Fahrers. Der Gestaltung gelingt es daher eindrucksvoll, einer derartigen Maschine eine neue Formensprache und Funktionalität zu verleihen. Auch durch ihre Nutzerfreundlichkeit werden neue Standards für die Zukunft definiert.

Statement by the jury

Form and functionality are merged in this forest machine into a highly impressive unity. All elements are carefully matched to lend this type of work machine an entirely new appearance. Showcasing a well thought-out operator environment, the harvester offers a high degree of ergonomics and safety. A particularly innovative solution has been realised in the cabin that gives the operator a good all-round vision at all times.

Begründung der Jury

Form und Funktionalität verbinden sich bei dieser Forstmaschine zu einer beeindruckenden Einheit. Alle Elemente sind sorgfältig aufeinander abgestimmt und geben einem solchen Arbeitsgerät eine völlig neue Anmutung. Mit einem gut durchdacht gestalteten Arbeitsumfeld bietet der Harvester ein Höchstmaß an Ergonomie und Sicherheit. Eine besonders innovative Lösung ist dabei die sich drehende Kabine, die dem Fahrer jederzeit eine gute Rundumsicht ermöglicht.

Designer portrait
See page 78
Siehe Seite 78

R 926 Compact
Excavator
Bagger

Manufacturer
Liebherr-France SAS, Colmar, France
In-house design
Jonas Lorenz
Web
www.liebherr.com

The R 926 Compact is a short-tail excavator which offers the operator a considerable space inside the cabin, in spite of its compact construction. The operator's cabin conforms to high ergonomic standards. The service and maintenance points at the side are easily accessible. The hinged maintenance doors with gas springs form a protective roof when open. The Liebherr Litronic system assures good performance, high productivity and low fuel consumption.

Statement by the jury
With its well-considered ergonomic design, a high level of functionality and robust appearance, the R 926 Compact makes an impression of reliability.

Der R 926 Compact ist ein Kurzheckbagger, der dem Maschinenführer trotz kompakter Bauart viel Bewegungsfreiheit bietet. Der Kabinenarbeitsplatz entspricht hohen ergonomischen Standards. Die Service- und Wartungspunkte an der Seite sind bequem erreichbar. Die gasfederunterstützten Wartungsklappen bilden geöffnet ein schützendes Dach. Das Liebherr-Litronic-System garantiert Leistungsfähigkeit, hohe Produktivität und geringen Kraftstoffverbrauch.

Begründung der Jury
Mit seiner durchdachten ergonomischen Gestaltung, hoher Funktionalität und einer robusten Erscheinung hinterlässt der R 926 Compact einen zuverlässigen Eindruck.

Linde P60-P80/W08
Electro Tractor /
Load Transporter
Elektro-Schlepper /
Plattformwagen

Manufacturer
Linde Material Handling GmbH, Aschaffenburg, Germany
Design
Porsche Engineering Group GmbH (Stefan Stark), Weissach, Germany
Web
www.linde-mh.com
www.porscheengineering.com

The Linde P60-80/W08 Tractor is a compact, powerful load hauler with a small turning radius. Thanks to a sophisticated steel construction, it combines the robustness of a warehouse vehicle with the agility of a scooter. The strongly curved front profile offers a safety zone and provides at the same time spacious legroom. The long wheelbase increases driving safety and comfort. The driver's cab offers automotive ergonomics with an extensive number of settings.

Statement by the jury
This tractor gains merit with a dynamic form and excellent ergonomics, thus setting standards in its vehicle category.

Der Linde P60-80/W08-Schlepper ist eine kompakte, kraftvolle Zugmaschine mit kleinem Wenderadius. Dank einer ausgefeilten Stahlkonstruktion kombiniert er die Robustheit eines Lagerfahrzeugs mit der Agilität eines Scooters. Die stark gewölbte Front bietet eine Sicherheitszone und zugleich große Beinfreiheit. Der lange Radstand erhöht Fahrsicherheit und -komfort. Der Fahrerarbeitsplatz bietet automobilkonforme Ergonomie mit vielen Einstellmöglichkeiten.

Begründung der Jury
Dieser Schlepper punktet mit einer dynamischen Formgebung und ausgezeichneter Ergonomie und setzt damit Maßstäbe in seiner Fahrzeugkategorie.

Coguaro4
Mine Mixer
Betonmischer

Manufacturer
CIFA S.p.A., Senago, Italy
In-house design
Web
www.cifa.com

The Coguaro4 has been specially designed for the transport of concrete in difficult environments such as mines or tunnels. When designing, attention was paid to combine ergonomics, functionality and technology with a concise exterior. The mixer is equipped with an innovative hydraulic transmission, a hydraulic pump and two hydraulic motors on the axles, thus assuring high traction. Thanks to compact dimensions, four-wheel drive and steering as well as an infra-red rear view camera, the Coguaro4 remains manoeuvrable even under difficult conditions.

Der Coguaro4 ist speziell für den Transport von Beton in anspruchsvollen Umgebungen wie Minen oder Tunneln konzipiert. Bei der Gestaltung wurde darauf geachtet, Ergonomie, Funktionalität und Technologie mit einem prägnanten Äußeren zu verbinden. Der Mischer ist mit einem innovativen Hydraulikantrieb, einer Hydraulikpumpe und zwei Hydraulikmotoren an den Achsen ausgestattet, die eine hohe Traktion gewährleisten. Dank kompakter Abmessungen, Allradantrieb und -lenkung sowie einer Infrarot-Rückfahrkamera bleibt der Coguaro4 auch unter schwierigen Bedingungen manövrierbar.

Statement by the jury
This concrete mixer for extreme conditions makes by its autonomous appearance a powerful and confident impression.

Begründung der Jury
Dieser Betonmischer für Extrembedingungen zeigt sich durch sein eigenständiges Erscheinungsbild kraftvoll und selbstbewusst.

361

Ammann ARX 90
Articulated Tandem Roller
Knickgelenkte Tandemwalze

Manufacturer
Ammann Schweiz AG, Langenthal, Switzerland
Design
Hammer und Runge Designerpartnerschaft,
Neuss, Germany
Web
www.ammann-group.com
www.hammer-runge.de

The ARX 90 is part of the new design line of the company. Characteristic for the hydraulic tandem roller is its concise lines running between frame, tanks and cabin. The spacious system cabin has been newly conceived from an ergonomic viewpoint. Thanks to a revolving and slidable seat unit, the driver has a good overview; furthermore, all controls including an incorporated monitor in the steering wheel are combined at the driver seat. The "ACEpro" vibration system measures the stiffness and temperature of the asphalt and regulates both frequency as well as amplitude automatically.

Das Design der ARX 90 ist Teil der neuen Gestaltungslinie des Unternehmens. Charakteristisch für die hydraulische Tandemwalze ist der markante Linienverlauf zwischen Rahmen, Tanks und Kabine. Die geräumige Systemkabine wurde unter ergonomischen Aspekten neu konzipiert. Dank einer dreh- und verschiebbaren Sitzeinheit hat der Fahrer einen guten Überblick, zudem sind alle Funktionen inklusive eines ins Lenkrad integrierten Monitors am Fahrerplatz konzentriert. Das Vibrationssystem „ACEpro" misst Verdichtung und Temperatur des Asphalts und regelt sowohl Frequenz als auch Amplitude automatisch.

Statement by the jury
The ARX 90 is logically designed from an ergonomic viewpoint and at the same time displays a distinctive, independent appearance.

Begründung der Jury
Die ARX 90 ist konsequent unter ergonomischen Gesichtspunkten gestaltet und zeigt zugleich ein markantes, eigenständiges Erscheinungsbild.

Ammann RAMMAX 1575
Remote Controlled Trench Roller
Ferngesteuerte Grabenwalze

Manufacturer
Ammann Schweiz AG,
Langenthal, Switzerland
Design
Hammer und Runge Designerpartnerschaft,
Neuss, Germany
Web
www.ammann-group.com
www.hammer-runge.de

This remote controlled trench roller has a maintenance-free articulated swivel joint which assures optimal traction and ground contact. The solar-powered remote control uses infrared technology for safety reasons so that the vehicle can only be operated in visual contact; thanks to a second receiver, however, hindrances can be circumvented without trouble. Both engine cowlings can be fully opened for servicing and maintenance.

Statement by the jury
With its functional, compact and robust form, the Rammax 1575 imparts the impression of strength and reliability.

Diese ferngesteuerte Grabenwalze hat ein wartungsfreies Knick-Pendelgelenk, das permanenten Bodenkontakt für optimale Traktion und Verdichtung garantiert. Die solarbetriebene Fernsteuerung nutzt aus Sicherheitsgründen Infrarot-Technik, sodass die Bedienung der Maschine nur bei Sichtkontakt möglich ist; dank eines zweiten Empfängers können Hindernisse jedoch störungsfrei umfahren werden. Die beiden Maschinenhauben lassen sich für Service und Wartung vollständig öffnen.

Begründung der Jury
Mit ihrer funktionalen, kompakten und robusten Formgebung vermittelt die Rammax 1575 den Eindruck von Stärke und Zuverlässigkeit.

Ammann APH 110-95
High Performance Plate Compactor
Hydraulischer Hochleistungsverdichter

Manufacturer
Ammann Schweiz AG,
Langenthal, Switzerland
Design
Hammer und Runge Designerpartnerschaft,
Neuss, Germany
Web
www.ammann-group.com
www.hammer-runge.de

Due to the combination of the water-cooled three-cylinder diesel engine with an innovative, hydraulic, 3-shaft vibration regulator, the APH 110-95 achieves an unusual compression power for a hand-guided machine. In spite of its great weight, the vehicle is well manoeuvrable. Thanks to a vibration-insulated handlebar, the operator is able to work fatigue-free. The engine cowling can be opened without tools, making servicing and maintenance easier.

Statement by the jury
A high level of functionality and performance characterises this high-performance compactor, which attracts attention due to its distinctive lines.

Der Verdichter APH 110-95 erreicht durch die Kombination des wassergekühlten Dreizylinder-Dieselmotors mit einem innovativen, hydraulischen 3-Wellen-Vibrationserreger eine für handgeführte Maschinen ungewöhnlich hohe Verdichtungsleistung. Trotz ihres hohen Gewichts ist die Maschine sehr beweglich. Dank einer schwingungsentkoppelten Deichsel ermöglicht sie ermüdungsfreies Arbeiten. Die werkzeuglos öffnende Motorhaube erleichtert Service und Wartung.

Begründung der Jury
Eine hohe Funktionalität und Leistungsfähigkeit kennzeichnen diesen Hochleistungsverdichter, der mit einer markanten Linienführung auf sich aufmerksam macht.

BT Levio P Series
Electric Powered Pallet Truck
Elektrischer Niederhubwagen

Manufacturer
Toyota Material Handling Europe,
Mjölby, Sweden
In-house design
Magnus Oliveira Andersson
Web
www.toyota-forklifts.eu

The energy-efficient pallet trucks of the BT Levio P series incorporate simple, safe operation, high comfort and great manoeuvrability. A robust, ergonomic design facilitates all-round visibility for safe working. Foldable side guards and driver flip-up platform, a small turning circle and a height-adjustable tiller arm contribute furthermore to the driving convenience and the efficiency of this compact and simultaneously powerful pallet truck.

Statement by the jury
The BT Levio P series is characterised by high ergonomic quality. The trucks are furthermore agile and nimble and can be manoeuvred well in small spaces.

Die energieeffizienten Niederhubwagen der BT Levio P-Serie sind auf eine einfache, sichere Bedienung mit großem Komfort und hoher Wendigkeit ausgelegt. Eine robuste, ergonomische Gestaltung ermöglicht Rundumsicht für einen sicheren Einsatz. Klappbare Seitenschutzwände und Fahrerplattform, ein kleiner Wendekreis und eine höhenverstellbare Deichsel tragen zusätzlich zu Fahrkomfort und Effizienz dieser kompakten und zugleich kraftvollen Niederhubwagen bei.

Begründung der Jury
Die BT Levio P-Serie zeichnet sich durch eine hohe ergonomische Qualität aus. Zudem sind die Wagen flink und wendig und lassen sich so auch auf engem Raum gut manövrieren.

Mitsubishi EDiA EX
Electric Counterbalance Truck
Elektro-Gegengewichtsstapler

Manufacturer
Rocla Oy, Järvenpää, Finland
In-house design
Kero Uusitalo, Mikko Hietanen
Web
www.rocla.com
www.mitforklift.com

EDiA EX combines a robust construction with operator comfort and safety. The cabin is easy to access, and, due to fingertip levers, an inclined dashboard and a sitting position which offers more all-round visibility, it is very ergonomic. The innovative Sensitive Drive System also assures close coordination between user commands and reaction of the forklift. Thanks to the dual drive of the front axles and all-wheel steering, it is stable and agile.

Statement by the jury
This 4-wheel electric forklift truck fulfils high demands of ergonomics and functionality and thus facilitates intuitive, fluid operation.

EDiA EX vereint eine robuste Bauweise mit Bedienkomfort und Sicherheit. Der Arbeitsplatz ist leicht zugänglich und mit den in die Armlehne integrierten Fingertipp-Hebeln, einem abgewinkelten Armaturenbrett und einer Sitzposition, die mehr Rundumsicht bietet, sehr ergonomisch. Das innovative Sensitive-Drive-System sorgt zudem für eine feine Abstimmung zwischen Benutzerbefehlen und Reaktion des Staplers. Dank Doppelantrieb an den Vorderachsen und Allradlenkung ist er stabil und wendig.

Begründung der Jury
Dieser 4-Rad-Elektrostapler erfüllt hohe Ansprüche an Ergonomie und Funktionalität und ermöglicht so eine intuitive, flüssige Bedienung.

Sandvik DD422i Development Drill Rig
Sandvik DD422i Vortriebsbohrwagen

Manufacturer
Sandvik Mining, Tampere, Finland
In-house design
Web
www.mining.sandvik.com

The new drill rig for hard rock mining has been developed for intelligent exploration drilling in underground work. The DD422i is equipped with an innovative drilling and boom control system, an ergonomic cabin as well as an improved carrier and offers a high level of automation. The design is based on a new, modular system platform, thanks to which its economic efficiency, productivity and safety have been improved.

Der neue Vortriebsbohrwagen für den Hartgesteinsabbau wurde speziell für intelligentes Aufschlussbohren im Tiefbau entwickelt. Der DD422i ist mit einem innovativen Bohr- und Ausleger-Kontrollsystem, einer ergonomischen Kabine und einem überarbeiteten Träger ausgestattet und bietet einen hohen Automatisierungsgrad. Die Gestaltung basiert auf einer neuen modularen Systemplattform, dank derer Wirtschaftlichkeit, Produktivität und Sicherheit verbessert werden konnten.

Statement by the jury
The DD422i convinces with high functionality. The underlying modular system contributes to greater safety and cost-effectiveness.

Begründung der Jury
Der DD422i überzeugt mit hoher Funktionalität. Das zugrunde liegende modulare System trägt zu mehr Sicherheit und Wirtschaftlichkeit bei.

Bullet 1000
Motorcycle Lamps
Motorradleuchten

Manufacturer
Kellermann GmbH,
Aachen, Germany

In-house design
Johannes Leugers,
Guido Kellermann

Web
www.kellermann-online.com

Impressive light scenario

The bullet motorcycle lamp showcases a familiar classical form that was used often in the era of streamline design. Featuring a strong sense of visual presence and plasticity in itself, it also achieves a similar distinctive effect in the field of motorcycle design. The design of the Bullet 1000 succeeds in delivering an exciting redefinition of this form by combining it with modern LED technology. At the centre of the design for this lamp series is the eye-catching and easily recognisable design element of a ring around the indicator, which further enhances the signalling effect of the lamps. The visual appearance is marked by a red ring around the rear and brake lights at the back, and a yellow ring for the position lights in front. Together they achieve an overall visual scenario of spectacular appearance that blends in well with the design vocabulary of the lamps. Alongside a pure indicator in the front, the sophisticated concept comprises a rear indicator with rear and break lights as well as a front indicator with position light. The motorcycle itself, its position and driving trajectory, is thus easily visible in daily traffic. The impressive elegance of these lamps is further enhanced by the use of high-quality materials; all metal parts are available in either polished chrome, satin chrome or satin black finish. The reinterpretation of a familiar design idiom paired with meaningfully employed contemporary lighting technology thus emerges as a lamp series with enticing qualities.

Eindrucksvolles Lichtszenario

Die Bullet-Leuchte hat eine bekannte klassische Form, welche besonders in der Ära des Streamline-Designs verwendet wurde. Durch ihre visuelle Präsenz und Plastizität erzielt sie auch im Einsatz bei Motorrädern eine prägnante Wirkung. Der Gestaltung der Bullet 1000 gelingt eine spannende Neudefinition dieser Form in Verbindung mit zeitgemäßer LED-Technologie. Ein auffälliges und gut wiedererkennbares gestalterisches Element dieser Leuchtenserie für Motorräder ist ein Leuchtring um den Blinker, der zugleich die Signalwirkung erhöht. Bei den Rücklichtern und den Positionsleuchten vorne prägt ein roter bzw. gelber Leuchtring das Erscheinungsbild. In der Gesamtheit entsteht ein spektakulär anmutendes Leuchtbild, das sich gut mit der Formensprache dieser Leuchten ergänzt. Das ausgeklügelte Konzept umfasst neben einem einfachen Blinker einen hinteren Blinker mit Rück- und Bremslicht sowie einen vorderen Blinker mit Positionsleuchte. Das Motorrad selbst sowie alle seine Aktivitäten sind im Straßenverkehr deshalb gut sichtbar. Die eindringliche Eleganz dieser Leuchten wird bekräftigt durch den Einsatz hochwertiger Materialien; alle Metallteile sind wahlweise hochglanzverchromt, seidenmatt verchromt oder schwarz seidenmatt lackiert. Die Neuinterpretation einer bekannten Formensprache im Einklang mit einer sinnvoll genutzten zeitgemäßen Lichttechnologie brachte somit eine Leuchtenserie mit bestechenden Eigenschaften hervor.

Statement by the jury

Thanks to a distinctive design vocabulary, the design of the Bullet 1000 lamp series merges the classic form of the bullet lamp with state-of-the-art LED technology. The resulting functionality is thus perfectly implemented to the last detail. The ring around the indicator possesses outstanding signalling effect, featuring in red for the rear and brake lights at the back and yellow for the indicator and position light at the front. Made of high-quality materials, these motorcycle lamps are both durable and highly elegant.

Begründung der Jury

Mittels einer prägnanten Formensprache vereint die Gestaltung der Serie Bullet 1000 die klassische Form der Bullet-Leuchte mit modernster LED-Technik. Perfekt umgesetzt auch im Detail ist die damit einhergehende Funktionalität. Eine sehr gute Signalwirkung hat ein Leuchtring um den Blinker, beim Blinker hinten mit Rück- und Bremslicht in Rot und beim Blinker vorn mit Positionslicht in Gelb. Durch die Fertigung aus hochwertigen Materialien sind diese Leuchten für Motorräder ebenso elegant wie langlebig.

Designer portrait
See page 80
Siehe Seite 80

Headlamp for Audi TT3
Scheinwerfer für den Audi TT3

Manufacturer
Automotive Lighting Reutlingen
GmbH, Reutlingen, Germany

Design
AUDI AG, Ingolstadt, Germany

Web
www.al-lighting.com
www.audi.com

reddot award 2015
best of the best

Matrix for the street

The shape of the headlamp plays a decisive role in lending a car its distinctive character. The lamps take centre stage in most automotive design processes and are part of the planning process from an early stage. The headlamp for the top version Audi TT3 impresses with a design idiom that is both futuristic and sporting in appearance. It is perfectly matched to enhance the adaptive function of the so-called "Matrix" light beam: a highly effective, glare-free high beam. The headlamps are equipped with an e-light projection module forming the low beam, while three reflectors bring the segmented glare-free high beam into operation. Also outstanding is the fact that the "Matrix" high beam is operational in automatic mode and activated by means of a camera located behind the wind screen. The high beam function is electronically controlled and the integrated LED chips are individually addressed, which ensures a very precise light distribution. The technological innovation of this headlamp is complemented with sophisticatedly designed elements such as a dynamic, linear-shaped daytime running light and a progressive turn signal. Forming an impressive unity of design and light technology, the headlamp emerges as a distinctive eye-catcher that also defines the character of the Audi TT3 with its particular lighting performance.

Matrix für die Straße

Die Form der Scheinwerfer eines Autos trägt entscheidend zur Wirkung des Fahrzeugs bei. Sie spielen deshalb im Gestaltungsprozess eine wichtige Rolle und werden in vielerlei Hinsicht in die Planung einbezogen. Die Scheinwerfer für die Top-Version des Audi TT3 beeindrucken mit einer futuristischen wie sportlichen Formensprache. Diese ist perfekt darauf abgestimmt, die adaptive Funktion des „Matrix"-Lichts zum Ausdruck zu bringen: ein sehr effektives, blendfreies Dauerfernlicht. Die Scheinwerfer sind mit einem E-Light-Projektionsmodul ausgestattet, welches das Abblendlicht kreiert, während drei Reflektoren das segmentierte, blendfreie Dauerfernlicht realisieren. Bemerkenswert ist zudem, dass das „Matrix"-Fernlicht in einem automatischen Modus funktionsbereit ist und mithilfe einer hinter der Windschutzscheibe platzierten Kamera geschaltet wird. Die Fernlichtfunktion ist elektronisch geregelt und die integrierten LED-Chips sind einzeln adressierbar, was zu einer sehr präzisen Lichtverteilung führt. Die lichttechnologische Innovation dieser Scheinwerfer geht einher mit gestalterisch sehr gut akzentuierten Elementen wie einem dynamisch-linear geformten Tagfahrlicht und einem progressiv anmutenden Fahrtrichtungsanzeiger. In einer beeindruckenden Einheit von Design und Lichttechnologie entstand ein faszinierender Scheinwerfer, der durch seine besondere Lichtleistung zugleich den Charakter des Audi TT3 definiert.

Statement by the jury

This full LED headlamp for the Audi TT3 successfully merges all elements into a symbiotic unity. With its highly impressive design, it serves as a dynamically shaped daytime running light as well as a progressive turn signal. This headlamp delivers a highly effective lighting performance thanks to its adaptive function of the glare-free "Matrix" high beam. With its futuristic design it underlines the distinctive character and expression of the Audi TT3.

Begründung der Jury

Bei diesem Voll-LED-Scheinwerfer für den Audi TT3 wurden alle Elemente in einer symbiotischen Einheit zusammengefasst. Sehr eindrucksvoll gestaltet sind ein dynamisch fließend geformtes Tagfahrlicht sowie ein progressiv anmutender Fahrtrichtungsanzeiger. Dieser Scheinwerfer bietet eine hocheffektive Lichtleistung durch seine adaptive Funktion des blendfreien „Matrix"-Dauerfernlichts. Mit seinem futuristischen Design unterstreicht er die markante Ausdruckskraft des Audi TT3.

Designer portrait
See page 82
Siehe Seite 82

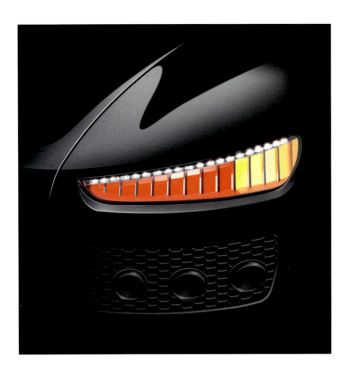

Rear Lamp
Rückleuchte
Exterior Automotive Lighting
Externe Fahrzeugbeleuchtung

Manufacturer
Automotive Lighting Reutlingen GmbH,
Reutlingen, Germany
In-house design
Web
www.al-lighting.com

The design of this exterior rear lamp is based on organic LEDs (OLEDs) which can be used not only in the interior of the car, but also for the rear lighting. The flat and thin modules emit a homogenous light over their entire surface, do not overheat and are energy-saving. The light sources, which are red and yellow, follow elegantly the outline contour of the car body and thus integrate harmoniously in the lines of the car. When switched off, the OLED elements convince with their very elegant silvery mirror look.

Statement by the jury
Thanks to the use of OLEDs with this type of rear lighting, forms can be realised which until now were not possible.

Die Gestaltung dieser externen Rückleuchte basiert auf organischen LEDs (OLEDs), die nicht nur im Fahrzeuginnenraum, sondern auch bei der Heckbeleuchtung eingesetzt werden können. Die flachen und dünnen Module geben mit ihrer gesamten Fläche ein homogenes Licht ab, überhitzen nicht und sparen Energie. Die Lichtquellen in Rot und Gelb folgen elegant der Außenkontur der Karosserie und fügen sich damit harmonisch in die Linienführung des Fahrzeugs ein. Im ausgeschalteten Zustand überzeugen die OLED-Elemente mit ihrem sehr eleganten, spiegelähnlichen Erscheinungsbild.

Begründung der Jury
Dank des Einsatzes von OLEDs können bei dieser Art der Heckbeleuchtung Formen realisiert werden, die so bislang nicht möglich waren.

Marksman Pegasus
Car Emergency Torch
Auto-Notfalltaschenlampe

Manufacturer
PF Concept, Roelofarendsveen, Netherlands
In-house design
Joeri van der Leeden
Web
www.pfconcept.com

This car emergency torch combines several functions in one elegant housing. The multifunctional device consists of an emergency hammer, a belt cutter, a torch and LEDs located at the side which can be set optionally as white working light or a flashing red warning light. Thanks to silicon-coated magnets on the outside, the Marksman Pegasus can be attached to any metal surface of the car.

Statement by the jury
With its reduced design and clever functions, the Marksman Pegasus assures that car drivers in emergencies have everything important at hand with one grasp.

Diese Notfalltaschenlampe vereint mehrere Funktionen in einem eleganten Gehäuse. Das Multifunktionsgerät beinhaltet Nothammer, Gurtschneider, Taschenlampe und seitlich angebrachte LEDs, die wahlweise als weißes Arbeitslicht oder als rot blinkendes Warnlicht eingesetzt werden können. Dank silikonbeschichteter Magneten an der Außenseite kann die Marksman Pegasus an jeder Metallfläche des Autos befestigt werden.

Begründung der Jury
Mit ihrer reduzierten Gestaltung und cleveren Funktionen sorgt die Marksman Pegasus dafür, dass Autofahrer im Notfall mit einem Griff alles Wichtige zur Hand haben.

Rechargable 12 V Car Light
Torch
Taschenlampe

Manufacturer
VARTA Consumer Batteries GmbH & Co.
KGaA, Ellwangen, Germany
In-house design
Web
www.varta-consumer.com

This mini pocket torch with elegant aluminium case is equipped with a 3.6 Volt Ni-MH 80 mAh rechargeable battery, which can be recharged simply from the cigarette lighter. With a length of 6.5 cm and a diameter of 2 cm, it fits in any compartment or on the dashboard and is thus always within reach. Fitted with a bright LED, the pocket torch shines as far as 27 metres.

Statement by the jury
This small car pocket torch gains merit with a minimalist cylindrical design and its small size, which makes it the ideal companion for those on the go.

Diese Mini-Taschenlampe mit elegantem Aluminiumgehäuse ist mit einem 3,6 Volt, 80 mAh Ni-MH-Akku ausgestattet und kann einfach mit dem Zigarettenanzünder wieder aufgeladen werden. Mit einer Länge von 6,5 cm und einem Durchmesser von 2 cm passt sie in jedes Fach oder auf das Armaturenbrett und ist so immer griffbereit. Mit einer hellen LED ausgestattet, leuchtet die Taschenlampe 27 Meter weit.

Begründung der Jury
Diese kleine Auto-Taschenlampe punktet mit einem minimalistischen zylindrischen Design und ihrer geringen Größe, die sie zum idealen Begleiter für unterwegs macht.

SAFETY TORCH OPTI-ON
Sicherheitslampe

Manufacturer
Lifehammer Products, Zoetermeer, Netherlands
In-house design
Bart Verbakel, Dalibor Borjanin
Design
FLEX/the INNOVATIONLAB B.V., Delft, Netherlands
Web
www.lifehammerproducts.com

Safety Torch Opti-On is a pocket torch and signal lamp in one. A simple press on the top of the torch is enough to alternate from one mode to the other. Thanks to a telescope system, the lamp extends as signal light and thus becomes a light source which is visible from far away. A reserve battery system assures that the torch is sure to function in an emergency. The safety torch can be attached to the centre console of the car so that it is always at hand, and by use of the Smart Suction Strap it is easily mounted on the bodywork or windows.

Safety Torch Opti-On ist zugleich Taschenlampe und Signallicht. Ein einfacher Druck auf die Oberseite der Leuchte genügt, um von einem Modus in den anderen zu wechseln. Dank eines Teleskopsystems schiebt sich die Leuchte als Signallicht weiter heraus und wird so zu einer weithin sichtbaren Lichtquelle. Ein Reserve-Batteriesystem gewährleistet, dass die Lampe im Notfall auch tatsächlich funktioniert. Die Sicherheitslampe lässt sich griffbereit an der Mittelkonsole des Fahrzeugs befestigen und mittels Smart Suction Strap einfach an Karosserie oder Fenstern anbringen.

Statement by the jury
With its clever functionality and a robust and compact design, the safety torch offers a simple solution in difficult situations.

Begründung der Jury
Mit ihrer cleveren Funktionalität und einer robusten und kompakten Gestaltung bietet die Sicherheitslampe in schwierigen Situationen eine einfache Lösung.

371

Corner Warning Assisting Device
Warnsystem für Kurven

Manufacturer
Automotive Research Testing Center, Changhua, Taiwan
In-house design
Web
www.artc.org.tw

This product identifies the light from vehicle headlights and thus recognises when a vehicle is approaching. The warning system is used in corners or bends which are poorly visible and automatically warns oncoming traffic of danger by means of a light signal. By this means, it increases safety in winding roads and helps avoiding accidents caused by bad visibility or insufficient lighting.

Dieses Produkt identifiziert das Licht von Autoscheinwerfern und erkennt dadurch, ob ein Fahrzeug naht. Das Warnsystem wird in schlecht einsehbaren Ecken oder Kurven installiert und warnt den Gegenverkehr mittels Leuchtsignal automatisch vor der Gefahr. Auf diese Weise erhöht es etwa die Sicherheit in Serpentinen und hilft, Unfälle zu vermeiden, die auf schlechte Sicht oder eine unzureichende Beleuchtung zurückzuführen zu sind.

Statement by the jury
Due to an almost figurative design, this warning system gives the impression of an attentive watchman and thus creates trust in its functionality.

Begründung der Jury
Durch eine beinahe figürliche Gestaltung bekommt dieses Warnsystem die Anmutung eines aufmerksamen Wächters und schafft so Vertrauen in seine Funktion.

X500
Dashcam
Autokamera

Manufacturer
Thinkware, Seongnam, South Korea
In-house design
Jae Hoon Lim, Seung Ho Lee
Web
www.thinkware.com

The X500 is a car dashboard camera which attracts attention due to its elegant design. For the compact, slim case, black is combined with silver. The colour contrast is further emphasised by the dissimilar haptic of the materials used. The black case has a fine pattern, giving it an appearance of soft leather, whereas the operating and function elements made of brushed aluminiun radiate cool precision.

Statement by the jury
The compact dashboard camera X500, with its slim case and an elegant design, is able to merge harmoniously into the car interior.

Die X500 ist eine Autokamera, die durch ihre elegante Gestaltung auf sich aufmerksam macht. Für das kompakte, schlanke Gehäuse wurde Schwarz mit Silber kombiniert. Der Farbkontrast wird durch die unterschiedliche Haptik der verwendeten Materialien zusätzlich betont. Das schwarze Gehäuse zeigt ein feines Muster, das ihm eine weiche Lederanmutung verleiht, während die aus gebürstetem Aluminium gefertigten Bedien- und Funktionselemente kühle Präzision ausstrahlen.

Begründung der Jury
Die kompakte Autokamera X500 fügt sich mit ihrem schlanken Gehäuse und einer eleganten Gestaltung harmonisch in das Wageninnere ein.

Black Prime
Dashcam
Autokamera

Manufacturer
Thinkware, Seongnam, South Korea
In-house design
Hyun Ju Lee, Seung Ho Lee
Web
www.thinkware.com

This LCD car dashcam films what is happening in front of and behind the car at the same time. The front view is recorded in full HD resolution, the rear view in HD resolution. The camera lens is designed as the centre of attention and is additionally emphasised by a U-shaped frame of high-gloss chromium, which also protects it from vibration. With its leather appearance, the camera blends harmoniously with the car interior.

Statement by the jury
With the Black Prime, its consistent high-quality design and precise engineering indicate its high technical merit.

Diese LCD-Autokamera filmt gleichzeitig, was vor und hinter dem Auto passiert. Die Vorderperspektive wird in Full-HD-Auflösung, die Rückperspektive in HD-Auflösung aufgezeichnet. Das Kameraobjektiv ist gestalterisch in den Mittelpunkt der Kamera gerückt und wird durch eine U-förmige Umrahmung aus hochglänzendem Chrom sowohl zusätzlich betont als auch vor Erschütterungen geschützt. Mit ihrer Lederanmutung fügt sich die Kamera harmonisch ins Fahrzeuginterieur ein.

Begründung der Jury
Die Black Prime kommuniziert mit ihrer durchgehend hochwertigen Gestaltung und präzisen Verarbeitung ihren hohen technischen Anspruch.

QXD900View
Dashcam
Autokamera

Manufacturer
Thinkware, Seongnam, South Korea
In-house design
Hyun Ju Lee, Seung Ho Lee
Web
www.thinkware.com

The QXD900View records by day and by night all images from a front and rear perspective with 1080P full HD resolution. Its premium impression is created by the use of high-quality materials. The front imparts in look and feel the impression of a soft leather coating, whereas diamond-cut aluminium is used for the D-shaped covering round the lens, thus giving the impression of precision engineering.

Statement by the jury
This dashcam captivates by a successful symbiosis of sophisticated technology and elegant, impressive aesthetics.

Die QXD900View zeichnet bei Tag und bei Nacht alle Bilder der Front- und Rückperspektive mit 1080P-Full-HD-Auflösung auf. Ihre Premiumanmutung entsteht durch die Verwendung hochwertiger Materialien. Die Vorderseite vermittelt visuell und haptisch den Eindruck einer weichen Lederbeschichtung, während für die D-förmige Abdeckung um die Linse herum diamantgeschnittenes Aluminium verwendet wird und so den Eindruck technischer Präzision vermittelt.

Begründung der Jury
Diese Autokamera besticht durch eine gelungene Symbiose aus anspruchsvoller Technik und einer eleganten, wertigen Ästhetik.

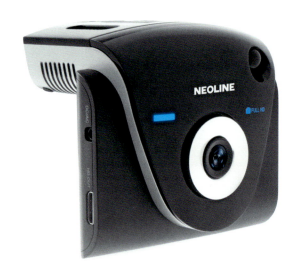

Neoline X-COP 9700
Car Video Recorder & Radar Detector
Auto-Videorekorder & Radardetektor

Manufacturer
Neoline, Moscow, Russia
In-house design
JH Yong
Web
www.neoline.ru

Neoline X-COP 9700 is a hybrid device for vehicles, combining radar detector and video recorder, whereby both are merged with one user interface. Via an intuitive, user-friendly interface, the driver is given access as required to each variably adjustable function of the device. With its purist, elegant housing, the Neoline X-COP 9700 blends harmoniously with the vehicle ambience.

Statement by the jury
This device presents a clear, functional design and convinces with a high degree of user-friendliness.

Neoline X-COP 9700 ist ein Hybridgerät für Fahrzeuge, das Radardetektor und Videorekorder kombiniert und beide auf einer Benutzeroberfläche zusammenführt. Über eine intuitive, bedienfreundliche Schnittstelle erhält der Fahrer Zugriff auf die je nach Anforderung variabel einstellbaren Funktionen des Geräts. Mit seinem puristisch-eleganten Gehäuse fügt sich der Neoline X-COP 9700 harmonisch in die Fahrzeugumgebung ein.

Begründung der Jury
Dieses Gerät zeigt eine klare, sachliche Gestaltung und überzeugt mit einer hohen Benutzerfreundlichkeit.

CAR GUARD
Car Emergency Mobile Power
Auto-Notbatterie

Manufacturer
Shenzhen Bifoxs Industrial Design Co., Ltd.,
Shenzhen, China
In-house design
Shaoshuai Cheng, Jiang Zhu, Jianhe Liu,
Guocheng Xiao
Web
www.bifoxs.com

With a weight of 500 grams, this D28 car power bank is a very light, mobile car emergency battery for petrol or diesel driven cars, which assures starting aid when the battery is weak. The compact unit is rechargeable, can also be used as a pocket torch, SOS signal lamp and, thanks to its integrated connection points, as charger for mobile end devices such as smartphones or tablet PCs.

Statement by the jury
Car Guard gains merit due to its sensible multifunctional properties and an extremely compact design.

Mit einem Gewicht von 500 Gramm ist die D28 Auto-Power-Bank eine sehr leichte, mobile Auto-Notbatterie für Benziner oder Diesel, die Starthilfe gewährt, wenn die Autobatterie schwach ist. Das kompakte, wieder aufladbare Gerät dient darüber hinaus als Taschenlampe, SOS-Signallicht und dank integrierter Anschlüsse auch als Ladestation für mobile Endgeräte wie Smartphones oder Tablet-PCs.

Begründung der Jury
Car Guard punktet mit einer sinnvollen Multifunktionalität und einer äußerst kompakten Gestaltung.

In-Vehicle Infotainment System (IVI System)
Tablet-Based Detachable IVI System
Tablet-basiertes Infotainmentsystem

Manufacturer
ASUSTek Computer Inc.,
Taipei, Taiwan
UniMax Electronics Inc. (ASUS Group),
Taipei, Taiwan
In-house design
Asus Design Center
UniMax Electronics Inc.
Web
www.asus.com
www.unimax.com.tw

In-Vehicle Infotainment (IVI) System is a car dashboard system to integrate a 7" LTE tablet. It provides in-car wireless charging and wireless video-audio communication between the tablet and the dashboard. The tablet serves as IVI's interactive interface when docked into the vehicle dashboard, allowing users to access personalised content and a wide range of the Android-based infotainment services (e.g. navigation, internet radio, etc.).

Statement by the jury
The IVI System is characterised by intelligent functionality and purist design which focuses on the user interface.

Das In-Vehicle Infotainment (IVI) System ist ein System für das Armaturenbrett zur Integration eines 7"-LTE-Tablets. Es erlaubt drahtloses Laden und die drahtlose Übertragung von Video- oder Audiodateien zwischen Tablet und Instrumententafel. In seiner Armaturenbrett-Halterung platziert, dient das Tablet als interaktive Benutzeroberfläche, die Nutzern den Zugriff auf persönliche Inhalte ebenso ermöglicht wie auf diverse Android-basierte Infotainment-Dienste (z. B. Navigation, Internetradio etc.).

Begründung der Jury
Das IVI-System zeichnet sich durch eine intelligente Funktionalität und eine puristische Gestaltung aus, die die Bedienoberfläche in den Mittelpunkt rückt.

Camino Smart Holder
Portable Terminal Holder
Halterung für mobile Geräte

Manufacturer
TCM Inc., Cheongju, South Korea
In-house design
Uijun Kwon
Web
www.caminolife.co.kr
Honourable Mention

The Camino Smart Holder is a holder for mobile devices, which was developed taking into account the special demands of car drivers towards such an equipment. The portable product is very flat when folded and thus does not impede the view when mounted on the dashboard. For use, it is folded upright and holds smartphones and other devices in a well visible and secure manner.

Statement by the jury
Thanks to a compact and plain design, the holder integrates discreetly in the car environment when not in use and yet remains ready for use at all times.

Der Camino Smart Holder ist eine Halterung für mobile Geräte, die unter spezieller Berücksichtigung der Ansprüche von Autofahrern an eine solche Halterung entwickelt wurde. Das tragbare Produkt ist zusammengeklappt sehr flach und schränkt so auch auf dem Armaturenbrett positioniert die Sicht nicht ein. Für den Gebrauch wird es vertikal hochgeklappt und hält Smartphone o. Ä. gut sichtbar und sicher fest.

Begründung der Jury
Dank einer kompakten, schlichten Gestaltung fügt sich die Halterung bei Nichtgebrauch dezent in die Wagenumgebung ein und bleibt doch die ganze Zeit einsatzbereit.

MAK Milano
Alloy Wheels
Alufelgen

Manufacturer
MAK S.p.A., Carpenedolo, Italy
Design
After Design srl (Gianfranco Melegari), Castelleone, Italy
Web
www.makwheels.it
www.afterdesign.it

MAK Milano is a universal wheel which is compatible with many cars, not least because of its simple, elegant design. Ten slim spokes, arranged in pairs, rise powerfully from the hub and express lightness and dynamics. The spokes feature a slight angle in the middle, setting the MAK Milano apart from other wheels. A double joint in the spokes at the outer rim assures good resistance.

Statement by the jury
This wheel impresses with a well-conceived construction, imparting at the same time an impression of lightness and dynamics.

MAK Milano ist eine Universalfelge, die nicht zuletzt dank ihrer schlichten, eleganten Gestaltung mit vielen Fahrzeugen kompatibel ist. Zehn schmale, paarweise angeordnete Speichen steigen kraftvoll aus der Felgenmitte auf und drücken Leichtigkeit und Dynamik aus. Die Speichen haben in der Mitte einen leichten Winkel, wodurch sich die MAK Milano von anderen Felgen unterscheidet. Eine Doppelverbindung der Speichen zum äußeren Rand sorgt für Widerstandsfähigkeit.

Begründung der Jury
Diese Felge beeindruckt mit einer durchdachten Konstruktion und einer Gestaltung, die zugleich leicht und dynamisch anmutet.

Sealant Tire
Car Tyre
Autoreifen

Manufacturer
Kumho Tire, Yongin, South Korea
In-house design
Web
www.kumhotire.com

The Sealant Tire is a self-sealing tyre containing a sealant which, in case of a puncture from a sharp object, assures that air does not escape. In this so-called "K-Seal" technology, the tyre consists of a gel layer inside which penetrates the damaged spot in the tyre and thus seals it automatically. By this means, driving can continue normally and safely.

Statement by the jury
As a self-sealing tyre, the Sealant Tire increases safety and convenience when driving since in case of a puncture, there is no need to change the wheel immediately.

Der Sealant Tire ist ein selbstversiegelnder Reifen. Er enthält Dichtmittel, das im Falle einer Beschädigung des Reifens durch einen spitzen Gegenstand dafür sorgt, dass die Luft nicht entweicht. Bei dieser sogenannten „K-Seal"-Technologie befindet sich eine durchgehende Gelschicht im Reifen, die in den beschädigten Bereich des Reifens eindringt und ihn so automatisch abdichtet, damit die Fahrt normal und sicher fortgesetzt werden kann.

Begründung der Jury
Als selbstversiegelnder Reifen erhöht der Sealant Tire Sicherheit und Komfort beim Autofahren, da bei einer Reifenpanne kein sofortiger Radwechsel nötig wird.

Smart Control AW02
Car Tyre
Autoreifen

Manufacturer
Hankook Tire, Seoul, South Korea
In-house design
Haeun Koog
Web
www.hankooktire.com

This all-position winter tyre for tractors and busses has been designed as a multipurpose tyre. After a period of wear as winter tyre, it can be further used as summer tyre. Due to its wide tread design, the tyre assures good road contact and fuel efficiency. 3D grooves provide high traction and mileage performance on icy and snowy roads.

Statement by the jury
The Smart Control AW02 gains merit with a sustainable concept, allowing its use first as winter tyre and later as summer tyre.

Dieser All-Position-Winterreifen für Zugmaschinen und Busse ist als Mehrzweckreifen konzipiert. Nach der ersten Abnutzung kann der Winterreifen als Sommerreifen weiter eingesetzt werden. Der Reifen gewährleistet durch sein breites Laufflächendesign guten Bodenkontakt und Kraftstoffeffizienz. 3D-Rillen sorgen für eine hohe Traktion und Laufleistung auf vereisten und verschneiten Straßen.

Begründung der Jury
Der Smart Control AW02 punktet mit einem nachhaltigen Konzept, das es ermöglicht, ihn zunächst als Winterreifen und später als Sommerreifen zu nutzen.

N'FERA SUR4
Car Tyre
Autoreifen

Manufacturer
Nexen Tire Coporation, Yangsan, South Korea
In-house design
Seung-Il Choi, Soung-Tae Shin
Web
www.nexentire.co.kr

The N'Fera Sur4 is suitable not only for roads, but also for race tracks and furthermore provides good adhesion even at high speeds. The tread design is inspired by the eagle and expresses sharpness, strength and aggression. Cornering and mileage have been enhanced by various technologies, among these "Profile Optimisation System" technology and "Hydro Evacuation Simulation" for greater traction on wet roads.

Statement by the jury
Tread pattern of this tyre, reminiscent of eagle wings, was optimised by use of advanced technologies and conveys the impression of speed and precision.

Der N'Fera Sur4 ist sowohl für die Straße als auch für Rennstrecken geeignet und bietet auch bei Höchstgeschwindigkeiten hohe Haftung. Die Profilgestaltung ist von einem Adler inspiriert und drückt Schärfe, Stärke und Aggressivität aus. Das Verhalten in Kurven und die Laufleistung wurden durch verschiedene Technologien verbessert, darunter die „Profile Optimization System"-Technologie und „Hydro Evacuation Simulation" für eine höhere Traktion bei Nässe.

Begründung der Jury
Die an Adlerschwingen erinnernde Profilgestaltung dieses Reifens wurde unter Einsatz zeitgemäßer Technologien optimiert und vermittelt den Eindruck von Schnelligkeit und Präzision.

ROADIAN AT PRO RA8
Car Tyre
Autoreifen

Manufacturer
Nexen Tire Coporation, Yangsan, South Korea
In-house design
Seung-Il Choi, Myung-Kwon Kim
Web
www.nexentire.co.kr

The distinctive tread design of this tyre for pickup cars is inspired by the form of a dinosaur tooth and conveys strength. This design element also has been applied in the sidewall and assures good traction and good braking performance, even in difficult weather conditions. By means of simulator technologies, it was possible to enhance driving conditions further; an HD steel belt system increases wear life of the tyre.

Statement by the jury
With its concise, powerful dinosaur tooth pattern, this tyre reflects the strength of the pickup cars for which it was developed.

Die markante Profilgestaltung dieses Reifens für Pick-ups ist von der Form eines Dinosaurierzahns inspiriert und kommuniziert Stärke. Dieses Designelement wurde auch in der Seitenwand eingesetzt und sorgt selbst bei schwierigen Wetterbedingungen für eine hohe Traktion und ein gutes Bremsverhalten. Mithilfe von Simulationstechnologien konnte das Fahrverhalten zusätzlich verbessert werden, ein HD-Stahlgürtelsystem verlängert die Lebensdauer des Reifens.

Begründung der Jury
Mit seinem prägnanten, kraftvollen Saurierzahnmuster spiegelt dieser Reifen die Stärke der Pick-ups wider, für die er entwickelt wurde.

CRUGEN HP91
Car Tyre
Autoreifen

Manufacturer
Kumho Tire, Yongin, South Korea
In-house design
Web
www.kumhotire.com

The outer side of the Crugen HP91 summer tyre for four-wheel drive SUVs consists of a silicate reinforced tread compound. The compound is formed to an asymmetric tread profile that, in combination with a continuous rib, assures directional stability and good steering response. Furthermore, the road holding is enhanced, and rolling resistance is reduced to the extent that it is similar to that of a saloon car.

Statement by the jury
Thanks to an optimal tread design using silicate, this SUV tyre achieves a clear reduction of rolling resistance.

Die Außenseite des Sommerreifens Crugen HP91 für Allrad-SUVs besteht aus einer mit Silikat verstärkten Laufflächenmischung. Die Mischung ist zu einem asymmetrischen Laufflächenprofil geformt, das in Kombination mit einer durchgehenden Rippe für Richtungsstabilität und gutes Lenkverhalten sorgt. Zudem wird die Bodenhaftung verbessert und der Rollwiderstand soweit verringert, dass er mit dem eines Pkw vergleichbar ist.

Begründung der Jury
Dank einer optimierten Laufflächengestaltung unter Einsatz von Silikat erreicht dieser SUV-Reifen eine deutliche Verringerung des Rollwiderstands.

SOLUS HS51
Car Tyre
Autoreifen

Manufacturer
Kumho Tire, Yongin, South Korea
In-house design
Web
www.kumhotire.com
Honourable Mention

Solus HS51 is an integrated product which meets the performance demands of the European market. The tyre runs quietly, behaves well on motorways and, thanks to widely formed transverse grooves, also offers high steering stability and good grip on wet road surfaces. The newly developed tyre tread minimises frictional and impact energy, contributes to improving noise levels and prevents block deforming due to the strengthened tread profile.

Statement by the jury
Thanks to precision tread profile, the Solus HS51 contributes to safe and low-noise handling of the car.

Solus HS51 ist ein integriertes Produkt, das die Leistungsanforderungen des europäischen Markts erfüllt. Der Reifen läuft leise, zeigt ein gutes Verhalten auf der Autobahn und bietet dank breit ausgeformter Querrillen zudem hohe Lenkstabilität und gute Haftung auf nassem Untergrund. Das neu entwickelte Reifenprofil minimiert die Reibungs- und Aufprallenergie, trägt zur Rauschleistungsverbesserung bei und verhindert durch das verstärkte Laufflächenprofil eine Blockverformung.

Begründung der Jury
Dank einer präzisen Profilgestaltung trägt der Solus HS51 zu einem sicheren und geräuscharmen Fahrverhalten bei.

Winter i*cept evo2
Car Tyre
Autoreifen

Manufacturer
Hankook Tire, Seoul, South Korea
In-house design
Soojin Son
Web
www.hankooktire.com

The Winter i*cept evo2 is an ultra-performance winter tyre of differentiated design. On the outer side, the tread has an ice cube block profile with lattice structure, providing good adhesion when cornering. On the inside, sharp grooves maximise the block edge length and assure good grip in snow. In the centre, wide grooves speed up water drainage, and a widened cross section facilitates fast driving on snowy and wet roads.

Statement by the jury
This winter tyre convinces by a profile design with winter pattern to achieve improved adhesion on ice and snow.

Der Winter i*cept evo2 ist ein differenziert gestalteter Ultra-Performance-Winterreifen. Außen zeigt das Profil ein eiswürfelförmiges Muster mit Netzstruktur, das eine gute Kurvenhaftung ermöglicht. Innen maximieren scharfe Rillen die Kantenlänge der Blocks und sorgen für gute Griffigkeit bei Schnee. In der Mitte beschleunigen breite Rillen die Wasserableitung, und ein verbreitertes Querprofil ermöglicht schnelles Fahren auf verschneiten und nassen Straßen.

Begründung der Jury
Dieser Winterreifen überzeugt mit einer Profilgestaltung, die sich winterlicher Muster bedient, um eine verbesserte Haftung bei Eis und Schnee zu erreichen.

JBL GTR Amplifiers
Car Audio Amplifiers
Audio-Verstärker fürs Auto

Manufacturer
Harman International Industries,
Northridge, USA
Design
LDA, Irvine, USA
Web
www.harman.com
www.lumdesign.com

This Bluetooth-enabled, aggressively styled GTR amplifier features the so-called Clari-Fi sound restoration technology to improve audio playback of compressed music files from smartphones and other band-limited sources. In Party Mode, multiple passengers can share playlists from their devices. Hands-free equipment, Clari-Fi function and Party Mode are remotely controlled.

Statement by the jury
A sharply pronounced, powerful design indicates the performance strength of the JBL GTR Amplifiers at first glance.

Dieser Bluetooth-fähige, aggressiv anmutende GTR-Verstärker bietet eine sogenannte Clari-Fi-Klangwiederherstellungstechnologie, um die Audiowiedergabe komprimierter Musikdateien beispielsweise von Smartphones zu verbessern. Im Partymodus können mehrere Personen Playlists von ihren Geräten teilen. Freisprecheinrichtung, Clari-Fi-Funktion und Partymodus werden über eine Fernbedienung gesteuert.

Begründung der Jury
Eine scharf akzentuierte, kraftvolle Gestaltung transportiert die Leistungsstärke der JBL GTR Amplifiers auf den ersten Blick.

Infinity Kappa Amplifiers
Car Audio Amplifiers
Audio-Verstärker fürs Auto

Manufacturer
Harman International Industries,
Northridge, USA
Design
LDA, Irvine, USA
Web
www.harman.com
www.lumdesign.com

The linear, partially open design of the Kappa amplifier exposes the heatsinks and enhances thermal dissipation. LED level displays indicate the instantaneous power output from each channel. The hands-free system, the Clari-Fi function for enhancing digital music and the Party Mode, which allows up to three persons to share their playlists, are all remotely controlled. Dynamic Bass Optimization (DBO) allows fine bass tuning.

Statement by the jury
The Infinity Kappa Amplifier convinces with a clear, geometrical form which imparts the impression of high technical precision.

Die geradlinige, partiell offene Gestaltung des Kappa-Verstärkers gewährt einen Blick auf die Kühlkörper und verbessert die Wärmeableitung. LED-Pegelanzeigen teilen die aktuelle Systemleistung mit. Die Fernbedienung steuert die Freisprecheinrichtung, die Clari-Fi-Funktion zur Verbesserung von digitaler Musik und den Partymodus, der es ermöglicht, dass bis zu drei Personen ihre Playlists teilen. Eine dynamische Bass-Optimierung (DBO) erlaubt feines Bass-Tuning.

Begründung der Jury
Der Infinity Kappa Amplifier überzeugt mit einer klaren, geometrischen Formgebung, die den Findruck hoher technischer Präzision vermittelt.

Interlock 7000
Alcohol Ignition Interlock Device
Atemalkoholgesteuerte Wegfahrsperre

Manufacturer
Dräger Safety AG & Co. KGaA,
Lübeck, Germany
In-house design
Design
Design3 GmbH, Hamburg, Germany
Web
www.draeger.com
www.design3.de
Honourable Mention

The Interlock 7000 is a breath alcohol controlled immobiliser which prevents a driver who has consumed alcohol from starting his vehicle. The high-quality, minimalist design is orientated towards smartphones so that the Interlock 7000 integrates discreetly in the vehicle interior. By this means, handling is made uncomplicated; at the same time unwanted observation is avoided, and the driver is protected from stigmatisation.

Statement by the jury
With a formal language which simulates modern communication devices, design of this breath alcohol controlled immobiliser contributes to its acceptance.

Das Interlock 7000 ist eine atemalkohol-gesteuerte Wegfahrsperre, die alkoholisierte Fahrer daran hindert, ihr Fahrzeug zu starten. Die hochwertige, minimalistische Gestaltung ist an Produkten wie Smartphones orientiert, sodass sich das Interlock 7000 zurückhaltend ins Autointerieur integriert. Auf diese Weise erlaubt es eine unkomplizierte Handhabung, gleichzeitig vermeidet es ungewollte Aufmerksamkeit und schützt den Fahrer vor Stigmatisierung.

Begründung der Jury
Mit einer Formensprache, die an moderne Kommunikationsgeräte angelehnt ist, trägt die Gestaltung zur Akzeptanz dieser atem-alkoholgesteuerten Wegfahrsperre bei.

Pisa Safety Hammer / Car Charger
Pisa Nothammer / Autoladegerät

Manufacturer
Shenzhen Triumph Innovate Brand Industrial Design Co., Ltd., Shenzhen, China
In-house design
Huang Gaoxiang
Web
www.tidesign.cn

This multifunctional device combines an emergency hammer with a fast car charger for iPad Air or other tablet PCs. When designing the cylindrical housing, it was possible to dispense with visible contacts and springs, resulting in a minimalist device. The stainless steel surface emphasises the purist impression and contributes to the durability of the product. An integrated LED eases its usability in the dark.

Statement by the jury
Pisa combines a USB car charger with an emergency hammer in a purist housing and thus offers a practical enhanced functionality.

Dieses multifunktionale Gerät kombiniert einen Nothammer mit einem schnellen Autoladegerät für das iPad Air oder andere Tablet-PCs. Bei der Gestaltung des zylindrischen Gehäuses konnte auf sichtbare Kontakte und Federn verzichtet werden, sodass ein sehr minimalistisches Gerät entstanden ist. Die Edelstahloberfläche betont die puristische Anmutung und trägt zur Langlebigkeit des Produkts bei. Eine integrierte LED erleichtert die Benutzung bei Dunkelheit.

Begründung der Jury
Pisa kombiniert ein USB-KFZ-Ladegerät mit einem Nothammer in einem puristischen Gehäuse und bietet so eine praktische erweiterte Funktionalität.

Side Cases SH36
Seitenkoffer SH36
Motorcycle Side Case
Motorradkoffer

Manufacturer
NAD s.l., SHAD,
Mollet del Vallès (Barcelona), Spain
In-house design
Anton Puigdomènech
Web
www.shad.es

SH36 is a motorcycle side case set which is produced in various colours and versions and matches a variety of motorcycle models. Enlarged capacity, a well-considered frame structure and optimised materials give the cases stability and lightness at the same time. The innovative 3P mounting fixes the SH36 at three points, transfers its weight towards the motorcycle's centre of gravity and thus provides stability and safety.

Statement by the jury
These side cases are distinguished not only by an elegant aerodynamic design, but also by a well-conceived mounting system.

SH36 ist ein Motorradkoffer-Set, das mit unterschiedlichen Farben und Ausführungen zu verschiedenen Motorrädern passt. Das vergrößerte Volumen, eine durchdachte Rahmenstruktur und optimierte Materialien verleihen den Koffern gleichzeitig Stabilität und Leichtigkeit. Die innovative 3P-Halterung fixiert die SH36 an drei Punkten, verlagert ihr Gewicht in Richtung des Schwerpunkts des Motorrads und bietet so mehr Stabilität und Sicherheit.

Begründung der Jury
Diese Seitenkoffer zeichnen sich sowohl durch ein elegantes, aerodynamisches Design als auch durch eine durchdachte Fixierungsvorrichtung aus.

Akrapovič Evolution Line (Titanium) with Carbon Fiber Tailpipes and Diffusor for the BMW M3/M4
Akrapovič Evolution Line (Titan) mit Endrohren und Diffusor aus Carbon für den BMW M3/M4

Exhaust System
Abgasanlage

Manufacturer
Akrapovič d.d., Ivančna Gorica, Slovenia
In-house design
Web
www.akrapovic.com

The Evolution Line exhaust system is elaborately manufactured of high-quality titanium, partially by casting processes. It gives the BMW M3/M4 more power and torque, at the same time reducing the weight by 11 kg (28 per cent). Thanks to the relatively bigger pipes compared to the series model, the exhaust gas pressure has been reduced. When designing, great emphasis was also laid on a deep, sporty sound. The appearance of the system is characterised by the concise tailpipes of carbon fibre and an elegantly formed carbon diffuser.

Die Evolution-Line-Abgasanlage wird aufwändig und teils im Gussverfahren aus hochwertigem Titan gefertigt. Sie verleiht dem BMW M3/M4 mehr Leistung und Drehmoment bei gleichzeitiger Gewichtsreduzierung um 11 kg (28 Prozent). Dank der im Vergleich zur Serienversion größeren Rohre wird der Abgasdruck verringert. Bei der Gestaltung wurde zudem großer Wert auf einen tiefen, sportlichen Klang gelegt. Das Erscheinungsbild der Anlage wird durch die prägnanten Endrohre aus Carbon sowie einen elegant geformten Carbon-Diffusor geprägt.

Statement by the jury
Materials, sound and technologies work together for the Evolution Line and produce not only sophisticated functionality, but also a very aesthetic overall image.

Begründung der Jury
Materialien, Sound und Technologien wirken bei der Evolution Line zusammen und erzeugen eine ausgereifte Funktionalität und ein sehr ästhetisches Gesamtbild.

Akrapovič Slip-On Line (Carbon) for the Kawasaki Ninja H2
Akrapovič Slip-On Line (Carbon) für die Kawasaki Ninja H2
Exhaust System
Abgasanlage

Manufacturer
Akrapovič d.d., Ivančna Gorica, Slovenia
In-house design
Igor Akrapovič
Web
www.akrapovic.com

This slip-on exhaust assembly is based on an innovative concept system. It consists of light-weight carbon fibre and high-quality titanium and thus reduces the weight of the motorcycle. The distinctive exhaust system matches the lines of the Kawasaki Ninja H2 and facilitates full exploitation of the machine's power. The new form is a further development of the hexagonal Akrapovič exhaust system, which was already presented in 2005. When designing the exhaust system, properties such as volume, ground clearance, performance and sound were taken into account.

Diese Slip-on-Auspuffanlage basiert auf einer innovativen Konzeptanlage. Sie besteht aus leichtgewichtigem Carbon und hochwertigem Titan und reduziert so das Gewicht des Motorrads. Der markante Auspuff passt sich der Linienführung der Kawasaki Ninja H2 an und ermöglicht es, die Leistung der Maschine voll auszuschöpfen. Die neue Form ist eine Weiterentwicklung des sechskantigen Akrapovič-Auspuffs, der bereits 2005 vorgestellt wurde. Bei der Gestaltung des Schalldämpfers wurden Punkte wie Volumen, Bodenfreiheit, Fahrverhalten und Klang mit einbezogen.

Statement by the jury
With its aerodynamic design, considered in detail, the Slip-On Line blends harmoniously with the distinctive look of the Kawasaki Ninja H2.

Begründung der Jury
Mit ihrer bis ins Detail durchdachten, aerodynamischen Gestaltung fügt sich die Slip-On Line harmonisch in das markante Erscheinungsbild der Kawasaki Ninja H2 ein.

Careliner
Horse Trailer
Pferdeanhänger

Manufacturer
Bücker Trailer GmbH, Emsdetten, Germany
In-house design
Web
www.buecker-trailer.com

The Careliner is a completely re-designed horse trailer of polyester construction which, due to a very high dividing line between the top section and the body, allows for various innovations, among them a body-sized entrance door and flush mounted windows. The generously dimensioned interior with LED lighting offers ample headroom for the horses and space for up to three saddles. In the dark, an innovative new rear light assists with manoeuvring as well as loading and unloading the horses.

Statement by the jury
A purist exterior design and dynamic lines lend an innovative appearance to this horse trailer.

Der Careliner ist eine neu konzipierte Generation von Pferdeanhängern mit einem Polyesteraufbau, der durch die sehr hoch angesetzte Trennlinie zwischen Haube und Korpus eine Reihe von Innovationen ermöglicht, darunter eine körperhohe Einstiegstür und flächenbündige Fensterscheiben. Der großzügige Innenraum mit LED-Beleuchtung bietet Pferden viel Kopffreiheit sowie Platz für bis zu drei Sättel. Eine neuartige Heckleuchte erleichtert bei Dunkelheit das Rangieren und das Auf- und Abladen der Pferde.

Begründung der Jury
Eine klare, puristische Flächengestaltung und eine dynamische Linienführung verleihen diesem Pferdeanhänger ein innovatives Erscheinungsbild.

easydriver
Manoeuvring Device
Rangierantrieb

Manufacturer
REICH GmbH, Regel- und Sicherheitstechnik, Eschenburg, Germany
Design
RRo industrial design B.V. (Hans Roelvink), Rotterdam, Netherlands
industrialpartners GmbH (Jens Arend), Frankfurt/Main, Germany
Web
www.reich-easydriver.com
www.rro.nl
www.industrialpartners.de

Easydriver systems are manoeuvring devices by means of which caravans and trailers can be easily moved. Thanks to light, powerful motors, an efficient gear and a newly developed electronic control, single or tandem axle caravans can be manoeuvred into small spaces and over rough terrain or turned on the spot using the remote control. The housing is of a special material which protects the easydriver from corrosion.

Statement by the jury
The easydriver earns merit with an intuitively understandable remote control which makes the complex functionality of the manoeuvring system simple to operate.

Easydriver sind Rangierhilfen, mit denen Wohnwagen und Anhänger einfach bewegt werden können. Dank leichter, leistungsstarker Motoren, einem effizienten Getriebe und einer neu entwickelten elektronischen Steuerung können Ein- oder Tandemachser mithilfe einer Fernbedienung auch in kleine Lücken und auf unwegsamem Gelände rangiert oder auf der Stelle gedreht werden. Das Gehäuse ist aus einem speziellen Material gefertigt, das den easydriver vor Korrosion schützt.

Begründung der Jury
Der easydriver punktet mit einer intuitiv verständlichen Fernsteuerung, die die komplexe Funktionalität des Rangiersystems einfach bedienbar macht.

Mercury Marine FourStroke Engine
Outboard Engine
Außenbordmotor

Manufacturer
Mercury Marine, Fond du Lac, USA
Design
Designworks, a BMW Group Company, Newbury Park, USA
Web
www.mercurymarine.com
www.bmwgroupdesignworks.com

The outboard engines of the Mercury Marine FourStroke Engine series are characterised by an elegant, dynamic design and balanced proportions and are compatible with various types of boats. Noise and vibration are minimised by the means of a sealed, thermally bonded engine cowling and a new exhaust with innovative clutch design which provide a more gentle and quiet boating experience.

Statement by the jury
The new Mercury Marine FourStroke engine convinces due to a contemporary, dynamic design and a quieter functionality with less vibration.

Die Außenbordmotoren der Mercury Marine FourStroke-Engine-Serie sind durch eine elegante, dynamische Gestaltung und ausgewogene Proportionen gekennzeichnet und mit verschiedenen Bootstypen kompatibel. Schall und Vibrationen werden durch eine abgedichtete, thermisch verbundene Motorhaube, und einen neuen Auspuff mit innovativem Kupplungsdesign minimiert, sodass die neuen Motoren ein sanfteres und ruhigeres Bootsfahrterlebnis bieten.

Begründung der Jury
Die neue Mercury Marine FourStroke Engine überzeugt durch eine zeitgemäße, dynamische Gestaltung und eine ruhigere, vibrationsärmere Funktionsweise.

Evinrude E-TEC G2
Outboard Engine
Außenbordmotor

Manufacturer
BRP Inc., Valcourt, Canada
In-house design
Web
www.brp.com

For this new generation of the Evinrude E-Tec engines, the engines have been upgraded from a functional as well as from an aesthetic viewpoint. The Evinrude E-Tec G2 two-stroke outboard engine with integrated electronic control reduces emissions by up to 75 per cent, and fuel consumption is 15 per cent less than that of its predecessors. Compact mounting, vertical alignment and powerful proportions assure a clear rear view. With over 400 possible colour and material combinations, engine design can be attuned to every boat.

Für diese neue Generation der Evinrude E-Tec-Motoren wurden die Motoren sowohl unter funktionalen als auch ästhetischen Gesichtspunkten überarbeitet. Der Zweitaktaußenbordmotor Evinrude E-Tec G2 mit integrierter elektronischer Steuerung reduziert die Emissionen um bis zu 75 Prozent und verbraucht 15 Prozent weniger Kraftstoff als seine Vorgänger. Kompakte Montagemöglichkeiten, eine vertikale Ausrichtung und kraftvolle Proportionen sorgen für eine klare Heckansicht. Mit mehr als 400 möglichen Farb- und Materialkombinationen lässt sich das Design des Motors auf jedes Boot abstimmen.

China Airlines Premium Business Class
Aircraft Interior
Flugzeug-Inneneinrichtung

Manufacturer
B/E Aerospace, Winston-Salem,
North Carolina, USA

Design
China Airlines, Taipei, Taiwan
Ray Chen International,
Taipei, Taiwan

Web
www.china-airlines.com
www.beaerospace.com

reddot award 2015
best of the best

Imperial aesthetics

Aircraft interior have a decisive influence on how passengers feel and what they perceive within an aircraft. Against this backdrop, the interior for the China Airlines Premium Business Class impresses with a highly refined concept that virtually stages the act of flying. As a result of a two-year long intensive development endeavour, the interior is inspired by the aesthetics of the Song Dynasty, a vibrant era that defined culture in Ancient China from 960 to 1279. In order to gain an in-depth insight into passenger experiences, the design process comprised several workshops and prototyping sessions with frequent business class flyers, leading to a consistently implemented interior based on the concept "Back to Nature". The interior showcases a variety of traditional elements from the Song Dynasty culture that lend it an outstandingly organic and elegant appearance. In order to deliver a maximum of comfort and convenience, these elements fuse with state-of-the-art technological amenities to form a homogeneous unity. Alongside architectonically impressive details such as a leisure and recreation space as well as an artfully arranged lavatory, the design centres on the customised B/E Super Diamond seats. These are perfectly adapted to passenger needs and fascinate with their functionality and ergonomics. The ambiance of the aircraft interior thus delivers the promise to give guests a memorable flying experience that touches the senses.

Kaiserliche Ästhetik

Die Innenausstattung eines Flugzeugs hat einen entscheidenden Einfluss darauf, wie man sich als Fluggast fühlt. Das Interieur für die China Airlines Premium Business Class beeindruckt vor diesem Hintergrund mit einem sehr feinsinnigen Konzept, welches das Fliegen geradezu inszeniert. Es ist inspiriert von der Ästhetik der Song-Dynastie, die von 960 bis 1279 das damalige chinesische Kaiserreich beherrschte, und ist das Ergebnis einer zweijährigen, intensiven Entwicklung. Um die Wünsche der Passagiere en détail in Erfahrung zu bringen, war der Gestaltungsprozess begleitet von Unternehmungen wie Workshops oder Prototyping-Sessions mit Fluggästen der Business Class. Auf stimmige Art und Weise wurde dann das Konzept „Back to Nature" umgesetzt. Im Interieur kommen traditionelle Elemente der Kultur der Song-Dynastie zum Einsatz, die zu einer bemerkenswert eleganten und organischen Anmutung führen. Um dem Fluggast zudem ein Höchstmaß an Komfort zu geben, verschmelzen diese mit den Annehmlichkeiten moderner Technologie zu einer homogenen Einheit. Eindrucksvolle architektonische Details bieten ein Erholungs- und Freizeitbereich sowie ein kunstvoll gestalteter Wasch- und Toilettenraum. Den Mittelpunkt bilden jedoch die B/E Super Diamond-Sitze. Sie sind perfekt den Bedürfnissen angepasst, werden maßgefertigt und begeistern durch ihre Ergonomie und Funktionalität. Das Ambiente dieser Flugzeugeinrichtung steht daher für ein unvergessliches Erlebnis und ein Design, welches die Sinne berührt.

In a fascinating manner, the design of the China Airlines Premium Business Class merges traditional Chinese symbolism from the historical Song Dynasty with state-of-the-art technology. The result is a charming contrast with refined highlights that make the aircraft interior stand out among otherwise purely technically functional interiors. Passengers gain the feeling of being at home surrounded by a highly comfortable environment. The cabin seat is a perfect complement to the interior and impresses with its comfort and sophisticated ergonomics.

In der Gestaltung der China Airlines Premium Business Class vereint sich auf faszinierende Weise die traditionelle Symbolik der historischen Song-Dynastie mit einer hochtechnologischen Ausstattung. Es entsteht ein reizvoller Kontrast, der raffinierte Akzente setzt und sich von einer rein technisch orientierten Anmutung löst. Der Fluggast hat das Gefühl, zu Hause und sehr komfortabel aufgehoben zu sein. Der Kabinensitz fügt sich hier perfekt ein. Er beeindruckt mit seinem Komfort und seiner ausgefeilten Ergonomie.

Designer portrait
See page 84
Siehe Seite 84

China Airlines New Sky Lounge
Airline Lounge
Loungebereich im Flugzeug

Manufacturer
AIM Aviation Ltd, Dorset, Great Britain
Design
Ray Chen International, Taipei, Taiwan
Web
www.china-airlines.com

This new lounge zone for Business Class offers a pleasant refuge for air travellers. The interior design displays aesthetic elements from the Song dynasty and combines harmony with the branded image of the airline. Dividing walls show the natural grain of the persimmon tree and exude peace and comfort. Reading lights and a display of up-to-date books appeal especially to passengers with literary interests. The Sky Lounge is thus a place of culture and relaxation at the same time.

Dieser neue Loungebereich für die Business Class bietet Flugreisenden einen schönen Rückzugsort. Das Interieurdesign zitiert ästhetische Elemente aus der Song-Dynastie und verbindet sie harmonisch mit dem Markenbild der Fluglinie. Die Trennwände zeigen die Maserung des Dattelpflaumenbaums und vermitteln Ruhe und Gemütlichkeit. Leseleuchten und Auslagen mit aktuellen Büchern sprechen insbesondere literarisch interessierte Reisende an. So ist die Sky Lounge zugleich ein Ort der Kultur und der Entspannung.

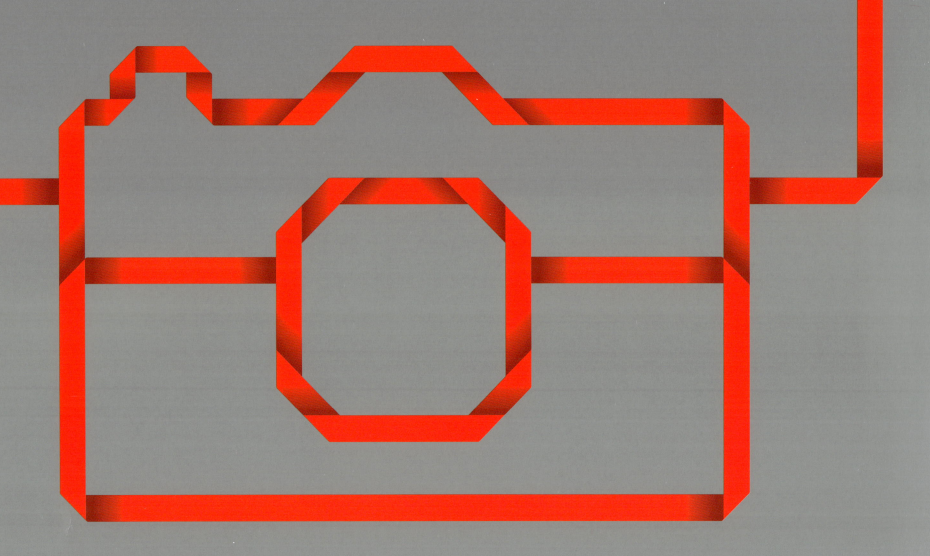

Amplifiers	Abspielgeräte
Bluetooth speakers	Beamer
Camera accessories	Bluetooth-Boxen
Cameras	Digitale Bilderrahmen
Digital photo frames	Fernbedienungen
Headphones	Hi-Fi-Systeme
Hi-fi systems	Home Cinema
Home cinema	Kameras
MP3 / MP4 players	Kamerazubehör
and accessories	Kopfhörer
Players	Lautsprecher
Projectors	MP3-/MP4-Player
Radios	und -Zubehör
Remote controls	Radios
Sound systems	Soundsysteme
Speakers	TV
TVs	Verstärker

Consumer electronics and cameras
Unterhaltungselektronik und Kameras

Blackmagic Studio Camera
Digital Film Camera
Digitale Filmkamera

Manufacturer
Blackmagic Design Pty Ltd,
Melbourne, Australia

In-house design
Blackmagic Design Pty Ltd

Web
www.blackmagicdesign.com

reddot award 2015
best of the best

The art of miniaturisation

The field of live productions in studios makes enormous demands on man and material. Many elements need to work smoothly together until the audience can watch a perfectly orchestrated show. The Blackmagic Studio Camera is a live broadcast camera, whose innovative design blends a compact body with a very large viewfinder. This very lightweight studio camera with slim proportions and a folding sunshade embodies a carefully considered study in miniaturisation. All necessary elements are fundamentally well thought through, so that they are in a perfectly balanced relationship. An essential aspect in connection with the light weight is the portability of the camera. Utilising the reusable packaging insert, the camera is well protected and easy to transport, while up to four cameras can be packed comfortably into a single road case. Designed specifically for multi-camera live productions such as talk shows, news or sporting events, the Blackmagic Studio Camera captures high-resolution Ultra HD footage. A further innovative feature is its ability to be controlled remotely via Blackmagic ATEM live production switchers. Along with aperture control, users can simultaneously colour balance multiple cameras in real-time using colour correction software. This ensures high quality and consistent footage across an entire live production.

Die Kunst der Miniaturisierung

Der Bereich der Liveaufnahmen im Studio stellt enorme Anforderungen an Mensch und Material. Bis die Zuschauer eine perfekte Sendung sehen, müssen viele Elemente ineinandergreifen. Die Blackmagic Studio Camera ist eine Live-Broadcast-Filmkamera, deren innovative Gestaltung ein kleines Gehäuse mit einem sehr großen Suchermonitor vereint. Diese sehr leichte Studiokamera verkörpert mit schlank ausgeführten Proportionen und einer faltbaren Gegenlichtblende eine sorgfältig erwogene Studie der Miniaturisierung. Alle nötigen Elemente wurden so grundlegend durchdacht, dass sie sich in einem perfekt austarierten Verhältnis befinden. Ein wesentlicher Aspekt ist die mit dem geringen Gewicht einhergehende Transportmöglichkeit der Kamera. Unkompliziert lässt sie sich, gut geschützt, in einer wiederverwendbaren Transportverpackung mitführen, wobei bis zu vier Kameras in einem Flightcase-Koffer verstaut werden können. Die speziell für Live-Mehrkamera-Produktionen wie Talkshows, Nachrichten- oder Sportsendungen konzipierte Blackmagic Studio Camera nimmt hochauflösende Ultra-HD-Bilder auf; ein innovatives Leistungsmerkmal ist zudem die Möglichkeit der Fernsteuerung über einen Live-Produktionsmischer der Blackmagic ATEM-Serie. Die Kameraleute können neben der Blendensteuerung mithilfe einer Farbkorrektursoftware auch die gleichzeitige Farbbalance mehrerer Kameras in Echtzeit vornehmen. Dies gewährleistet hochqualitative, durchgängige Aufnahmen für die gesamte Liveproduktion.

Statement by the jury

The Blackmagic Studio Camera offers versatile new possibilities in the field of live productions. It is affordable, very light and integrates extremely important features for high-quality live recordings such as the remote control live production switchers. Its design excellently succeeds in fitting all necessary elements for this type of camera in a considerably smaller body. All details and proportions are perfectly tuned to each other.

Begründung der Jury

Die Blackmagic Studio Camera bietet vielseitige neue Möglichkeiten im Bereich der Liveproduktion. Sie ist kostengünstig, sehr leicht und integriert für hochqualitative Liveaufnahmen äußerst wichtige Features wie die Fernsteuerung über Live-Produktionsmischer. Ihrer Gestaltung gelingt es auf exzellente Weise, die für diese Art von Kameras notwendigen Elemente in einem wesentlich kleineren Gehäuse unterzubringen. Ihre Details und Proportionen sind dabei perfekt aufeinander abgestimmt.

Designer portrait
Working
See page 76
Siehe Seite 76

Blackmagic URSA
Digital Film Camera
Digitale Filmkamera

Manufacturer
Blackmagic Design Pty Ltd, Melbourne, Australia
In-house design
Web
www.blackmagicdesign.com

The Blackmagic URSA is a high-end digital cinema camera designed specifically for the production of feature films and documentaries, as well as for news gathering. The user can personally upgrade the image sensor and lens mount and thus adapt the camera to future technical innovations. Its durable, CNC-precision-machined aluminium casing and uncompromising aesthetic design characterise the camera as a professional tool whose performance and reliability can be trusted, even in difficult production environments.

Die Blackmagic URSA ist eine speziell für die Produktion von Spielfilmen, Dokumentationen und die Erfassung von Nachrichten konzipierte digitale High-End-Kinofilmkamera. Der Nutzer kann Bildsensor und Objektivanschluss eigenhändig aufrüsten und damit an künftige technische Entwicklungen anpassen. Ihr widerstandsfähiges CNC-präzisionsgefrästes Aluminiumgehäuse und ihre kompromisslose Ästhetik charakterisieren die Kamera als Profiwerkzeug, auf dessen Leistung und Zuverlässigkeit man auch in schwierigen Produktionsumgebungen vertrauen kann.

Statement by the jury
The URSA impresses with its outstanding equipment and future-oriented construction style with upgrade options for the user.

Begründung der Jury
Die URSA beeindruckt durch ihre hervorragende Ausstattung und die zukunftsfähige Bauweise mit Upgrademöglichkeiten durch den Benutzer.

RE
Point-of-View Photo/Video Camera
Subjektive Foto/Video-Kamera

Manufacturer
HTC Corporation, Smart Phone, New Taipei, Taiwan
In-house design
Web
www.htc.com

Simplicity and fun were the two major objectives in designing this camera. It automatically turns on as soon as the user picks it up, and it shoots both photos and videos without a viewfinder. The simple push of a button is enough to take pictures, whereas pressing the button a little longer starts the video recording function. The wide-angle lens captures everything located in the visual field of the user. The camera can even be used underwater. It directly streams all data to a smartphone and automatically stores this data in the cloud.

Einfachheit und Spaß bei der Bedienung waren Ziele bei der Gestaltung dieser Kamera. Sie schaltet sich automatisch ein, sobald der Benutzer sie in der Hand hält, und nimmt Fotos und Videos ohne Sucher auf. Mit einem einfachen Tastendruck werden Fotos geschossen, wird der Knopf länger gedrückt, startet die Videoaufnahme. Das Weitwinkelobjektiv erfasst alles, was sich im Blickfeld des Benutzers befindet. Auch unter Wasser lässt sich die Kamera verwenden. Damit die Aufnahmen leicht angesehen und geteilt werden können, streamt die Kamera direkt alle Daten auf das Smartphone und sichert sie automatisch in der Cloud.

393

Leica T
Camera
Kamera

Manufacturer
Leica Camera AG,
Wetzlar, Germany

Design
AUDI AG, Audi Industrial Design
München, Munich, Germany

Web
www.leica-camera.com
www.audi.com

reddot award 2015
best of the best

Pure precision

Leica cameras have reached a virtually legendary status in photography. The 35 mm film format, which is associated with Leica, offered new freedom and possibilities for international journalism as well as art photography. Continuing the form language of the classic Leica camera, the Leica T fascinates with a purism that seamlessly ties in with the familiar product design and its message. At the same time, it also embodies the maxims of Audi Design in an impressive manner. With clean lines and sleek surfaces reduced to the essential form, this camera powerfully visualises its purpose: to serve a pure, creative photography and the best image quality. In order to furthermore achieve the nearly legendary durability of a Leica camera, the compact body is manufactured by hand with innovative precision techniques from a single block of aluminium. This camera is exceptionally solid and rests perfectly in the hand. Complementing the purist design of the camera, the operating concept of the Leica T is equally modern. The clear user interface of the touchscreen display at the heart of the camera is easy to understand, with four ergonomically positioned haptic control elements that are intuitive in their operation. Consistently following the Leica system concept, the Leica T series also comprises camera accessories in four trendy colours as well as functional bags. In a highly convincing way, this camera thus successfully continues the Leica legend.

Pure Präzision

Die Kameras von Leica haben im Bereich der Fotografie eine nahezu mythische Bedeutung erlangt. Das mit ihnen verbundene Kleinbildformat bot neue Freiheiten für den internationalen Journalismus wie auch die künstlerische Fotografie. Die Formensprache der klassischen Leica-Kameras weiterführend, fasziniert die Leica T durch einen Purismus, der nahtlos an das bekannte Produktdesign und die damit verbundene Aussage anknüpft. Auf beeindruckende Weise verkörpert sie dabei auch die Maximen von Audi Design. Mit klaren Linien und glatten Oberflächen auf die essenzielle Form reduziert, visualisiert diese Kamera eindringlich ihre Bestimmung: einer puren, kreativen Fotografie und dem besten Bildergebnis dienen zu wollen. Um zudem die geradezu legendäre materielle Langlebigkeit einer Leica-Kamera zu erreichen, wird das kompakte Gehäuse aus nur einem Stück Aluminium von Hand und unter Einsatz innovativer Präzisionstechnik gefertigt. Diese Kamera ist außerordentlich solide und liegt perfekt in der Hand. Im Einklang mit ihrer puristischen Gestaltung wurde auch das Bedienkonzept der Leica T zeitgemäß entwickelt. Ihre übersichtliche Oberfläche mit einem zentralen Touchscreen ist einfach zu verstehen, wobei sich vier ergonomisch angeordnete, haptische Bedienelemente vom Nutzer intuitiv bedienen lassen. Den Leica-Systemgedanken konsequent fortführend, schließt das Konzept der Leica T auch Kamerazubehör in vier Trendfarben sowie funktionale Taschen mit ein. Auf eine sehr überzeugende Weise gelingt es so, an den Mythos Leica anzuknüpfen.

Statement by the jury

The design of the Leica T realises the perfect reduction to the essential. Its clean lines and sleek surfaces visualise a design dedicated to a purist, creative photography. The compact camera rests exceptionally well in the hand, while a hand-made aluminium housing makes it durable and solid. With a clearly structured touchscreen, the Leica T allows for an intuitive operation in order to achieve the best results in photography.

Begründung der Jury

Die Gestaltung der Kamera Leica T verwirklicht die perfekte Reduktion auf das Wesentliche. Ihre klare Linienführung und glatten Oberflächen visualisieren die Ausrichtung an einer puristischen, der kreativen Fotografie verschriebenen Form. Ausgesprochen kompakt liegt sie in der Hand des Nutzers, ein handgefertigtes Aluminiumgehäuse macht sie solide und langlebig. Mittels eines klar und übersichtlich strukturierten Touchscreens ermöglicht die Leica T eine intuitive Bedienung für beste Fotoergebnisse.

Designer portrait
See page 86
Siehe Seite 86

Leica M Edition 60
Digital Camera
Digitalkamera

Manufacturer
Leica Camera AG, Wetzlar, Germany
Design
AUDI AG, Konzept Design München,
Munich, Germany
Web
www.leica-camera.com
www.audi.com

The Leica M Edition 60 is a limited edition which clearly shows the formal connection between the Audi and Leica design language. The features are reduced to the essential functions needed for taking pictures: shutter speed, aperture, focusing and ISO sensitivity. For this reason, the display has been replaced by the ISO selector dial. The visible metal components of both camera and lens are made of stainless steel, a resistant and durable material guaranteeing long-term value.

Statement by the jury
Made from premium materials, the Leica M Edition 60 offers uncompromisingly reduced functionality and places the responsibility for the picture back in the hands of the photographer.

Die Leica M Edition 60 ist eine limitierte Sonderedition, die deutlich die formale Verbindung der beiden Designsprachen von Audi und Leica zeigt. Ihre Ausstattung ist auf die wesentlichen Funktionen, die zum Fotografieren benötigt werden, reduziert: Zeit, Blende, Entfernung und ISO-Empfindlichkeit. Aus diesem Grund findet sich anstelle eines Displays lediglich das ISO-Einstellrad. Die sichtbaren Metallteile von Kamera und Objektiv bestehen aus Edelstahl, der gleichermaßen widerstandsfähig wie beständig ist und einen langen Werterhalt garantiert.

Begründung der Jury
Aus hochwertigen Materialien hergestellt, bietet die Leica M Edition 60 einen kompromisslos reduzierten Funktionsumfang und legt die Verantwortung für das Bild wieder in die Hände des Fotografen.

Nikon 1 S2
Digital Camera
Digitalkamera

Manufacturer
Nikon Corporation, Tokyo, Japan
In-house design
Chikara Fujita
Web
www.nikon.com

The S2 provides an entry into the world of system cameras with interchangeable lenses. Its autofocus measures the picture contrast and additionally works with the phase-change process. With this functionality, the camera rapidly focuses on scenes in motion, allowing for up to 20 pictures per second. Both the camera and the appropriate lens are small and comfortably lightweight at 230 grams and 83 grams respectively. The minimalist form underscores its convenient handling.

Statement by the jury
The S2 digital camera is compact and thus within easy reach in any situation. Its rounded edges lend the device a congenial appearance.

Die S2 bietet den Einstieg in die Welt der Systemkameras mit Wechselobjektiv. Ihr Autofokus misst den Bildkontrast und arbeitet zusätzlich mit dem Phase-Change-Verfahren. Dadurch stellt sich die Kamera sehr schnell auf sich bewegende Motive scharf und ermöglicht bis zu 20 Bilder pro Sekunde. Die Kamera und das dazu passende Objektiv sind klein und mit 230 bzw. 83 Gramm angenehm leicht. Ihre minimalistische Form unterstreicht die einfache Handhabung.

Begründung der Jury
Die Digitalkamera S2 ist kompakt und dadurch in jeder Situation schnell zur Hand. Mit ihren abgerundeten Kanten zeigt sie ein sympathisches Äußeres.

Nikon 1 J4
Digital Camera
Digitalkamera

Manufacturer
Nikon Corporation, Tokyo, Japan
In-house design
Tatsuya Kobayashi
Web
www.nikon.com

The J4 is a versatile system camera with interchangeable lenses. Thanks to its high-speed autofocus system and frame rate of up to 20 pictures per second, it captures scenes in motion and image series with high visual sharpness. Its seamless housing is compact and lightweight, and its fast-response touchscreen is used to shoot still images and videos as well as to set important functions. As the camera is equipped with Wi-Fi, images can be directly shared with others.

Die J4 ist eine vielseitige Systemkamera mit Wechselobjektiv, die dank ihres schnellen Autofokussystems und einer Bildrate von bis zu 20 Bildern pro Sekunde auch bewegte Szenen und Serien in hoher Schärfe festhält. Ihr nahtlos konzipiertes Gehäuse ist kompakt und leicht. Über den reaktionsschnellen Touchscreen werden Fotos und Filme aufgenommen und wichtige Funktionen eingestellt. Da die Kamera mit Wi-Fi ausgestattet ist, lassen sich Aufnahmen direkt mit anderen teilen.

Statement by the jury
The seamless form gives this camera a very elegant appearance and also highlights its easy handling.

Begründung der Jury
Die nahtlose Form verleiht der Kamera eine überaus elegante Anmutung und weist zudem auf die einfache Bedienbarkeit hin.

397

X100T
Digital Camera
Digitalkamera

Manufacturer
FUJIFILM Corporation, Tokyo, Japan
In-house design
Haruka Ikegame
Web
www.fujifilm.com

The X100T premium compact camera offers a large number of innovative functions. One example is the hybrid viewfinder that enables simple switching between optical and electronic operation. When the optical sensor is used, the detail view can be faded in electronically so that the photographer is able to focus on the image for optimum sharpness. Control elements like buttons and dials have been arranged ergonomically, incorporating the feedback given by photographers.

Statement by the jury
The X100T puts the control of taking pictures into the hands of the photographer, yet it provides support with intelligent functions and matured technologies.

Die Premium-Kompaktkamera X100T bietet eine Vielzahl an innovativen Funktionen. Ein Beispiel ist der Hybridsucher, der das einfache Umstellen zwischen optischem und elektronischem Sucher erlaubt. Wird der optische Sucher verwendet, lässt sich der Bildausschnitt elektronisch einblenden, damit der Fotograf sehr genau fokussieren kann, bis das Bild scharf ist. Bedienelemente wie Tasten und Drehräder wurden nach Rückmeldung von Fotografen ergonomisch angeordnet.

Begründung der Jury
Die X100T überlässt dem Fotografen die Kontrolle über seine Aufnahmen, unterstützt ihn dabei aber mit intelligenten Funktionen und ausgereiften Technologien.

X30
Digital Camera
Digitalkamera

Manufacturer
FUJIFILM Corporation, Tokyo, Japan
In-house design
Keita Kamei
Web
www.fujifilm.com

The X30 is a high-grade compact camera with a magnesium die-cast top and bottom and aluminium control dials. A special feature is the electronic OLED viewfinder with 2.36 million pixels and a viewing angle of 25 degrees. As the viewfinder brightness adapts to the ambient light, the user can select the suitable image detail, even in bright sunlight. The foldable display facilitates the taking of pictures from unusual perspectives.

Statement by the jury
Metal components render this camera durable and underline its classic design.

Die X30 ist eine hochwertige Kompaktkamera mit einem Magnesiumgehäuse und Einstellrädern aus Aluminium. Eine Besonderheit ist der elektronische OLED-Sucher mit einem Bildwinkel von 25 Grad und 2,36 Millionen Bildpunkten. Da sich die Helligkeit des Sucherbildes dem Umgebungslicht anpasst, kann der Benutzer selbst bei Sonnenschein den passenden Bildausschnitt wählen. Das klappbare Display erleichtert Aufnahmen aus ungewöhnlichen Perspektiven.

Begründung der Jury
Die Metallkomponenten machen die Kamera widerstandsfähig und unterstreichen ihre klassische Anmutung.

X-T1
Digital Camera
Digitalkamera

Manufacturer
FUJIFILM Corporation, Tokyo, Japan
In-house design
Masazumi Imai
Web
www.fujifilm.com

The X-T1 is a premium-class digital camera with an X-Trans CMOS II sensor in APS-C size enabling excellent images. The real-time viewfinder works with virtually no delay, and the 0.77x magnification makes taking pictures particularly comfortable. The camera is splash-proof, dust-proof as well as cold-resistant and thus suitable for use in difficult conditions. The large, swivelling display allows pictures to be taken from nearly any position.

Statement by the jury
The casing of the X-T1 with its classic, robust design conceals high-grade technology, turning the camera into a reliable device for demanding photographers.

Die X-T1 ist eine Digitalkamera der Premiumklasse mit einem X-Trans-CMOS-II-Sensor in APS-C-Größe, der exzellente Aufnahmen ermöglicht. Der Echtzeitsucher arbeitet nahezu ohne Verzögerung, die Suchervergrößerung von 0,77x macht das Fotografieren besonders komfortabel. Die Kamera ist spritzwasserfest, staubgeschützt und kältebeständig und daher auch unter schwierigen Bedingungen einsetzbar. Dazu ermöglicht das große, schwenkbare Display Aufnahmen aus beinahe jeder Position heraus.

Begründung der Jury
In dem klassischen, robusten Gehäuse der X-T1 verbergen sich hochwertige Technologien, die die Kamera zu einem zuverlässigen Gerät für anspruchsvolle Fotografen machen.

DSC-RX100M3
Digital Camera
Digitalkamera

Manufacturer
Sony Corporation, Tokyo, Japan
In-house design
Yasuaki Isonaga, Shin Miyashita
Web
www.sony.net

This compact camera shows a slim aluminium casing conveying a high-quality impression. It includes a large 1" sensor, an electronic viewfinder and a premium zoom lens with a focal length ranging from 24–70 mm, thus enabling both wide-angle and portrait shots. The display can be folded out and rotated by 180 degrees so that pictures may also be taken from an unusual viewing angle. Moreover, the camera records videos in full HD.

Diese Kompaktkamera zeigt ein schlankes Aluminiumgehäuse, das einen wertigen Eindruck vermittelt. Sie verfügt über einen großen 1"-Sensor, einen elektronischen Sucher und ein hochwertiges Zoom-Objektiv mit einer Brennweite von 24–70 mm, das Weitwinkel- wie auch Porträt-Aufnahmen ermöglicht. Da das Display ausgeklappt und um 180 Grad gedreht werden kann, können Motive auch aus einem ungewöhnlichen Blickwinkel fotografiert werden. Zudem nimmt die Kamera Videos in Full-HD auf.

FinePix XP80
Digital Camera
Digitalkamera

Manufacturer
FUJIFILM Corporation, Tokyo, Japan
In-house design
Haruka Ikegame
Web
www.fujifilm.com

The FinePix XP80 is a robust camera complying with the IP 68 standard. It is waterproof down to a depth of 15 metres, is impact-resistant when dropped from a height of up to 1.75 metres, can withstand cold temperatures down to minus 10 centigrade, and is dustproof as well. With its 5x optical zoom, a wide angle of 28 mm and image stabilisation, the camera allows sophisticated images to be made even under extreme conditions. In addition, Wi-Fi is integrated to enable smooth communication with a smartphone.

Statement by the jury
With the FinePix XP80, taking pictures is fun in any situation. The camera is compact, powerful and has a robust design.

Die FinePix XP80 ist eine robuste Kamera, die den IP-68-Standard erfüllt. Sie ist wasserdicht bis 15 Meter Tiefe, stoßfest bis zu einer Fallhöhe von 1,75 Metern, kältebeständig bis minus 10 Grad Celsius sowie staubdicht. Mit ihrem fünffach optischen Zoom, einem Weitwinkel von 28 mm und Bildstabilisierung ermöglicht sie auch unter extremen Bedingungen anspruchsvolle Aufnahmen. Darüber hinaus ist Wi-Fi für die reibungslose Kommunikation mit einem Smartphone integriert.

Begründung der Jury
Mit der FinePix XP80 macht das Fotografieren in jeder Situation Spaß. Die Kamera ist kompakt, leistungsstark und besitzt ein robustes Design.

Coolpix S6900
Digital Compact Camera
Digitale Kompaktkamera

Manufacturer
Nikon Corporation, Tokyo, Japan
In-house design
Toshiko Odashima, Kenichi Soejima
Web
www.nikon.com

The Coolpix S6900 is particularly suitable for shooting self-portraits. It includes a built-in camera stand and a front shutter release button. Its 3" touchscreen can be swivelled and tilted, offering a high degree of flexibility during operation. NFC and Wi-Fi functionality facilitate the simple transfer of the captured images to other devices like smartphones or tablets. Moreover, the wide-angle lens makes it easy to shoot self-portraits within a group setting.

Statement by the jury
This camera offers well-thought-out functions that make it easy for the user to shoot selfies. It also features a modern and stylish design.

Die Coolpix S6900 eignet sich besonders für die Aufnahme von Selbstporträts. Sie besitzt einen integrierten Kamerafuß sowie einen Auslöser an der Vorderseite. Ihr 3"-Touchscreen ist dreh- und neigbar und bietet damit eine hohe Flexibilität bei der Aufnahme. NFC- und Wi-Fi-Funktionen ermöglichen die einfache Übertragung der Fotos auf andere Geräte wie Smartphones oder Tablets. Durch das Objektiv mit Weitwinkelabdeckung sind auch Selbstaufnahmen in der Gruppe problemlos möglich.

Begründung der Jury
Die Kamera bietet durchdachte Funktionen, mit denen die Aufnahme von Selfies erleichtert wird. Passend dazu ist sie modern und stilvoll gestaltet.

Instax Wide 300
Instant Camera
Sofortbildkamera

Manufacturer
FUJIFILM Corporation, Tokyo, Japan
In-house design
Hiroyuki Sakai
Web
www.fujifilm.com

The Instax Wide 300 is an instant camera using the film format 62 x 92 mm, which is of particular advantage when taking group or landscape photographs. The camera has a lateral grip enabling it to be held securely with the right hand. It also features tiltable shutter releases that can be pushed naturally with the index finger. With these properties, reliable handling is possible in any situation.

Statement by the jury
The Instax Wide 300 allows uncomplicated, spontaneous images to be made in analogue style. It offers sophisticated technology in an ergonomic shape.

Die Instax Wide 300 ist eine Sofortbildkamera, die Breitfilm im Format 62 x 92 mm verwendet, was besonders bei Gruppen- oder Landschaftsaufnahmen von Vorteil ist. An ihrem seitlichen Griff kann die Kamera mit der rechten Hand sicher festgehalten werden, der neigbare Auslöser wird ganz natürlich mit dem Zeigefinger gedrückt. Dank dieser Merkmale ist die zuverlässige Handhabung in jeder Situation möglich.

Begründung der Jury
Die Instax Wide 300 ermöglicht unkomplizierte, spontane Aufnahmen im analogen Stil. Dabei bietet sie ausgereifte Technik in einer ergonomischen Form.

Cine Lens ZK12x25
Kino-Objektiv

Manufacturer
FUJIFILM Corporation, Tokyo, Japan
In-house design
Hiroyuki Sakai
Web
www.fujifilm.com

The ZK12x25 is a long zoom cinema lens for film shots with a focal length ranging from 25 to 300 mm and continuous 4K resolution. The photographer can thus optimally adjust the zoom in many situations, which is of particular advantage for documentaries and nature scenes. A large font is used to indicate the operation status on the display so that it can be read easily, even under poor lighting conditions.

Statement by the jury
Thanks to the large zoom range, the lens is highly versatile. As it delivers high-resolution images across the entire spectrum, it is ideal for professionals.

Das ZK12x25 ist ein langes Kino-Zoomobjektiv für Filmaufnahmen mit einer Brennweite von 25 bis 300 mm bei einer durchgehenden 4K-Auflösung. Dadurch kann der Fotograf den Zoom in zahlreichen Situationen optimal anpassen, was vor allem bei Dokumentarfilmen oder Naturaufnahmen große Vorteile bietet. Für die Betriebsanzeige auf dem Display wird eine große Schrift verwendet, so kann sie auch bei schlechter Beleuchtung gut gelesen werden.

Begründung der Jury
Das Objektiv lässt sich dank seines großen Zoombereichs vielseitig einsetzen. Da es über das gesamte Spektrum hochauflösende Bilder liefert, eignet es sich besonders für Profis.

Commax Bulb Cam
Video Security System
Videoüberwachungssystem

Manufacturer
Commax, Seongnam, South Korea
In-house design
Jaesang Park, Yonghee Han
Web
www.commax.co.kr

The IP-based video security system Commax Bulb Cam is powered via a standard light-bulb fitting, which makes the device easy and quick to install. The system includes four modules: camera lens, body, power and mounting. This enables a wide range of combinations for any environment and application. In addition, the modules are characterised by a uniform design.

Das IP-basierte Videoüberwachungssystem Commax Bulb Cam wird über eine standardmäßige Glühbirnenfassung mit Strom versorgt. Dadurch kann das Gerät sehr einfach und schnell installiert werden. Das System besteht aus vier Modulen: Kameralinse, Gehäuse, Stromquelle und Befestigung. Dadurch entstehen viele verschiedene Kombinationsmöglichkeiten für jede Umgebung und jede Anwendung. Die Module zeichnen sich zudem durch ein einheitliches Design aus.

Statement by the jury
The Commax Bulb Cam is astonishingly easy to mount thanks to its lamp socket. Its discreet aesthetics blend in with any interior design.

Begründung der Jury
Die Commax Bulb Cam lässt sich durch den Lampensockel verblüffend einfach montieren. Ihre dezente Ästhetik fügt sich in jedes Ambiente.

CAMERA_module_PT

EXTENSION_module_L

POWER_module_WIFI

Dahua Full-Range PTZ Camera
Security Camera
Überwachungskamera

Manufacturer
Zhejiang Dahua Technology Co., Ltd., Hangzhou, China
In-house design
Jixiang Jin, Cao Li
Web
www.dahuasecurity.com

This security camera features a 36x wide-angle zoom lens. It can be rotated at high speed by 360 degrees horizontally and between 45 and 90 degrees vertically. The compact aluminium-alloy housing is water-resistant and dustproof. With the help of the integrated infrared LED light compensation technology, the camera can also record images in poor weather conditions and even in complete darkness.

Diese Überwachungskamera verfügt über einen 36-fach Weitwinkelzoom. Sie kann mit hoher Geschwindigkeit um 360 Grad waagerecht und um 45–90 Grad senkrecht gedreht werden. Das kompakte Gehäuse mit Aluminiumlegierung ist wasser- und staubdicht. Mithilfe der integrierten Infrarot-LED-Kompensationstechnologie kann die Kamera auch bei schlechten Wetterverhältnissen und sogar bei völliger Dunkelheit Bilder aufnehmen.

Statement by the jury
The unique rounded cover that frames the security camera is a special feature, conveying a high degree of individuality.

Begründung der Jury
Die ausgefallene abgerundete Blende, die die Überwachungskamera umrahmt, stellt eine Besonderheit dar und verleiht ihr ein hohes Maß an Individualität.

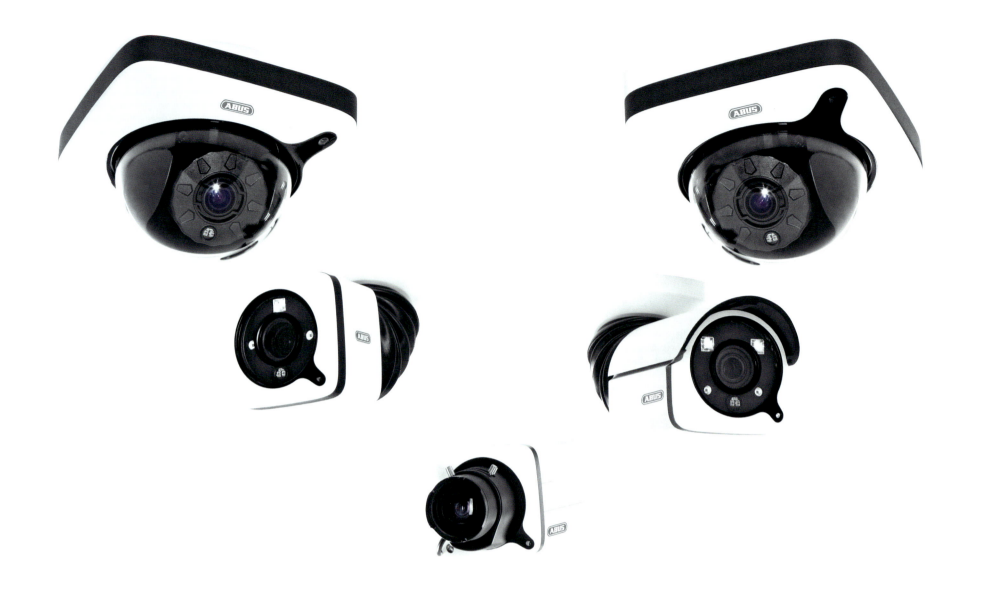

IP Camera Series
IP-Kameraserie

Manufacturer
ABUS Security-Center GmbH & Co. KG, Affing, Germany
In-house design
Web
www.abus.com

With its five construction styles and nine different models, this IP camera series offers numerous application options. The devices convince with an optimum ratio between image resolution and bandwidth required, as well as a high degree of image usability. Moreover, they are compatible with existing video surveillance systems and easy to install and configure. Integrated mounts and concealed cable routing are part of a stringent functional design which characterises the entire range.

Die IP-Kameraserie bietet mit ihren fünf Bauarten und neun Modellen zahlreiche Anwendungsmöglichkeiten. Die Geräte überzeugen durch ein optimales Verhältnis zwischen Bildauflösung und beanspruchter Bandbreite sowie eine hohe Bildnutzbarkeit. Darüber hinaus sind sie kompatibel mit bereits bestehenden Videoüberwachungsanlagen und einfach zu installieren und zu konfigurieren. Integrierte Halterungen und verdeckte Kabelführungen sind Teil eines stringenten, funktionalen Designs, das die gesamte Serie auszeichnet.

Statement by the jury
This camera series offers an optimal solution for any purpose. Thanks to a unified design language, all versions of the camera are readily recognisable as part of the same product family.

Begründung der Jury
Die Kameraserie bietet für jeden Zweck die passende Lösung. Dabei sind aufgrund der einheitlichen Formensprache alle Varianten stets als Teil der Produktfamilie erkennbar.

Arlo™
Smart Home Security Camera
Smart-Home-Sicherheitskamera

Manufacturer
NETGEAR, Inc., San Jose, USA
In-house design
John Ramones, Jenny Ouk
Design
Enlisted Design (Beau Oyler, Julian Bagirov,
Charles Bates, Jared Aller), Oakland, USA
Web
www.netgear.com
www.enlisteddesign.com

Arlo is a Wi-Fi security camera capable
of transmitting and recording HD videos.
The small and versatile device features
a magnetic ball mount and thus enables
the monitoring of every possible angle
in the room. Since the device is weather-
proof and equipped with infrared functions, it can be used both indoors and
outdoors for round-the-clock registering
of movement and image recording. A
single battery charge lasts for up to six
months of operation time.

Arlo ist eine Wi-Fi-Sicherheitskamera, die
HD-Videos überträgt und aufzeichnet. Das
kleine und vielseitige Gerät wird per
Magnet an einer kugelförmigen Halterung
montiert und überwacht dadurch jeden
Winkel im Raum. Da sie wetterfest und mit
Infrarot-Funktionen ausgestattet ist, kann
sie im Innen- und Außenbereich angebracht
werden und rund um die Uhr Bewegung
erfassen und Bilder aufzeichnen. Eine einzige Akkuladung reicht für eine Betriebsdauer von sechs Monaten.

Statement by the jury
This small camera considerably facilitates the monitoring of both home and
yard as it can be integrated virtually
anywhere.

Begründung der Jury
Diese kleine Kamera erleichtert die Überwachung von Haus und Grundstück erheblich, da sie überall eingesetzt werden kann.

Dahua SD21 Series
Security Camera
Überwachungskamera

Manufacturer
Zhejiang Dahua Technology Co., Ltd., Hangzhou, China
In-house design
Jixiang Jin, Cao Li
Web
www.dahuasecurity.com

The SD21 series offers full HD dome cameras with a 3x optical zoom that is used in a broad range of public settings. They can be integrated into networks in different ways, for instance via smartphone, PC or smart terminal. Their high-performance lenses are housed in a chassis made of special material, improving the transmission of infrared light. In this way, the cameras record high-quality images under any level of illumination. The lenses are rotatable and thus can be pointed to any spot in the room. The devices are easily installed and dust-proof as well.

Die SD21-Serie bietet Full-HD-Dome-Kameras mit dreifachem optischem Zoom, die in zahlreichen öffentlichen Bereichen eingesetzt werden. Sie lassen sich auf unterschiedliche Weise in Netzwerke einbinden, z. B. über Smartphone, PC oder Smart Terminal. Die leistungsfähigen Linsen sind in einem Gehäuse aus speziellem Material untergebracht, das die Übertragung von Infrarotlicht verbessert. Dadurch nehmen die Kameras Bilder in hoher Qualität bei jeder Helligkeit auf. Da die Linsen drehbar sind, lässt sich zudem jede Stelle im Raum überwachen. Die Geräte sind leicht zu installieren und überdies staubdicht.

FLEXIDOME IP panoramic 7000 MP
Security Camera
Überwachungskamera

Manufacturer
Bosch Sicherheitssysteme GmbH, Grasbrunn,
Germany
Design
TEAMS Design GmbH, Esslingen, Germany
Web
www.boschsecurity.com
www.teamsdesign.com

The Flexidome IP panoramic 7000 MP
video security camera has a discreet design and is easily installed. With a 360-
or 180-degree viewing angle, it captures
a whole range in a single image and, at a
rate of 30 frames per second, it registers
fast-moving objects in high detail. The
camera recognises suspicious events live
as well as in subsequent analysis, thus
facilitating the review of surveillance
videos.

Statement by the jury
The camera offers intelligent functions
through an inconspicuous, friendly design and thus makes video surveillance
considerably easier.

Die Videoüberwachungskamera Flexidome
IP panoramic 7000 MP ist dezent gestaltet
und lässt sich einfach installieren. Sie erfasst mit einer 360- oder 180-Grad-Sicht
einen ganzen Bereich in einem Bild und
liefert mit einer Bildrate von 30 Bildern pro
Sekunde die detailgetreue Erfassung von
sich schnell bewegenden Objekten. Die Kamera erkennt verdächtige Ereignisse sowohl live als auch in nachträglicher Analyse
und erleichtert dadurch die Auswertung
der Überwachungsvideos.

Begründung der Jury
Die Kamera bietet intelligente Funktionen in
einem unauffälligen, freundlichen Design
und vereinfacht so die Videoüberwachung
erheblich.

Rhea Smart Cam
Security Camera
Überwachungskamera

Manufacturer
LG Innotek, Ansan, South Korea
In-house design
Kihyun Jeong
Web
www.lginnotek.com

The Rhea IP security camera detects
movement in interior areas around the
clock. Recorded images are transferred
to mobile phones via Wi-Fi or stored
in the memory of the device. Since the
camera is very small, it can be positioned
nearly anywhere. As such, it may easily
be installed on a table, a wall or a ceiling
and directed towards any spot in the
room. With its smooth, rounded shape,
it also renders a friendly impression.

Statement by the jury
The attractive exterior of the camera is
shown to an advantage especially in
indoor settings, which increases the acceptance of this monitoring device.

Die IP-Überwachungskamera Rhea erfasst
rund um die Uhr Bewegungen im Innenbereich. Die Aufnahmen werden im Gerät gespeichert oder per Wi-Fi an Mobiltelefone
übertragen. Da die Kamera so klein ist,
findet sie an vielen Orten einen Platz. Sie
lässt sich beispielsweise leicht an Tisch,
Wand oder Decke montieren und auf jeden
beliebigen Punkt im Raum ausrichten. Mit
ihrer glatten runden Form macht sie zudem
einen freundlichen Eindruck.

Begründung der Jury
Das sympathische Äußere der Kamera
kommt in Innenräumen besonders gut zur
Geltung und erhöht die Akzeptanz für
dieses Überwachungsgerät.

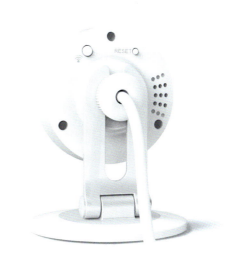

Beseye Pro
Cloud-Based Home Security Camera
Cloud-basierte
Heim-Überwachungskamera

Manufacturer
Beseye Cloud Security Co., Ltd., Taipei, Taiwan
In-house design
Pei-Yun Tsai
Web
www.beseye.com

Beseye Pro combines intelligent home monitoring with decorative art. The camera, whose form is reminiscent of a water drop, is secured in the mounting with a neodymium magnet and can be rotated by 360 degrees. This allows it to be easily positioned in the desired direction. The camera also houses modern technologies, such as infrared LEDs, cloud-based intelligent video analyses and SSL encryption.

Statement by the jury
This security camera impresses with its sculptural, smooth design and clever magnetic mounting.

Beseye Pro kombiniert intelligente Heim-überwachung mit dekorativer Kunst. Die Kamera, deren Form an einen Wassertropfen erinnert, wird von einem Neodym-Magnet in der Halterung fixiert und lässt sich um 360 Grad drehen. So kann sie einfach in jede gewünschte Position ausgerichtet werden. Außerdem beherbergt die Kamera moderne Technologien, wie Infrarot-LEDs, Cloud-basierte intelligente Videoanalysen und SSL-Verschlüsselung.

Begründung der Jury
Die Überwachungskamera punktet mit ihrer skulpturalen, sanften Gestaltung und der cleveren Befestigung mittels Magnethalterung.

SNH-E6110BN
Security Indoor Camera
Sicherheitskamera für
den Innenbereich

Manufacturer
Samsung Techwin, Seongnam, South Korea
In-house design
Hojin Park, Hyunkyu Park
Web
www.samsungtechwin.co.kr

The SNH-E6110BN is a slim and compact full HD home security camera. Installation is quick and easy as the metal bracket is equipped with hinges. Once mounted to the wall, the bracket on the back of the camera housing is folded in so that the camera hangs flush against the wall. Configuration and subsequent monitoring are done via an app installed on Wi-Fi-capable smart devices. The camera also features two-way audio and thus allows the user to hold a conversation.

Statement by the jury
The camera mount is an intelligent solution in that the foldable bracket allows for unobtrusive mounting directly on the wall.

Die SNH-E6110BN ist eine schlanke und kompakte Full-HD-Überwachungskamera für den Heimbereich. Sie wird schnell und einfach installiert, da der Halter mit Gelenken ausgestattet ist. Einmal an der Wand montiert, wird der Halter an der Rückseite des Kameragehäuses zusammengeklappt, sodass die Kamera flach an der Wand hängt. Die Konfigurierung und spätere Überwachung erfolgt über eine App von Wi-Fi-fähigen Smartgeräten aus. Ebenso besitzt die Kamera eine Zwei-Wege-Audio-Funktion und ermöglicht dadurch eine Unterhaltung.

Begründung der Jury
Die Befestigung der Kamera ist intelligent gelöst, denn dank des klappbaren Halters lässt sie sich unauffällig direkt auf der Wand montieren.

Photo Sensor
Fotosensor
Security Camera
Überwachungskamera

Manufacturer
LG Innotek, Ansan, South Korea
In-house design
Kihyun Jeong
Web
www.lginnotek.com

The Photo Sensor is a security camera with a discreet design that automatically detects intruders and transfers still images to the user via Wi-Fi. It distinguishes persons from animals to minimise false alarms. It also sends an alert message to the user when detached from the wall or opened. The camera enables different viewing angles to monitor preferred areas.

Der Fotosensor ist eine dezent gestaltete Überwachungskamera, die Eindringlinge automatisch erkennt und Standbilder per Wi-Fi an das Mobiltelefon des Benutzers übermittelt. Dabei unterscheidet er zwischen Tieren und Menschen, um Fehlalarme zu minimieren. Auch wenn er von der Wand genommen oder geöffnet wird, sendet er einen Alarm an den Benutzer. Zudem lassen sich unterschiedliche Betrachtungswinkel einstellen, um die gewünschten Bereiche zu überwachen.

Statement by the jury
The Photo Sensor has a discreet, friendly shape so that it can be installed unobtrusively in interior spaces.

Begründung der Jury
Der Fotosensor besitzt eine zurückhaltende, freundliche Form, sodass er in Innenräumen unauffällig installiert werden kann.

Withings Home
All-in-One Security Camera with Air Quality Sensors
Überwachungskamera mit Sensoren zur Kontrolle der Raumluftqualität

Manufacturer
Withings, Issy-les-Moulineaux, France
Design
Elium Studio, Paris, France
Web
www.withings.com
www.eliumstudio.com

Withings Home is an intelligent security camera with a 135-degree, wide-angle, 12x zoom and high-quality night vision. The camera detects both motion and sound, which triggers smart alerts. Users can access live streaming and video footage directly from their smartphone application. The camera also has embedded air quality sensors to monitor indoor pollution.

Withings Home ist eine intelligente Überwachungskamera mit einem zwölffach Weitwinkel-Zoom von 135 Grad und ausgezeichneter Nachtsichtqualität. Die Kamera registriert sowohl Bewegungen als auch Geräusche, woraufhin intelligente Benachrichtigungen ausgelöst werden. Über eine Smartphone-App können Nutzer direkt auf Live-Bilder und Videoaufzeichnungen zugreifen. Darüber hinaus ist die Kamera mit Sensoren ausgestattet, mit denen die Luftqualität im Wohnbereich überwacht werden kann.

Statement by the jury
With its cylindrical shape and bright colouring, the camera has a highly congenial appearance. It can be set up easily anywhere, yet remains comfortably discreet.

Begründung der Jury
Mit ihrer zylindrischen Form und der hellen Farbgebung wirkt die Kamera ausgesprochen sympathisch. Sie lässt sich leicht überall aufstellen und bleibt dabei angenehm diskret.

Canary
Security System
Sicherheitssystem

Manufacturer
Canary, New York, USA
In-house design
James Krause
Web
www.canary.is

Canary was designed with the idea to make an uncomplicated, inexpensive and intelligent security system available that anyone can install at home. All necessary functions are accommodated in one single device: it records audio and video, registers motion and even measures temperature, humidity and air quality. Moreover, it is equipped with a learning function and adapts over time to the events in the room being monitored.

Statement by the jury
Canary has a very friendly appearance, which makes it easier to approach the product and emphasises its simple operation.

Canary wurde mit dem Gedanken entwickelt, ein unkompliziertes, preiswertes und intelligentes Sicherheitssystem anzubieten, das jeder selbst zu Hause installieren kann. Alle notwendigen Funktionen sind in einem einzigen Gerät untergebracht: Es nimmt Audio und Video auf, registriert Bewegung und misst dazu noch Temperatur, Luftfeuchtigkeit und Luftqualität. Zudem ist es mit einer Lernfunktion ausgestattet und passt sich mit der Zeit an die Vorgänge im überwachten Raum an.

Begründung der Jury
Canary besitzt eine überaus freundliche Anmutung, die die Annäherung an das Produkt erleichtert und die einfache Bedienung unterstreicht.

Water Wolf HD UW1
Underwater Fishing Camera
Unterwasserkamera
zum Angeln

Manufacturer
Svendsen Sport, Gadstrup, Denmark
Design
MOEF A/S (Martin Holmberg, Michael Trøst), Copenhagen, Denmark
Web
www.svsshop.com
www.moef.dk

The Water Wolf HD UW1 underwater camera enables the recording of images during fishing in salt- and freshwater. It is mounted to the fishing line with a steel bar and can be used for different types of fishing, such as artificial lure fishing or surface float fishing. The mount is conceived in such a way that the camera is always secured in the right position underwater. Its slim shape results in low water resistance so that the camera glides through the water smoothly.

Statement by the jury
A high degree of functionality is concealed behind the plain form, as the camera masters divergent situations during fishing and can be used extremely versatilely.

Die Unterwasserkamera Water Wolf HD UW1 ermöglicht Aufnahmen beim Angeln in Salz- und Süßwasser. Sie wird mithilfe einer Stahlstange an der Angelschnur befestigt und kann bei verschiedenen Formen des Angelns wie Kunstköderangeln oder Angeln mit Schwimmern eingesetzt werden. Die Befestigung ist so konzipiert, dass die Kamera stets in der richtigen Position im Wasser liegt. Zudem bietet ihre schlanke Form nur einen geringen Wasserwiderstand, sodass die Kamera ruhig durch das Wasser gleitet.

Begründung der Jury
Hinter der schlichten Form verbirgt sich eine hohe Funktionalität, denn die Unterwasserkamera meistert die unterschiedlichsten Situationen beim Angeln und kann vielseitig eingesetzt werden.

Handheld 3D Scanner

Manufacturer
Industrial Technology Research Institute, Hsinchu, Taiwan
In-house design
Design
Gearlab, Taipei, Taiwan
Web
www.itri.org.tw
www.gearlab.com.tw

This handheld 3D scanner features a patented IR depth camera and 3D software which allows for continuous 3D stitching and fast model reconstruction without needing a position-tracking device. Furthermore, it has balanced weight distribution to reduce stress on the wrists. An ergonomically fitting handle grip at a specific angle also provides comfortable and precise adjustment towards the scanning target. A streamlined status indicator provides vital information during operation.

Dieser 3D-Handheld-Scanner ist mit einer patentierten IR-Tiefenkamera und 3D-Software ausgestattet, was die kontinuierliche 3D-Abtastung und schnelle Modellrekonstruktion ohne gesondertes Gerät zur Positionsbestimmung ermöglicht. Zudem ist die Gewichtsverteilung ausgewogen, was Belastungen für das Handgelenk reduziert. Ein ergonomisch geformter, speziell gewinkelter Handgriff gewährleistet die präzise Ausrichtung auf das zu scannende Objekt. Ein optimiertes Statusmeldesystem bietet wichtige Informationen bei der Bedienung.

Statement by the jury
The 3D scanner is conceived in such a way that it reduces physical stress during work to a minimum and allows for precise handling.

Begründung der Jury
Der 3D-Scanner ist so konzipiert, dass er präzise zu handhaben ist und die körperlichen Belastungen während der Arbeit auf ein Minimum reduziert.

Motorola Scout 5000
Dog Collar with Smart Functions
Hundehalsband mit Smart-Funktionen

Manufacturer
Binatone Electronics International Limited, Hong Kong
In-house design
Chow Hung Pong, Valentino
Web
www.binatoneglobal.com

The Motorola Scout 5000 is a dog collar with smart functions. It features a 4K Wi-Fi camera so that dog owners can use their smartphone to see what the dog sees or record the action in HD quality. The dog can be located via the GPS tracking system, and a motion sensor measures the dog's activity. The user can engage in two-way audio communication with their pet via a speaker built into the collar and will receive bark notifications when the dog passes the geo fence.

Motorola Scout 5000 ist ein Hundehalsband mit Smart-Funktionen. In das Halsband ist eine 4K-Wi-Fi-Kamera integriert, sodass Hundebesitzer auf ihrem Smartphone sehen, was ihr Hund sieht, oder eine Videoaufnahme in HD-Qualität machen können. Über das GPS-System lässt sich der Hund orten, ein Bewegungssensor misst die Aktivität des Vierbeiners. Der Nutzer kann über den im Halsband integrierten Lautsprecher mit seinem Haustier in beide Richtungen kommunizieren. Dadurch hört er, wenn der Hund bellt, und er erhält Benachrichtigungen, wenn der Hund den Geo-Zaun durchbricht.

Statement by the jury
The Motorola Scout 5000 enriches the coexistence of owner and pet and also contributes to the dog's safety.

Begründung der Jury
Motorola Scout 5000 bereichert das Zusammenleben von Haustier und Besitzer und trägt zur Sicherheit des Hundes bei.

iSHOXS ProMount
Mounting System for Action Cameras
Halterungssystem für Action-Kameras

Manufacturer
Tormaxx GmbH, Mönchengladbach, Germany
In-house design
Hans Hubert Koch, Armin Dilling
Web
www.tormaxx.de

The iSHOXS ProMount is designed to attach action cameras to pipes, for instance bicycle frames. With the easily adjustable diameter range from 20–42 mm and compatibility with round, oval and angular profiles, a wide range of applications is possible. The mount is machined from aluminium and features a high-quality anodised surface. It offers multiple adjustment options, a quick-release fastener and is simple to handle.

Statement by the jury
This camera mount is very versatile as it can be easily and securely attached to a large number of pipe-like component parts.

Mit dem iSHOXS ProMount werden Action-Kameras an Rohren von z. B. Fahrrad-rahmen befestigt. Durch den einfach zu verstellenden Durchmesserbereich von 20–42 mm und die Kompatibilität mit runden, ovalen und eckigen Profilen ergibt sich ein großer Anwendungsbereich. Die Halterung ist aus Aluminium gefräst und hochwertig eloxiert. Sie bietet vielfältige Verstellmöglichkeiten, einen Schnell-verschluss und ist leicht zu handhaben.

Begründung der Jury
Die Kamerahalterung ist sehr vielseitig, da sie sich an zahlreichen rohrähnlichen Bauteilen einfach und sicher befestigen lässt.

iSHOXS Hell Rider
Mounting System for Action Cameras
Halterungssystem für Action-Kameras

Manufacturer
Tormaxx GmbH, Mönchengladbach, Germany
In-house design
Hans Hubert Koch
Web
www.tormaxx.de

The iSHOXS Hell Rider is a camera mount specifically designed for use on motor-cycles. The chain system adapts to all conceivable profiles and also enables mounting in areas that are difficult to reach. It offers high holding force and absorbs vibrations, which is indispensable for action-camera mounts on motor-cycles, quad bikes and other motorised vehicles. In this way, the picture and sound quality of the videos is enhanced.

Statement by the jury
This mounting system meets high demands with regard to attachment and stabilisation, thus bringing action cameras into new fields of application.

Der iSHOXS Hell Rider ist eine speziell für den Einsatz an Motorrädern entworfene Kamerahalterung. Das Kettensystem passt sich allen denkbaren Profilen an und ermöglicht die Montage auch an schwer erreichbaren Stellen. Dabei bietet es eine hohe Haltekraft und absorbiert Vibrationen, was unabdingbar ist für eine Action-Kamera-Halterung an Motorrädern, Quads und anderen motorisierten Fahrzeugen. Dadurch wird die Bild- und Tonqualität von Videoaufnahmen verbessert.

Begründung der Jury
Dieses Halterungssystem meistert erhöhte Anforderungen hinsichtlich Befestigung und Stabilisierung und bringt Action-Kameras damit in neue Bereiche.

iSHOXS Shark
Mounting System for Action Cameras
Halterungssystem für Action-Kameras

Manufacturer
Tormaxx GmbH, Mönchengladbach, Germany
In-house design
Hans Hubert Koch
Web
www.tormaxx.de

The iSHOXS Shark is designed in an open C construction and features a wide clamping range of up to 65 mm on oval and angular profiles. Its construction type is aligned to achieve the highest possible flexibility in all applications. Since it is saltwater-resistant, the mounting system is also suitable for use in maritime environments. Silicone inserts ensure a solid grip on softer materials and reduce vibrations and shocks.

Statement by the jury
The iSHOXS Shark is a robust all-rounder which can be attached virtually anywhere and is also deployable under extreme conditions.

Der iSHOXS Shark ist in offener C-Konstruk-tion gestaltet und verfügt über einen gro-ßen Klemmbereich von bis zu 65 mm bei ovalen und eckigen Profilen. Seine Bauwei-se ist auf die größtmögliche Flexibilität bei allen Anwendungen ausgerichtet. So ist das Halterungssystem seewassertauglich und eignet sich damit auch für den Einsatz im maritimen Bereich. Silikon-Inlets garantie-ren an weicheren Materialien festen Halt und reduzieren Vibrationen und Stöße.

Begründung der Jury
Der iSHOXS Shark ist ein robuster Alles-könner, der sich überall befestigen lässt und auch unter extremen Bedingungen einsetzbar ist.

SwiftCam G3
Handheld Gimbal for GoPro Action Cameras
Handgimbal für GoPro-Action-Kameras

Manufacturer
SwiftCam Technologies Group Company Limited, Hong Kong
In-house design
Jacky Cheung, Kenneth Choi
Web
www.swiftcam.com

The SwiftCam G3 is a slenderly designed, handheld gimbal for GoPro action cameras with three-axis stabilisation, a 360-degree horizontal pan, as well as joystick control for panning left and right or tilting up and down. The user can release or set each axis independently from the others so that everything from completely dynamic motion to full stabilisation of the camera is possible. The mount is equipped with a magnetic locking system and may thus be effortlessly stowed away.

SwiftCam G3 ist ein schlank gestalteter Handgimbal für GoPro-Action-Kameras mit 3-Achsen-Stabilisierung, 360-Grad-Horizontalschwenk sowie einer Joystick-Steuerung für Rechts-/Linksschwenk und Neigung. Der Benutzer kann jede Achse unabhängig voneinander lösen oder fixieren. Somit ist von einer komplett dynamischen Bewegung bis zur vollständigen Stabilisierung der Kamera alles möglich. Da die Halterung mit einem Magnetverschlusssystem ausgestattet ist, lässt sie sich leicht verstauen.

Statement by the jury
The mount has an elegant and functional appearance. With its multiple adjustment options, the user enjoys a maximum degree of flexibility with regard to camera positioning.

Begründung der Jury
Die Halterung zeigt sich elegant und funktional. Mit ihren vielfältigen Verstellmöglichkeiten kann der Nutzer seine Kamera sehr flexibel einsetzen.

XPRO Fluid Head
XPRO Fluid-Kopf

Manufacturer
Manfrotto, Cassola, Italy
In-house design
Design
Momodesign, Milan, Italy
Web
www.manfrotto.com
www.momodesign.com

The XPRO Fluid is a tripod head whose fluidity can be adjusted according to specific preferences. This gives the photographer the opportunity to select between hard and soft fluidity according to the desired tilting motion speed and equipment being supported. The body is fashioned from aluminium and technopolymer, resulting in a strongly reduced weight of only 700 grams and a payload of up to 4 kg.

Statement by the jury
The tripod head is pleasingly lightweight and, thanks to the adjustable fluidity, offers enhanced comfort when shooting photos and videos.

Der XPRO Fluid ist ein Stativkopf, dessen Fluidität sich nach Belieben einstellen lässt. So hat der Fotograf die Möglichkeit, je nach gewünschter Geschwindigkeit der Neigebewegung und der verwendeten Ausrüstung zwischen harter und weicher Fluidität zu wählen. Dank des Gehäuses aus Aluminium und Polymer-Kunststoff ist der Kopf besonders leicht und wiegt lediglich 700 Gramm bei einer maximalen Tragkraft von 4 kg.

Begründung der Jury
Der Stativkopf ist erfreulich leicht und bietet dank der anpassbaren Fluidität viel Komfort bei der Aufnahme von Fotos und Videos.

XPRO Geared Head
XPRO Getriebeneiger

Manufacturer
Manfrotto, Cassola, Italy
In-house design
Design
Momodesign, Milan, Italy
Web
www.manfrotto.com
www.momodesign.com

The XPRO geared tripod head was specifically designed for outdoor use and weighs merely 750 grams with a payload of 4 kg. The technopolymer body makes the tripod head robust and lightweight. Micrometric adjustment allows photographers to easily bring the camera into the desired position on all three axes. The adjustment knobs are ergonomically designed and convenient to handle.

Statement by the jury
This tripod head impresses with its option to precisely adjust the camera to the desired position. This offers flexibility in the selection of image details.

Der Getriebeneiger XPRO wurde besonders für die Verwendung im Freien entwickelt und wiegt lediglich 750 Gramm bei einer Tragkraft von 4 kg. Das Gehäuse aus Technopolymer macht den Stativkopf robust und leicht. Mit der mikrometrischen Feinjustierung können Fotografen ihre Kamera über alle drei Achsen einfach in die gewünschte Position bringen. Die Einstellknöpfe sind ergonomisch gestaltet und lassen sich bequem bedienen.

Begründung der Jury
Der Getriebeneiger beeindruckt mit der Möglichkeit, die Kameraposition äußerst genau einstellen zu können. Das bietet Flexibilität bei der Wahl des Bildausschnitts.

XPRO Ball Head
XPRO Kugelkopf

Manufacturer
Manfrotto, Cassola, Italy
In-house design
Web
www.manfrotto.com

The XPRO is an innovative, lightweight ball head designed to meet the needs and capabilities of a large number of photographers. The head weighs merely 530 grams, thanks to its hollow sphere and magnesium body, yet it has an impressive payload of up to 8 kg. A central locking system made up of three individual wedges guarantees secure latching. The head is available with either the 200PL quick-release plate or the advanced Top Lock system.

Statement by the jury
The XPRO ball head has a coherent design that offers comprehensive functionality at a low weight and high carrying capacity.

XPRO ist ein innovativer, leichter Kugelkopf, der den Bedürfnissen und Fähigkeiten vieler Fotografen gerecht wird. Der Kopf wiegt dank seiner Hohlkugel und dem Magnesiumgehäuse lediglich 530 Gramm und hat dabei eine beeindruckende Tragkraft von bis zu 8 kg. Ein zentrales Verriegelungssystem aus drei individuellen Keilen garantiert eine sichere Arretierung. Der Kopf ist mit der Schnellwechselplatte 200PL oder dem fortschrittlichen Top-Lock-System erhältlich.

Begründung der Jury
Der XPRO Kugelkopf ist schlüssig gestaltet und bietet eine umfassende Funktionalität bei geringem Gewicht und hoher Tragkraft.

Befree Carbon Tripod
Befree Carbonstativ

Manufacturer
Manfrotto, Cassola, Italy
In-house design
Design
Momodesign, Milan, Italy
Web
www.manfrotto.com
www.momodesign.com

The Befree carbon tripod is very light-weight and compact. Its legs, which are made of 100 per cent carbon, provide a high degree of rigidity at a weight of merely 1.1 kg. Thanks to the tripod's special mechanism and the quick-release adapter for its ball head, the legs can be reduced to a compact pack size with a length of only 40 cm. This consider-ably facilitates transport with the pad-ded carry case included.

Statement by the jury
With its carbon legs, this tripod is very lightweight and stable, making it a reli-able and versatile travelling companion for photographers.

Das Carbonstativ Befree ist sehr leicht und kompakt. Seine Beine bestehen zu 100 Prozent aus Carbon und bieten eine hohe Steifigkeit bei einem Gewicht von lediglich 1,1 kg. Der besondere Mechanis-mus des Stativs sowie der Schnellwech-seladapter seines Kugelkopfes ermöglichen es, dass sich die Beine auf ein kompaktes Packmaß von nur 40 cm Länge zusammen-legen lassen. Dadurch wird der Transport in der mitgelieferten gepolsterten Trageta-sche erheblich erleichtert.

Begründung der Jury
Das dank seiner Carbonbeine sehr leichte und stabile Stativ ist ein zuverlässiger und vielseitiger Reisebegleiter für Fotografen.

Compact Action
Tripod
Stativ

Manufacturer
Manfrotto, Cassola, Italy
In-house design
Web
www.manfrotto.com

The Compact Action tripod was designed for entry-level DSLR users who like to shoot a lot of photos and videos. The ergonomic joystick head is securely locked with a rotary knob, while the grip is comfortable to hold and allows im-pressive still images and smooth movies to be captured with little effort. More-over, the camera can be easily attached to the circular quick-release plate.

Statement by the jury
This tripod is easy to handle, even for novices. It offers all functions needed for the safe and comfortable shooting of photos and videos.

Das Stativ Compact Action wurde speziell für die Besitzer von Einsteiger-DSLR-Kame-ras konzipiert, die gerne viele Fotos machen und Filme aufnehmen. Der ergonomische Joystick-Kopf wird über ein Drehrad sicher verriegelt, der Griff liegt gut in der Hand und ermöglicht beeindruckende Fotos und fließende Filmaufnahmen mit geringem Aufwand. Zudem lässt sich die Kamera auf der runden Schnellwechselplatte mühelos befestigen.

Begründung der Jury
Das Stativ ist besonders für Einsteiger ein-fach zu bedienen. Es bietet alle Funktionen für das sichere, bequeme Fotografieren und Filmen.

Off-Road Tripod
Off-Road Stativ

Manufacturer
Manfrotto, Cassola, Italy
In-house design
Design
Momodesign, Milan, Italy
Web
www.manfrotto.com
www.momodesign.com

Off-Road is optimally suited for hikers who value convenient transport and comfortable operation. It is the lightest tripod in the Manfrotto range and, with a diameter of 5 cm, is very compact and easy to stow. The aluminium legs close telescopically with a simple twist-to-lock mechanism. With the universal 1/4" camera attachment and the quick wheel, the camera can be conveniently attached and detached.

Statement by the jury
This tripod allows for quick attachment of the camera in a secure position when-ever the photographer wants to shoot pictures or videos outdoors.

Off-Road ist bestens für Wanderer geeig-net, für die ein einfacher Transport und eine komfortable Bedienung wichtig sind. Es ist das leichteste Stativ aus dem Manfrotto-Sortiment und ist mit einem Durchmesser von 5 cm sehr kompakt und leicht zu verstauen. Die Beine aus Alu-minium schließen teleskopartig mit einem einfachen Drehriegelverschluss. Dank der universellen 1/4"-Kamerabefestigung und dem Schnellrad kann die Kamera bequem befestigt und abgenommen werden.

Begründung der Jury
Mit diesem Stativ ist die Kamera schnell in sicherer Position befestigt, wenn der Foto-graf im Freien Bilder schießen oder filmen möchte.

Cardboard
Virtual Reality Viewer for Mobile Phones
Virtual-Reality-Erweiterung für Mobiltelefone

Manufacturer
Google, Mountain View, USA
In-house design
Web
www.google.com

You can also find this product in
Dieses Produkt finden Sie auch in
Working
Page 343
Seite 343

Cardboard is a low-cost virtual reality headset for mobile phones. It allows the user to easily explore 3D environments and consume VR content and media. It is made from a single sheet of cardboard and shipped in a flat envelope, reducing manufacturing as well as transport and storage costs. The design templates are freely available on the Internet, which considerably simplifies both industrial manufacture and replication by individuals.

Cardboard ist eine kostengünstige Virtual-Reality-Erweiterung für Mobiltelefone, die es dem Benutzer erlaubt, leicht in 3D-Umgebungen einzutauchen und VR-Inhalte und -Medien zu nutzen. Sie wird aus einem einzigen Stück Karton hergestellt und als flacher Umschlag versandt. Dadurch sind sowohl die Herstellungs- als auch die Transport- und Lagerkosten gering. Die Pläne sind für jedermann im Internet zugänglich, was sowohl die industrielle Herstellung als auch den Nachbau durch Einzelpersonen erheblich vereinfacht.

Statement by the jury
The Cardboard VR viewer implements the concept of simplification in an outstanding way. It uses fewer resources and, thanks to the open-source philosophy, becomes a part of the community.

Begründung der Jury
Cardboard setzt das Konzept der Vereinfachung hervorragend um. Die Erweiterung verbraucht wenige Ressourcen und wird durch die Open-Source-Philosophie Teil der Gemeinschaft.

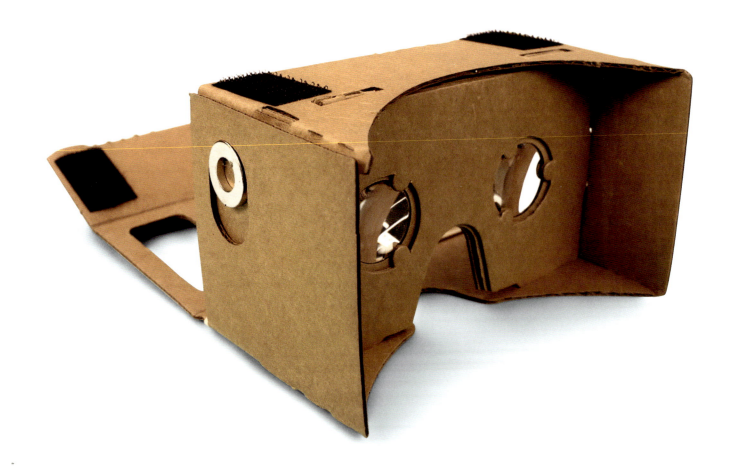

Arrowscope
Jewellery Experience Equipment
Diamanten-Prüfausstattung

Manufacturer
Goldway Technology Ltd., Hong Kong
In-house design
Carol Cheng
Design
Dr. Tom Kong, Dr. Wai Keung Yeung, Martin Liu, Ian Kwan
Web
www.goldwayltd.com

The Arrowscope allows customers to take a detailed look at the diamonds they are interested in buying. They can, for instance, find out whether a diamond shows the perfect cut, thus belonging to the category of so-called Hearts & Arrows diamonds. In such a case, the Arrowscope shows an even pattern of eight arrows and eight hearts. With the built-in LEDs, the Arrowscope lights up the cut; in addition, it simulates the sparkling effect engendered by the LEDs and a rotating platform. The related images can be streamed wirelessly to a tablet.

Mit dem Arrowscope können Kunden Diamanten beim Kauf im Detail betrachten. Dadurch lässt sich beispielsweise herausfinden, ob der Diamant den perfekten Schliff aufweist und zu den sogenannten Hearts-and-Arrows-Diamanten gehört. Das Arrowscope liefert in diesem Fall ein regelmäßiges Muster aus acht Pfeilen und acht Herzen. Durch die eingebauten LEDs beleuchtet das Arrowscope den Schliff; zusätzlich simuliert das Gerät den Glitzereffekt über die LEDs und eine rotierende Plattform. Die Aufnahmen können kabellos auf ein Tablet gestreamt werden.

Statement by the jury
The Arrowscope offers a new way of looking at high-grade jewellery. Not only does it serve to establish quality; it also adds an entertaining component to the inspection process.

Begründung der Jury
Das Arrowscope ermöglicht eine neue Sichtweise auf hochwertige Schmuckstücke. Es dient nicht nur der Begutachtung ihrer Qualität, sondern bereichert die Betrachtung um eine unterhaltsame Komponente.

77EG9900
OLED TV

Manufacturer
LG Electronics Inc.,
Seoul, South Korea

In-house design
Byung-Mu Huh, Min-Ji Seo,
Young Kyoung Kim, Sang-Ik Lee

Web
www.lg.com

reddot award 2015
best of the best

Luxurious flexibility

Based on technological advances, TV sets have undergone enormous developments over the years providing maximum viewing comfort by boasting ultra-high-definition images. Against this backdrop, the flexible 77EG9900 4K OLED TV impresses with a concept that offers users a novel type of adjusting the unit to both their individual needs and viewing preferences. This design approach has been facilitated by a 77" display that is not only light and ultra-thin but also flexible and bendable. The display thus allows for free variable adjustment of its plane and curvature, while the size of the display and the actual viewing distance are also taken into account for an optimum ratio between the two, depending on the viewing angle. Interestingly, the bending process of the display itself is almost imperceptible to the naked eye. Thanks to a slim bezel with a width of only 9 mm, the housing of the OLED display can pliably curve or flatten, whereas the curvature is already achieved by applying only minimal force. Users can comfortably adjust the screen's curvature by remote control, selecting from a flat plane to a maximum curvature of 4000R. This high-definition TV set also fascinates users with an overall design of strict lines and a successful reduction to essential elements. The design merges outstanding luxurious comfort with the subliminal experience evoked when users subtly adjust the display curvature to enhance their individual needs and desires.

Luxuriöse Flexibilität

Durch den technologischen Fortschritt hat der Fernseher in den vergangenen Jahren eine enorme Entwicklung hin zu überaus komfortablen Geräten mit sehr hoher Auflösung vollzogen. Der biegbare 4K-OLED-Fernseher 77EG9900 beeindruckt vor diesem Hintergrund mit einem Konzept, das dem Nutzer eine neue Art der Anpassung des Fernsehgerätes an seine individuellen Bedürfnisse und Sehgewohnheiten erlaubt. Möglich wurde seine Gestaltung durch ein sehr flaches und leichtes 77"-Display, das zugleich biegsam und flexibel ist. Es lässt sich in seiner Fläche und Krümmung völlig frei variieren. Je nachdem, wo man sitzt, können die Bildschirmgröße und der jeweilige Betrachtungsabstand so in das optimale Verhältnis gebracht werden. Bemerkenswert ist dabei, dass dieser Vorgang der Krümmung mit bloßem Auge kaum zu sehen ist. Dank einer schlanken, nur 9 mm breiten Einfassung ist auch das Gehäuse biegsam, wobei die Wölbung des OLED-Bildschirms schon durch eine nur minimale Krafteinwirkung erreicht wird. Der Nutzer kann ihn von einer planen Fläche bis hin zu einer maximalen Krümmung von 4000R auch per Fernsteuerung komfortabel einstellen. Dieses hochauflösende TV-Gerät fasziniert den Betrachter außerdem durch eine strikte Linienführung und eine gelungene Reduktion auf die wesentlichen Elemente. Seine Gestaltung verknüpft ausgesprochen luxuriösen Komfort mit den sinnlichen Eindrücken der fast unmerklichen Anpassung des Bildschirms an die Wünsche des Nutzers.

Statement by the jury

The flexible 77EG9900 4K OLED TV and its innovation of a freely and independently adjustable display curvature offers users entirely new possibilities. It even allows adjusting the curvature comfortably via remote control and thus bringing the display size and the individual viewing distance into an optimal ratio. The slim, purist and perfectly implemented form of the device is an enhancement to almost any interior environment.

Begründung der Jury

Der flexible 4K-OLED-Fernseher 77EG9900 bietet dem Nutzer mit der Innovation einer freien und eigenständigen Anpassung von Fläche und Krümmung völlig neue Möglichkeiten. Äußerst komfortabel kann er die Wölbung dabei auch aus der Entfernung anpassen und damit die Bildschirmgröße und den Betrachtungsabstand in ein jeweils optimales Verhältnis bringen. Die gestalterisch perfekt umgesetzte puristische und schlanke Form dieses Gerätes bereichert jedes Ambiente.

Designer portrait
See page 88
Siehe Seite 88

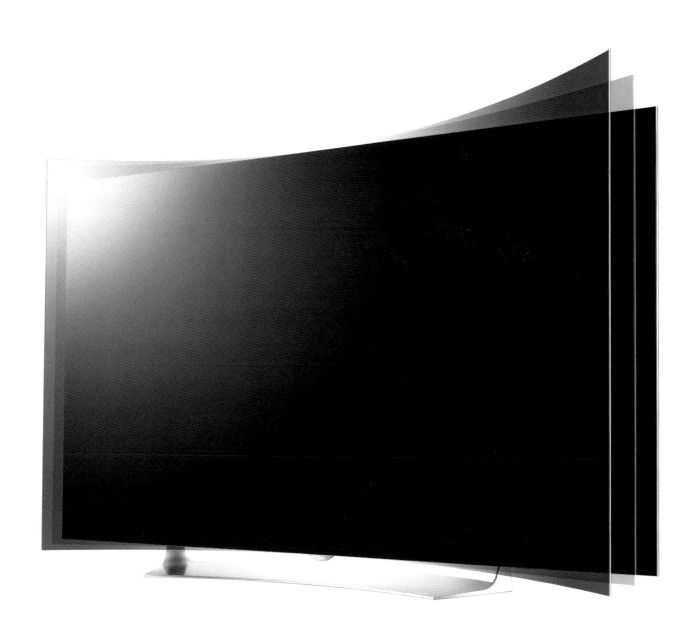

65EF9800
OLED TV

Manufacturer
LG Electronics Inc., Seoul, South Korea
In-house design
Yoo-Seok Kim, Youn-Soo Kim,
Yoon-Kyeong Lee, Byung-Mu Huh
Web
www.lg.com

This 4K OLED TV displays a clear functional and optical separation of screen and speakers. The speakers are connected wirelessly and are located below the very thin screen. They can also function as a stand and fill the space between screen and footprint. With their plain horizontal and vertical lines, the speakers foster an appealing contrast with the extremely smooth screen.

Statement by the jury
With their striking design, the speakers distinctly contrast with the screen, resulting in the television's unique appearance.

Bei dem 4K-OLED-Fernseher sind Bildschirm und Lautsprecher funktional wie optisch deutlich voneinander getrennt. Die Lautsprecher sind kabellos angebunden und befinden sich unterhalb des sehr dünnen Bildschirms. Sie können auch als Sockel fungieren und füllen den Raum zwischen Bildschirm und Standfläche. Dabei bilden sie mit ihren einfachen horizontal und vertikal verlaufenden Linien einen ansprechenden Kontrast zu dem äußerst glatten Bildschirm.

Begründung der Jury
Die Lautsprecher heben sich durch ihre auffällige Gestaltung deutlich vom Bildschirm ab und sorgen dadurch für ein eigenständiges Erscheinungsbild des Fernsehers.

65EG9600
OLED TV

Manufacturer
LG Electronics Inc., Seoul, South Korea
In-house design
Byung-Mu Huh, Sun-Ha Park, Sang-Ik Lee
Web
www.lg.com

Transparency is the design concept of this HD OLED TV with a curved screen. The stand is made from fine metal and the vertical connection to the screen is transparent, resulting in a feather-light appearance of the screen, which appears to be floating above the stand. The stand solution with its reduced design and clear structure creates a certain tension with the curvature of the screen.

Statement by the jury
The particular appeal of this television arises through interplay between the functionally designed stand with its combination of glass and metal and the curved, thin screen.

Transparenz heißt das Designkonzept dieses HD-OLED-Fernsehers mit gekrümmtem Bildschirm. Der Fuß ist aus feinem Metall gefertigt, die vertikale Verbindung mit dem Bildschirm ist transparent ausgeführt. Dadurch wirkt der Bildschirm federleicht und scheint über dem Fuß zu schweben. Die reduziert gestaltete Standlösung mit ihrer klaren Struktur steht dabei in einer gewissen Spannung zu der gewölbten Linie des Bildschirms.

Begründung der Jury
Das Zusammenspiel von dem sachlich gestalteten Fuß mit seiner Kombination aus Glas und Metall und dem geschwungenen dünnen Bildschirm macht den besonderen Reiz des Fernsehers aus.

UG8700
UHD TV

Manufacturer
LG Electronics Inc., Seoul, South Korea
In-house design
Yongho Lee, Sangwon Yoon, Kyongtae Han
Web
www.lg.com

The UG8700 UHD TV has a curved screen conveying a realistic cinematic experience to the viewer. Thanks to a sanding and buffing process, its aluminium frame has been structured in such a way that the metal shows elegant contrasts. The silhouette of the stand made from thin, metal-plated bands highlights the curvature of the screen, making the polished metal finish appear softer.

Statement by the jury
The soft line of the stand harmonises excellently with the curved screen. In addition, the metal elements give rise to sophisticated design accents.

Der UHD-Fernseher UG8700 besitzt einen gewölbten Bildschirm, der dem Zuschauer ein realistisches Kinoerlebnis vermittelt. Sein Aluminiumrahmen wurde über Aufrauen und Schleifen so strukturiert, dass das Metall elegante Kontraste zeigt. Die Silhouette des dünnen, aus Metallplattenbändern gefertigten Ständers unterstreicht die Wölbung des Bildschirms. Dabei erscheint seine polierte Metalloberfläche umso weicher.

Begründung der Jury
Die sanfte Linie des Ständers harmoniert vorzüglich mit der Krümmung des Bildschirms. Darüber hinaus setzen die Metallelemente edle Akzente.

Alpha 1
Curved TV

Manufacturer
Top Victory Electronics Co., Ltd., New Taipei City, Taiwan
In-house design
Sun-Cheng Lin, Yii-Hong Wu, Hsiao-Jung Hsu,
Chao-Yang Chou, Chia-Wei Hu
Web
www.tpv-tech.com

The Alpha 1 is a curved TV with a very slim design. Its shape was inspired by the Greek letter α and embodies harmony and relaxation. The base encases the speaker, which appears to be rolled towards the front in the shape of coiled paper, and it also enables secure positioning. Due to the curved construction of both screen and speaker, the viewer becomes the point of focus visually and acoustically.

Alpha 1 ist ein geschwungener TV-Bildschirm in einer sehr schmalen Bauweise. Seine Form wurde vom griechischen Buchstaben α inspiriert und verkörpert Harmonie und Entspannung. Der Fuß beherbergt zum einen den Lautsprecher, der wie nach vorne aufgerolltes Papier wirkt, zum anderen sorgt er für einen sicheren Stand. Durch die Krümmung von Bildschirm und Lautsprecher rückt der Betrachter visuell und akustisch in den Mittelpunkt.

Statement by the jury
The slim screen has a lightweight and filigree appearance. The unique stand underlines its self-sufficiency and also facilitates improved sound development.

Begründung der Jury
Der schlanke Bildschirm wirkt leicht und filigran. Der besondere Fuß unterstreicht nicht nur seine Eigenständigkeit, sondern trägt auch zur besseren Klangentfaltung bei.

Philips 9100
4K Ultra HD LED TV

Manufacturer
TP Vision, Amsterdam, Netherlands
In-house design
Web
www.tpvision.com

Equipped with high-end technology, the very thin 9100 4K Ultra HD LED TV offers outstanding performance. With the selection of materials, the iconic stand and the nearly invisible bezel, it harmoniously blends into modern living interiors. The four-sided Ambilight intensifies the viewing experience with its all-round lighting. Environmental concerns are taken into account, for the device works very efficiently and economically.

Statement by the jury
The softly curved stand is distinguished from the straight line of the screen, thus fostering a charming contrast.

Bestückt mit High-End-Technologie, bietet der sehr flache LED-Fernseher 9100 4K Ultra HD eine hervorragende Leistung. Mit der Materialwahl, dem ikonischen Fuß und dem nahezu unsichtbaren Rahmen fügt er sich harmonisch in eine moderne Einrichtung ein. Zudem wird das Fernseherlebnis durch die Rundumbeleuchtung mit vierseitigem Ambilight intensiver. Das Gerät arbeitet sehr effizient und sparsam, womit es dem Umweltgedanken Rechnung trägt.

Begründung der Jury
Der sanft geschwungene Fuß setzt sich von der geraden Linie des Bildschirms ab und schafft dadurch einen reizvollen Kontrast.

Philips 8900 Curved
4K Ultra HD LED TV

Manufacturer
TP Vision, Amsterdam, Netherlands
In-house design
Web
www.tpvision.com

The 8900 4K Ultra HD LED TV combines choice materials in an unmistakable form. A particularly striking design feature is the chromium-plated stand which runs like a curved band below the screen. The three-sided Ambilight guarantees an impressive television-viewing experience, which is complemented by the enlarged screen and the new Android platform with pre-installed apps and access to online content.

Statement by the jury
This 4K television impresses with its premium materials and extraordinary stand shape, thus enriching any living space.

Der LED-Fernseher 8900 4K Ultra HD kombiniert ausgesuchte Materialien in einer unverwechselbaren Form. Besonders markant ist der verchromte Standfuß, der wie ein geschwungenes Band unter dem Bildschirm verläuft. Das dreiseitige Ambilight garantiert ein beeindruckendes Fernseherlebnis, das durch den vergrößerten Bildschirm und die Plattform Android mit vorinstallierten Apps und Zugriff auf Online-Inhalte komplettiert wird.

Begründung der Jury
Dieser 4K-Fernseher beeindruckt durch die hochwertigen Materialien und die ungewöhnliche Form des Fußes. Er wird damit zur Bereicherung für jeden Wohnraum.

Philips 8700 Curved
LED TV

Manufacturer
TP Vision, Amsterdam, Netherlands
In-house design
Web
www.tpvision.com

The 8700 LED TV combines a curved silhouette with a high degree of innovation, aiming to deliver excellent sound in compact dimensions. The three-sided Ambilight enlarges the screen optically, while the Android platform facilitates easy interaction and operation. Premium components like the polished-chrome stand underline the excellent quality of this TV.

Statement by the jury
With its softly curved screen and airy stand solution, the device renders an unobtrusive and light impression.

Der 8700 LED-TV verbindet eine geschwungene Silhouette mit einem hohen Maß an Innovation, um exzellenten Sound in kompakten Abmessungen zu liefern. Ambilight ist an drei Seiten integriert, was den Bildschirm optisch vergrößert. Die Plattform Android erleichtert Interaktion und Bedienung. Hochwertige Komponenten wie der Standfuß aus poliertem Chrom unterstreichen die ausgezeichnete Qualität dieses Fernsehers.

Begründung der Jury
Mit seinem sanft gekrümmten Bildschirm und der luftigen Standlösung wirkt das Gerät unaufdringlich und leicht.

Philips 8600
LED TV

Manufacturer
TP Vision, Amsterdam, Netherlands
In-house design
Web
www.tpvision.com

The 8600 LED TV accommodates leading-edge technology in a design communicating high performance and blending harmoniously into modern interiors. The four-sided Ambilight enlarges the screen and turns the watching of television into a cinematic experience. The wireless speakers are detachable and can thus be flexibly situated anywhere in the room. Refined materials in dark chrome and top-quality speakers convey an impressive sound performance.

Statement by the jury
This TV features an intelligent sound solution, providing a unique viewing experience and also offering a wide range of placement options.

Der 8600 LED-TV bringt führende Technologie in einem Design unter, das hohe Leistung kommuniziert und in moderne Einrichtungen passt. Das vierseitige Ambilight vergrößert den Bildschirm und macht aus dem Fernseh- ein Kinoerlebnis. Die kabellosen Lautsprecher sind abnehmbar, lassen sich also beliebig im Raum platzieren. Die edlen Materialien in Dunkelchrom und die hochwertigen Lautsprecher vermitteln eine beeindruckende Sound-Performance.

Begründung der Jury
Der Fernseher besitzt eine intelligente Soundlösung, die zum einen das Fernseherlebnis zu einem besonderen macht und zum anderen diverse Möglichkeiten bei der Platzierung bietet.

Philips 7100
LED TV

Manufacturer
TP Vision, Amsterdam, Netherlands
In-house design
Web
www.tpvision.com

The 7100 LED TV is a full HD television with extraordinary design details. The chromium-plated stand lends the nearly frameless screen a sort of lightness, while the two-sided Ambilight enhances the cinematic experience. The circumferential edge finds its continuation in the stand so that screen and stand appear as if made from one single piece. The new Android Google platform offers the user extensive interaction options.

Statement by the jury
Glass and chromium-plated metal characterise the elegant look of this television and emphasise its high technical quality.

Der 7100 LED-TV ist ein Full-HD-Fernseher mit außergewöhnlichen Gestaltungsdetails. So verleiht der verchromte Standfuß dem nahezu rahmenlosen Bildschirm Leichtigkeit, während das zweiseitige Ambilight das Kinoerlebnis verstärkt. Der den Bildschirm umlaufende Rand findet seine Fortführung im Standfuß, sodass Bildschirm und Fuß wie aus einem Guss erscheinen. Die Android-Google-Plattform bietet dem Benutzer weitreichende Möglichkeiten bei der Interaktion.

Begründung der Jury
Glas und verchromtes Metall prägen die elegante Erscheinung dieses Fernsehers und unterstreichen seine hohe technische Qualität.

CooCaa A55
TV

Manufacturer
Skyworth, Shenzhen, China
In-house design
Yu Tian, Xiaohui Zhang, Zhiyong Chen, Shuxiao Wei, Yulan Sheng, Feixiang Fang, Tingming Ren, Mingming Wang, Liyuan Peng, Huaming Gao, Yonghong Diao, Xiaochen Wan
Web
www.skyworth.com

This 55" flat screen TV is no longer controlled with a remote control, but with a mobile phone. The signal is transferred wirelessly via HDMI, which allows the unit to do without any wire-based interfaces, thus facilitating free interaction for the user. A characteristic feature of the TV is its very narrow bezel, which highlights the size of the screen. Thanks to the intelligent user interface, games can also be displayed on this television.

Statement by the jury
The screen is self-sufficient, operating without any wired connectivity. The type of control is flexible thanks to the possibility of integrating one's own smartphone.

Der 55"-Flachbildfernseher wird nicht mehr mit einer Fernbedienung, sondern mit einem Mobiltelefon gesteuert. Das Signal wird kabellos per HDMI übertragen. Dadurch kommt er ohne Schnittstellen aus, die verkabelt werden müssen, dies ermöglicht dem Nutzer die freie Interaktion. Der Fernseher zeichnet sich durch einen sehr schlanken Rahmen aus, der die Größe des Bildschirms unterstreicht. Auch Spiele können dank des intelligenten User Interfaces auf dem Bildschirm wiedergegeben werden.

Begründung der Jury
Der Bildschirm steht ganz für sich selbst, denn er funktioniert ohne Anbindung über Kabel. Die Art der Steuerung ist flexibel, da der Nutzer sein eigenes Smartphone verwenden kann.

BRAVIA™ X9500B Series
4K LED HD TV

Manufacturer
Sony Visual Products Inc., Tokyo, Japan
In-house design
Sony Corporation (Hiroaki Yokota)
Web
www.sony.net

Televisions from the Bravia X9500B series provide sharp images in high-definition 4K quality. Their slim, minimalist design underlines the superior picture quality. A thin sheet of film envelops the screen and bezel, creating a seamless connection between both components so that pictures appear to float in space. Thanks to their elegant contouring, the devices embody the appeal of exquisite furnishings. All cables can be stowed away inside the stand, maintaining a clear and discreet overall impression.

Fernseher der Serie Bravia X9500B liefern scharfe Bilder in hochauflösender 4K-Qualität. Ihre flache, minimalistische Gestaltung unterstreicht die hervorragende Darstellung der Inhalte. Eine dünne Folie umgibt sowohl den Bildschirm als auch den Rahmen, sodass beide Komponenten nahtlos ineinander übergehen und Bilder im Raum zu schweben scheinen. Durch ihre eleganten Konturen erwecken die Geräte den Eindruck von erlesenem Mobiliar. Kabel können innerhalb des Ständers untergebracht werden, wodurch der aufgeräumte Eindruck erhalten bleibt.

Statement by the jury
The minimalist design of these 4K TVs allows for unobtrusive placement in interior spaces, which even holds true for the larger representatives of this series.

Begründung der Jury
Die minimalistische Gestaltung dieser 4K-Fernseher trägt dazu bei, dass sich selbst die größeren Vertreter der Serie harmonisch in den Raum einfügen.

BRAVIA™ X9000C Series
4K LED HD TV

Manufacturer
Sony Visual Products Inc., Tokyo, Japan
In-house design
Sony Corporation (Daisuke Shiono)
Web
www.sony.net

These 4K-compatible LCD TVs have a very slim screen with a thickness of only 4.9 mm at the thinnest part. The stand made of extruded aluminium is similarly slender, yet sturdy enough to hold the screen securely. When mounted, the screen is nearly flush with the wall and extends only 4 cm outwards. This is made possible by practical, space-saving structures on the back accommodating the cables, providing sufficient ventilation and offering an intelligent solution for the stand attachment.

Diese 4K-kompatiblen LCD-Fernsehgeräte besitzen einen äußerst schlanken Bildschirm, der an der dünnsten Stelle nur 4,9 mm dick ist. Der aus extrudiertem Aluminium gefertigte Ständer ist ähnlich flach und trotzdem robust genug, um den Bildschirm sicher zu tragen. Bei der Wandmontage hängt der Bildschirm mit einem Abstand von etwa 4 cm dicht an der Wand. Möglich wird dies durch praktische, platzsparende Strukturen an der Rückseite, die Kabel aufnehmen, für die Belüftung sorgen und eine intelligente Lösung für die Ständerbefestigung bieten.

BOE Alta
TV

Manufacturer
Beijing BOE Multimedia Technology Co., Ltd., Beijing, China
Design
BOE In-house design, LUNAR Europe GmbH, Munich, Germany
Web
www.boe.com
www.lunar-europe.com

The sculptural design of the Alta TV is based on plain lines, embodying the vision of a very slim screen. The display is supported by an aluminium element that houses the electronic components and speakers. The purist design is continued in the selection of materials: the frameless glass panel is attached to a lightweight aluminium housing with magnets and integrated screw fasteners. In addition, the aluminium remote control and media box offer simple connectivity and easy access to a number of connectors.

Die skulpturale Gestaltung des Fernsehers Alta basiert auf schlichten Linien und verkörpert die Vision eines sehr dünnen Bildschirms. Das Display wird von einem Aluminium-Element gestützt, in dem Elektronik und Lautsprecher integriert sind. Das puristische Design setzt sich in der Wahl des Materials fort: Das rahmenlose Glaspaneel wird von einem leichten Aluminiumgehäuse über Magnete und eine integrierte Schraubbefestigung gehalten. Die Fernbedienung und Mediabox aus Aluminium ermöglichen die einfache Vernetzung und den leichten Zugang zu einer Vielzahl an Anschlüssen.

LSPX–W1S
4K Ultra Short Throw Projector
4K-Ultrakurzdistanz-Projektor

Manufacturer
Sony Corporation, Tokyo, Japan

In-house design
Yusuke Tsujita

Web
www.sony.net

reddot award 2015
best of the best

New horizons

Watching TV is usually confined to the place where the TV is located. This 4K projector changes the way people watch videos and at the same time allows for a new kind of spatial flexibility. The ultra short throw projector casts an up to 147" large image onto the wall in any ordinary room. With its extremely high resolution, it provides users with a viewing experience in cinema quality wherever they desire. The 4K projector itself, as well as all its peripheral components, combine this innovative way of image generation with a highly sophisticated design. With horizontally flowing lines, the projector, the stereo speakers and the complementary AV cabinet modules harmoniously blend into any environment. In addition, the aesthetically appealing form, reminiscent of stacked slates, offers many advantages. It improves ventilation as well as the sound quality of the speakers, and allows for wireless/IR connectivity. A further functional and rather interesting detail is that at start-up, a hatch on top folds upwards for projection, while keeping the opening of the projector itself concealed. This new experience of watching videos without a fixed screen presents itself to users in an extremely flexible and harmonious way, allowing them to fully concentrate on what is shown.

Neue Horizonte

Das Fernsehen ist normalerweise stets auf den Ort fixiert, an dem der Fernseher steht. Dieser 4K-Projektor verändert die Sehgewohnheiten und ermöglicht dabei zugleich eine neue Form der räumlichen Flexibilität. Mit einer ultrakurzen Bildwurfweite kann das Gerät in jedem gewöhnlichen Raum ein bis zu 147" großes Bild an die Wand projizieren. Durch dessen zudem enorm hohe Auflösung bietet sich dem Nutzer so ein Erlebnis in Kinoqualität, wo immer er dies gerade wünscht.
Bei dem 4K-Projektor selbst, wie auch den Komponenten in seiner Peripherie, verbindet sich diese innovative Art der Bilderzeugung mit einer sehr durchdachten Gestaltung. Der Projektor, die Stereo-Lautsprecher sowie die dazugehörigen AV-Schrankmodule fügen sich mit ihren horizontal fließenden Linien harmonisch in die jeweilige Umgebung ein. Die ästhetisch ansprechende, an gestapelte Schieferplatten erinnernde Form bietet zusätzlich viele Vorteile. Sie verbessert die Belüftung, die Tonwiedergabe über die Lautsprecher und die Möglichkeiten für Drahtlos-/Infrarot-Verbindungen. Ein funktionales und ausgesprochen interessantes Detail ist außerdem, dass sich beim Hochfahren für die Projektion eine Klappe an der Oberseite öffnet, während die Projektoröffnung selbst verborgen bleibt. Überaus flexibel und harmonisch bietet sich dem Betrachter diese neue Art des Fernsehens, die ohne festen Bildschirm neue Horizonte erschließt. Er kann sich ganz auf das konzentrieren, was gezeigt wird.

Statement by the jury

The 4K ultra short throw projector and the associated method of image projection reinterpret the way we watch TV in an exciting way. With its ultra short throw technology, it projects an up to 147" large image onto an ordinary wall. This high-resolution projector thus replaces conventional screens as viewing devices. It stands out with both a functional and aesthetically appealing design, which perfectly reflects its technical potential.

Begründung der Jury

Der 4K-Ultrakurzdistanz-Projektor und die damit verbundene Art der Bildprojektion interpretieren auf spannende Weise das Fernsehen neu. Mit einer ultrakurzen Bildwurfweite projiziert er ein bis zu 147" großes Bild an eine beliebige Wand. Dadurch ersetzt dieses hochauflösende Gerät übliche Bildschirme als Betrachtungsmedium. Es besticht mit einer funktionalen wie ästhetisch ansprechenden Gestaltung, die seine technischen Möglichkeiten auf perfekte Weise widerspiegelt.

Designer portrait
See page 90
Siehe Seite 90

leaf
Remote Control
Fernbedienung

Manufacturer
ruwido austria gmbh,
Neumarkt, Austria

In-house design
Ferdinand Maier

Design
Zeug Design Ges.m.b.H.
(Erwin Weitgasser,
Detlev Magerer),
Salzburg, Austria

Web
www.ruwido.com
www.zeug.at

Elegant individualist

The remote control "leaf" has been specifically designed and manufactured to navigate modern TV user interfaces in the context of the home. As such, it is picked up and held in the hand frequently. With a design following the maxim of usability, "leaf" is a remote control with a fascinatingly new appearance. At first glance, it catches the eye particularly due to its exceptionally thin product architecture. Measuring only 2.97 mm, the remote control is well balanced in its centre of gravity. A form language that appeals to the emotions of the user is combined with a successful reduction to the essential operating elements. These present themselves in a clearly arranged and easy-to-understand way. The "leaf" remote control also introduces users to a future-oriented multimodality, supporting data transmission via infrared and Bluetooth Low Energy. With integrated voice recognition as well as key-based interaction, it allows exciting user experiences with the control of modern user interfaces. The appeal of this remote control is further enhanced by the use of high-quality aluminium and carbon materials, which ensure both durability and a pleasant feel. With its innovative, sensuous design and smart, sophisticated user interface it stirs up emotions of the user each and every day.

Eleganter Individualist

Die Fernbedienung „leaf" wurde speziell für die Navigation umfangreicher TV-Angebote im privaten Wohnumfeld gestaltet und entwickelt. Als solche wird sie entsprechend oft in die Hand genommen. Dank eines an der Maxime der Nutzerfreundlichkeit orientierten Designs entstand mit „leaf" eine Fernbedienung, die durch eine faszinierend neue Anmutung besticht. Auf den ersten Blick zieht sie den Betrachter vor allem durch ihr ungewöhnlich dünnes Erscheinungsbild in den Bann. Dieses misst nur 2,97 mm, wobei die Fernbedienung in ihrem Schwerpunkt genau austariert ist. Eine die Sinne ansprechende Formgebung verbindet sich mit einer gelungenen Reduktion auf die wesentlichen Bedienelemente. Diese bieten sich übersichtlich angeordnet und in ihrer jeweiligen Bedeutung verständlich dar. Die Fernbedienung „leaf" eröffnet dem Nutzer zudem zukunftsweisende Multimodalität, unterstützt von den Übertragungsmechanismen Infrarot und Bluetooth Low Energy. Mittels integrierter Spracherkennung sowie des tastenbasierten Interaktionsmechanismus ermöglicht sie spannende Erlebnisse in der Bedienung moderner User-Interfaces. Bekräftigt wird die Wirkung dieser Fernbedienung durch den Einsatz der hochwertigen Materialien Aluminium und Carbon, die sie langlebig und haptisch angenehm machen. Mit ihrer innovativen, sinnlichen Formensprache und einer klug durchdachten Bedienoberfläche emotionalisiert sie Nutzer täglich aufs Neue.

Statement by the jury

The design of "leaf" combines the maxim of usability with subtle aesthetics. Skilfully reduced to the essential elements, the remote control is extremely thin and feels very light in the hand. The interface with voice recognition and an intuitive, key-based interaction mechanism is comfortable in its use. Its multimodality as well as the support of data transmission via infrared and Bluetooth Low Energy are highly visionary.

Begründung der Jury

Die Gestaltung von „leaf" verknüpft die Maxime der Nutzerfreundlichkeit mit einer feinsinnigen Ästhetik. Diese gekonnt auf die wichtigsten Elemente reduzierte Fernbedienung ist extrem dünn, sodass sie sehr leicht in der Hand liegt. Komfortabel ist ihr Interface mit Spracherkennung sowie einem intuitiv bedienbaren, tastenbasierten Interaktionsmechanismus. Ihre multimodale Ausrichtung sowie die Unterstützung der Übertragungsmechanismen Infrarot und Bluetooth Low Energy sind dabei zukunftsweisend.

Designer portrait
See page 92
Siehe Seite 92

mic

ruwido

TS4
TV Remote Control
TV-Fernbedienung

Manufacturer
Grundig Intermedia GmbH,
Nuremberg, Germany
In-house design
Web
www.grundig.com

With its reduced design, this remote control displays a clear and elegant appearance. The balanced key size allows for comfortable handling. The central key element has a surface with different textures, making it easy to find menu functions. The upper shell of the remote control is colour-coded, while the lower shell is smoothly shaped and rounded, providing a comfortable grip. Length, centre of mass and weight are perfectly balanced.

Statement by the jury
The TS4 has an ergonomic design enabling very comfortable handling. At the same time, it maintains an elegant character due to its reduced design.

Die Fernbedienung wirkt durch ihre reduzierte Gestaltung übersichtlich und elegant. Durch die ausgewogene Tastengröße ist sie komfortabel zu bedienen. Das zentrale Tastenelement besitzt eine Oberfläche mit unterschiedlicher Haptik, wodurch Menüfunktionen leicht zu finden sind. Die Oberschale der Fernbedienung hebt sich farblich ab, die Unterschale ist weich geformt und abgerundet, dadurch liegt sie angenehm in der Hand. Länge, Schwerpunkt und Gewicht sind perfekt ausbalanciert.

Begründung der Jury
Die TS4 ist ergonomisch gestaltet, was die Handhabung sehr angenehm macht. Zugleich bewahrt sie sich einen eleganten Charakter durch ihre reduzierte Gestaltung.

Phantom
Set-Top Box

Manufacturer
MitraStar Technology Corporation,
Hsinchu, Taiwan
In-house design
Konica Lee, Yung-Mao Lai,
Yung-Jung Peng, Yu-Ning Chang
Web
www.mitrastar.com.tw

The Phantom set-top box allows the user to enjoy multimedia content from different providers. Its transparent and lightweight design blends into any living-room environment. The elegant and well-structured remote control contributes to a perfectly relaxed user experience. The remote control is not only aesthetically appealing but also designed ergonomically with optimally placed control buttons and a slight curvature.

Statement by the jury
With the transparency of its lower shell, this set-top box has an independent appearance and can be placed in such a way that it appears to be floating.

Mit der Set-Top-Box Phantom lassen sich Multimedia-Inhalte von verschiedenen Anbietern genießen. Ihre transparente und leichte Erscheinung fügt sich in jedes Wohnzimmer ein. Eine elegante und gut strukturierte Fernbedienung vervollständigt das entspannte Benutzererlebnis. Die Fernbedienung ist nicht nur ästhetisch auffallend, sondern mit den optimal platzierten Tasten und einer leichten Krümmung auch ergonomisch gestaltet.

Begründung der Jury
Mit ihrer transparenten unteren Gehäusehälfte wirkt die Set-Top-Box eigenständig und kann so aufgestellt werden, als würde sie schweben.

NEEO
Home Automation System
Hausautomatisierungssystem

Manufacturer
NEEO AG, Solothurn, Switzerland
In-house design
Oliver Studer, Raphael Oberholzer
Web
www.neeo.com

NEEO is a smart home automation system consisting of a remote control and a central unit with which a large number of devices in the house can be controlled. High-grade aluminium and acrylic glass lend the system an elegant appearance. The device is silent in operation and, thanks to its plain design, blends perfectly with many living environments. The reduced design also underscores the intuitive user guidance, making the system easily accessible for everyone.

Statement by the jury
NEEO cleverly conceals its functionality as a control centre behind a discreet, albeit elegant design.

NEEO ist ein Smart-Home-System, bestehend aus Fernbedienung und Zentraleinheit, mit dem sich zahlreiche Geräte im Haus steuern lassen. Hochwertiges Aluminium und Acrylglas verleihen dem System eine elegante Anmutung. Das Gerät arbeitet geräuschlos und fügt sich dank seiner schlichten Form ideal in Wohnumgebungen ein. Zudem unterstreicht die reduzierte Gestaltung die intuitive Benutzerführung, die das System für jedermann leicht zugänglich macht.

Begründung der Jury
NEEO verbirgt geschickt seine Funktion als Zentrale für die Steuerung von Geräten hinter einer zurückhaltenden, wenngleich eleganten Gestaltung.

HMS (Home Media Server)
Set-Top Box

Manufacturer
Humax, South Korea
In-house design
Humax Design Team
Web
www.humaxdigital.com

The high-performance Home Media Server has four tuners, offering users a strong degree of flexibility when watching TV, recording programs and streaming content. Media content is stored on the 2.5" HDD and can be played back with devices such as smartphones, tablets or computers. The leather-like stitch design, available in different colours, protects the device from scratches. It therefore addresses the diverse needs and wishes of consumers, as well as their lifestyle.

Statement by the jury
The Home Media Server is optimally equipped for recording and streaming television programmes to other devices. Its novel surface structure is functional and decorative.

Der leistungsfähige Home Media Server besitzt vier Tuner und bietet dem Benutzer damit eine große Flexibilität beim Fernsehen, Aufnehmen und Streamen. Inhalte werden auf der 2,5"-HDD-Festplatte gespeichert und stehen für Geräte wie Smartphones, Tablets oder Computer zur Verfügung. Die lederähnliche Oberfläche in unterschiedlichen Farben schützt das Gerät vor Kratzern. Dadurch passt es zu den vielfältigen Bedürfnissen und Lebensstilen der Konsumenten.

Begründung der Jury
Der Home Media Server ist bestens für das Aufnehmen und Streamen von Fernsehsendungen auf andere Geräte ausgestattet. Seine neuartige Oberfläche ist funktional und dekorativ.

Mi Box Plus
Internet TV Set-Top Box

Manufacturer
Xiaomi Corporation, Beijing, China
In-house design
Cheng Zhaopeng
Web
www.mi.com

The Mi Box Plus is a high-definition Internet TV Box that provides content such as movies, apps, and television series and shows. It also streams content from mobile devices. A remote control allows for comfortable operation of the box. The curved top surface and the hard edge give the device a sculptural look, so that it is almost perceived as a piece of art rather than a technical object. The contrast between the matt surface and shiny logo highlights the delicate surface finish and high-quality workmanship.

Statement by the jury
This set-top box takes a less-technical approach to providing multimedia content. In this way, it blends naturally and harmoniously into the existing interior.

Mi Box Plus ist eine hochauflösende Internet-TV-Box, über die sich Inhalte wie Filme, Apps, TV-Serien und -Shows beziehen oder von einem mobilen Endgerät streamen lassen. Eine Fernbedienung erlaubt die komfortable Bedienung der Box. Die kurvige Oberseite und die harte Kante verleihen dem Gerät ein skulpturales Aussehen, sodass es weniger als technisches Objekt, sondern fast als Schmuckstück wahrgenommen wird. Der Kontrast zwischen der matten Oberfläche und dem glänzenden Logo unterstreicht die feine Oberflächentextur und hohe Verarbeitungsqualität.

Begründung der Jury
Die Set-Top-Box stellt multimediale Inhalte für Fernsehgeräte in einer wenig technischen Form bereit. Sie wird dadurch wie selbstverständlich zum Teil der Einrichtung.

Vodafone TV Center 2000
Set-Top Box

Manufacturer
Vodafone D2 GmbH, Eschborn, Germany
Design
Ziba Munich (Oliver Lang, Thor Unbescheid), Planegg, Germany
Web
www.vodafone.de
www.ziba.com

The TV Center 2000 is a 4K-capable set-top box with remote control. Thanks to its satellite and IPTV receiver, Internet access and internal hard drive, the device can play back all kinds of accessible content. Its striking design is characterised by large radii and slightly curved surface progressions. The remote control continues the clear design language of the set-top box. Its buttons are grouped according to functions and contrast with the body with regard to haptics.

Statement by the jury
This set-top box and accompanying remote control are, in terms of form and function, optimally coordinated with one another. With their rounded design, they have a pleasantly unobtrusive appearance.

Das TV Center 2000 ist eine 4K-fähige Set-Top-Box mit Fernbedienung, die dank Satelliten- und IPTV-Receiver, Internetzugang und einer Festplatte alle verfügbaren Inhalte wiedergeben kann. Ihre markante Gestalt ist durch große Radien und weiche, leicht gewölbte Flächenverläufe gekennzeichnet. Die Fernbedienung folgt der klaren Formensprache der Set-Top-Box. Ihre Tasten sind nach Funktionen gruppiert und setzen sich haptisch vom Gehäuse ab.

Begründung der Jury
Set-Top-Box und Fernbedienung sind in Form und Funktion gut aufeinander abgestimmt. Dank ihrer runden Linienführung wirken sie angenehm unauffällig.

Mi Box Mini
Internet TV Set-Top Box

Manufacturer
Xiaomi Corporation, Beijing, China
In-house design
Shen Xijie
Web
www.mi.com

The Mi Box Mini is a small set-top box capable of reproducing HD videos with a resolution of up to 1080p. Since it is very compact, it can easily be taken along while travelling. With its quad-core processor and fast graphics unit, it guarantees smooth playback of videos and games. The box includes a foldable plug, an HDMI port and a Bluetooth remote control. Its design pursues the approach "restrained yet active" which is reflected in the selection of six vibrant colours.

Statement by the jury
This small, colourful set-top box brings videos and games to the TV in a refreshing way, and it blends well with an active, mobile lifestyle.

Die Mi Box Mini ist eine kleine Set-Top-Box, die HD-Videos mit bis zu 1080p wiedergeben kann. Da sie so kompakt ist, eignet sie sich besonders zum Mitnehmen auf Reisen. Mit ihrem Vierkern-Prozessor und der schnellen Grafikeinheit gibt sie Videos und Spiele flüssig wieder. Die Box besitzt einen ausklappbaren Stecker, einen HDMI-Anschluss und eine Bluetooth-Fernbedienung. Ihre Gestaltung folgt dem Ansatz „zurückhaltend, aber aktiv", was sich in den sechs leuchtenden Farben widerspiegelt, in denen sie erhältlich ist.

Begründung der Jury
Diese kleine, farbenfrohe Set-Top-Box bringt Videos und Spiele in einer erfrischenden Art auf den Fernseher und passt gut zu einem aktiven, mobilen Lebensstil.

TB 01
Mobile TV Tuner
Mobiler TV-Tuner

Manufacturer
Huawei Technologies Co., Ltd, Shenzhen, China
In-house design
Mihoko Hotta, Yuan Lei
Web
consumer.huawei.com/en/

The TB 01 TV tuner enables smartphones and tablets to receive TV programmes while on the move. It also offers additional functions like video recording, Internet connection via tethering and the charging of 3,560 mAh batteries. With its eye-catching colour scheme, the slim, handy casing in high-gloss black and matt yellow attracts attention.

Statement by the jury
The tuner's appeal results from the interplay of plain shapes and striking colouring. Pleasing in particular are the smooth surfaces and straight lines.

Mit dem TV-Tuner TB 01 werden Smartphones und Tablets so erweitert, dass sie unterwegs Fernsehprogramme empfangen können. Zudem bietet der Tuner weitere Funktionen wie die Aufnahme von Videos, die Internetanbindung über Tethering und das Laden von 3.560-mAh-Akkus. Das handliche, schlanke Gehäuse zieht mit seiner gewagten Farbkombination von glänzendem Schwarz und mattem Gelb die Aufmerksamkeit auf sich.

Begründung der Jury
Der Tuner bezieht seinen Reiz aus dem Zusammenspiel von schlichter Form und auffälliger Farbgebung. Dabei gefallen besonders die glatte Oberfläche und klaren Linien.

Google Chromecast
Digital Streaming Device
Digitales Streaming-Gerät

Manufacturer
Google, Mountain View, USA
Design
Whipsaw Inc, San Jose, USA
Web
www.google.com/chromecast
www.whipsaw.com

Google Chromecast is a digital media player in the form of a dongle that connects to the HDMI port of a television. It streams audio and video content directly to the TV via Wi-Fi or LAN from mobile devices or Internet service providers. A smartphone or tablet is used for navigation. As the device is very small, it fits into any TV without attracting attention, and it can be easily taken along. Its shape, tapered toward the front, guarantees that the dongle does not collide with other connectors.

Statement by the jury
Chromecast is extremely small and offers full streaming functionality. Its intelligent design facilitates convenient handling.

Google Chromecast ist ein digitaler Media-Player in Form eines Dongles für den HDMI-Anschluss eines Fernsehers. Es streamt Audio- und Videoinhalte von Mobilgeräten oder Internetanbietern über Wi-Fi oder LAN direkt auf den Bildschirm. Navigiert wird mit dem eigenen Smartphone oder Tablet. Da der Media-Player so klein ist, passt er in jeden Fernseher, ohne aufzufallen, und kann einfach mitgenommen werden. Seine sich nach vorn verjüngende Form garantiert, dass der Dongle nicht mit anderen Steckern kollidiert.

Begründung der Jury
Chromecast ist äußert klein und bietet die volle Streaming-Funktionalität. Seine intelligente Gestaltung erleichtert die Handhabung.

Smart OTT Box
Set-Top Box

Manufacturer
Humax, South Korea
In-house design
Humax Design Team
Web
www.humaxdigital.com

The Smart OTT Box allows users to enjoy online media content anywhere and anytime. It transfers movies, media library programmes, Internet content and music, as well as files from a home media server or mobile device, to a large television screen. The top cover of the box is available in different wood styles, which blend well into most interiors, offering options for personalising the product. The rounded design and accents in white also lend the device a natural appeal.

Statement by the jury
The Smart OTT Box conceals its technical character in a natural form and thus becomes part of the interior design.

Die Smart OTT Box erlaubt Nutzern, Inhalte von Online-Medien überall und jederzeit zu genießen. Filme, Mediathek-Programme, Internetinhalte und Musik bringt sie ebenso auf den großen Fernsehschirm wie Dateien von einem Home Media Server oder einem mobilen Gerät. Die obere Abdeckung der Box ist in unterschiedlichen Holzdekoren erhältlich, die sich gut in die Umgebung einfügen und Möglichkeiten zur Personalisierung des Produkts bieten. Die abgerundete Gestalt und Akzente in Weiß verleihen dem Gerät eine natürliche Ausstrahlung.

Begründung der Jury
Die Smart OTT Box verbirgt ihren technischen Charakter in einer natürlich anmutenden Form und wird dadurch zu einem Teil der Einrichtung.

Nixplay Edge
Wi-Fi Cloud Photo Frame
Wi-Fi-Cloud-Bilderrahmen

Manufacturer
Creedon Technologies, Hong Kong
In-house design
Web
www.nixplay.com
www.creedontechnologies.com

The Nixplay Edge Wi-Fi cloud photo frame features a 13.3" high-resolution IPS display in 16:9 format. Its photo processing components are located behind a curved gloss back, contrasted by the matt rubberised housing. The frame's slim profile is accentuated by an elegant stand, which achieves even weight distribution and allows for positioning in both landscape and portrait formats. Built-in motion and infrared sensors support all optical functions with customised flat lenses.

Der Wi-Fi-Cloud-Bilderrahmen Nixplay Edge besitzt ein hochauflösendes 13,3"-IPS-Display im 16:9-Format. Seine bildverarbeitenden Komponenten befinden sich hinter einer gerundeten, glänzenden Rückseite, die einen Kontrast zu dem matt gummierten Gehäuse bildet. Betont wird das schlanke Profil des Rahmens durch einen eleganten Ständer, der für eine gleichmäßige Gewichtsverteilung sorgt und das Aufstellen im Quer- sowie Hochformat ermöglicht. Eingebaute Bewegungs- und Infrarotsensoren unterstützen mit maßgeschneiderten flachen Linsen alle optischen Funktionen.

Statement by the jury
With its plain and elegant appearance, the Nixplay Edge blends harmoniously into any living environment. The motion and infrared sensors offer an intuitive user experience.

Begründung der Jury
Dank seiner schlichten und eleganten Erscheinung fügt sich die Nixplay Edge in jedes Wohnumfeld harmonisch ein. Die Bewegungs- und Infrarotsensoren bieten ein intuitives Nutzererlebnis.

Philips Flat Micro
DCM2260W/BTM2280W
Stereo System

Manufacturer
Gibson Innovations, Hong Kong
In-house design
Web
www.gibson.com

This all-in-one stereo system can be placed on a shelf or mounted to a wall. Its slim shape is space-saving and thus ideal for rooms with limited wall space. With its elegant modern design and use of soft fabrics, the stereo integrates well into any home interior. All controls and the CD slot are concentrated in one area so that the casing wraps entirely around the system, fostering a seamless appearance.

Statement by the jury
The design of this stereo system is flat and unobtrusive, allowing it to be placed nearly anywhere. It also blends harmoniously with many interior styles.

Das All-in-One-Stereosystem kann auf ein Regal gestellt oder an der Wand angebracht werden. Aufgrund seiner schlanken Form braucht es wenig Platz und ist somit ideal für Räume mit begrenzter Wandfläche. Mit seiner eleganten, modernen Gestaltung und der Verwendung von weichen Stoffen passt es gut in jedes Zuhause. Alle Kontrollen und das CD-Laufwerk befinden sich in einem Bereich, wodurch das Gehäuse das System durchgehend umschließt und eine nahtlose Form schafft.

Begründung der Jury
Das Stereosystem ist flach und unaufdringlich gestaltet. Damit findet es nicht nur problemlos einen Platz, sondern passt zu vielen Einrichtungsstilen.

Philips Soundbar Speaker
HTL6145C
Lautsprecher

Manufacturer
Gibson Innovations, Hong Kong
In-house design
Web
www.gibson.com

With a thickness of only 3 cm, the HTL6145C soundbar is very slim, just like the complementing subwoofer. A metal grille wraps around the front and extends to both ends, providing full-range sound. Touch controls in the centre top strip blend seamlessly with the metal foil extending along the entire top surface. This results in a harmonious, seamless overall shape with optimum dimensions. The soundbar can be mounted to a wall or placed anywhere in the room.

Statement by the jury
Both soundbar and subwoofer are very slim and thus blend well with flat-screen TVs, which are becoming ever more slender as well.

Die Soundbar HTL6145C ist mit einer Dicke von nur 3 cm ebenso wie der ergänzende Subwoofer sehr schlank. Ein Gitter aus Metall umschließt ihre Vorderseite bis zu beiden Enden und sorgt für einen vollen Klang. Touch-Regler im Mittelstreifen der Oberseite verschmelzen mit der Metallfolie, die sich über die gesamte Oberseite zieht. Dadurch ergibt sich eine harmonische, nahtlose Gesamtform mit optimalen Abmessungen. Die Soundbar kann an der Wand befestigt oder im Raum aufgestellt werden.

Begründung der Jury
Soundbar und Subwoofer sind sehr schlank und passen damit gut zu den immer flacher werdenden TV-Bildschirmen.

Philips Soundbar
HTL7140
Soundbar

Manufacturer
Gibson Innovations, Hong Kong
In-house design
Web
www.gibson.com

The HTL7140 is based on the concept of accommodating a 5.1 system in a single soundbar accompanied by a wireless subwoofer. The angled tweeters at the ends play back the rear channels and produce convincing surround sound. The soundbar is only 38 mm high and 55 mm thick, enabling it to be placed under any television. The chassis is made of sandblasted, anodised aluminium and conveys superior quality.

Statement by the jury
This soundbar with a very slim design renders a highly elegant impression, while simultaneously offering convincing 5.1 surround sound.

Hinter der HTL7140 steht das Konzept, ein 5.1-System in einer einzigen Soundbar mit kabellosem Subwoofer unterzubringen. Die angewinkelten Hochtöner an den Enden geben die hinteren Kanäle wieder und sorgen für einen überzeugenden Raumklang. Die Soundbar ist nur 38 mm hoch und 55 mm dick, daher lässt sie sich unter jedem Fernseher aufstellen. Das Gehäuse besteht aus sandgestrahltem, eloxiertem Aluminium und vermittelt eine hohe Qualität.

Begründung der Jury
Die sehr flach gestaltete Soundbar wirkt überaus elegant und bietet zugleich einen überzeugenden 5.1-Raumklang.

Philips Soundstage HTL5130
Sound System
Soundsystem

Manufacturer
Gibson Innovations, Hong Kong
In-house design
Web
www.gibson.com

With its three powerful speaker drivers, the HTL5130 sound system produces space-filling volume. They are arranged in such a way that the centre speaker points directly towards the user, while the other drivers direct sound around the user. The subwoofer is integrated to attain a coherent overall solution. The textured circular surface reflects sound distribution, and the metal grilles wrapping around the entire front length convey spatial sound reproduction.

Statement by the jury
Thanks to the integrated subwoofer, this sound system is compact and displays a coherent design. The pattern on the top surface has an independent and symbolic appearance.

Das Soundsystem HTL5130 erzielt mit seinen drei kraftvollen Treibern raumfüllenden Klang. Sie sind so angeordnet, dass der mittlere Lautsprecher direkt zum Benutzer zeigt, während die anderen Treiber den Klang um den Benutzer herum abstrahlen. Der Subwoofer ist integriert, um eine stimmige Gesamtlösung zu erreichen. Die kreisförmig texturierte Oberfläche spiegelt die Verteilung des Klangs wider. Auch die über die gesamte Vorderseite verlaufenden Metallgitter vermitteln die räumliche Klangwiedergabe.

Begründung der Jury
Das Soundsystem ist dank des integrierten Subwoofers kompakt und zeigt eine schlüssige Gestaltung. Das Muster auf der Oberseite wirkt eigenständig und symbolhaft.

Philips Fidelio XS1 Soundstage
Speaker
Lautsprecher

Manufacturer
Gibson Innovations, Hong Kong
In-house design
Web
www.gibson.com

The slim Fidelio XS1 Soundstage offers high sound quality and, at a height of just 41 mm, fits snugly under the television without impairing the viewing experience. Its stable, polished glass surface reflects ambient light, and the speaker cover with rounded edges gives it a floating appearance. The slim shape provides sufficient space for connectors and the control panel along the side.

Statement by the jury
The Fidelio XS1 Soundstage has a very flat and reduced design, harmonising with any modern television.

Die schlanke Fidelio XS1 Soundstage bietet eine hohe Klangqualität und passt mit nur 41 mm Höhe problemlos unter den Fernseher, ohne das Fernseherlebnis zu beeinträchtigen. Ihre polierte, stabile Glasoberfläche reflektiert das Umgebungslicht, die Lautsprecherabdeckung mit den abgerundeten Kanten verleiht ihr eine schwebende Anmutung. Die schlanke Form bietet genügend Platz für Anschlüsse und das Bedienteil an der Seite.

Begründung der Jury
Die Fidelio XS1 Soundstage ist sehr flach und reduziert gestaltet. So harmoniert sie mit jedem modernen Fernsehgerät.

Philips Wireless Speaker BT7500
Kabelloser Lautsprecher

Manufacturer
Gibson Innovations, Hong Kong
In-house design
Web
www.gibson.com

The BT7500 wireless Bluetooth speaker is particularly suited for home use, allowing users to quickly and easily stream music from their devices. The curved front enhances acoustic performance and space-filling sound perception. The metal base and aluminium volume wheel lend the speaker a high-quality appearance. It can connect to multiple devices via Bluetooth or NFC.

Statement by the jury
The speaker is a user-friendly device for the living room. All control elements are easily accessible, and the forward-facing shape renders an inviting impression.

Der kabellose Bluetooth-Lautsprecher BT7500 eignet sich besonders für zu Hause, da Nutzer damit schnell und einfach Musik von ihren Geräten streamen können. Die gebogene Vorderseite unterstützt die akustische Leistung und raumfüllende Klangwahrnehmung. Der Sockel aus Metall und der Lautstärkeregler aus Aluminium verleihen dem Lautsprecher eine hochwertige Anmutung. Per Bluetooth oder NFC lassen sich mit ihm mehrere Geräte verbinden.

Begründung der Jury
Der Lautsprecher ist ein nutzerfreundliches Gerät fürs Wohnzimmer. Die Bedienelemente sind leicht erreichbar, die nach vorn gewandte Form wirkt einladend.

HS9
Array Speaker
Array-Lautsprecher

Manufacturer
LG Electronics Inc., Seoul, South Korea
In-house design
Hee-Su Yang, He-Won Kihl, Jun-Ki Kim
Web
www.lg.com

The HS9 array speaker streams music wirelessly via Wi-Fi or Bluetooth from compatible devices. The system simulates 7.1 sound with a power of 700 watts. The speakers are located in a casing made of punched metal (SECC) whose surface is aesthetically complemented by transverse grooves. This emphasises the conciseness of sound reproduction and also lends the speaker a sturdy appearance.

Statement by the jury
With the finely structured metal finish and colour-distinguished front, the speaker renders a reliable and sophisticated impression.

Der Array-Lautsprecher HS9 streamt Musik kabellos per Wi-Fi oder Bluetooth von kompatiblen Geräten. Das System simuliert einen 7.1-Klang bei einer Leistung von 700 Watt. Die Lautsprecher sind in einem Gehäuse aus Stanzmetall (SECC) untergebracht, dessen Oberfläche durch quer laufende Rillen ästhetisch abgerundet wird. Dies unterstreicht zum einen die Prägnanz der Tonwiedergabe und verleiht dem Lautsprecher zum anderen eine massive Anmutung

Begründung der Jury
Durch die fein strukturierte Metalloberfläche und die farblich abgesetzte Vorderseite wirkt der Lautsprecher verlässlich und edel.

Harman Kardon Omni Bar with Subwoofer
Soundbar

Manufacturer
Harman International Industries, Northridge, USA
In-house design
Harman Design Team
(Damian Mackiewicz, Hyungwoo Yoon)
Web
www.harman.com

The Harman Kardon Omni Bar is a compact soundbar with Wi-Fi and Bluetooth in a premium, classic design. It offers sound in HD studio quality, thus guaranteeing an impressive audio experience. Since it can be placed on a shelf or mounted to the wall, it blends into nearly any room. Audio signals are transmitted wirelessly both to the subwoofer and from the television and can be streamed to one or all of the speakers connected to the system.

Statement by the jury
As the Omni Bar is a completely wireless system with a discreet design, it offers a wide variety of placement options.

Die Harman Kardon Omni Bar ist eine kompakte Soundbar mit Wi-Fi und Bluetooth in klassisch-hochwertiger Anmutung. Sie bietet Sound in HD-Studioqualität und ermöglicht dadurch ein eindrucksvolles Klangerlebnis. Da sie aufgestellt oder an der Wand montiert werden kann, passt sie in nahezu jeden Raum. Audiosignale werden kabellos sowohl zum Subwoofer als auch vom Fernseher übertragen und an einzelne oder alle vernetzten Lautsprecher übermittelt.

Begründung der Jury
Da die Omni Bar ganz auf kabellose Übertragung setzt und zudem dezent gestaltet ist, bietet sie einen großen Spielraum bei der Platzierung.

JBL Boost TV
Soundbar

Manufacturer
Harman International Industries, Northridge, USA
In-house design
Harman Design Team
(Joshua Fischer, Irfan Kachwalla)
Web
www.harman.com

The Boost TV is a compact soundbar with a width of less than 40 cm. It is covered with a high-quality black fabric and thus harmonises well with modern TVs and interior furnishings. Two bass ports are located at the outer ends, visually completing the design. The soundbar can be easily controlled with a mobile device, and the user can switch back and forth between TV sound and mobile device.

Statement by the jury
With its compact and rounded shape, the Boost TV soundbar renders an unobtrusive and friendly impression. It also convinces with its user-friendly operation.

Boost TV ist eine kompakte Soundbar mit einer Breite von weniger als 40 cm. Sie ist mit einem hochwertigen schwarzen Gewebe beschichtet und harmoniert dadurch gut mit modernen Fernsehgeräten und Wohnzimmereinrichtungen. An den äußeren Enden sind zwei Bass-Ports untergebracht, die die Form visuell abschließen. Die Soundbar lässt sich einfach mit einem Mobilgerät ansteuern, wobei der Benutzer zwischen TV-Sound und Mobilgerät wechseln kann.

Begründung der Jury
Die Soundbar Boost TV wirkt mit ihrer kompakten und rundlichen Form unaufdringlich und freundlich. Zudem überzeugt sie mit ihrer nutzerfreundlichen Bedienung.

SRS-X99
Bluetooth Speaker
Bluetooth-Lautsprecher

Manufacturer
Sony Corporation, Tokyo, Japan
In-house design
Design Center Europe, Takahiro Naito
Web
www.sony.net

This Bluetooth speaker is equipped for the reproduction of high-resolution audio and thus offers outstanding sound quality. With its minimalist design, the speaker blends well with most interiors. The three different materials used for the surfaces significantly contribute to this impression. It features seven speakers and comprehensive interface options for wireless and wired connections. The unit can be controlled via smartphone, tablet or remote control. In addition, several buttons on the device offer access to all essential functions.

Dieser Bluetooth-Lautsprecher ist für die Wiedergabe von High-Resolution Audio gerüstet und bietet damit eine ausgezeichnete Klangqualität. Mit seiner minimalistischen Gestaltung passt er gut in die meisten Interieurs. Dazu tragen nicht zuletzt die drei verschiedenen Werkstoffe der Oberflächen bei. Er besitzt sieben Lautsprecher und umfangreiche Schnittstellen für die kabellose und kabelgebundene Verbindung. Steuern lässt er sich per Smartphone, Tablet oder Fernbedienung, zudem bieten diverse Tasten am Gerät den Zugang zu den wesentlichen Funktionen.

Statement by the jury
The SRS-X99 speaker displays clear structures with straight lines and smooth surfaces. In this way, it harmoniously blends with nearly any interior-design style.

Begründung der Jury
Der Lautsprecher SRS-X99 zeigt klare Strukturen mit gerader Linienführung und glatten Oberflächen. Dadurch ordnet er sich nahezu jedem Einrichtungsstil unter.

Harman Kardon Esquire 2
Portable Conferencing System with Speaker
Tragbares Konferenzsystem mit Lautsprecher

Manufacturer
Harman International Industries, Northridge, USA
In-house design
Harman Design Team
(Alexander Demin, Hyungwoo Yoon)
Web
www.harman.com

The Esquire 2 portable conferencing system features high-performance microphones with noise cancellation, enabling clear conversations even in loud environments. The four integrated speakers deliver harmonious sound that lends phone calls the feel of personal conversations in a room. In addition, the device is a mobile Bluetooth speaker and charging station. Thanks to its slim shape and reduced weight, it fits into any case or handbag.

Statement by the jury
The Esquire 2 conferencing system evinces a reduced design, yet it renders a precious and reliable impression with its ceramic grille and leather finishing.

Das tragbare Konferenzsystem Esquire 2 besitzt leistungsstarke Mikrofone mit Geräuschunterdrückung, die auch in lauten Umgebungen klare Gespräche ermöglichen. Die vier integrierten Lautsprecher sorgen für eine harmonische Klangwiedergabe, sodass Telefonate wie persönliche Gespräche in einem Raum klingen. Zudem ist das Gerät ein mobiler Bluetooth-Lautsprecher und eine Ladestation. Dank seiner schlanken Form und seinem geringen Gewicht passt es problemlos in jede Handtasche.

Begründung der Jury
Das Konferenzsystem Esquire 2 ist reduziert gestaltet. Mit seinem Keramikgitter und der Lederoberfläche wirkt es edel und zuverlässig.

Castor
Speaker
Lautsprecher

Manufacturer
Iriver, Seoul, South Korea
In-house design
Iriver Design Group
Design
Metal Sound Design (Kukil Yu), Seoul, South Korea
Web
www.iriver.com
www.ilmsd.com

Castor is a two-way speaker with the aspiration to reproduce sound in original quality. The design process is based on the philosophy that sound is inversely proportional to weight. The speaker was thus manufactured from metal components to avoid any unnecessary vibrations that could impair sound quality. At the same time, it guarantees powerful and clear bass reproduction.

Statement by the jury
With its solid, massive appearance, the speaker attracts attention and promises dynamic sound development.

Castor ist ein Zwei-Wege-Lautsprecher mit dem Anspruch, Sound in Originalqualität wiederzugeben. Dem Gestaltungsprozess liegt die Philosophie zugrunde, dass der Klang sich umgekehrt proportional zum Gewicht verhält. Deshalb wurde der Lautsprecher aus Metall gefertigt, denn es ist das ideale Material, um unnötige Vibrationen zu vermeiden, die den Klang beeinträchtigen könnten. Gleichzeitig gewährleistet es die kraftvolle und klare Wiedergabe der Bässe.

Begründung der Jury
Der Lautsprecher zieht mit seiner soliden, wuchtigen Anmutung die Aufmerksamkeit auf sich und verspricht eine dynamische Klangentfaltung.

H3
Multiroom Speaker
Netzwerk-Lautsprecher

Manufacturer
LG Electronics Inc., Seoul, South Korea
In-house design
Seon-Kyu Kim, Yuon-Ui Chong
Web
www.lg.com

This multiroom speaker shows an inconspicuous functional design and blends with any interior furnishings without creating an intrusive effect. It can be connected via Wi-Fi, Bluetooth and NFC and additionally features integrated streaming services like Spotify, Deezer or Napster. A music library is automatically created via an app so that the user can easily access his or her favourite tunes. In addition, two speakers can be interconnected for stereo sound.

Statement by the jury
The speaker provides outstanding network connectivity and intelligent functions for easily and flexibly streaming music at home.

Dieser Netzwerk-Lautsprecher zeigt eine unauffällig funktionale Gestaltung und passt zu jeder Einrichtung, ohne aufdringlich zu wirken. Er lässt sich über Wi-Fi, Bluetooth und NFC verbinden und hat zudem Streamingdienste wie Spotify, Deezer oder Napster integriert. Per App wird automatisch eine Musikbibliothek angelegt, damit der Benutzer seine Lieblingsmusik leicht finden kann. Darüber hinaus können zwei Lautsprecher für Stereoklang verbunden werden.

Begründung der Jury
Der Lautsprecher bietet eine hervorragende Netzanbindung und intelligente Funktionen, um Musik zu Hause einfach und flexibel zu streamen.

ODIN
Wireless Speaker
Kabelloser Lautsprecher

Manufacturer
Clint Digital, Ballerup, Denmark
In-house design
Phillip Bodum
Web
www.clintdigital.com

The Odin wireless speaker delivers powerful and space-filling sound with its two woofers and two tweeters, making it perfectly suited to placement in a living room. Its design is plain and characterised by clear lines; the aluminium and steel-grille enclosure lends the speaker a high-grade impression. The buttons on top allow for simple and comfortable handling, while purple LEDs serve as indicators for the speaker status.

Statement by the jury
With its unique, triangular shape, the speaker fits into any room corner. The selection of materials and the design convey a premium look and feel.

Der kabellose Lautsprecher Odin bietet durch seine zwei Tieftöner und zwei Hochtöner einen kraftvollen und raumfüllenden Klang, sodass er problemlos in einem Wohnzimmer aufgestellt werden kann. Seine Gestaltung ist schlicht und von klaren Linien bestimmt, das Gehäuse aus Aluminium und Stahlgitter verleiht ihm eine hochwertige Anmutung. Die Tasten auf der Oberseite ermöglichen eine einfache und komfortable Bedienung, lila LEDs geben eine Rückmeldung über den Status des Lautsprechers.

Begründung der Jury
Der Lautsprecher passt mit seiner eigenwilligen dreieckigen Form in jede Ecke im Raum. Die Materialwahl und die Formgebung vermitteln einen hochwertigen Eindruck.

FREYA
Wireless Speaker
Kabelloser Lautsprecher

Manufacturer
Clint Digital, Ballerup, Denmark
In-house design
Phillip Bodum
Web
www.clintdigital.com

The Freya wireless speaker has a plain and simple design. It is small and lightweight so that it can be carried effortlessly from room to room or used outdoors. In connection with a second speaker, it delivers space-filling stereo sound. With its smooth, cylindrical shape and calm colouring, the speaker blends harmoniously into any space, whether it is an office, living room or bedroom. The buttons on top are easily accessible and comfortably handled.

Statement by the jury
Freya convinces with its monolithic design, thanks to which it can be unobtrusively integrated into any living or working environment.

Der kabellose Lautsprecher Freya ist schlicht und einfach gestaltet. Er ist klein und leicht, sodass er problemlos von Raum zu Raum getragen oder im Freien genutzt werden kann. In Verbindung mit einem zweiten Lautsprecher liefert er raumfüllenden Stereoklang. Dank seiner glatten Zylinderform und der ruhigen Farbgebung passt der Lautsprecher in jeden Raum, egal ob Büro, Wohnzimmer oder Schlafzimmer. Die Tasten an der Oberseite sind leicht zugänglich und bequem zu bedienen.

Begründung der Jury
Freya überzeugt mit seiner monolithischen Gestaltung, dank derer er sich unauffällig in jede Büro- und Wohnumgebung einfügt.

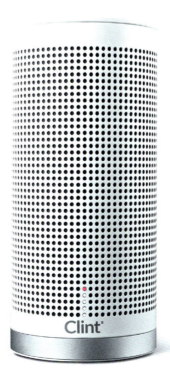

M3
Wi-Fi Speaker
Kabelloser Lautsprecher

Manufacturer
Shenzhen GGMM Industrial Co., Ltd., Shenzhen, China
In-house design
GGMM International Co., Ltd.
Web
www.ggmm.com

The M3 digital speaker supports three connect modes: AUX, Bluetooth and Wi-Fi. Its housing is coated with a textured leather surface available in different colours, and the front is overlaid with a brushed aluminium panel, making the speaker suitable for a large variety of interior styles. It is operated via a single, multifunctional knob located on top of the speaker.

Der digitale Lautsprecher M3 unterstützt drei Verbindungsmodi: AUX, Bluetooth und Wi-Fi. Sein Gehäuse ist mit einer strukturierten Lederoberfläche überzogen, die in mehreren Farben erhältlich ist. Die Front ist mit gebürstetem Aluminium verkleidet. Dadurch eignet sich der Lautsprecher für eine Vielzahl von Einrichtungsstilen. Der Lautsprecher wird über einen einzigen multifunktionalen Knopf bedient, der an der Oberseite angebracht ist.

Statement by the jury
With its metal front and prominent positioning of the multifunctional knob, the M3 renders a sophisticated and solid impression. The leather surface provides a warm contrast and adapts well to any living environment.

Begründung der Jury
Mit seiner Metallfront und dem prominent angebrachten Drehknopf wirkt der M3 edel und solide. Dazu bildet die Lederoberfläche einen warmen Kontrast und stellt die Verbindung zum Wohnbereich her.

M4
Wi-Fi Speaker
Kabelloser Lautsprecher

Manufacturer
Shenzhen GGMM Industrial Co., Ltd., Shenzhen, China
In-house design
GGMM International Co., Ltd.
Web
www.ggmm.com

The M4 is a portable Wi-Fi/Bluetooth speaker for indoor and outdoor use. The wooden housing with a leather finish is available in various colours and thus suits different lifestyles. Practical for outdoor use are the built-in rechargeable battery and adjustable leather carry straps. The speaker can be operated via AUX, Bluetooth and Wi-Fi connectivity. Two or more devices may be linked to form a network to achieve even deeper sound.

M4 ist ein tragbarer Wi-Fi/Bluetooth-Lautsprecher, der sich drinnen wie draußen verwenden lässt. Da sein Holzgehäuse mit Lederüberzug in unterschiedlichen Farben erhältlich ist, passt er zu verschiedenen Lebensstilen. Praktisch für die Verwendung im Freien sind der eingebaute Akku und der verstellbare Trageriemen aus Leder. Ansteuern lässt sich der Lautsprecher über AUX, Bluetooth und Wi-Fi. Zwei oder mehr Geräte können in einem Netzwerk verbunden werden, um noch mehr Klangfülle zu erzielen.

Statement by the jury
The combination of the materials wood and leather creates a cosy effect and highlights the speaker's indoor use. At the same time, it renders the device robust enough for outdoors.

Begründung der Jury
Die Materialkombination Holz/Leder wirkt wohnlich und stellt den Bezug zur Nutzung in Wohnräumen her. Gleichzeitig macht sie den Lautsprecher robust genug für die Verwendung im Freien.

449

Celia&Perah P Series
Wireless Speaker
Kabelloser Lautsprecher

Manufacturer
EGOS Technology Corp., Taipei, Taiwan
In-house design
Web
www.celia-perah.com

The speakers of the Celia&Perah P series combine the warm sound of wood with state-of-the-art wireless Bluetooth technology. The casing consists of natural wood with elegant curves, giving the devices a refined appearance. With two-way treble technology per audio channel, they deliver authentic, high-quality sound. Like the casing, the carry strap is made from natural materials, rounding off the overall impression. With 20 hours of battery life, the speakers are supplied with sufficient power, making them ideally suited for outdoor use.

Die Lautsprecher der Serie Celia&Perah P verbinden den warmen Klang von Holz mit der neuesten kabellosen Bluetooth-Technologie. Ihr Gehäuse besteht aus elegant gebogenem Naturholz, was den Geräten eine ursprüngliche Anmutung verleiht. Mit ihrer Zwei-Wege-Hochtöner-Technologie pro Audio-Kanal liefern sie einen unverfälschten Klang in hoher Qualität. Wie die Gehäuse so sind auch die Tragegurte aus Naturmaterialien hergestellt und runden dadurch das Gesamtbild ab. Mit einer Akkulaufzeit von 20 Stunden bieten die Lautsprecher eine ausdauernde Leistung und eignen sich dadurch hervorragend für die Verwendung im Freien.

Statement by the jury
This speaker series displays a successful symbiosis of natural materials and modern equipment. It thus fits very well into rural environments.

Begründung der Jury
Die Lautsprecher der Serie zeigen eine gelungene Symbiose von Naturmaterialien und moderner Ausstattung. Sie fügen sich dadurch hervorragend in ländliche Umgebungen ein.

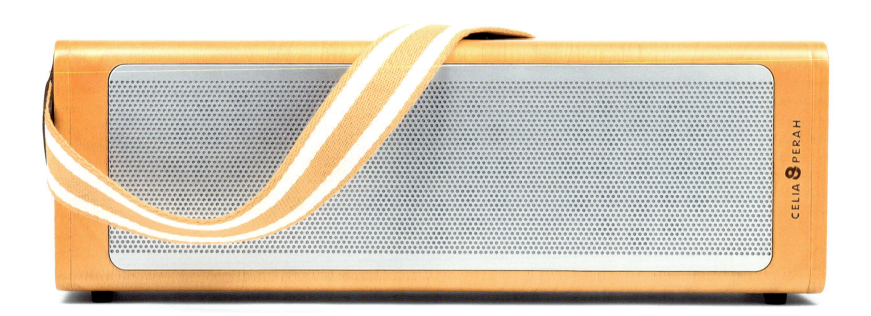

Mu-so
Wireless Music System
Kabelloses Musiksystem

Manufacturer
Naim Audio, Salisbury, Great Britain
In-house design
Web
www.naimaudio.com

The Mu-so wireless music system features a wooden enclosure wrapped in a layer of anodised aluminium, giving the system a sophisticated appearance and increasing sound quality. A milled and anodised aluminium heat sink runs along the entire rear face, efficiently dissipating heat and adding to the aesthetic appearance of the product. The heat sink also houses the Wi-Fi antenna. The illuminated base made from premium optical acrylic, the curved front and the elegant rotary control knob with intuitive handling lend the product a strong visual appeal.

Das kabellose Musiksystem Mu-so besitzt ein aus Holz gefertigtes Gehäuse mit einer Hülle aus eloxiertem Aluminium, was dem Gerät eine edle Anmutung verleiht und seine Stabilität sowie Klangqualität erhöht. Über die gesamte Rückseite erstreckt sich ein aus Aluminium gefräster, eloxierter Kühlkörper, der die entstehende Wärme effizient ableitet und die Ästhetik des Produktes unterstützt. Der Kühlkörper beherbergt zudem die Wi-Fi-Antenne. Die beleuchtete Basis aus hochwertigem Acryl, die geschwungene Front und der elegante, intuitiv bedienbare Drehregler verleihen dem Produkt einen besonderen visuellen Reiz.

Wireless Speaker System
Kabelloses Lautsprechersystem

Manufacturer
Raumfeld, Berlin, Germany
In-house design
Web
www.raumfeld.com

The Wi-Fi speakers enable audio streaming in any room. Advanced audio technology converts lossless digital files to produce genuine hi-fi sound. With elegant enclosures fashioned from real wood and aluminium, along with high-quality audio components and an intuitive user interface, they offer unlimited music enjoyment. The system comprises compact all-in-one speakers and also stereo floorstanding speakers to meet any need.

Statement by the jury
With regard to furnishings and design, the speaker system does not make any compromises. It meets the most sophisticated demands and is freely expandable.

Die Wi-Fi-Lautsprecher ermöglichen Audio-Streaming in jedem Raum. Fortschrittliche Audiotechnologie wandelt digitale Daten verlustfrei um und produziert auf diese Weise echten Hi-Fi-Sound. Mit ihren eleganten Gehäusen aus Echtholz und Aluminium, hochwertigen Audiokomponenten und der intuitiven Benutzerschnittstelle bieten sie unbegrenzten Musikgenuss. Das System umfasst kompakte All-in-one- und Stereo-Standlautsprecher für jeden Bedarf.

Begründung der Jury
Das Lautsprechersystem geht hinsichtlich Ausstattung und Gestaltung keine Kompromisse ein. Es bedient jeden Anspruch und ist problemlos erweiterbar.

Airstream S200
Wireless Speaker
Kabelloser Lautsprecher

Manufacturer
Monitor Audio, Essex, Great Britain
In-house design
Emily Hesslegrave, Charles Minett
Web
www.monitoraudio.co.uk

The Airstream S200 is a premium-class speaker system. Featuring a simple design, its footprint is so small that it requires minimal space. At the same time, it delivers surprisingly powerful sound. Bluetooth and Airplay facilitate the simple playback of stored or streamed content from any mobile device or PC. The innovative, twisted driver array ensures uniform sound distribution throughout the entire room.

Statement by the jury
With its independent form, the Airstream S200 can be placed even on small surfaces and provides a unique design highlight.

Der Airstream S200 ist ein Lautsprechersystem der Premiumklasse. Wegen seiner einfachen Gestalt ist seine Auflagefläche gering, sodass er nur wenig Platz benötigt. Gleichzeitig bietet er einen erstaunlich kraftvollen Klang. Bluetooth und Airplay gewährleisten eine einfache Wiedergabe von gespeicherten oder zu streamenden Inhalten von jedem mobilen Gerät oder PC. Die innovativ verdrehte Treiberanordnung sorgt dafür, dass der Klang im ganzen Raum verteilt wird.

Begründung der Jury
Dank seiner eigenständigen Form passt der Airstream S200 selbst auf kleine Flächen und setzt gestalterische Akzente.

Harman Kardon Wireless HD Audio System
Kabelloses HD-Audiosystem

Manufacturer
Harman International Industries, Northridge, USA
In-house design
Harman Design Team
(Damian Mackiewicz, Alexander Demin)
Web
www.harman.com

This wireless HD audio system combines speaker and audio adapter so that music and streaming content can be played back around the entire house without loss of quality. The Omni 10 and Omni 20 speakers with Bluetooth and Firecast play back sound in HD quality, while the Adapt turns the existing hi-fi system into an additional component part of the wireless HD audio system. With the WPS button on each product, installation of the system is child's play.

Statement by the jury
Speaker and adapter are perfectly matched with regard to design and convince both as a stand-alone solution and a network system.

Dieses kabellose HD-Audiosystem kombiniert Lautsprecher und Audio-Adapter, damit Musik und Streaming-Inhalte ohne Qualitätsverlust im gesamten Haus abgespielt werden können. Die Lautsprecher Omni 10 und Omni 20 mit Bluetooth und Firecast geben Klang in HD-Qualität wieder, während die Omni Adapt die Musikanlage zum Bestandteil des drahtlosen HD-Audiosystems macht. Dank der WPS-Taste auf jedem Produkt ist die Installation kinderleicht.

Begründung der Jury
Lautsprecher und Adapter sind gestalterisch perfekt aufeinander abgestimmt und überzeugen sowohl als Einzellösung wie auch als System.

Tranquil Moments Bedside Speaker
Bluetooth Speaker for Bedrooms
Bluetooth-Lautsprecher für Schlafzimmer

Manufacturer
Brookstone, Merrimack, USA
In-house design
Chris Petersen
Web
www.brookstone.com

The Tranquil Moments Bedside Speaker is a Bluetooth device particularly suitable for use in the bedroom. It is capable of playing 12 different sounds that, based on scientific findings, help with relaxing and falling asleep. At first, the user selects a sound by simply turning the dial on top, then he presses the button to start the 30-minute sleep enhancement mode. Outside of the bedroom the device can be used like any conventional Bluetooth speaker.

Der Tranquil Moments Bedside Speaker ist ein Bluetooth-Lautsprecher, der insbesondere für die Verwendung im Schlafzimmer konzipiert wurde. Er kann zwölf unterschiedliche Klänge abspielen, die nach wissenschaftlichen Erkenntnissen ausgewählt wurden, um das Einschlafen zu erleichtern. Der Benutzer wählt durch einfaches Drehen des oberen Teils einen Sound aus, dann drückt er auf den Knopf, um das 30-minütige Einschlafprogramm zu starten. Außerhalb des Schlafzimmers lässt sich das Gerät wie ein herkömmlicher Bluetooth-Lautsprecher verwenden.

Statement by the jury
The speaker impresses with its intuitive operating concept and congenial design. Thanks to its rubberised bottom, it rests safely on any surface.

Begründung der Jury
Der Lautsprecher besticht durch sein intuitives Bedienkonzept und die freundliche Formgebung. Dank seiner gummierten Unterseite steht er sicher auf jeder Oberfläche.

453

TransSpeakers
Portable Bluetooth Speaker
Tragbarer Bluetooth-Lautsprecher

Manufacturer
Topneer International Ltd., Hong Kong
In-house design
Honda Hu, Jimmy Wan
Web
www.topneer.com

The TransSpeakers portable Bluetooth speaker comprises four speaker columns that can be individually adjusted to suit one's personal preferences. They serve as speakers but also as a stand for smartphone or tablet, thus providing an optimal view of the device. The high-performance driver and passive radiator on each column deliver superb sound quality. Extended columns can be easily folded to a compact size for on-the-road use. The speaker's silicone cover protects it from scratches, facilitates cleaning and underscores its aesthetic appeal.

Der tragbare Bluetooth-Lautsprecher TransSpeakers besteht aus vier Lautsprechersäulen, die individuell angepasst werden können, um den persönlichen Vorlieben zu entsprechen. Sie dienen sowohl als Lautsprecher wie auch als Halter für Smartphone oder Tablet und bieten so eine optimale Sicht auf das Gerät. Die Hochleistungstreiber und der Passivstrahler an jeder Säule liefern einen ausgezeichneten Klang. Für unterwegs können die Säulen einfach zu einem kompakten Würfel zusammengefaltet werden. Die Silikonhülle des Lautsprechers schützt ihn vor Kratzern, erleichtert die Reinigung und unterstreicht seine Ästhetik.

P7
Portable Speaker
Tragbarer Lautsprecher

Manufacturer
LG Electronics Inc., Seoul, South Korea
In-house design
Seon-Kyu Kim, Jun-Ki Kim, Sang-Hoon Yoon
Web
www.lg.com

The P7 portable speaker has a compact design and connects to other devices via Bluetooth. When two speakers are interconnected, they provide stereo sound due to a special mode. The metal frame with grille creates the impression that the entire speaker enclosure is made of metal. The round metal key on top has a minimalist aesthetic design and offers access to numerous device functions.

Statement by the jury
Thanks to its reduced design and compact dimensions, this speaker can be unobtrusively placed in any living environment.

Der kompakte, tragbare Lautsprecher P7 verbindet sich per Bluetooth mit anderen Geräten. Werden zwei Lautsprecher zusammengeschaltet, bieten sie über einen speziellen Modus Stereoklang. Der Metallrahmen mit Gitter sorgt dafür, dass auch das Lautsprechergehäuse wie aus Metall wirkt. Die runde Taste aus Metall auf der Oberseite erscheint in minimalistischer Ästhetik und bietet Zugang zu den zahlreichen Funktionen des Geräts.

Begründung der Jury
Dieser Lautsprecher lässt sich dank seiner reduzierten Gestaltung und kompakten Maße in jedem Wohnraum unauffällig aufstellen.

JBL GO
Portable Bluetooth Speaker
Tragbarer Bluetooth-Lautsprecher

Manufacturer
Harman International Industries, Northridge, USA
In-house design
Harman Design Team
(Joshua Fischer, Irfan Kachwalla)
Web
www.harman.com

The JBL GO is a portable full-range speaker with Bluetooth connectivity, AUX connector and hands-free functionality. Due to its compact dimensions, it fits into any hand. With the integrated carabiner hook, the speaker can be attached to bags or articles of clothing. Its buttons are located along the top surface so as to be easily reached and operated with the thumb. The speaker is available in eight vivid colours.

Statement by the jury
Despite its reduced design, the JBL GO expresses a high degree of independence. The angular form and vibrant colours make it an eye-catching companion in any environment.

JBL GO ist ein tragbarer Komplettlautsprecher mit Bluetooth, AUX-Anschluss und Freisprechfunktion. Dank seiner kompakten Größe passt er in jede Hand. Mit dem integrierten Karabinerhaken kann der Lautsprecher an Taschen oder der Kleidung befestigt werden. Seine Tasten sind auf der oberen Seite so positioniert, dass sie problemlos mit dem Daumen erreicht und bedient werden können. Der Lautsprecher ist in acht lebhaften Farben verfügbar.

Begründung der Jury
Trotz seiner reduzierten Gestaltung besitzt der JBL GO eine hohe Eigenständigkeit. Mit seiner kantigen Form und den kräftigen Farben wird er in jeder Umgebung zu einem auffälligen Begleiter.

JBL Xtreme
Portable Bluetooth Speaker
Tragbarer Bluetooth-Lautsprecher

Manufacturer
Harman International Industries, Northridge, USA
In-house design
Harman Design Team
(Joshua Fischer, Fenoson Zafimahova)
Web
www.harman.com

The JBL Xtreme is a portable, high-performance speaker with a noise-cancelling speakerphone and a waterproof casing. All electronic connections are located behind the cover on the back, yet the well-protected keys on top are still easily accessible. The shape of the casing allows for secure carrying in the hand. In addition, the speaker includes two passive radiators visualising the sound while music is played.

Statement by the jury
The JBL Xtreme is splash-proof so that it can be used in bathrooms or near swimming pools.

Der JBL Xtreme ist ein leistungsfähiger tragbarer Lautsprecher mit Freisprechfunktion mit Geräuschunterdrückung und einem wasserfesten Gehäuse. Alle elektronischen Anschlüsse befinden sich hinter der Abdeckung auf der Rückseite, während die Tasten auf der oberen Seite gut geschützt und trotzdem leicht zu erreichen sind. Das Gehäuse ist so geformt, dass es beim Tragen sicher in der Hand gehalten werden kann. Zudem besitzt der Lautsprecher zwei Passivboxen, die den Sound während der Musikwiedergabe visualisieren.

Begründung der Jury
Der JBL Xtreme ist unempfindlich gegen Wasserspritzer, sodass er auch im Bad oder Schwimmbad eingesetzt werden kann.

AeroTwist
Portable Bluetooth Speaker
Tragbarer Bluetooth-
Lautsprecher

Manufacturer
Jarre Technologies, Hong Kong
In-house design
Kateryna Sokolova
Web
www.jarre.com
Honourable Mention

This ring-shaped Bluetooth speaker can be opened and thus easily attached to a bag or bicycle. It weighs less than 1 kg and includes four speakers delivering clear sound with 40 watts of power. An additional feature is the microphone with integrated noise cancellation. The AeroTwist is available in seven vivid colours to match a modern lifestyle.

Statement by the jury
The unconventional shape of the AeroTwist turns out to be practical as it allows for attachment to loops and eyelets.

Der ringförmige Bluetooth-Lautsprecher lässt sich öffnen und dadurch leicht an einer Tasche oder einem Fahrrad befestigen. Er wiegt weniger als 1 kg und besitzt vier Lautsprecher, die einen klaren Klang bei einer Leistung von 40 Watt liefern. Auch ein Mikrofon mit integrierter Geräuschunterdrückung ist mit an Bord. Der AeroTwist ist in sieben lebhaften Farben erhältlich, die zu einem modernen Lebensstil passen.

Begründung der Jury
Die eigenwillige Form des AeroTwist erweist sich als praktisch, da sie die Befestigung an Schlaufen und Ösen ermöglicht.

Powergo
Portable Bluetooth Speaker
Tragbarer Bluetooth-
Lautsprecher

Manufacturer
Platform2 International Limited, Hong Kong
In-house design
Michael Tse, Rex Chan
Web
www.iuidesign.com

Powergo is a portable Bluetooth speaker with additional functions that are of particular use for business people. It has a high-capacity powerbank for charging smartphones and other mobile devices. It can be used as a speaker for a phone and includes a detachable Bluetooth headset for hands-free conversation. The casing is made of elegant aluminium and is so compact that it fits into any pocket or briefcase.

Statement by the jury
Powergo offers clever functions for mobile business people in a handy device. Its aluminium casing renders a durable, high-grade impression.

Powergo ist ein tragbarer Bluetooth-Lautsprecher mit Zusatzfunktionen, die besonders für Geschäftsleute nützlich sind. So besitzt er einen Akku, um Smartphones und andere mobile Geräte aufzuladen. Er kann als Lautsprecher fürs Telefon verwendet werden und wird mit einem abnehmbaren Bluetooth-Headset für das freie Sprechen geliefert. Das Gehäuse besteht aus elegantem Aluminium und ist so kompakt, dass es in Akten- oder Hosentaschen passt.

Begründung der Jury
Powergo bietet in einem handlichen Gerät clevere Funktionen für mobile Geschäftsleute. Mit seinem Aluminiumgehäuse wirkt es wertig und robust.

X6
Portable Bluetooth Speaker
Tragbarer Bluetooth-
Lautsprecher

Manufacturer
Gibson Innovations, Hong Kong
In-house design
Charlie Bolton
Web
www.gibson.com

The X6 portable Bluetooth speaker has a rigid extruded aluminum casing which gives the product a high quality, robust appearance. The semi elliptical profile randomizes the direction of sound waves for natural sound reproduction. Up to five devices can be paired with the X6, and a rechargeable battery supplies power for up to eight hours of music playback. The speaker also includes USB and audio connections, as well as a microphone.

Statement by the jury
This speaker impresses with its independent design, which is both functional and stylish, and thus enriches any room.

Der tragbare Bluetooth-Lautsprecher X6 hat ein festes, extrudiertes Aluminiumgehäuse, das dem Produkt ein hochqualitatives, robustes Aussehen verleiht. Das halbelliptische Profil randomisiert die Richtung der Schallwellen für eine natürliche Klangreproduktion. Bis zu fünf Geräte können mit dem X6 gekoppelt werden, und ein Akku sorgt für die Stromversorgung für bis zu acht Stunden Musikwiedergabe. Der Lautsprecher besitzt zudem USB- und Audioanschlüsse sowie ein Mikrofon.

Begründung der Jury
Der Lautsprecher beeindruckt durch seine eigenständige Gestaltung, die funktional wie stilvoll ist und damit jeden Raum bereichert.

Stellé Audio Pillar®
Portable Bluetooth Speaker
Tragbarer Bluetooth-Lautsprecher

Manufacturer
Stellé Audio, Newport Beach, USA
In-house design
Wayne Ludlum
Web
www.stelleaudio.com

The Stellé Audio Pillar speaker bridges the gap between music, design and technology. Accordingly, it features an elegant, smooth cylindrical shape, offers impressive 360-degree sound and connects to any compatible device via Bluetooth. With its large number of designs available in different wood, metal and colour varieties, it blends with any interior style, both at home and in the office.

Statement by the jury
With designs catering to any taste and thanks to its slim shape, this speaker blends perfectly into any interior, like a piece of furniture.

Der Lautsprecher Stellé Audio Pillar schließt die Lücke zwischen Musik, Design und Technologie. Entsprechend zeigt er sich in einer eleganten, glatten Zylinderform, bietet einen beeindruckenden 360-Grad-Sound und lässt sich dank Bluetooth mit jedem kompatiblen Gerät verbinden. Mit seinen diversen Dekoren, die in unterschiedlichen Holz-, Metall- und Farbvarianten erhältlich sind, passt er zu jedem Einrichtungsstil zu Hause wie auch im Büro.

Begründung der Jury
Dank seiner Dekore für jeden Geschmack und der schlanken Form fügt sich der Lautsprecher wie ein Möbelstück bestens in jedes Interieur ein.

Hoop
Bluetooth Speaker
Bluetooth-Lautsprecher

Manufacturer
Lexon Design, Boulogne, France
In-house design
Valentina Del Ciotto, Simone Spalvieri
Web
www.lexon-design.com

The Hoop is a portable, high-performance Bluetooth speaker based on the idea of sharing music on the go. Its two useful accessories expand the range of use: thanks to a practical silicone rubber cord and elastic hook, it is easy to carry and can be hung anywhere. The speaker is made of steel and ABS with a rubber finish, making it durable and convenient to use in everyday life.

Statement by the jury
Hoop can be easily taken along and used anywhere. Its accessories show a surprisingly simple design and offer many options.

Hoop ist ein leistungsstarker, tragbarer Bluetooth-Lautsprecher, der auf der Idee beruht, Musik unterwegs zu teilen. Seine beiden nützlichen Zubehörteile erweitern das Anwendungsspektrum: Dank des praktischen Silikon-Gummibands und des elastischen Hakens lässt er sich leicht tragen und überall aufhängen. Der Lautsprecher besteht aus ABS mit einem Kunststoffüberzug und Stahl, ist daher strapazierfähig und alltagstauglich.

Begründung der Jury
Hoop lässt sich leicht mitnehmen und überall einsetzen. Das Zubehör ist verblüffend einfach gestaltet und bietet viele Möglichkeiten.

Fine
Bluetooth Speaker
Bluetooth-Lautsprecher

Manufacturer
Lexon Design, Boulogne, France
In-house design
Pauline Deltour
Web
www.lexon-design.com

The Fine speaker has the shape of a lipstick and is similarly handled as well: to switch it on, it has to be turned once. The volume is likewise controlled via the turning mechanism. Its surface is made of aluminium with delicate grooves to enhance sound quality and also provide a better grip when turning the device. The speaker is charged via USB and, with a diameter of only 6.5 cm, it is very compact.

Statement by the jury
With its grooved aluminium structure and operation via fine rotary movements, this speaker displays both industrial and feminine aesthetics.

Der Lautsprecher Fine hat die Form eines Lippenstifts und lässt sich auch ähnlich bedienen: Um ihn anzuschalten, muss einmal daran gedreht werden. Die Lautstärke wird ebenfalls über die Drehung geregelt. Seine Oberfläche besteht aus gerilltem Aluminium, was für einen besseren Klang sorgt und zudem den Grip bei der Drehbewegung erhöht. Der Lautsprecher wird über USB aufgeladen und ist mit einem Durchmesser von nur 6,5 cm sehr kompakt.

Begründung der Jury
Mit seiner gerillten Aluminiumstruktur und der Bedienung über feine Drehbewegungen besitzt dieser Lautsprecher eine industrielle wie feminine Ästhetik.

Cube
Bluetooth Speaker
Bluetooth-Lautsprecher

Manufacturer
John Lewis, London, Great Britain
Design
AlexHammond Design, London, Great Britain
Web
www.johnlewis.com
www.alexhammond.co.uk

The Bluetooth speaker Cube has a compact cuboid shape with a solid appearance. As the bass speaker is integrated into the foot, the table or shelf on which the unit rests turns into a resonance surface, enhancing bass performance. The built-in rechargeable battery supplies power for up to six hours and provides enough energy to charge connected mobile devices via a USB port. In addition, Cube can also be used as speaker for mobile phones.

Statement by the jury
Circular elements, such as front grille and control buttons, are gracefully integrated into the cube shape of the speaker.

Der Bluetooth-Lautsprecher Cube besitzt eine kompakte, solide wirkende Würfelform. Da der Basslautsprecher im Fuß integriert ist, wird die Oberfläche, auf der der Lautsprecher steht, zur Resonanzfläche, die die Bassleistung verstärkt. Der Akku garantiert eine Betriebsdauer von bis zu sechs Stunden und bietet genügend Energie, um per USB-Kabel angeschlossene Mobilgeräte aufzuladen. Zudem kann Cube als Lautsprecher für Mobiltelefone verwendet werden.

Begründung der Jury
Die kreisförmigen Elemente wie Frontgitter und Bedienknöpfe sind harmonisch in die Würfelform des Lautsprechers integriert.

DrumBass X BT
Bluetooth Stereo Speaker Set
Bluetooth Stereo-Lautsprecher

Manufacturer
Lifetrons Switzerland AG, Dicken, Switzerland
In-house design
Lifetrons Switzerland Design Team
Web
www.lifetrons.ch

These Bluetooth stereo speakers in pocket format provide powerful bass and clear sound with a good reproduction of mid- and high-range frequencies. At the top, a small LED light circle indicates charging or playback status. In addition, a voice alert system informs the user of the current status, such as "low battery" or "power on". The speakers can be connected via Bluetooth or an AUX port.

Statement by the jury
This practical speaker not only has a sophisticated look but also provides the user with all important information for trouble-free operation.

Diese Bluetooth-Stereolautsprecher im Taschenformat bieten kräftige Bässe und einen klaren Sound mit einer guten Wiedergabe mittlerer und hoher Tonlagen. An der Oberseite befindet sich ein kleiner LED-Lichtkreis, der den Lade- oder Wiedergabestatus anzeigt. Zusätzlich gibt eine Sprachnachricht Auskunft über den gegenwärtigen Status wie „niedrige Batterie" oder „Power on". Neben Bluetooth bieten die Lautsprecher auch einen AUX-Anschluss.

Begründung der Jury
Diese handliche Lautsprecher sehen nicht nur edel aus, sondern geben dem Benutzer auch alle wichtigen Informationen für eine problemlose Bedienung.

SoundLink® Colour
Bluetooth® Speaker
Bluetooth-Lautsprecher

Manufacturer
Bose Corporation, Framingham, USA
In-house design
Web
www.bose.com

The SoundLink Colour Bluetooth speaker weighs merely 570 grams and is very compact. It can be easily transported and set up anywhere. Two efficient transducers, combined with dual-opposing passive radiators, deliver natural, full-range sound. The speaker connects wirelessly to smartphones, tablet PCs or other Bluetooth-capable devices. Voice prompts support the user in setting up the speaker and identifying the connected source device.

Statement by the jury
This speaker delivers top-quality audio, is very user-friendly and, thanks to its small dimensions, also renders a good impression on the go.

Der SoundLink Colour Bluetooth Speaker wiegt nur 570 Gramm und ist sehr kompakt. Daher lässt er sich leicht transportieren und überall aufstellen. Zwei effiziente Schallwandler, kombiniert mit zwei gegenüberliegenden Passivstrahlern liefern einen naturgetreuen Klang. Der Lautsprecher verbindet sich kabellos mit Smartphones, Tablet-PCs und anderen Bluetooth-fähigen Geräten. Sprachansagen unterstützen dabei, den Lautsprecher einzurichten und das verbundene Gerät zu erkennen.

Begründung der Jury
Der Lautsprecher liefert Spitzenklang, ist überaus benutzerfreundlich und macht dank seiner geringen Maße auch unterwegs einen guten Eindruck.

Ultimate Ears Megaboom
Bluetooth Speaker
Bluetooth-Lautsprecher

Manufacturer
Ultimate Ears, Newark, California, USA
Design
NONOBJECT, Palo Alto, USA
Web
www.ultimateears.com
www.nonobject.com

The Ultimate Ears Megaboom wireless speaker is lightweight, fits comfortably in the hand and impresses with its good sound. It is covered in a protective skin which, over time, is designed to take on an individual patina. Since the speaker is IPX7 certified and thus waterproof, the skin can simply be wet-cleaned. Control functions are reduced to the bare essentials so that users can enjoy music without distraction. The rechargeable battery supplies power for 20 hours and is completely recharged within two and a half hours.

Der kabellose Lautsprecher Ultimate Ears Megaboom ist leicht, liegt bequem in der Hand und besticht durch seinen guten Sound. Er ist mit einer schützenden Haut verkleidet, die mit der Zeit eine individuelle Patina ansetzt. Da der Lautsprecher IPX7-zertifiziert und damit wasserfest ist, lässt sich diese Haut auch einfach feucht reinigen. Die Bedienung ist auf das Wesentliche reduziert, damit der Benutzer seine Musik ohne Ablenkung genießen kann. Der Akku liefert Energie für 20 Stunden und ist innerhalb von zweieinhalb Stunden wieder vollständig aufgeladen.

Statement by the jury
With its clear and plain form, robust design and striking colours, this speaker is optimally equipped to appeal to a young target group.

Begründung der Jury
Mit seiner klaren Form, der robusten Gestaltung und den frechen Farben ist dieser Lautsprecher für seine junge Zielgruppe optimal ausgestattet.

Eton Rukus Xtreme
Solar-Powered Bluetooth Speaker
Solarbetriebener Bluetooth-Lautsprecher

Manufacturer
Etón Corporation, Palo Alto, USA
Design
Whipsaw Inc, San Jose, USA
Web
www.etoncorp.com
www.whipsaw.com

The Eton Rukus Xtreme Bluetooth speaker features a solar panel generating enough energy for the entire day. In addition, the internal rechargeable battery provides power for another 12 hours. Its robust housing consists of shock- and splash-proof plastic and rubber, making the speaker particularly suitable for outdoor use. The handle at the narrower end of the speaker serves as carrying aid or as an attachment for a backpack or tent. When the speaker is connected to a smartphone via USB, it charges the device with its internal rechargeable battery.

Statement by the jury
From the practical handle all the way to the energy supply via solar panel, the Eton Rukus Xtreme is optimally suited for outdoor use.

Der Bluetooth-Lautsprecher Eton Rukus Xtreme besitzt ein Solarpanel, das genügend Energie für den ganzen Tag erzeugt. Zusätzlich liefert der interne Akku Strom für weitere zwölf Stunden. Sein robustes Gehäuse besteht aus stoßfestem Kunststoff und Gummi, was ihn für den Einsatz im Freien besonders geeignet macht. Der Griff am schlankeren Ende des Lautsprechers dient als Tragehilfe oder zur Befestigung am Rucksack oder in einem Zelt. Wird er mit einem Smartphone über USB verbunden, lädt er dieses über seinen internen Akku auf.

Begründung der Jury
Der Eton Rukus Xtreme ist optimal für die Verwendung im Freien geeignet, angefangen beim praktischen Griff bis hin zur Energieversorgung über ein Solarpanel.

Cast-Fi 7
HDMI Docking Speaker
HDMI-Dockinglautsprecher

Manufacturer
TVLogic, Aurender, Seoul, South Korea
Design
JEI Design Works (Kyuhong Han, Chul Jin), Seoul, South Korea
Web
www.aurender.com
www.designjei.com

This docking speaker with a monitor has a high-grade appearance, as the enclosure consists of a full-metal frame and an aluminium grille on the front. The frame not only dominates the aesthetic appearance but also contributes to optimum sound development. Control features on the device are reduced to a minimum since the functions of the speaker are accessed via remote control or smartphone. This design gives the user considerably more freedom while operating the speaker.

Statement by the jury
With its metal components, the Cast-Fi 7 renders a sophisticated and massive impression. The bevelled front conveys independence and creates harmonious synergy between image and sound.

Der Dockinglautsprecher mit Monitor besitzt eine wertige Anmutung, denn das Gehäuse besteht aus Metall, während das Frontgitter in Aluminium ausgeführt ist. Dabei bestimmt der Rahmen nicht nur die Ästhetik, sondern trägt auch zur Entfaltung des Klangs bei. Die Steuerung am Gerät ist auf ein Minimum reduziert, da der Zugriff auf die Funktionen mit Fernbedienung oder Smartphone erfolgt. Dadurch erhält der Benutzer wesentlich mehr Freiheit bei der Bedienung des Lautsprechers.

Begründung der Jury
Durch die Metallkomponenten zeigt sich der Cast-Fi 7 edel und massiv. Die abgeschrägte Vorderseite verleiht ihm Eigenständigkeit und schafft eine harmonische Verbindung zwischen Bild und Ton.

Flow [v1000]
DAC and Headphone Amplifier
D/A-Wandler und Kopfhörerverstärker

Manufacturer
TVLogic, Aurender, Seoul, South Korea
Design
JEI Design Works (Chul Jin), Seoul, South Korea
Web
www.aurender.com
www.designjei.com

The Flow amplifier was milled from a single piece of aluminium, making it not only highly compact but giving it a precious appearance as well. The basic design concept of warm music flowing out of the cold metal is reflected by its wave-shaped top. Thanks to the amplifier's small dimensions, it can be easily used en route and placed on any surface. The device is user-friendly in operation, as all important functions are controlled by the large rotary knob.

Statement by the jury
Soft lines along the top of the device simplify access to this technical product. In addition, the control concept with a rotary knob and easily accessible lateral buttons is a convincing feature.

Der Verstärker Flow wurde aus einem Stück Aluminium gefräst. Dadurch ist er nicht nur sehr kompakt, sondern wirkt überaus edel. Das ihm zugrunde liegende Designkonzept von warmer Musik, die aus kaltem Metall herausfließt, wird durch seine geschwungene Oberseite angedeutet. Dank seiner geringen Maße lässt er sich leicht mitnehmen und platzieren. Auch die Bedienung ist nutzerfreundlich, denn alle wichtigen Funktionen werden über einen großen Drehknopf gesteuert.

Begründung der Jury
Sanfte Linien an der Oberseite vereinfachen den Zugang zu diesem technischen Produkt. Zudem überzeugt das Bedienkonzept mit Drehknopf und leicht erreichbaren Tasten an der Seite.

Reed Muse 3C
Turntable
Plattenspieler

Manufacturer
Tonearms.lt, UAB, Kaunas District, Lithuania
In-house design
Loreta Kudirkaite, Vidmantas Triukas
Design
Alfonsas Vaura, Kaunas, Lithuania
Web
www.reed.lt

The Muse 3C turntable is housed in a massive aluminium enclosure. With only a few moves, it can be swiftly converted from belt drive to friction drive. The turntable disc velocity is very consistent as it is stabilised by a quartz-based control system. Using an electronic level for measuring slope and tilt angle, the turntable can be precisely aligned without needing any additional tools.

Statement by the jury
The Muse 3C impresses with an independent design that conveys its special quality and underscores the high-grade technical equipment.

Der Plattenspieler Muse 3C ist in einem massiven Aluminiumgehäuse untergebracht. Er lässt sich mit wenigen Handgriffen von Riemen- auf Reibradantrieb umbauen. Die Drehzahl des Plattentellers ist sehr gleichmäßig, da sie mit einem quarzgesteuerten System stabilisiert wird. Durch eine elektronische Einstellhilfe, die Abweichungen von der Horizontalen in zwei Ebenen misst, kann der Plattenspieler beim Aufstellen ohne zusätzliche Werkzeuge präzise ausgerichtet werden.

Begründung der Jury
Der Muse 3C beeindruckt durch seine eigenständige Gestaltung. Sie vermittelt seine besondere Qualität und unterstreicht die hochwertige technische Ausstattung.

Double Matrix Professional Sonic
Record Cleaning Machine
Plattenreinigungsmaschine

Manufacturer
clearaudio electronic GmbH,
Erlangen, Germany
In-house design
Web
www.clearaudio.de

The Double Matrix Professional Sonic cleans vinyl records very gently and intensively through vibration. Users have the possibility of operating the whole cleaning process at the touch of a single button, as well as the ability to control each step in the process with a user-friendly operating panel. The cleaner is nearly silent in operation, easily handled and maintained. Exchange of the microfibre cleaning strips, for instance, is accomplished in just a few steps. The LED level indicator communicates the amount of cleaning fluid remaining at any given time.

Statement by the jury
This record cleaning machine shows its professional character in form and function, yet it is also refreshingly easy to operate and user-friendly.

Die Double Matrix Professional Sonic reinigt Vinylschallplatten besonders schonend und intensiv durch Vibration. Sie lässt dem Anwender die Wahl zwischen manueller Steuerung der einzelnen Reinigungsschritte und voll automatisierten Reinigungsabläufen mit nur einem Knopfdruck. Dabei arbeitet sie nahezu lautlos, ist leicht zu bedienen und zu warten, wie beispielsweise beim Austausch der Mikrosamtstreifen, der in wenigen Schritten erfolgt. Die LED-Füllstandsanzeige informiert jederzeit über die verbleibende Reinigungsflüssigkeit.

Begründung der Jury
Diese Plattenreinigungsmaschine zeigt ihren professionellen Charakter in Form und Funktion, ist dabei aber erfreulich einfach zu bedienen.

Noxon Nova
Audio System
Audiosystem

Manufacturer
NOXON Vertriebsgesellschaft mbH,
Nettetal, Germany
Design
design networx,
Mülheim an der Ruhr, Germany
Web
www.noxonradio.de
www.design-networx.de

The Noxon Nova 2.1 sound system offers comprehensive functions such as Internet radio, DAB+, UKW and music playback via Bluetooth, as well as streaming from the cloud and USB. It can be operated using the typical rotary knob, the remote control and also an app for Android and iOS. The timeless design, premium materials such as aluminium and wood, simple operation and impressive sound turn this audio system into a device that leaves nothing to be desired.

Statement by the jury
Noxon Nova is an all-rounder with a timeless design. It can be used to receive and play music from any source.

Das 2.1-Soundsystem Noxon Nova bietet umfassende Funktionen wie Radioempfang über Internet, DAB+ und UKW sowie Musikwiedergabe über Bluetooth, Streaming aus der Cloud und USB. Es kann über den typischen Drehknopf, die Fernbedienung wie auch über eine App für Android und iOS bedient werden. Die zeitlose Gestaltung, hochwertige Materialien wie Aluminium und Holz, die einfache Bedienung und der eindrucksvolle Klang machen das Audiosystem zu einem Gerät, das keine Wünsche offen lässt.

Begründung der Jury
Noxon Nova ist ein zeitlos gestalteter Alleskönner, denn mit ihm lässt sich Musik von jeder Quelle empfangen und abspielen.

MEDION® LIFE® P85100
Internet Radio

Manufacturer
Medion AG, Essen, Germany
In-house design
Web
www.medion.com

The Medion Life P85100 is a DNLA Internet radio that is comfortably controlled using a smartphone or tablet. In the tradition of classical consumer electronics, this compact music cube delivers excellent digital sound that is true to nature. The straightforward design language emphasises the high quality of the technology integrated into the cube.

Statement by the jury
This Internet radio derives its uniqueness from the contrasting surfaces and clever combination of the geometric forms of the circle and the cube.

Das Medion Life P85100 ist ein DNLA-Internetradio, das sich bequem von einem Smartphone oder Tablet aus bedienen lässt. In der Tradition klassischer Unterhaltungselektronik produziert dieser kompakte Musikwürfel einen ausgezeichneten digitalen sowie naturgetreuen Klang. Die geradlinige Formensprache unterstreicht die hohe Qualität der Technologie, die sich in dem Würfel befindet.

Begründung der Jury
Dieses Internetradio bezieht seine Eigenständigkeit aus den kontrastierenden Oberflächen und der geschickten Kombination der geometrischen Formen Kreis und Würfel.

SuperSystem
Digital Radio
Digitalradio

Manufacturer
Revo Technologies Ltd, Lanark, Great Britain
In-house design
David Baxter, Jeremy Offer
Web
www.revo.co.uk

The SuperSystem is reminiscent of a traditional radio but houses modern digital technology in its interior. It receives a large number of Internet radio stations, streams music from smart devices or computers via an app, and can access millions of songs in the cloud. All this is included in a wooden enclosure with an aluminium front, fostering a sophisticated and durable appearance. In this way, radio gains new relevance in the digital world.

Statement by the jury
With high-grade materials, such as wood and aluminium, and its classic shape, the SuperSystem successfully invokes the age of analogue radio.

Das SuperSystem erinnert zwar an ein traditionelles Radio, birgt aber in seinem Inneren moderne digitale Technik. Es empfängt eine Vielzahl von Internet-Radiostationen, streamt per App Musik von Smartgeräten oder einem Computer und kann auf Millionen von Titeln in der Cloud zugreifen. All das befindet sich in einem Holzgehäuse mit Aluminiumfront, die das Gerät edel und langlebig wirken lassen. So gewinnt das Radio eine neue Relevanz in der digitalen Welt.

Begründung der Jury
Mit seinen wertigen Materialien wie Holz und Aluminium und der klassischen Form ist das SuperSystem eine gelungene Referenz an das analoge Radiozeitalter.

Pop
Digital Radio Range
Digitalradio-Reihe

Manufacturer
Pure (Imagination Technologies), Kings Langley, Great Britain
In-house design
Web
www.pure.com

Pop is a digital radio range with the aspiration to reinvent radio for a new listening experience. Its essential design features were derived from findings about how people use and interact with the medium of radio. The result is a series with three products varying in size. All of them are characterised by easy access through the large button on top. Once the button is pressed, the radio will automatically tune in to the last station listened to, with the volume previously set. Bluetooth connectivity and a rechargeable battery complement the equipment.

Statement by the jury
Pop offers an uncomplicated radio listening experience in an appealing, compact design. The radio is available in three sizes and blends into any environment.

Pop ist eine Digitalradio-Reihe mit dem Anspruch, das Radio für den Hörer neu erlebbar zu machen. Seine wesentlichen Gestaltungsmerkmale wurden daraus abgeleitet, wie Menschen ein Radio verwenden und mit ihm interagieren wollen. Das Ergebnis ist eine Reihe mit drei Produkten unterschiedlicher Größe. Allen gemeinsam ist der leichte Zugang über den großen Knopf an der Oberseite. Einmal gedrückt, schaltet das Gerät auf den letzten eingestellten Sender in der zuvor gewählten Lautstärke. Bluetooth und ein Akku runden die Ausstattung ab.

Begründung der Jury
Pop bietet ein unkompliziertes Radioerlebnis in einem ansprechenden kompakten Design. Das in drei Größen erhältliche Gerät passt in jede Umgebung.

AK Junior
Portable Audio Player
Tragbarer Audio-Player

Manufacturer
Iriver, Seoul, South Korea
In-house design
Baekjin Seong, Eunkyung Lee
Design
Metal Sound Design (Kukil Yu), Seoul, South Korea
Web
www.iriver.com
www.ilmsd.com

The AK Junior audio player offers hi-fi technology in a compact, portable form. The display is situated at the top in a raised position while the edges are flattened. This provides a comfortable grip and allows the player to appear generally slimmer than it actually is. The aluminium volume wheel on the side is slightly elevated, and thus easily recognisable and comfortably operated. The overall handling has been simplified in such a way that it conforms to the habits of smartphone users, thereby allowing for intuitive operation. This concept is also reflected in the plain design of the player.

Der Audio-Player AK Junior bietet Hi-Fi-Technologie in einer kompakten tragbaren Form. Der Bildschirm liegt erhaben auf der Oberseite, während die Ränder abgeflacht sind. Dadurch lässt sich der Player bequem anfassen, zudem wirkt das Gerät insgesamt schlanker, als es eigentlich ist. Auch der Aluminium-Lautstärkeregler an der Seite ist hervorgehoben und dadurch einfach zu erkennen und zu bedienen. Insgesamt wurde die Handhabung derart vereinfacht, dass sie den Gewohnheiten von Smartphone-Benutzern entspricht und demnach intuitiv erfolgen kann. Dieses Konzept spiegelt sich auch in der schlichten Anmutung des Players wider.

Statement by the jury
The AK Junior is reduced to the bare essentials and has a coherent, uniform appearance. Another convincing feature is the state-of-the-art operating concept.

Begründung der Jury
Der AK Junior ist auf das Wesentliche reduziert und wirkt wie aus einem Guss. Auch sein zeitgemäßes Bedienkonzept überzeugt.

AK120 II
Portable Audio Player
Tragbarer Audio-Player

Manufacturer
Iriver, Seoul, South Korea
In-house design
Jungbeom Han, Baekjin Seong
Web
www.iriver.com

The AK120 II audio player was designed with the objective of creating the perfect combination of performance and design. The volume knob was placed all the way at the top so that the space for the rechargeable battery could be maximised. The knob remains accessible and easy to operate, even when the player is held with both hands. The aluminium casing with its large glass display has a sophisticated appearance, provides a pleasant tactile feel and underscores the quality of the player.

Der Audio-Player AK120 II wurde mit der Maßgabe entworfen, Leistung und Design perfekt zu kombinieren. Damit der Akku möglichst viel Platz hat, wurde der Lautstärkeregler an der obersten Stelle platziert. Dadurch ist er zudem leicht zugänglich und zu bedienen, selbst wenn das Gerät mit beiden Händen gehalten wird. Das Gehäuse aus Aluminium mit dem großen Glasdisplay wirkt edel, bietet eine angenehme Haptik und unterstreicht die Qualität des Players.

Statement by the jury
The AK120 II is a player with excellent workmanship, which leaves nothing to be desired for music lovers with a high appreciation of quality.

Begründung der Jury
Der AK120 II ist ein vorzüglich verarbeiteter Player, der bei Musikliebhabern mit Sinn für Qualität keine Wünsche offen lässt.

NW-ZX2
Digital Media Player
Digitaler Media-Player

Manufacturer
Sony Corporation, Tokyo, Japan
In-house design
Shogo Yashiro
Web
www.sony.net

The NW-ZX2 media player combines a range of advanced audio technologies to reproduce media files with top quality. The compact body is made of milled aluminium for maximum rigidity. The soft texture on the back provides pleasant haptics and enhances comfortable operation as well. Another detail accentuating the high quality of this player is the range of precisely manufactured aluminium buttons.

Der Media-Player NW-ZX2 vereinigt in sich eine Reihe fortschrittlicher Audiotechnologien, um Mediendateien in bester Qualität wiederzugeben. Das kompakte Gehäuse besteht aus gefrästem Aluminium und besitzt dadurch eine hohe Steifigkeit. Die weiche Textur der Rückseite sorgt für eine angenehme Haptik und macht zusätzlich die Bedienung komfortabler. Ein weiteres Detail, das die Wertigkeit dieses Players unterstreicht, sind die präzise gefertigten Aluminiumtasten.

AK500N
Network Audio Player
Netzwerk-Audioplayer

Manufacturer
Iriver, Seoul, South Korea
In-house design
Baekjin Seong, Eunkyung Lee
Design
Metal Sound Design (Kukil Yu), Seoul, South Korea
Web
www.iriver.com
www.ilmsd.com

The AK500N network audio player shows a striking three-dimensional front, which lends it a high level of autonomy. It was designed for music lovers who place value on space-filling sound in hi-fi quality. When the player is connected to high-end amplifiers and speakers, it reproduces sound in concert quality. The player also enables the comfortable copying and storage of music CDs; data are simply saved in its SSD memory. As the player is exclusively operated via a rechargeable battery, it offers an optimum signal-to-noise ratio. Thanks to the foldable touch display, the device is very user-friendly.

Der Netzwerk-Audioplayer AK500N zeigt eine markante dreidimensionale Front, die ihm eine hohe Eigenständigkeit verleiht. Er wurde für Musikliebhaber entworfen, die Wert auf einen raumfüllenden Klang in Hi-Fi-Qualität legen. Wird der Player an High-End-Verstärker und -Lautsprecher angeschlossen, gibt er Klang in Konzertqualität wieder. Daneben ermöglicht er das komfortable Auslesen und Speichern von Musik-CDs, deren Daten auf einer SSD gespeichert werden. Da der Player ausschließlich über einen Akku betrieben wird, bietet er einen optimalen Signal-Rausch-Abstand. Dank der Touchbedienung mit klappbarem Display ist er sehr benutzerfreundlich.

DDA-LA20RC
Portable Headphone Amplifier
Tragbarer Kopfhörerverstärker

Manufacturer
Deff Corporation, Osaka, Japan
In-house design
Taku Yahara
Web
www.deff.co.jp

This portable headphone amplifier was created with the aspiration to convey high sound quality solely through the design itself. Its round shape rests nicely in the hand and, at the same time, it offers surface area for an intuitive four-way control unit. The device is small, for instead of containing a rechargeable battery it is powered by the connected audio device. Except for the clip, the amplifier is entirely made of aluminium. It is available in black, white and silver and thus blends well with most device and cable colours.

Der tragbare Kopfhörerverstärker wurde mit dem Anspruch geschaffen, schon allein durch die Erscheinung eine hohe Klang-qualität zu vermitteln. Seine runde Form schmeichelt der Hand und bietet gleichzeitig die Fläche für eine intuitive Vier-Wege-Steuerung. Das Gerät ist klein, da es keinen Akku besitzt, sondern den Strom vom angeschlossenen Audiogerät bezieht. Bis auf den Clip besteht das ganze Gerät aus Aluminium. Es ist in Schwarz, Weiß und Silber erhältlich und passt dadurch zu den gängigen Geräte- und Kabelfarben.

Statement by the jury
With its round shape and high-grade aluminium casing, the headphone amplifier lies comfortably in the hand and is an eye-catcher on the table as well.

Begründung der Jury
Mit seiner runden Form und dem wertig anmutenden Aluminiumgehäuse liegt der Kopfhörerverstärker angenehm in der Hand und wird auch auf einem Tisch zum Blickfang.

S900
Bluetooth Noise Cancelling NFC Headset
Bluetooth-NFC-Headset mit Geräuschunterdrückung

Manufacturer
Shenzhen Rapoo Technology Co., Ltd., Shenzhen, China
In-house design
Qingjie Hu
Web
www.rapoo.com

The S900 is a high-end Bluetooth headset with efficient noise cancellation in elegant design. Its carefully machined aluminium construction lends it a modern appeal, and the leather earcups guarantee high wearing comfort. Connectivity is established via NFC, while Bluetooth 4.1 enables reliable wireless signal transfer in top quality. Lossless aptX audio encoding and decoding produces clear, realistic sound.

Statement by the jury
Aluminium and leather give this headset a precious appearance. In terms of wearing comfort, the product fully convinces as well.

Das S900 ist ein High-End-Bluetooth-Headset mit wirkungsvoller Störgeräusch-unterdrückung im eleganten Design. Seine sorgfältige Ausführung in Aluminium verleiht ihm eine moderne Anmutung, die Ohrmuscheln aus Leder sorgen für einen hohen Tragekomfort. Die Kopplung erfolgt über NFC, Bluetooth bietet eine zuverlässige kabellose Signalübertragung in erstklassiger Qualität. Die verlustfreie aptX-Audiocodierung gewährleistet einen klaren, realistischen Klang.

Begründung der Jury
Aluminium und Leder verleihen diesem Headset eine wertige Anmutung. Auch in puncto Tragekomfort kann das Produkt vollends überzeugen.

QuietComfort® 25 Acoustic Noise Cancelling® Headphones
Kopfhörer mit Geräusch-unterdrückung

Manufacturer
Bose Corporation, Framingham, USA
In-house design
Web
www.bose.com

These headphones offer a unique combination of noise reduction, high quality sound and comfort for music and calls. They feature measurement microphones, both inside and outside the earcups, which register unwanted noise. A precise equal and opposite noise cancellation signal is calculated and distracting noise is thus immediately cancelled out, even in demanding environments. The materials selected, the headband shape and the soft earpads all result in high wearing comfort.

Statement by the jury
Sophisticated technical features and well-conceived design contribute to an all-around successful premium product.

Der Kopfhörer bietet eine einzigartige Kombination aus Lärmreduzierung, Spitzenklang und Komfort für Musik und Telefonate. Er besitzt im Inneren der Hörmuscheln sowie außen Messmikrofone, die unerwünschte Geräusche registrieren. Ein präzises Schall-Gegenschall-Signal wird errechnet, und Störgeräusche werden so selbst unter schwierigen Bedingungen sofort ausgeblendet. Für einen hohen Tragekomfort sorgen Materialien und Form des Kopfbügels sowie die weichen Ohrpolster.

Begründung der Jury
Die ausgefeilten technischen Eigenschaften und das Design machen diesen Kopfhörer zu einem rundum gelungenen hochwertigen Produkt.

SoundLink® On-Ear Bluetooth® Headphones
Bluetooth-Kopfhörer

Manufacturer
Bose Corporation, Framingham, USA
In-house design
Web
www.bose.com

The SoundLink on-ear Bluetooth headphones deliver natural, powerful sound, offer new functions and display an innovative design. The wireless connection allows for complete freedom when listening to music, watching movies or taking phone calls. Made from impact-resistant materials, the headphones are particularly suitable for an active lifestyle. They are operated intuitively via the control keys located on the earcup. A practical feature for safe storage is the folding mechanism.

Statement by the jury
The headphones impress with their reduced weight and comfortable wearing characteristics. This is accompanied by dynamic sound with powerful bass and clear treble.

Der Bluetooth-Kopfhörer SoundLink bietet naturgetreuen, kraftvollen Klang, neue Funktionen und ein innovatives Design. Die kabellose Verbindung ermöglicht zudem die völlige Freiheit beim Musikhören, Videoschauen oder Telefonieren. Der Kopfhörer ist aus strapazierfähigen Materialien gefertigt und dadurch besonders für einen aktiven Lebensstil geeignet. Er wird intuitiv über die Steuertasten der Hörmuschel bedient. Für die Aufbewahrung praktisch ist der Faltmechanismus.

Begründung der Jury
Der Kopfhörer besticht durch sein leichtes Gewicht und komfortable Trageeigenschaften. Dazu gesellt sich ein dynamischer Klang mit kraftvollen Bässen und klaren Höhen.

aëdle VK-1 Legacy edition
Headphones
Kopfhörer

Manufacturer
aëdle SAS, Bois-Colombes, France
In-house design
Baptiste Sancho, Raphaël Lebas de Lacour
Design
Eugeni Quitllet, Barcelona, Spain
FLOZ Studio, Paris, France
Web
www.aedle.net

The VK-1 Legacy edition headphones are characterised by their timelessly iconic design. Each set of headphones is hand-assembled in France and thus unique. Their exclusive quality is underlined by high-grade materials: the drivers are fashioned from titanium-neodymium, the earcups are CNC-milled from a single block of aluminium, and the ear cushions are covered with genuine lambskin leather. Each detail is manufactured with the objective of achieving the highest possible degree of robustness and reliability.

Statement by the jury
The headphones embody high quality and exclusivity in all respects. The design is unobtrusive, yet still characterised by timeless elegance.

Der Kopfhörer VK-1 Legacy edition lebt von seiner zeitlos ikonischen Gestaltung. Jeder einzelne Kopfhörer wird in Frankreich handgefertigt und ist damit ein Unikat. Unterstrichen wird seine exklusive Qualität durch hochwertige Materialien: Die Treiber bestehen aus Titan-Neodymium, die Ohrmuscheln wurden aus einem Block Aluminium gefräst, und die Ohrpolster wurden aus Lammleder genäht. Jedes Detail wird nach der Maßgabe gefertigt, eine größtmögliche Robustheit und Zuverlässigkeit zu erzielen.

Begründung der Jury
Der Kopfhörer verkörpert in jeder Hinsicht Hochwertigkeit und Exklusivität. Die Formensprache ist dabei unaufdringlich und doch von einer zeitlosen Eleganz.

AKG Y50
Headphones
Kopfhörer

Manufacturer
Harman International Industries, Northridge, USA
In-house design
Harman Design Team (Rafael Czaniecki)
Web
www.harman.com

The AKG Y50 has a stainless-steel headband, which, due to its self-contained design, does not compromise sound quality. The other components are fashioned from premium materials as well. The earcups are made of aluminium, and the hinges consist of fibreglass-reinforced polycarbonate. Ear and head cushions are made from memory foam to enhance the durability and maximise the wearing comfort of these headphones.

Statement by the jury
All headphone components are manufactured from high-grade materials. This ensures comfortable use and reliability.

Der AKG Y50 besitzt einen Kopfbügel aus Edelstahl, der dank des geschlossenen Gehäusedesigns keine Kompromisse bei der Klangqualität eingeht. Auch die anderen Komponenten bestehen aus ausgewählten Materialien. So sind die Hörmuscheln aus Aluminium gefertigt und die Gelenke aus mit Glasfaser verstärktem Polykarbonat. Die Polsterungen bestehen aus Memory Schaum, was den Tragekomfort des Kopfhörers erhöht und zur Langlebigkeit beiträgt.

Begründung der Jury
Für jede Komponente dieses Kopfhörers wurden hochwertige Materialien verwendet. Dies gewährleistet eine bequeme Nutzung und Zuverlässigkeit.

AKG N60NC
Headphones
Kopfhörer

Manufacturer
Harman International Industries, Northridge, USA
In-house design
Harman Design Team (Rafael Czaniecki, Gabriel Djohar)
Web
www.harman.com

The AKG N60NC is distinguished by a high-quality headband with soft ear cushions made of fine leather. The comfortable fit and reduced weight provide high wearing comfort so that the headphones can be worn for many hours on end. Moreover, passive and active noise cancellation guarantees outstanding sound. Thanks to the 3D-axis folding mechanism, the headphones are easily stowed away for transport.

Statement by the jury
With their smooth design, these headphones render a comfortable impression. The metal finishing fosters a premium exterior appearance.

Der AKG N60NC zeichnet sich durch einen hochwertigen Kopfbügel mit weichen Polsterungen aus feinem Leder aus. Die bequeme Passform und das geringe Gewicht bieten einen hohen Komfort, sodass sich der Kopfhörer über viele Stunden tragen lässt. Darüber hinaus sorgen die passive und aktive Geräuschunterdrückung für einen hervorragenden Klang. Dank des 3D-Faltmechanismus kann der Kopfhörer einfach transportiert und aufbewahrt werden.

Begründung der Jury
Durch seine glatte Gestaltung wirkt der Kopfhörer bequem. Die Metallanmutung verleiht ihm ein hochwertiges Äußeres.

Phiaton Chord MS 530
Headphones
Kopfhörer

Manufacturer
Phiaton Corporation, Irvine, USA
Design
TEAGUE (Youjin Nam, Tad Toulis, Devin Liddell, Clement Gallois, John Mabry, Benoit Collette, Roger Jackson, Ashley Newcomer), Seattle, USA
Web
www.phiaton.com
www.teague.com

With a dual-chamber structure and new audio technology, the Phiaton Chord MS 530 headphones facilitate a studio-grade sound experience and are thus particularly suitable for demanding music lovers. The equipment includes Bluetooth 4.0, switchable active noise cancellation, a built-in rechargeable battery and an additional wired option. Genuine perforated leather and machined aluminium components guarantee elegance and high wearing comfort.

Statement by the jury
The high-grade materials selected underscore the sound quality of these headphones and also convey a solid, valuable impression.

Der Kopfhörer Phiaton Chord MS 530 ermöglicht dank seiner Zwei-Kammern-Struktur und einer neuen Audio-Technologie ein Klangerlebnis auf Studio-Niveau und eignet sich damit besonders für anspruchsvolle Musikliebhaber. Zur Ausstattung zählen Bluetooth 4.0, eine zuschaltbare aktive Geräuschunterdrückung, ein eingebauter Akku sowie ein optionales Kabel. Durch perforiertes Echtleder und gefrästes Aluminium bietet der Kopfhörer Eleganz und einen hohen Tragekomfort.

Begründung der Jury
Die hochwertigen Materialien unterstreichen die Klangqualität dieses Kopfhörers und vermitteln darüber hinaus einen soliden, wertigen Eindruck.

Philips A1 Professional DJ Headphones
Headphones
Kopfhörer

Manufacturer
Gibson Innovations, Hong Kong
In-house design
Web
www.gibson.com

The A1 Professional DJ headphones were developed for DJs who want to compose and mix music on stage as well as at home. They consist of durable materials and transfer classic headphone elements into a contemporary form. Both lightweight and robust, they offer comfort and flexibility. The cushions are interchangeable and allow for different uses. If the DJ wants to listen to samples during production work, he can manually fold an earcup upward for quick pickup to the ear.

Statement by the jury
The headphones render a high-quality, reliable impression and convey the degree of professionalism required for DJ'ing.

Der A1 Professional DJ Kopfhörer wurde für DJs entwickelt, die auf der Bühne wie auch zu Hause Musik komponieren und mischen möchten. Er besteht aus langlebigen Materialien und überträgt klassische Kopfhörerelemente in eine zeitgemäße Form. Da er sowohl leicht als auch robust ist, bietet er Komfort und Flexibilität. Die Polster sind austauschbar und gestatten unterschiedliche Verwendungen. Möchte der DJ während der Produktion Samples anhören, kann er eine Ohrmuschel schnell hochklappen und ans Ohr führen.

Begründung der Jury
Der Kopfhörer macht einen hochwertigen, verlässlichen Eindruck und vermittelt dabei die Professionalität, die für den DJ-Bereich angemessen ist.

Philips Fidelio NC1
Noise Cancelling Headphones
Kopfhörer mit Geräusch-unterdrückung

Manufacturer
Gibson Innovations, Hong Kong
In-house design
Web
www.gibson.com

The Fidelio NC1 are high-performance headphones with active noise cancellation for hi-fi music playback on the go. The lightweight earcups are made of aluminium and guarantee controlled noise cancellation. The soft-foam ear cushions covered in fabric and leather offer enhanced wearing comfort. The headphones can be folded up in different ways and thus fit into any bag.

Statement by the jury
The Fidelio NC1 headphones stand for music enjoyment in top quality. With the aluminium design accents they have a self-contained appearance.

Der Fidelio NC1 ist ein Hochleistungskopfhörer mit aktiver Geräuschunterdrückung für die Hi-Fi-Musikwiedergabe unterwegs. Die leichten Ohrmuscheln bestehen aus Aluminium und gewährleisten die kontrollierte Geräuschunterdrückung. Dabei sorgen die Ohrpolster aus weichem Schaumstoff mit ihrem Bezug aus Stoff und Leder für einen hohen Tragekomfort. Da sich der Kopfhörer auf unterschiedliche Arten zusammenklappen lässt, passt er in jede Tasche.

Begründung der Jury
Der Kopfhörer Fidelio NC1 steht für Musikgenuss in erstklassiger Qualität und zeigt zudem mit Akzenten aus Aluminium eine eigenständige Gestaltung.

Sennheiser URBANITE
Headphones
Kopfhörer

Manufacturer
Sennheiser electronic GmbH & Co. KG,
Wedemark, Germany
In-house design
Design
designaffairs GmbH, Munich, Germany
Web
www.sennheiser.com
www.designaffairs.com

The Urbanite series is a range of closed headphones designed in a robust urban style. They come in six individual designs with colour-coordinated, fabric-covered headbands and high-grade, stainless-steel hinges. The folding mechanism and exchangeable cable guarantee comfortable handling, even when on the go. Soft earpads ensure insulation from external noise. The Urbanite is available in both over-ear and on-ear versions.

Statement by the jury
The Urbanite offers a high degree of comfort and useful details en route. In addition, the headphones cleverly combine materials such as stainless steel and fabric in different colours.

Die geschlossenen Kopfhörer der Urbanite-Reihe wurden in einem robusten urbanen Stil gestaltet. Die Serie besteht aus sechs individuellen Designs mit farblich abgestimmten, stoffüberzogenen Kopfbügeln und hochwertigen Gelenken aus Edelstahl. Der Faltmechanismus und das austauschbare Kabel garantieren auch unterwegs eine bequeme Handhabung. Für einen von Umgebungsgeräuschen abgeschirmten Hörgenuss sorgen die weichen Ohrpolster. Der Urbanite ist als Over-Ear-und On-Ear-Variante erhältlich.

Begründung der Jury
Der Urbanite bietet einen hohen Komfort und nützliche Details für unterwegs. Zudem kombiniert er geschickt die Materialien Edelstahl und Stoff in unterschiedlichen Farben.

Tt eSPORTS SHOCK 3D 7.1
Gaming Headset

Manufacturer
Thermaltake Technology Co., Ltd.,
Taipei, Taiwan
In-house design
Web
www.thermaltakecorp.com

Thanks to the latest 3D surround sound technology, the Tt eSports Shock 3D 7.1 gaming headset allows users to immerse themselves deeply into the gaming world. The microphone automatically mutes when pivoted upward. Four equaliser settings serve to fine-tune the sound for games, music or movies. Buttons on the earcup enable the user to turn the 3D 7.1 surround sound on and off and to change equaliser settings quickly and conveniently at any time.

Statement by the jury
This gaming headset is easily configured and has a comfortable fit. Its dynamic appearance is optimally attuned to gamers' demands.

Das Gaming-Headset Tt eSports Shock 3D 7.1 lässt Spieler dank verbessertem virtuellem 3D-Surround-Sound tief in die Spielwelt eintauchen. Das Mikrofon wird automatisch stumm gestellt, wenn es hochgeklappt ist. Über vier Equalizer-Einstellungen wird der Ton fein geregelt und auf Spiele, Musik oder Filme abgestimmt. Mit Reglern an der Ohrmuschel kann der Anwender den 3D-7.1-Surround-Sound an- und ausschalten und die Equalizer-Einstellungen schnell und bequem zu jeder Zeit anpassen.

Begründung der Jury
Das Gaming-Headset lässt sich leicht konfigurieren und bequem tragen. Seine dynamische Erscheinung ist optimal auf die Anforderungen von Gamern abgestimmt.

Cronos AD
Gaming Headset

Manufacturer
Thermaltake Technology Co., Ltd.,
Taipei, Taiwan
In-house design
Web
www.thermaltakecorp.com

The Cronos AD gaming headset has an auto-adjusting headband and thus fits any head size and shape. In addition, it has a secure and comfortable fit, which is an advantage in the case of longer gaming sessions. Lycra padding, LED illumination and a retractable microphone are also integrated into the headset. As a special feature, the product is delivered with two removable cables: one for use with a PC and one for connecting the headset to mobile devices.

Statement by the jury
This headset offers outstanding wearing comfort thanks to the adjustable headbands and padded earcups. The two-cable solution and retractable microphone round out the overall favourable impression.

Das Gaming-Headset Cronos AD besitzt einen selbstjustierenden Kopfbügel und passt dadurch auf jede Kopfgröße und -form. Zudem sitzt es sicher und bequem, was für längere Gaming-Sessions von Vorteil ist. Die Polster aus Lycra, LED-Beleuchtung und das einziehbare Mikrofon sind weitere Merkmale des Headsets. Als Besonderheit wird das Produkt mit zwei abnehmbaren Kabeln geliefert: eins für die Benutzung am Computer und eins für den Anschluss an mobilen Geräten.

Begründung der Jury
Das Headset bietet dank der verstellbaren Kopfbügel und gepolsterten Ohrstücke hervorragende Trageeigenschaften. Die Zwei-Kabel-Lösung und das einziehbare Mikrofon runden den guten Eindruck ab.

RS 175
Digital Wireless Headphone System
Digitales Funkkopfhörersystem

Manufacturer
Sennheiser electronic GmbH & Co. KG, Wedemark, Germany
Design
Brennwald Design, Kiel, Germany
Web
www.sennheiser.com
www.brennwald-design.de

The RS 175 headphone system offers excellent wireless sound with impressive bass and surround modes for an extraordinary listening experience. The novel shape of the earpiece makes geometrical and ergonomic reference to the human ear and highlights the technical innovativeness. All functions can be intuitively located and operated from the right earpiece. The transmitter is both a docking station and a stylish storage solution.

Statement by the jury
The headphones and the transmitter form a harmonious, self-contained unit, underscoring the excellent sound qualities of the system.

Das Kopfhörersystem RS 175 bietet einen ausgezeichneten kabellosen Klang mit beeindruckendem Bass und Surround-Sound-Modi für außergewöhnlichen Hörgenuss. Die neuartige Form der Hörmuschel nimmt in Geometrie und Ergonomie Bezug auf das menschliche Ohr und betont die technologische Innovation. Alle Funktionen können intuitiv an der rechten Hörmuschel ertastet und bedient werden. Der Sender ist gleichzeitig Ladestation und stilvolle Ablage.

Begründung der Jury
Kopfhörer und Sender bilden eine harmonische Einheit, die eigenständig wirkt und die exzellenten Klangeigenschaften des Systems unterstreicht.

RS 185
Digital Wireless Headphone System
Digitales Funkkopfhörersystem

Manufacturer
Sennheiser electronic GmbH & Co. KG, Wedemark, Germany
Design
Brennwald Design, Kiel, Germany
Web
www.sennheiser.com
www.brennwald-design.de

The RS 185 digital headphone system was designed for users with high audio and design demands. Their acoustic open construction and ergonomically optimised design with the characteristic earpiece shape enable precise, detailed sound reproduction and comfortable wear for many hours. With the manual level control for a precise sound reproduction, the RS 185 is the right choice for discerning users.

Statement by the jury
The wireless headphone system impresses with its generously designed earpieces promising high wearing comfort and outstanding sound.

Das digitale Funkkopfhörersystem RS 185 wurde für Benutzer entworfen, die hohe Ansprüche an Klang und Design stellen. Seine akustisch offene Bauweise und die ergonomisch optimierte Gestaltung mit der charakteristischen Ohrmuschelform ermöglichen eine präzise, detaillierte Klangwiedergabe und stundenlanges, entspanntes Tragen. Mit der manuellen Pegeleinstellung für eine präzise Klangwiedergabe ist der RS 185 die richtige Wahl für anspruchsvolle Benutzer.

Begründung der Jury
Das Funkkopfhörersystem beeindruckt mit seinen großzügig ausgelegten Ohrmuscheln, die hohen Komfort und eindrucksvollen Klang versprechen.

RS 195
Digital Wireless Headphone System
Digitales Funkkopfhörersystem

Manufacturer
Sennheiser electronic GmbH & Co. KG, Wedemark, Germany
Design
Brennwald Design, Kiel, Germany
Web
www.sennheiser.com
www.brennwald-design.de

Thanks to its ergonomically shaped earpieces, the RS 195 digital headphone system can be worn for many hours. All daily relevant operating functions can be intuitively controlled at the right earpiece. The additional control elements on the transmitter were designed as analogue dials for ergonomic reasons to guarantee best possible usability and visual feedback with regard to the switching status. Overall, it is very easy for the user to optimize the system for crystal clear speech and music.

Statement by the jury
The RS 195 has an ergonomic design and is equipped with clearly aligned control elements. Analogue dials are arranged in a sensible and user-friendly way.

Dank der ergonomisch geformten Ohrmuscheln lässt sich das digitale Funkkopfhörersystem RS 195 über mehrere Stunden bequem tragen. Alle wesentlichen Bedienfunktionen können an der rechten Hörmuschel intuitiv bedient werden. Die zusätzlichen Bedienelemente am Sender wurden aus ergonomischen Erwägungen als analoge Drehregler konzipiert, um eine bestmögliche Bedienbarkeit und visuelle Rückmeldung über den Status zu gewährleisten und das System für kristallklare Sprach- und Musikwiedergabe zu optimieren.

Begründung der Jury
Das RS 195 ist ergonomisch gestaltet und mit übersichtlichen Bedienelementen ausgestattet. Analoge Regler sind sinnvoll und benutzerfreundlich angeordnet.

Yurbuds Inspire & Focus 300 & 400 Series (powered by JBL)
Headphones
Kopfhörer

Manufacturer
Harman International Industries,
Northridge, USA
In-house design
Harman Design Team (Effrosini Karayiannis)
Web
www.harman.com

The headphones of this series are equipped with ear hooks guaranteeing a secure fit during all sorts of activities. The FlexSoft fitting technology ensures comfortable wear, even over extended periods of time. For simple transport, the headphones are equipped with a patented magnet system. In addition, they deliver full, bass-driven sound.

Statement by the jury
These headphones convince with their excellent fit. They are particularly suitable for athletes, a quality that is highlighted by their dynamic colouring.

Die Kopfhörer dieser Serie sind mit einem Ohrbügel ausgestattet, der bei allen Aktivitäten für einen sicheren Sitz sorgt. Zudem gewährleistet die FlexSoft-Passform, dass sie sich auch über einen längeren Zeitraum angenehm tragen lassen. Für den einfachen Transport sind die Kopfhörer mit einem patentierten Magnetsystem versehen. Darüber hinaus liefern sie einen satten, bassbetonten Klang.

Begründung der Jury
Die Kopfhörer überzeugen mit ihrer ausgezeichneten Passform. Sie eignen sich besonders für Sportler, was auch ihre dynamische Farbgebung unterstreicht.

Yurbuds Inspire & Focus 100 & 200 Series
Headphones
Kopfhörer

Manufacturer
Harman International Industries,
Northridge, USA
In-house design
Harman Design Team (Effrosini Karayiannis)
Web
www.harman.com

These in-ear headphones are available in two models – Focus features ear hooks, while Inspire does without them – thus covering a broad range of different demands. Both versions provide an optimum fit, guaranteeing convenient and secure wearing comfort. Ambient sound is not completely blocked, so that the user can still perceive the surrounding environment and stay safe.

Statement by the jury
With their soft earbuds, the headphones of this series are comfortable and yet provide a firm fit, always holding the headphones in place, even during intense workouts.

Der In-Ear-Kopfhörer ist wahlweise mit Ohrbügeln als Modell Focus, oder ohne Ohrbügel als Modell Inspire erhältlich. Dadurch deckt die Serie die unterschiedlichsten Ansprüche ab. Gemeinsam ist beiden Varianten die optimale Passform, die einen bequemen und sicheren Sitz garantiert. Umgebungsgeräusche werden nicht vollkommen abgeblockt, sodass der Benutzer sein Umfeld wahrnimmt und seine Sicherheit gewährleistet bleibt.

Begründung der Jury
Die Kopfhörer der Serie sitzen mit ihren weichen Ohrstöpseln komfortabel und doch so fest, dass sie auch beim Sport immer an ihrem Platz bleiben.

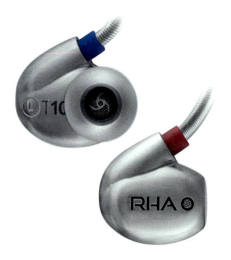

RHA T10i
In-Ear Headphones
In-Ear-Kopfhörer

Manufacturer
RHA, Glasgow, Great Britain
In-house design
Lewis Heath
Web
www.rha.co.uk

The T10i in-ear headphones combine pioneering technologies, materials and manufacturing processes to set a high standard in the field of portable audio devices. It features a handmade driver in an ergonomically shaped stainless-steel enclosure. The headphones also feature a mouldable over-ear hook consisting of a stainless steel spring and firm steel wire; the hooks can be adapted to the shape of any ear to ensure a high level of comfort.

Der In-Ear-Kopfhörer T10i vereint zukunfts- weisende Technologien, Materialien und Fertigungsprozesse und setzt so einen ho- hen Standard bei tragbaren Audiogeräten. Er verfügt über einen handgefertigten Trei- ber in einem ergonomisch geformten Edel- stahlgehäuse. Der Kopfhörer besitzt zudem formbare Ohrbügel, die aus einer Stahlfeder und festem Stahldraht bestehen. Dadurch können sie an jede Ohrform angepasst wer- den, um einen hohen Komfort zu garantie- ren.

Statement by the jury
With its stainless-steel driver enclosure and mouldable ear hooks made of steel components, the headphones convey premium quality and durability.

Begründung der Jury
Mit seinem Treibergehäuse aus Edelstahl und den formbaren Ohrbügeln aus Stahl- komponenten vermittelt der Kopfhörer eine hohe Wertigkeit und Langlebigkeit.

Jabra Sport Pulse Wireless
Wireless Sports Earbuds
Kabelloser Sportkopfhörer

Manufacturer
GN Netcom A/S, Jabra, Ballerup, Denmark
Design
Howl Designstudio (Jonas Samrelius), Stockholm, Sweden
Web
www.jabra.com
www.howlstudio.se

Jabra Sport Pulse Wireless are all-in-one earbuds with an integrated heart rate monitor. The heart rate is measured directly in the inner ear so that users can monitor their performance and establish personal training zones according to the heart rate. With the accompanying Jabra Sport Life app, individual goals can be defined based on distance, time or calorie consumption and the training session subsequently evaluated through diagrams. Furthermore, the earbuds are sweat- and weather-resistant.

Der Jabra Sport Pulse Wireless ist ein All-in-one-in-Ear-Kopfhörer mit integrier- tem Pulsmesser. Der Puls wird direkt im Innenohr gemessen, damit der Anwender während des Trainings die eigene Leistung überwachen und anhand des Pulswertes persönliche Trainingszonen erstellen kann. Mit der dazugehörigen Jabra Sport Life App kann er individuelle Ziele wie Entfernung, Zeit oder Kalorienverbrauch festlegen und anhand von Diagrammen das Training an- schließend auswerten. Zudem sind die Ohr- stöpsel schweiß- und wetterbeständig.

Statement by the jury
With the heart rate monitor feature, the Jabra Sport Pulse Wireless is a useful accessory for athletes. Thanks to the compact and lightweight construction, the headphones fit securely and com- fortably.

Begründung der Jury
Der Jabra Sport Pulse Wireless bietet mit dem Pulsmesser eine sinnvolle Ergänzung für Sportler. Dank seiner kompakten und leichten Bauweise sitzt er sicher und be- quem.

JBL® Reflect Aware™
Headphones
Kopfhörer

Manufacturer
Harman International Industries, Northridge, USA
In-house design
Harman Design Team (Joshua Fischer, Bochao Yang)
Web
www.harman.com

The JBL Reflect Aware headphones do not require a battery and thus count among the lightest headphones with noise can- cellation. They can be controlled via a microphone and are compatible with all smartphones thanks to a lightning con- nector and micro-USB converter. Reflec- tive cables contribute to better visibility of the user in dim lighting conditions. Moreover, the headphones are made of sweat-proof material and thus particu- larly suited for athletes.

Der JBL Reflect Aware benötigt keinen Akku und ist dadurch einer der leichtesten Kopf- hörer mit Geräuschunterdrückung. Er kann über das Mikrofon gesteuert werden und ist dank seines Lightning-Anschlusses und eines Mikro-USB-Adapters mit allen Smartphones kompatibel. Da der Kopfhörer reflektierende Kabel besitzt, wird der Be- nutzer bei Dunkelheit leichter gesehen. Zu- dem ist er aus schweißresistentem Material gefertigt und eignet sich daher besonders für Sportler.

Statement by the jury
These headphones are perfectly suited for users with an active lifestyle. They are lightweight, anatomically shaped and compatible with many devices.

Begründung der Jury
Der Kopfhörer ist bestens für Benutzer mit einem aktiven Lebensstil geeignet. Er ist leicht, anatomisch geformt und mit vielen Geräten kompatibel.

Normal Earphones
3D Printed Earphones
In-Ear-Kopfhörer aus dem
3D-Drucker

Manufacturer
Normal, New York, USA
Design
Impel Studio LLC, New York, USA
Web
www.nrml.com
www.impelstudio.com

Normals are premium earphones that are individually 3D printed to securely fit into each customer's ear. The concentric design language of the CNC'd aluminium housing harmonises with the pivoting coaxial cable connectors and the high-quality 14 mm dynamic driver inside. Each pair has a fully customisable design – from the colour choice of components to the cable length and the laser-engraving of the innovative pop-up carrying case.

Statement by the jury
Innovative 3D printing technology is intelligently used in creating these headphones to deliver a tailor-made product for every single customer.

Normals sind hochwertige In-Ear-Kopfhörer, die individuell mit einem 3D-Drucker produziert werden. Dadurch passen sie sicher in das Ohr eines jeden Nutzers. Die konzentrische Formensprache des CNC-gefrästen Aluminiumgehäuses harmoniert mit den drehbaren Koaxial-Steckverbindungen und den hochwertigen dynamischen 14-mm-Wandlern im Innern. Jedes Paar hat ein vollständig anpassbares Design, von der Farbwahl der Komponenten über die Kabellänge bis hin zur Lasergravur des Pop-up-Tragekästchens.

Begründung der Jury
Die innovative 3D-Drucktechnologie wird bei diesen Kopfhörern intelligent eingesetzt, um jedem Kunden ein maßgeschneidertes Produkt zu liefern.

KEF M100
Hi-Fi Earphones
Hi-Fi-Kopfhörer

Manufacturer
GP Acoustics (UK) Ltd, Kent, Great Britain
Design
GP Acoustics International Ltd, Product Planning & Design Department, Hong Kong
Web
www.kef.com

The M100 earphones are lightweight and robust. Their aluminium casing is machined with diamond-cut chamfered edges and contains a series of technologies for optimum sound enhancement. Each earpiece features a high-performance 10 mm neodymium driver to deliver excellent dynamics in all frequency ranges. The very thin silicone eartips and ergonomically shaped shells guarantee a comfortable fit without compromising sound quality.

Statement by the jury
These earphones are precisely machined with attention to detail and have a high-grade appearance. They not only deliver outstanding sound but also have a sophisticated look.

Der Kopfhörer M100 ist leicht und robust, sein Aluminiumgehäuse ist mit diamantgeschnittenen abgeschrägten Kanten gefertigt und beherbergt eine Reihe von Technologien zur Klangverbesserung. In beiden Ohrstöpseln arbeitet ein leistungsstarker 10-mm-Neodym-Treiber, der in allen Tonbereichen eine ausgezeichnete Dynamik bietet. Die sehr dünnen Ohreinsätze aus Silikon und die ergonomisch geformten Muscheln garantieren eine gute Passform ohne Klangverlust.

Begründung der Jury
Die Kopfhörer sind im Detail äußerst präzise und wertig ausgeführt. Dadurch liefern sie nicht nur einen hervorragenden Klang, sondern sehen auch edel aus.

BackBeat FIT
Sports Headset
Sport-Headset

Manufacturer
Plantronics, Inc., Santa Cruz, USA
In-house design
Web
www.plantronics.com

The BackBeat FIT is a Bluetooth stereo headset for athletes. Thanks to miniaturised components, a flexible neckband and wireless design, it is compact and comfortable. Easily accessible buttons on the headset enable a simple operation and the intuitive management of music and phone calls. Bright colours and the light-reflecting surface increase the visibility of athletes in the dark. The nano-coating protects the headset against moisture and makes it robust and durable.

Statement by the jury
This headset is an outstanding all-round device for athletes. It offers useful functions, an ergonomic fit and the highest possible degree of freedom during a workout.

Das BackBeat FIT ist ein Bluetooth-Stereo-Headset für Sportler. Dank verkleinerter Komponenten, flexiblem Nackenband und schnurlosem Design ist es kompakt und bequem. Leicht zugängliche Tasten am Headset ermöglichen die einfache Bedienung und das intuitive Management von Musik und Telefongesprächen. Leuchtende Farben und die lichtreflektierende Oberfläche sorgen dafür, dass Sportler bei Dunkelheit besser gesehen werden. Die Nanobeschichtung schützt das Headset gegen Feuchtigkeit und macht es robust und strapazierfähig.

Begründung der Jury
Das Headset ist ein hervorragendes All-roundgerät für Sportler. Es bietet nützliche Funktionen, eine ergonomische Passform und die größtmögliche Freiheit beim Training.

Sennheiser SPORTS Series
Headphones
Kopfhörer

Manufacturer
Sennheiser electronic GmbH & Co. KG, Wedemark, Germany
In-house design
Design
designaffairs GmbH, Munich, Germany
Web
www.sennheiser.com
www.designaffairs.com

The Sports series features four models and thus offers a suitable solution for any athlete. All headphones in this series come in an ergonomic design which conforms to the user's head shape whilst securing maximum freedom of movement. The reinforced, oval cables prevent tangle and clutter, while the cable sleeves also reduce structure-born noise. The included ear adapters have an antibacterial coating and are thus particularly hygienic.

Statement by the jury
The headphones in this series are optimally aligned to the needs of athletes in both indoor and outdoor environments.

Die Sports-Serie umfasst vier Modelle und bietet somit für jeden Sportler die passende Lösung. Alle Kopfhörer der Serie überzeugen durch ihr ergonomisches Design, das sich der Kopfform anpasst und maximale Bewegungsfreiheit garantiert. Die verstärkten ovalen Kabel verhindern Kabelgewirr und reduzieren dank einer speziellen Ummantelung Körperschall. Die mitgelieferten Ohr-Adapter sind antibakteriell beschichtet und damit besonders hygienisch.

Begründung der Jury
Die Kopfhörer der Serie sind optimal auf die Bedürfnisse von Sportlern abgestimmt, sowohl für den Indoor- als auch für den Outdoor-Bereich.

Philips ActionFit 2015
Sports Headphone Range
Sportkopfhörerserie

Manufacturer
Gibson Innovations, Hong Kong
In-house design
Web
www.gibson.com

The ActionFit 2015 headphone series is characterised by a unified, timeless design and offers an ergonomic shape in all versions. Thanks to premium materials and a special production technique, these headphones are durable, sweat-resistant and waterproof. Made from only a few component parts, they are feather-light and thus perfectly suited for athletes.

Statement by the jury
The series comprises different models, which are distinguished by a coherent shared design approach.

Die Serie ActionFit 2015 ist durch eine einheitliche, zeitlose Gestaltung gekennzeichnet und bietet in allen Varianten eine ergonomische Form. Dank der hochwertigen Materialien und einer speziellen Herstellungstechnik sind die Sportkopfhörer robust, schweißbeständig und wasserdicht. Da sie aus nur wenigen Komponenten bestehen, sind sie federleicht und damit hervorragend für Sportler geeignet.

Begründung der Jury
Die Serie umfasst unterschiedliche Modellformen, die sich durch einen schlüssigen gemeinsamen Gestaltungsansatz auszeichnen.

Alpine Filter Earplugs
Hearing Protection
Gehörschutz

Manufacturer
Alpine Hearing Protection, Soesterberg,
Netherlands
In-house design
Web
www.alpine.eu

This hearing protection provides natural,
even noise attenuation and a comfortable fit. Each of the six available versions
possesses its own specific attenuation
filter for different situations. The user
can thus rely upon correct and safe hearing protection at any given time. The
earplugs are made of a thermoplastic
material that shapes itself, using body
heat, to optimally fit the ear.

Statement by the jury
Thanks to their different attenuation
properties, the Alpine Filter Earplugs
offer an appropriate solution for every
situation.

Dieser Gehörschutz bietet eine naturge-
treue, flache Dämpfung und eine komfortable Passform. Jedes der sechs verfügbaren Modelle besitzt einen eigenen
spezifischen Dämpfungsfilter für verschiedene Situationen. So kann der Benutzer
sich jederzeit darauf verlassen, dass sein
Gehör auf die richtige Art geschützt wird.
Die Ohrstöpsel werden aus thermoplastischem Material hergestellt, das sich durch
Körperwärme dem Ohr optimal anpasst.

Begründung der Jury
Dank ihrer unterschiedlichen Dämpfungs-
eigenschaften bieten die Alpine Filter
Earplugs für jeden Nutzer in jeder Situation
die passende Lösung.

BSH-53TW
Wireless Earbuds
Kabelloser Kopfhörer

Manufacturer
Mavin Technology Inc., Chupei, Taiwan
In-house design
Web
www.mavintec.com
Honourable Mention

BSH-53TW are very small, wireless in-ear
headphones. They consist of two compact earpieces without a cable connection that deliver stereo sound like a Bluetooth headset. The built-in microphone
allows the headphones to also be used as
a hands-free headset for phone calls.
The rechargeable battery guarantees up
to five hours of playtime or up to four
hours of phone conversation. The control
buttons on the earbuds are easy to activate.

Statement by the jury
The Bluetooth headphones are highly
functional for everyday use and can be
worn unobtrusively and comfortably.

BSH-53TW ist ein sehr kleiner kabelloser
In-Ear-Kopfhörer. Er verfügt über zwei
kompakte Ohrstücke ohne Kabelverbindung,
die Stereosound wie ein Bluetooth-Headset
liefern. Dank des Mikrofons lässt sich der
Kopfhörer auch als Freisprechanlage für ein
Telefon nutzen. Der Akku gewährleistet
eine Spieldauer von fünf Stunden oder eine
Gesprächsdauer von vier Stunden. Die
Steuertasten am Ohrstück sind einfach zu
bedienen.

Begründung der Jury
Der Bluetooth-Kopfhörer hat einen guten
Funktionsumfang für die tägliche Nutzung.
Er ist unauffällig und angenehm leicht zu
tragen.

Xiaomi Piston
Earphones 3.0
Kopfhörer

Manufacturer
1more Inc., Shenzhen, China
In-house design
Web
www.1more.com

Thanks to the metal-composite
diaphragm, these headphones deliver
powerful bass and an extended high-
frequency range. Three tuning chambers
enhance low and high frequencies and
reduce distortion. The earplugs are
designed ergonomically and are adjusted
to the natural shape of the ear canal,
resulting in a comfortable and secure fit.
The Kevlar-coated wire prevents cable
breaks and enhances durability.

Statement by the jury
These in-ear headphones have an unobtrusive and functional appearance. They
offer everything music lovers need for
everyday use.

Dank der Membran aus Verbundmetall
bietet der Kopfhörer kraftvolle Bässe und
einen erweiterten Hochfrequenzbereich.
Drei Tuning-Kammern verbessern die niedrigen bis hohen Frequenztöne und reduzieren Verzerrungen. Die Ohrstöpsel sind
ergonomisch gestaltet und an die natürliche
Form des Gehörgangs angepasst. Dadurch
sitzen sie bequem und sicher. Das Kabel ist
mit Kevlar ummantelt, was Kabelbrüche
verhindert und die Langlebigkeit erhöht.

Begründung der Jury
Der In-Ear-Kopfhörer zeigt sich unauffällig
und funktional. Er bietet alles, was Musikliebhaber beim täglichen Gebrauch benötigen.

The jury 2015
International orientation and objectivity
Internationalität und Objektivität

The jurors of the Red Dot Award: Product Design
All members of the Red Dot Award: Product Design jury are appointed on the basis of independence and impartiality. They are independent designers, academics in design faculties, representatives of international design institutions, and design journalists.

The jury is international in its composition, which changes every year. These conditions assure a maximum of objectivity. The members of this year's jury are presented in alphabetical order on the following pages.

Die Juroren des Red Dot Award: Product Design
In die Jury des Red Dot Award: Product Design wird als Mitglied nur berufen, wer völlig unabhängig und unparteiisch ist. Dies sind selbstständig arbeitende Designer, Hochschullehrer der Designfakultäten, Repräsentanten internationaler Designinstitutionen und Designfachjournalisten.

Die Jury ist international besetzt und wechselt in jedem Jahr ihre Zusammensetzung. Unter diesen Voraussetzungen ist ein Höchstmaß an Objektivität gewährleistet. Auf den folgenden Seiten werden die Jurymitglieder des diesjährigen Wettbewerbs in alphabetischer Reihenfolge vorgestellt.

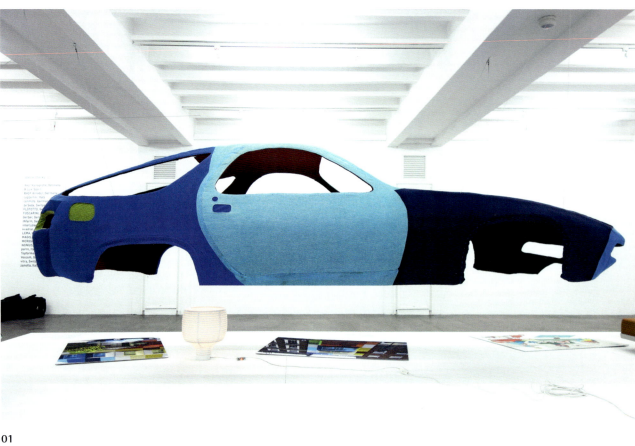

01

Prof.
Werner Aisslinger
Germany
Deutschland

The works of the designer Werner Aisslinger cover the spectrum of experimental and artistic approaches, including industrial design and architecture. He makes use of the latest technologies and has helped introduce new materials and techniques to the world of product design. His works are part of prestigious, permanent collections of international museums such as the Museum of Modern Art and the Metropolitan Museum of Art in New York, the Fonds National d'Art Contemporain in Paris, the Victoria & Albert Museum London, the Neue Sammlung Museum in Munich, and the Vitra Design Museum in Weil, Germany. In 2013, Werner Aisslinger opened his first solo show called "Home of the future" in the Berlin museum Haus am Waldsee. Besides numerous international awards he received the prestigious A&W Designer of the Year Award in 2014. Werner Aisslinger lives in Berlin and works for companies such as Vitra, Foscarini, Haier, Canon, 25hours Hotel Bikini and Moroso, for whom he has developed the "Hemp chair" together with BASF, which is the first ever biocomposite monobloc chair.

Die Arbeiten des Designers Werner Aisslinger umfassen das Spektrum experimenteller und künstlerischer Ansätze samt Industriedesign und Architektur. Er bedient sich modernster Technologien und hat dazu beigetragen, neue Materialien und Techniken in die Welt des Produktdesigns einzuführen. Seine Arbeiten sind Bestandteil bedeutender Museumssammlungen, darunter Museum of Modern Art und Metropolitan Museum of Art in New York, Fonds National d'Art Contemporain in Paris, Victoria & Albert Museum in London, Die Neue Sammlung in München und Vitra Design Museum in Weil. 2013 eröffnete Werner Aisslinger seine erste Solo-Show „Home of the future" im Berliner Museum „Haus am Waldsee". Neben zahlreichen internationalen Auszeichnungen wurde er 2014 als „A&W-Designer des Jahres" ausgezeichnet. Der in Berlin lebende Designer arbeitet u. a. für Unternehmen wie Vitra, Foscarini, Haier, Canon, 25hours Hotel Bikini und Moroso, für das er zusammen mit BASF den „Hemp chair", den weltweit ersten biologisch zusammengesetzten Monoblock-Stuhl, entwickelt hat.

01 Upcycling & tuning

The textile cover for Porsche 928 from the 1970s as a colourful reminder of the possible further uses of old cars. Upcycling and tuning are considered as a tool for prolonging the life cycles of products. Exhibition at Kölnischer Kunstverein 2014.

Textilhülle für den Porsche 928 aus den 1970ern als farbenfrohe Erinnerung an die weitere mögliche Nutzung alter Fahrzeuge. Upcycling und Tuning lassen sich als Werkzeuge verstehen, um den Lebenszyklus eines Produktes zu verlängern. Ausstellung im Kölnischen Kunstverein 2014.

02 Luminaire

Osram OLED, concept study 2014

Osram OLED, Konzeptstudie 2014

02

"Design is one of the most important professions of our time."

„Design ist eine der wichtigsten Professionen unserer Zeit."

What is, in your opinion, the significance of design quality in the industries you evaluated?
The design quality of the established brands is consistently high with a tendency towards a greater focus on product design, which is overall a positive trend.

How important is design quality in the global market?
Design quality is the ultimate distinguishing criterion. Particularly in product groups in which technical evolution creates a level playing field, design becomes the only relevant distinguishing criterion for products.

Do you see a dominating trend in this year's designs?
Innovative materials and new manufacturing methods are almost the norm in ambitious product developments. What is remarkable this year, however, is the sophistication with which these topics are combined with product design.

Wie würden Sie den Stellenwert von Designqualität in den von Ihnen beurteilten Branchen einschätzen?
Die Designqualität der etablierten Marken ist gleichbleibend hoch, mit Tendenz zu einem größeren Fokus auf das Produktdesign – alles in allem ein erfreulicher Trend.

Wie wichtig ist Designqualität im globalen Markt?
Designqualität ist das Unterscheidungskriterium schlechthin – gerade in Produktgruppen, in denen sich die technische Evolution nivelliert, wird Design zum einzigen relevanten Unterscheidungsfaktor für Produkte.

Sehen Sie eine Entwicklung im Design, die sich in diesem Jahr durchsetzt?
Innovative Materialien und neue Herstellungsmethoden sind bei ambitionierten Produktentwicklungen fast schon Standard – erstaunlich in diesem Jahr ist jedoch die „Sophistication", diese Themen mit dem Produktdesign zu verknüpfen.

01

Manuel
Alvarez Fuentes
Mexico
Mexiko

Manuel Alvarez Fuentes studied industrial design at the Universidad Iberoamericana, Mexico City, where he later served as Director of the Design Department. In 1975, he also received a Master of Design from the Royal College of Art, London. He has over 40 years of experience in the fields of product design, furniture and interior design, packaging design, signage and visual communications. From 1992 to 2012, he was Director of Diseño Corporativo, a product design consultancy. Currently, he is head of innovation at Alfher Porcewol in Mexico City. He has acted as consultant for numerous companies and also as a board member of various designers' associations, including as a member of the Icsid Board of Directors, as Vice President of the National Chamber of Industry of Mexico, Querétaro, and as Director of the Innovation and Design Award, Querétaro. Furthermore, Alvarez Fuentes has been senior tutor in Industrial Design at Tecnológico de Monterrey Campus Querétaro since 2009. In 2012 and 2013, he was president of the jury of Premio Quorum, the most prestigious design competition in Mexico.

Manuel Alvarez Fuentes studierte Industriedesign an der Universidad Iberoamericana, Mexiko-Stadt, wo er später die Leitung des Fachbereichs Design übernahm. 1975 erhielt er zudem einen Master of Design vom Royal College of Art, London. Er hat über 40 Jahre Erfahrung in Produkt- und Möbelgestaltung, Interior Design, Verpackungsdesign, Leitsystemen und visueller Kommunikation. Von 1992 bis 2012 war er Direktor von Diseño Corporativo, einem Beratungsunternehmen für Produktgestaltung. Heute ist Alvarez Fuentes Head of Innovation bei Alfher Porcewol in Mexiko-Stadt. Er war als Berater für zahlreiche Unternehmen sowie als Vorstandsmitglied in verschiedenen Designverbänden tätig, z. B. im Icsid-Vorstand, als Vizepräsident der mexikanischen Industrie- und Handelskammer im Bundesstaat Querétaro und als Direktor des Innovation and Design Award, Querétaro. Seit 2009 ist Alvarez Fuentes Senior-Dozent für Industriedesign am Tecnológico de Monterrey Campus Querétaro. 2012 und 2013 war er Präsident der Jury des Premio Quorum, des angesehensten Designwettbewerbs in Mexiko.

01 ELICA MAFdi 1
Kitchen hood, limited edition,
ceiling mount, island type.
The design of the hood is based on
Elica's patented, highly effective
Evolution ventilation system.
The steel carcase finish in porcelain
enamel (Peltre) is by Alfher Porcewol.
Küchenhaube, limitierte Auflage,
Deckenmontage, Insel-Typ.
Die Gestaltung der Haube basiert auf
dem patentierten, hocheffizienten
Ventilationssystem Elica Evolution.
Das Finish aus Porzellan-Emaille
(Peltre) ist von Alfher Porcewol.

02 ELICA MAFdi 1
Kitchen hoods, limited edition,
wall mounted.
The hood includes illumination
offered also in the Twin Evolution
system. Peltre finish colours:
orange, blue and black.
Küchenhauben, limitierte Auflage,
Wandmontage.
Die Haube bietet die gleiche Be-
leuchtung wie das „Twin Evolution"-
System. Farben des Peltre-Finishs:
Orange, Blau und Schwarz.

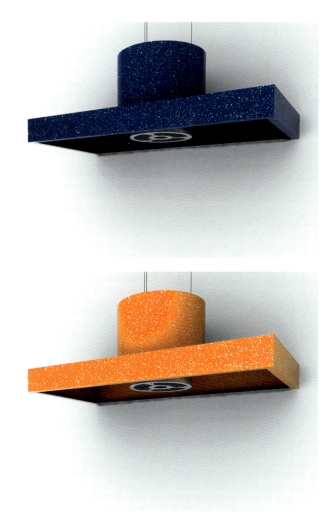

02

"My work philosophy is based on
the idea that design is not only
responsible for continuous innovation. Its main focus should be
the users, their happiness and
satisfaction."

„Meine Arbeitsphilosophie basiert
auf der Idee, dass Gestaltung
nicht nur für laufende Innovation
verantwortlich ist. Das Haupt-
augenmerk sollte vielmehr auf
den Benutzern, ihrem Glück und
ihrer Zufriedenheit liegen."

What are your sources of inspiration?
Living with a very talented and creative Mexican
architect, my father Augusto Alvarez, was a source
of inspiration at the beginning and later in my life.
Subsequently, the most reliable sources came from
a thorough observation of people.

**What do you take special notice of when you are
assessing a product as a jury member?**
I search for intelligent, simple, meaningful and
purposeful products; in many respects I try to
differentiate originality "per se" from true innovation.
I disregard those products that are only good-looking
but do not contribute to a better material world.

**What significance does winning an award in
a design competition have for the designer?**
It should mean a triggering motivation for his or her
creative work, to do better every time.

Was sind Ihre Inspirationsquellen?
Das Zusammenleben mit einem sehr talentierten und
kreativen mexikanischen Architekten, meinem Vater
Augusto Alvarez, war eine Inspirationsquelle für mich,
sowohl zu Beginn als auch später in meinem Leben.
Danach ergaben sich die zuverlässigsten Inspirations-
quellen aus der sorgfältigen Beobachtung von
Menschen.

**Worauf achten Sie besonders, wenn Sie ein
Produkt als Juror bewerten?**
Ich bin auf der Suche nach intelligenten, einfachen,
bedeutungsvollen und zweckmäßigen Produkten.
In vielerlei Hinsicht versuche ich, zwischen Originalität
per se und wahrer Innovation zu unterscheiden.
Ich ignoriere diejenigen Produkte, die lediglich gut
aussehen und keinen Beitrag zu einer besseren
materiellen Welt leisten.

**Welche Bedeutung hat die Auszeichnung in einem
Designwettbewerb für den Designer?**
Sie sollte einen Motivationsschub für die kreative
Arbeit des Designers darstellen, einen Ansporn, jedes
Mal etwas besser zu machen.

David Andersen
Denmark
Dänemark

David Andersen, born in 1978, graduated from the Glasgow School of Art and the Copenhagen Academy of Fashion and Design in 2003. In 2004, he was awarded "Best Costume Designer" and received the award "Wedding Gown of the Year" from the Royal Court Theatre in Denmark. He develops designs for ready-to-wear clothes, shoes, perfume, underwear and home wear and emerged as a fashion designer working as chief designer at Dreams by Isabell Kristensen as well as designing couture for artists and dance competitions under his own name. He debuted his collection "David Andersen" in 2007 and apart from Europe it also conquered markets in Japan and the US. In 2010 and 2011, the Danish Fashion Award nominated him in the category "Design Talent of the Year", and in 2010 as well as 2012, David Andersen received a grant from the National Art Foundation. He regularly shows his sustainable designs, which are worn by members of the Royal Family, politicians and celebrities, at couture exhibitions around the world. Furthermore, he is a guest lecturer at TEKO, Scandinavia's largest design and management college.

David Andersen, 1978 geboren, graduierte 2003 an der Glasgow School of Art und an der Copenhagen Academy of Fashion and Design. 2004 wurde er als „Best Costume Designer" sowie für das „Hochzeitskleid des Jahres" vom Royal Court Theatre in Dänemark prämiert. Er entwirft Konfektionskleidung, Schuhe, Parfüm, Unterwäsche und Heimtextilien und wurde als Modedesigner bekannt, als er als Chefdesigner bei Dreams von Isabell Kristensen sowie unter seinem Namen für Künstler und Tanzwettbewerbe arbeitete. Seine Kollektion „David Andersen", erschienen ab 2007, eroberte neben Europa Märkte in Japan und den USA. 2010 und 2011 nominierte ihn der Danish Fashion Award in der Kategorie „Design Talent of the Year" und 2010 sowie 2012 erhielt er ein Stipendium der National Art Foundation. David Andersen präsentiert seine nachhaltigen Entwürfe, die von Mitgliedern der Königsfamilie, Politikern oder Prominenten getragen werden, regelmäßig in Couture-Ausstellungen weltweit und ist zudem Gastdozent an der TEKO, der größten Hochschule für Design und Management Skandinaviens.

"My work philosophy is to be
true to myself."

„Meine Arbeitsphilosophie besteht
darin, mir selbst treu zu bleiben."

In your opinion, what will never be out of fashion?
Personality.

**Which of the projects in your lifetime arc you
particular proud of?**
The first dress I made for HRH Crown Princess Mary of
Denmark was a cerise-coloured dress she wore at the
royal wedding in Monaco. It was both thrilling and
challenging. Meeting her privately was one thing,
but knowing that this dress will be showcased in the
press all around the world was just nerve-racking.

**The Red Dot Award has been uncovering the
best designs for 60 years now. Which product
innovation would you like to see in the next
60 years, and why?**
I hope that we as designers will own up to our
responsibility for the environment through the
production of our products and that the focus will
shift from getting turnover to what actually happens
on the journey there.

Was ist Ihrer Meinung nach nie unmodern?
Persönlichkeit.

Auf welche Ihrer bisherigen Projekte sind Sie
besonders stolz?
Das erste Kleid, das ich für Ihre Königliche Hoheit
Kronprinzessin Mary von Dänemark entwarf, war ein
kirschrotes Kleid, das sie zur königlichen Hochzeit
in Monaco trug. Dieses Projekt war sowohl aufregend
als auch eine Herausforderung. Sie privat zu treffen,
war eine Sache, aber zu wissen, dass dieses Kleid von
der internationalen Presse weltweit zur Schau gestellt
werden würde, war schlicht nervenaufreibend.

Der Red Dot Award ermittelt seit 60 Jahren die
besten Gestaltungen. Welche Produktinnovation
würden Sie sich für die nächsten 60 Jahre
wünschen?
Ich hoffe, dass wir unserer Verantwortung als Designer
für die Umwelt durch die Herstellung unserer Produkte
gerecht werden und sich das Augenmerk weg von dem
Ziel des Umsatzes hin zu dem verschiebt, was auf dem
Weg dorthin tatsächlich passiert.

01

Prof. Martin Beeh
Germany
Deutschland

Professor Martin Beeh is a graduate in Industrial Design from the Darmstadt University of Applied Sciences in Germany and the ENSCI-Les Ateliers, Paris, and completed a postgraduate course in business administration. In 1995, he became design coordinator at Décathlon in Lille/France, in 1997 senior designer at Electrolux Industrial Design Center Nuremberg and Stockholm and furthermore became design manager at Electrolux Industrial Design Center Pordenone/Italy, in 2001. He is a laureate of several design awards as well as founder and director of the renowned student design competition "Electrolux Design Lab". In the year 2006 he became general manager of the German office of the material library Material ConneXion in Cologne. Three years later, he founded the design office beeh_innovation. He focuses on Industrial Design, Design Management, Design Thinking and Material Innovation. Martin Beeh lectured at the Folkwang University of the Arts in Essen and at the University of Applied Sciences Schwäbisch Gmünd. Since 2012, he is professor for design management at the University of Applied Sciences Ostwestfalen-Lippe in Lemgo/Germany.

Professor Martin Beeh absolvierte ein Studium in Industriedesign an der Fachhochschule Darmstadt und an der ENSCI-Les Ateliers, Paris, sowie ein Aufbaustudium der Betriebswirtschaft. 1995 wurde er Designkoordinator bei Décathlon in Lille/Frankreich, 1997 Senior Designer im Electrolux Industrial Design Center Nürnberg und Stockholm sowie 2001 Design Manager am Electrolux Industrial Design Center Pordenone/Italien. Er ist Gewinner diverser Designpreise und gründete und leitete den renommierten Designwettbewerb für Studierende, das „Electrolux Design Lab". Im Jahr 2006 wurde er General Manager der deutschen Niederlassung der Materialbibliothek „Material ConneXion" in Köln. Drei Jahre später gründete Martin Beeh das Designbüro beeh_innovation. Seine Schwerpunkte liegen in den Bereichen Industriedesign, Design Management, Design Thinking und Materialinnovation. Martin Beeh hatte Lehraufträge an der Folkwang Universität der Künste in Essen und an der Hochschule für Gestaltung Schwäbisch Gmünd. Seit 2012 ist er Professor für Designmanagement an der Hochschule Ostwestfalen-Lippe in Lemgo.

02

"My work philosophy is to aim for the best for and together with the customer, because every task is a new adventure."

„Meine Arbeitsphilosophie ist es, dass ich das Beste für und mit dem Kunden erreichen will, denn jede Aufgabe ist ein neues Abenteuer."

What are the main challenges in a designer's everyday life?
The movement between inspiration and stringency, order and creative chaos. Those who only follow a routine dry out and stop growing.

What impressed you most during the Red Dot judging process?
The concentrated competence of jurors and our interaction, the professionalism of the Red Dot team and the courage of the large number of companies that take part in the global competition.

What do you take special notice of when you are assessing a product as a jury member?
That the product is suitable for daily use, its ergonomic quality, and that there are no unnecessary frills. I also check if the design matches the brand, and I look for the product that stands out in a positive way.

What significance does winning an award in a design competition have for the manufacturer?
The role design plays in a company is strengthened and it is "officially" recognised as a successful tool.

Was sind große Herausforderungen im Alltag eines Designers?
Die Bewegung zwischen Inspiration und Stringenz, Ordnung und schöpferischem Chaos. Wer nur Routine macht, trocknet aus und wächst nicht mehr.

Was hat Sie bei der Red Dot-Jurierung am meisten beeindruckt?
Die geballte Kompetenz der Juroren und unser Zusammenspiel, die Professionalität des Red Dot-Teams und der Mut der vielen Unternehmen, die sich dem globalen Wettbewerb stellen.

Worauf achten Sie besonders, wenn Sie ein Produkt als Juror bewerten?
Auf Alltagstauglichkeit, Ergonomie, gestalterische Ordnung und den Verzicht auf Schnickschnack. Ich schaue auch, ob die Gestaltung zur Marke passt, und suche das Produkt, das sich positiv abhebt.

Welche Bedeutung hat die Auszeichnung in einem Designwettbewerb für den Hersteller?
Die Position von Design im Unternehmen wird gestärkt und „offiziell" als erfolgreiches Instrument anerkannt.

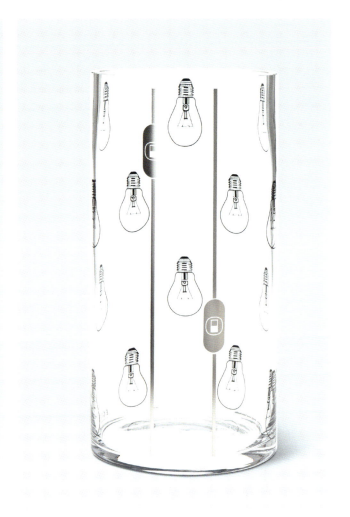

01

Dr Luisa Bocchietto
Italy
Italien

Luisa Bocchietto, architect and designer, graduated from the Milan Polytechnic. She has worked as a freelancer undertaking projects for local development, building renovations and urban planning. As a visiting professor she teaches at universities and design schools, she takes part in design conferences and international juries, publishes articles and organises exhibitions on architecture and design. Over the years, her numerous projects aimed at supporting the spread of design quality. She was a member of the Italian Design Council at the Ministry for Cultural Heritage, the Polidesign Consortium of the Milan Polytechnic, the CIDIC Italo-Chinese Council for Design and Innovation and the CNAC National Anti-Counterfeiting Council at the Ministry of Economic Development. From 2008 until 2014 she was National President of the ADI (Association for Industrial Design). Currently, she is a member of the Icsid Board, and since January 2015 she has been Editorial Director of PLATFORM, a new magazine about Architecture and Design.

Luisa Bocchietto, Architektin und Designerin, graduierte am Polytechnikum Mailand. Sie arbeitet freiberuflich und führt Projekte für die lokale Entwicklung, Gebäudeumbauten und Stadtplanung durch. Als Gastprofessorin lehrt sie an Universitäten und Designschulen, sie nimmt an Designkonferenzen und internationalen Jurys teil, veröffentlicht Artikel und betreut Ausstellungen über Architektur und Design. Ihre zahlreichen Projekte über die Jahre hinweg verfolgten das Ziel, die Verbreitung von Designqualität zu unterstützen. Sie war Mitglied des italienischen Rates für Formgebung am Kulturministerium, der Polidesign-Vereinigung des Polytechnikums Mailand, der CIDIC, der italienisch-chinesischen Vereinigung für Design und Innovation, und der CNAC, der Nationalen Vereinigung gegen Fälschungen am Ministerium für wirtschaftliche Entwicklung. Von 2008 bis 2014 war sie nationale Präsidentin der ADI, des Verbandes für Industriedesign. Aktuell ist sie Gremiumsmitglied des Icsid und seit Januar 2015 Chefredakteurin von PLATFORM, einem neuen Magazin über Architektur und Design.

02

"My work philosophy is to simplify, initiate sensible projects, generate quality and share joy."

„Meine Arbeitsphilosophie besteht darin, Dinge und Vorgänge zu vereinfachen, sinnvolle Projekte anzustoßen, Qualität zu schaffen und Freude zu teilen."

You are the editorial director of the new magazine "PLATFORM". What fascinates you about this job?
It allows me to explore the design aspects that I am more passionate about while at the same time keeping in touch with all the extraordinary people I have met during my ADI presidency.

The Red Dot Award has been uncovering the best designs for 60 years now. Which product innovation would you like to see in the next 60 years?
Products that are meant to improve people's lives – not only in an aesthetic and functional sense, but also in a social and ethical one.

What is the advantage of commissioning an external designer compared to having an in-house design team?
Working with an internal technical team means optimising in a linear way, while engaging with external designers entails the opportunity of a "lateral deviation", which often leads to unexpected and great results.

Sie sind die Chefredakteurin des neuen Magazins „PLATFORM". Was fasziniert Sie an dieser Rolle?
Sie erlaubt mir, die Gestaltungsaspekte, die mich besonders begeistern, näher zu untersuchen und zugleich mit all den außergewöhnlichen Menschen, die ich während meiner ADI-Präsidentschaft kennengelernt habe, in Kontakt zu bleiben.

Der Red Dot Award ermittelt seit 60 Jahren die besten Gestaltungen. Welche Produktinnovation würden Sie sich für die nächsten 60 Jahre wünschen?
Produkte, die darauf abzielen, das Leben zu verbessern – nicht nur in einem ästhetischen und funktionalen Sinn, sondern auch in einem sozialen und ethischen.

Worin liegt der Vorteil, einen externen Designer zu beauftragen, im Vergleich zu einem Inhouse-Designteam?
Mit einem internen technischen Team zu arbeiten, bedeutet, Dinge auf lineare Weise zu optimieren, wohingegen die Zusammenarbeit mit externen Designern die Gelegenheit für eine „laterale Abweichung" bietet, die oft zu unerwarteten und großartigen Ergebnissen führt.

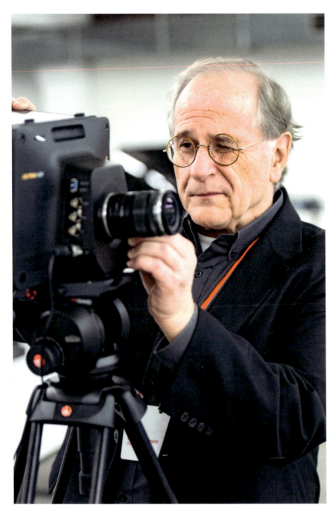

01

Gordon Bruce
USA
USA

Gordon Bruce is the owner of Gordon Bruce Design LLC and has been a design consultant for 40 years working with many multinational corporations in Europe, Asia and the USA. He has worked on a very wide range of products, interiors and vehicles – from aeroplanes to computers to medical equipment to furniture. From 1991 to 1994, Gordon Bruce was a consulting vice president for the Art Center College of Design's Kyoto programme and, from 1995 to 1999, chairman of Product Design for the Innovative Design Lab of Samsung (IDS) in Seoul, Korea. In 2003, he played a crucial role in helping to establish Porsche Design's North American office. For many years, he served as head design consultant for Lenovo's Innovative Design Center (IDC) in Beijing and he is presently working with Bühler in Switzerland and Huawei Technologies Co., Ltd. in China. Gordon Bruce is a visiting professor at several universities in the USA and in China and also acts as an author and design publicist. He recently received Art Center College of Design's "Lifetime Achievement Award".

Gordon Bruce ist Inhaber der Gordon Bruce Design LLC und seit mittlerweile 40 Jahren als Designberater für zahlreiche multinationale Unternehmen in Europa, Asien und den USA tätig. Er arbeitete bereits an einer Reihe von Produkten, Inneneinrichtungen und Fahrzeugen – von Flugzeugen über Computer bis hin zu medizinischem Equipment und Möbeln. Von 1991 bis 1994 war Gordon Bruce beratender Vizepräsident des Kioto-Programms am Art Center College of Design sowie von 1995 bis 1999 Vorsitzender für Produktdesign beim Innovative Design Lab of Samsung (IDS) in Seoul, Korea. Im Jahr 2003 war er wesentlich daran beteiligt, das Büro von Porsche Design in Nordamerika zu errichten. Über viele Jahre war er leitender Designberater für Lenovos Innovative Design Center (IDC) in Beijing. Aktuell arbeitet er für Bühler in der Schweiz und Huawei Technologies Co., Ltd. in China. Gordon Bruce ist Gastprofessor an zahlreichen Universitäten in den USA und in China und als Buchautor sowie Publizist tätig. Kürzlich erhielt er vom Art Center College of Design den Lifetime Achievement Award.

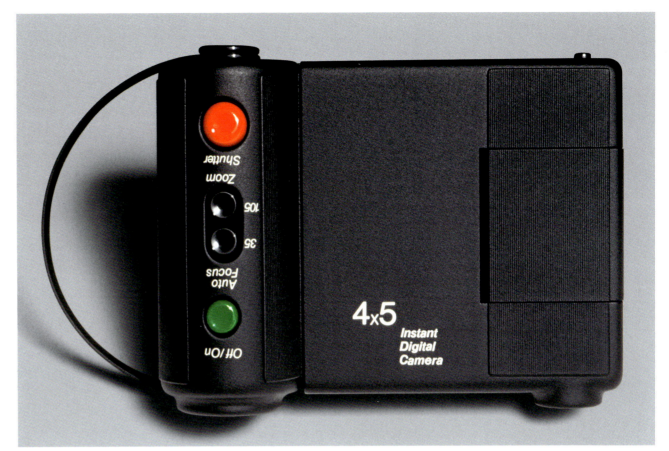

02

"My work philosophy is to
maintain a sense of curiosity.
Then, I conceptualise an idea
from my inquisitiveness, test it
and start the process over again."

„Meine Arbeitsphilosophie besteht
darin, Neugierde zu bewahren.
In einem zweiten Schritt konzipiere
ich eine Idee auf Basis meiner
Wissbegierde, teste sie und beginne
den Prozess von vorne."

Which IT project that you were involved in are you most proud of?
I was one of three designers who helped design the first "Massively Parallel Processor", a super computer produced by the Thinking Machines Company.
The unique design, using a series of black translucent boxes while accommodating many issues based on improved usability, resulted in a paradigm shift for large computers.

The Red Dot Award has been uncovering the best designs for 60 years now. Which product innovation would you like to see in the next 60 years, and why?
I would like to see a common theme in design move more and more towards Mother Nature's design.

What significance does winning an award in a design competition have for the designer?
Winning a design award from a very well-recognised design competition gives designers affirmation that they are making the right choices in their design decision processes.

Auf welches IT-Projekt, an dem Sie beteiligt waren, sind Sie besonders stolz?
Ich war einer von drei Designern, die dabei halfen, den ersten „Massively Parallel Processor", einen Super-Computer der Thinking Machines Company, zu gestalten. Das einzigartige Design, das aus einer Reihe lichtdurchlässiger Kästen besteht und gleichzeitig viele Aspekte verbesserter Benutzerfreundlichkeit beherbergt, begründete einen Paradigmenwechsel für große Computer.

Der Red Dot Award ermittelt seit 60 Jahren die besten Gestaltungen. Welche Produktinnovation würden Sie sich für die nächsten 60 Jahre wünschen, und warum?
Ich würde gerne einen einheitlichen Trend im Design-bereich sehen, der sich mehr und mehr zu einer Ge-staltung im Stil von Mutter Natur hinbewegt.

Welche Bedeutung hat die Auszeichnung in einem Designwettbewerb für den Designer?
Eine Auszeichnung in einem allgemein anerkannten Designwettbewerb zu gewinnen, gibt Designern die Bestätigung, dass sie die richtigen Entscheidungen im Designprozess getroffen haben.

01

Rüdiger Bucher
Germany
Deutschland

Rüdiger Bucher, born in 1967, graduated in political science from the Philipps-Universität Marburg and subsequently completed the postgraduate study course "Interdisciplinary studies on France" in Freiburg, Germany. While still at school he wrote for daily newspapers and magazines, before joining publishing house Verlagsgruppe Ebner Ulm in 1995, where he was in charge of "Scriptum. Die Zeitschrift für Schreibkultur" (Scriptum. The magazine for writing culture) for five years. In 1999 he became Chief Editor of Chronos, the leading German-language special interest magazine for wrist watches, with the same publishing house. During his time as Chief Editor, since 2005, Chronos has positioned itself internationally with subsidiary magazines and licensed editions in China, Korea, Japan and Poland. At the same time, Rüdiger Bucher established a successful corporate publishing department for Chronos. Since 2014, he has been Editorial Director of the business area "Watches" at the Ebner publishing house and besides Chronos he has also been in charge of the sister magazines concerning watches and classic watches as well as the New York-based "WatchTime".

Rüdiger Bucher, geboren 1967, studierte Politikwissenschaft an der Philipps-Universität Marburg und schloss daran den Aufbaustudiengang „Interdisziplinäre Frankreich-Studien" in Freiburg an. Schon als Schüler schrieb er für verschiedene Tageszeitungen und Zeitschriften, bevor er 1995 zum Ebner Verlag Ulm kam und dort fünf Jahre lang „Scriptum. Die Zeitschrift für Schreibkultur" betreute. Im selben Verlag wurde er 1999 Redaktionsleiter von „Chronos", dem führenden deutschsprachigen Special-Interest-Magazin für Armbanduhren. Während seiner Amtszeit als Chefredakteur, seit 2005, hat sich Chronos mit Tochtermagazinen und Lizenzausgaben in China, Korea, Japan und Polen international aufgestellt. Gleichzeitig baute Rüdiger Bucher für Chronos einen erfolgreichen Corporate-Publishing-Bereich auf. Seit 2014 verantwortet er als Redaktionsdirektor des Ebner Verlags im Geschäftsbereich „Uhren" neben Chronos auch die Schwestermagazine „Uhren-Magazin", „Klassik Uhren" sowie die in New York beheimatete „WatchTime".

01 **Chronos Special Uhrendesign**
Published once a year each
September since 2013.
Erscheint seit 2013 einmal
jährlich im September.

02 Chronos is present around the
globe with issues of differing
themes plus special issues.
Mit verschiedenen Ausgaben
und Sonderheften ist Chronos
rund um den Globus vertreten.

02

"My work philosophy is to inform
our readers competently, while at
the same time entertaining and
occasionally surprising them with
something new and unexpected."
„Meine Arbeitsphilosophie ist es,
den Leser kompetent zu informieren,
ihn dabei zu unterhalten und ihn
gelegentlich durch Neues, Unerwar-
tetes zu überraschen."

**What fascinates you about your job as Chief
Editor of Chronos?**
It is a very multi-faceted position. We distribute our
information using different channels and at the same
time we are closely in touch with what's happening
in a highly interesting industry, which is characterised
by a fascinating tension between past and future.

When did you first think consciously about design?
As a child I already preferred using specific glasses
and a specific cutlery set. Apart from the look, it was
above all the well-designed haptic qualities that
appealed to me.

**The Red Dot Award has been uncovering the
best designs for 60 years now. Which product
innovations would you like to see in the next
60 years, and why?**
A technology that does away with cables and at the
same time is non-hazardous to health.

**What significance does winning an award in
a design competition have for the manufacturer?**
An objective quality seal increases credibility, generates attention and enhances its profile.

**Was fasziniert Sie an Ihrem Beruf als Chef-
redakteur von Chronos?**
Er ist sehr vielseitig: Wir verbreiten unsere Informationen auf unterschiedlichsten Kanälen und sind zugleich
nah dran an einer hochinteressanten Branche, die sich
in Technik und Design in einem faszinierenden Span-
nungsfeld zwischen Vergangenheit und Zukunft bewegt.

**Wann haben Sie das erste Mal bewusst über Design
nachgedacht?**
Schon als Kind benutzte ich bestimmte Gläser und
bestimmtes Besteck lieber als andere. Neben der Optik
sprach mich vor allem eine gelungene Haptik an.

**Der Red Dot Award ermittelt seit 60 Jahren die
besten Gestaltungen. Welche Produktinnovation
würden Sie sich für die nächsten 60 Jahre
wünschen, und warum?**
Eine Technik, die es erlaubt, auf Kabel zu verzichten,
und dabei gesundheitlich unbedenklich ist.

**Welche Bedeutung hat die Auszeichnung in einem
Designwettbewerb für den Hersteller?**
Ein objektives Gütesiegel erhöht die Glaubwürdigkeit,
schafft Aufmerksamkeit und schärft sein Profil.

01

Wen-Long Chen
Taiwan
Taiwan

Wen-Long Chen has been CEO of the Taiwan Design Center since 2013 and has been designated as "Taiwan's Top Boss in Design". He has accumulated over 25 years of practical experience in design management and product design. In 1988, he founded Nova Design with support from Chinfon Trading Group, and has since led and completed hundreds of design projects. During his time at Nova Design, Wen-Long Chen led the company to tremendous growth with visionary design thinking. In response to clients' needs, he utilised the "Design System Competitiveness" as the core value of development, established a KMO (Knowledge Management Officer) system to oversee six branch offices in Asia, North America and Europe, and invested millions of dollars in the most advanced facilities. Today, Nova Design has over 200 employees worldwide and has won over 100 international design awards since 2006. Wen-Long Chen continues to propel the development of Taiwan's design industry using "Design x Knowledge Management", striving to shape Taiwan into a powerful international design force.

Wen-Long Chen ist seit 2013 CEO des Taiwan Design Centers und ist zu „Taiwan's Top Boss in Design" ernannt worden. Er verfügt über mehr als 25 Jahre praktische Erfahrung im Designmanagement und Produktdesign. 1988 gründete er mit Unterstützung der Chinfon Trading Group Nova Design und hat seitdem Hunderte Designprojekte durchgeführt. Während dieser Zeit führte und leitete er Nova Design mit visionärer Denkweise zu enormem Erfolg. Als Antwort auf die Bedürfnisse der Kunden nutzte er „Design System Competitiveness" als zentralen Entwicklungswert, etablierte das System „KMO" (Knowledge Management Officer), um sechs Zweigbüros in Asien, Nordamerika und Europa zu betreuen, und investierte mehrere Millionen Dollar in die fortgeschrittensten Einrichtungen. Heute hat Nova Design mehr als 200 Angestellte weltweit und erhielt seit 2006 über 100 internationale Designauszeichnungen. Wen-Long Chen treibt weiterhin die Entwicklung der taiwanesischen Designindustrie voran, indem er „Design x Knowledge Management" nutzt und danach strebt, Taiwan zu einer weltweit starken Designmacht zu formen.

01 SYM Fighter 4V 150

02 **Transformed Crane**
Commissioned by SANY Heavy
Industry Co., Ltd. and exhibited
at "bauma China 2012", the
International Trade Fair for
Construction Machinery,
Building Material Machines,
Construction Vehicles and
Equipment.
Transformierter Kran
In Auftrag gegeben von SANY
Heavy Industry Co., Ltd. und
ausgestellt auf der „bauma
China 2012", der internationalen
Messe für Baumaschinen, Bau-
materialmaschinen, Konstruktion,
Baufahrzeuge und Baugeräte.

02

"My work philosophy is: do the right thing right, which will eventually lead you to the right people and the right resources."

„Meine Arbeitsphilosophie lautet: Mache die richtige Sache richtig, denn das wird dich letztlich zu den richtigen Menschen und den richtigen Quellen führen."

Which product area will Taiwan have most success with in the next ten years?
Our efficient design industry will achieve success as a whole. There are both manufacturers with strong technology expertise and high adaptivity as well as a rich amount of creative talent in various fields.

Which country do you consider to be a pioneer in product design?
I would still consider Germany the pioneer in product design, due to the value they place on craftsmanship. Even though nowadays most breakthrough innovation comes from US-based start-ups, I believe in the end, it is hard to outperform Germany, both in terms of aesthetics and practicality.

What significance does winning an award in a design competition have for the designer?
While winning the award is proof of their career progress, the most beneficial effect comes from designers having the means to position themselves in the industry and determine their status.

In welchen Produktbereichen wird Taiwan in den nächsten zehn Jahren den größten Erfolg haben?
Unsere effiziente Designindustrie wird als Ganzes erfolgreich sein. Es gibt sowohl Hersteller mit großem Technologiewissen und hoher Adaptionsfähigkeit als auch eine Vielzahl an kreativen Talenten in den verschiedensten Bereichen.

Welches Land halten Sie für einen Pionier im Bereich Produktgestaltung?
Ich halte Deutschland nach wie vor für den Pionier im Produktdesign, weil es großen Wert auf handwerk-liches Können legt. Obwohl die meisten bahnbrechen-den Innovationen heutzutage von Start-up-Unter-nehmen aus den USA kommen, glaube ich, dass es letztlich schwer ist, Deutschland zu übertreffen, sowohl in puncto Ästhetik als auch in Praktikabilität.

Welche Bedeutung hat die Auszeichnung in einem Designwettbewerb für den Designer?
Während die Auszeichnung eine Bestätigung des Karrierefortschritts der Designer ist, besteht ihr hilf-reichster Effekt darin, dass sie ihnen ermöglicht, sich in der Branche zu positionieren und ihren eigenen Status zu ermitteln.

01

Vivian Wai Kwan Cheng
Hong Kong
Hongkong

On leaving Hong Kong Design Institute after 19 years of educational service, Vivian Cheng founded "Vivian Design" in 2014 to provide consultancy services and promote her own art in jewellery and glass. She graduated with a BA in industrial design from the Hong Kong Polytechnic University and was awarded a special prize in the Young Designers of the Year Award hosted by the Federation of Hong Kong Industries in 1987, and the Governor's Award for Industry: Consumer Product Design in 1989, after joining Lambda Industrial Limited as the head of the Product Design team. In 1995 she finished her Master degree and joined the Vocational Training Council teaching product design and later became responsible for, among others, establishing an international network with design-related organisations and schools. Vivian Cheng was the International Liaison Manager at the Hong Kong Design Institute (HKDI) and member of the Chartered Society of Designers Hong Kong, member of the Board of Directors of the Hong Kong Design Centre (HKDC) and board member of the Icsid from 2013 to 2015. Furthermore, she has been a panel member for the government and various NGOs.

Nach 19 Jahren im Lehrbetrieb verließ Vivian Cheng 2014 das Hong Kong Design Institute und gründete „Vivian Design", um Beratungsdienste anzubieten und ihre eigene Schmuck- und Glaskunst weiterzuentwickeln. 1987 machte sie ihren BA in Industriedesign an der Hong Kong Polytechnic University. Im selben Jahr erhielt sie einen Sonderpreis im Wettbewerb „Young Designers of the Year", veranstaltet von der Federation of Hong Kong Industries, sowie 1989 den Governor's Award for Industry: Consumer Product Design, nachdem sie bei Lambda Industrial Limited als Leiterin des Produktdesign-Teams angefangen hatte. 1995 beendete sie ihren Master-Studiengang und wechselte zum Vocational Training Council, wo sie Produktdesign unterrichtete und später u. a. für den Aufbau eines internationalen Netzwerks mit Organisationen und Schulen im Designbereich verantwortlich war. Vivian Cheng war International Liaison Manager am Hong Kong Design Institute (HKDI), Mitglied der Chartered Society of Designers Hong Kong, Vorstandsmitglied des Hong Kong Design Centre (HKDC) und Gremiumsmitglied des Icsid. Außerdem war sie Mitglied verschiedener Bewertungsgremien der Regierung und vieler Nichtregierungsorganisationen.

01 **AIR**
Casting in Silver
Silberguss

02 **FIRE**
Casting in Shibuichi
(metal alloy)
Shibuichi-Guss
(Metalllegierung)

02

"My work philosophy is: things I do today will be a bit better than yesterday's. I look into tomorrow, while enjoying today, and articulate what I experienced yesterday."

„Meine Arbeitsphilosophie ist: Dinge, die ich heute mache, werden ein bisschen besser als die gestrigen sein. Ich schaue auf das Morgen, genieße das Heute und spreche aus, was ich gestern erfahren habe."

When did you first think consciously about design?
I always wanted to dress only in the way I like since I was a small child. And I was told that I started to draw on walls at the age of three. So I believe I was born to appreciate art and design.

What do you take special notice of when you are assessing a product as a jury member?
I pay special attention to whether the design combines well with the choice of materials and the required technology, the craftsmanship, its positioning in the market, and that the user's emotional needs have been taken into account.

What is the advantage of commissioning an external designer compared to having an in-house design team?
An external designer can bring an outsider's view and provide the opportunity for a second party to challenge the product's functions and the way it operates. It might also help with the extension of product functions as well as the application of new materials and technology in different ways.

Wann haben Sie das erste Mal bewusst über Design nachgedacht?
Schon als Kleinkind wollte ich nur die Kleidung tragen, die mir gefiel. Und ich weiß von Erzählungen, dass ich bereits im Alter von drei Jahren begann, Wände zu bemalen. Daher glaube ich, dass ich dazu geboren bin, Kunst und Design zu würdigen.

Worauf achten Sie besonders, wenn Sie ein Produkt als Jurorin bewerten?
Ich achte besonders darauf, dass die Gestaltung gut zur Wahl der Materialien und Technologien, der Verarbeitung und der Marktpositionierung passt und dass die emotionalen Bedürfnisse des Benutzers berücksichtigt wurden.

Worin liegt der Vorteil, einen externen Designer zu beauftragen, im Vergleich zu einem Inhouse-Designteam?
Ein externer Designer kann eine Sicht von außen einbringen und Raum für eine zweite Seite schaffen, die die Funktionen und die Handhabung der Produkte hinterfragt. Zudem kann es dabei helfen, die Produktfunktionen zu erweitern und neue Materialien und Technologien auf unterschiedliche Arten einzubeziehen.

01

Datuk Prof. Jimmy Choo OBE
Malaysia / Great Britain
Malaysia / Großbritannien

Datuk Professor Jimmy Choo OBE is descended from a family of Malaysian shoemakers and learned the craft from his father. He studied at Cordwainers College, which is today part of the London College of Fashion. After graduating in 1983, he founded his own couture label and opened a shoe shop in London's East End whose regular customers included the late Diana, Princess of Wales. In 1996, Choo launched his ready-to-wear line with Tom Yeardye. Choo sold his share in the business in November 2001 to Equinox Luxury Holdings Ltd, charging them with the ongoing use of the label on the luxury goods market, while he continued to run his couture line. Choo now spends his time promoting design education. He is an ambassador for footwear education at the London College of Fashion, a spokesperson for the British Council in their promotion of British Education to foreign students and also spends time working with the non-profit programme, Teach For Malaysia. In 2003, Jimmy Choo was honoured for his contribution to fashion by Queen Elizabeth II who appointed him "Officer of the Order of the British Empire".

Datuk Professor Jimmy Choo OBE, der einer malaysischen Schuhmacher-Familie entstammt und das Handwerk von seinem Vater lernte, studierte am Cordwainers College, heute Teil des London College of Fashion. Nach seinem Abschluss 1983 gründete er sein eigenes Couture-Label und eröffnete ein Schuhgeschäft im Londoner East End, zu dessen Stammkundschaft auch Lady Diana gehörte. 1996 führte Choo gemeinsam mit Tom Yeardye seine Konfektionslinie ein und verkaufte seine Anteile an dem Unternehmen im November 2001 an die Equinox Luxury Holdings Ltd. Diese beauftragte er damit, das Label auf dem Markt für Luxusgüter fortzuführen, während er sich weiter um seine Couture-Linie kümmerte. Heute fördert Jimmy Choo die Designlehre. Er ist Botschafter für Footwear Education am London College of Fashion sowie Sprecher des British Council für die Förderung der Ausbildung ausländischer Studenten in Großbritannien und arbeitet für das gemeinnützige Programm „Teach for Malaysia". Für seine Verdienste in der Mode verlieh ihm Königin Elisabeth II. 2003 den Titel „Officer of the Order of the British Empire".

01 Red Sandal
In modern curves, made with
silk and a leather strap.
Rote Sandale
In modernen Kurven, hergestellt
aus Seide und einem Lederband.

02 Light Brown Sandal
Made with traditional woven
cotton from East Malaysia
(Pua Kumbu) with leather strap.
Leichte braune Sandale
Hergestellt aus traditioneller,
gewobener Baumwolle aus
Ostmalaysia (Pua Kumbu) mit
Lederband.

02

"My work philosophy is my
philosophy of life: always
move forward. Work hard.
Believe in yourself."
„Meine Arbeitsphilosophie ist
meine Lebensphilosophie:
Gehe immer vorwärts. Arbeite
hart. Glaube an dich selbst."

How does it feel to be a brand and a person of the same name?
I am very proud of what I have achieved with Jimmy Choo Couture. I no longer hold shares in Jimmy Choo London, but I am proud of co-founding a company that has become a global phenomenon.

Which country do you consider to be a pioneer in product design?
England has a rich history of producing leading design talent – and that continues to this day.

What impressed you most during the Red Dot judging process?
That professionals from around the world gather in one place, which gives us the opportunity to choose the most outstanding talent among the best.

What do you take special notice of when you are assessing a product as a jury member?
I always look for something that I haven't seen before – e.g. a new take on a well-known product. To catch my eye, the product also has to be of the highest quality, functional and durable.

Wie ist es, zugleich eine Marke und eine Person mit demselben Namen zu sein?
Ich bin sehr stolz auf das, was ich mit Jimmy Choo Couture erreicht habe. Ich besitze keine Anteile mehr an Jimmy Choo London, aber ich bin stolz darauf, ein Unternehmen mitgegründet zu haben, das sich zu einem globalen Phänomen entwickelt hat.

Welche Nation ist für Sie Vorreiter im Produktdesign?
England hat in der Vergangenheit eine Vielzahl führender Designtalente hervorgebracht – und das bis heute.

Was hat Sie bei der Red Dot-Jurierung am meisten beeindruckt?
Dass sich Profis aus aller Welt an einem Ort versammeln, was uns die Gelegenheit gibt, die hervorstechendsten Talente aus den Besten auszuwählen.

Worauf achten Sie besonders, wenn Sie ein Produkt als Juror bewerten?
Ich suche immer nach etwas, das ich vorher noch nicht gesehen habe – etwa eine neue Interpretation eines bekannten Produktes. Um meine Aufmerksamkeit zu erregen, muss es sowohl die höchsten Qualitätsstandards erfüllen als auch funktional und langlebig sein.

01

Vincent Créance
France
Frankreich

After graduating from the Ecole Supérieure de Design Industriel, Vincent Créance began his career in 1985 at the Plan Créatif Agency where he became design director and developed, among other things, numerous products for hi-tech and consumer markets, for France Télécom and RATP (Paris metro). In 1996 he joined Alcatel as Design Director for all phone activities on an international level. In 1999, he became Vice President Brand in charge of product design and user experience as well as all communications for the Mobile Phones BU. During the launch of the Franco-Chinese TCL and Alcatel Mobile Phones joint venture in 2004, Vincent Créance advanced to the position of Design and Corporate Communications Director. In 2006, he became President and CEO of MBD Design, one of the major design agencies in France, providing design solutions in transport design and product design. Créance is a member of the APCI (Agency for the Promotion of Industrial Creation), on the board of directors of ENSCI (National College of Industrial Creation), and a member of the Strategic Advisory Board for Strate College.

Vincent Créance begann seine Laufbahn nach seinem Abschluss an der Ecole Supérieure de Design Industriel 1985 bei Plan Créatif Agency. Hier stieg er 1990 zum Design Director auf und entwickelte u. a. zahlreiche Produkte für den Hightech- und Verbrauchermarkt, für die France Télécom oder die RATP (Pariser Metro). 1996 ging er als Design Director für sämtliche Telefonaktivitäten auf internationaler Ebene zu Alcatel und wurde 1999 Vice President Brand, zuständig für Produktdesign und User Experience sowie die gesamte Kommunikation für den Geschäftsbereich „Mobile Phones". Während des Zusammenschlusses des französisch-chinesischen TCL und Alcatel Mobile Phones 2004 avancierte Vincent Créance zum Design and Corporate Communications Director. 2006 wurde er Präsident und CEO von MBD Design, einer der wichtigsten Designagenturen in Frankreich, und entwickelte Designlösungen für Transport- und Produktdesign. Créance ist Mitglied von APCI (Agency for the Promotion of Industrial Creation), Vorstand des ENSCI (National College of Industrial Design) und Mitglied im wissenschaftlichen Beirat des Strate College.

02/03

"My work philosophy is: when risks appear, it becomes interesting!"

„Meine Arbeitsphilosophie lautet: Wenn Risiken auftauchen, wird es interessant!"

What are the main challenges in a designer's everyday life?
For a young one: to acquire experience in order to avoid big mistakes. For a senior one: to forget his or her experience in order to avoid big mistakes!

The Red Dot Award has been uncovering the best designs for 60 years now. Which product innovations would you like to see in the next 60 years, and why?
If everything we read about global warming etc. is true, it is obvious that product innovations have to find answers to these issues.

What significance does winning an award in a design competition have for the designer?
It's a recognition from his peers, always pleasant and helps to improve his image.

What significance does winning an award in a design competition have for the manufacturer?
It's a very good means to check the "non-rational" performance of their products. Because people also want pleasure, even from professional goods.

Was sind große Herausforderungen im Alltag eines Designers?
Für einen jungen Designer: Erfahrung sammeln, um große Fehler zu vermeiden. Für einen erfahrenen Designer: seine Erfahrung vergessen, um große Fehler zu vermeiden!

Der Red Dot Award ermittelt seit 60 Jahren die besten Gestaltungen. Welche Produktinnovationen würden Sie sich für die nächsten 60 Jahre wünschen, und warum?
Falls alles, was wir über die globale Erwärmung etc. lesen, wahr ist, ist offensichtlich, dass Produktinnovationen für diese Probleme Lösungen finden müssen.

Welche Bedeutung hat die Auszeichnung in einem Designwettbewerb für den Designer?
Sie ist eine Anerkennung seiner Partner, immer erfreulich und hilfreich, um sein Image zu verbessern.

Welche Bedeutung hat die Auszeichnung in einem Designwettbewerb für den Hersteller?
Es ist ein gutes Mittel, um die nicht-rationale Leistung seiner Produkte zu überprüfen. Denn Menschen wollen Wohlgefallen, selbst bei Fachprodukten.

01

Martin Darbyshire
Great Britain
Großbritannien

Martin Darbyshire is a founder and CEO of the internationally renowned design consultancy tangerine, working for clients such as Asiana, Azul, B/E Aerospace, Huawei, Nikon, The Royal Mint, Snoozebox.com and Virgin Australia. Before founding tangerine in 1989, he worked for Moggridge Associates and then in San Francisco at ID TWO (now IDEO). Martin Darbyshire is responsible for tangerine's commercial management, leading design projects and creating new business opportunities. Most notably, he led the multidisciplinary team that created both generations of the "Club World" business-class aircraft seating for British Airways – the world's first fully flat bed in business class which, since its launch in 2000, has remained the profit engine of the airline and transformed the industry. Besides, Martin Darbyshire has been a UK Trade and Investment ambassador for the UK Creative Industries sector, he is a recognised industry spokesperson, an advisor on design and innovation and was a board member of the Icsid.

Martin Darbyshire ist Gründer und CEO des international renommierten Designbüros tangerine, das für Kunden wie Asiana, Azul, B/E Aerospace, Huawei, Nikon, The Royal Mint, Snoozebox.com und Virgin Australia tätig ist. Bevor er tangerine 1989 gründete, arbeitete er für Moggridge Associates und danach in San Francisco bei ID TWO (heute IDEO). Martin Darbyshire verantwortet das kaufmännische Management von tangerine, wozu die Leitung von Designprojekten und die Entwicklung neuer Geschäftsmöglichkeiten gehört. Unter seiner Leitung stand insbesondere das multidisziplinäre Team, das beide Generationen der Business-Class-Sitze „Club World" für British Airways gestaltet hat – das weltweit erste komplett flache Bett in einer Business Class, das der Airline seit seiner Markteinführung im Jahr 2000 enorme Umsatzzahlen beschert und die Branche nachhaltig verändert hat. Darüber hinaus ist Martin Darbyshire für das Ministerium für Handel und Investition des Vereinigten Königreichs Botschafter für den Bereich der Kreativindustrie, ein anerkannter Sprecher der Branche sowie Berater für Design und Innovation und er war Gremiumsmitglied des Icsid.

01 tangerine's design work for Snoozebox.com created a premium hotel guest experience within a very small space (3.6m x 2m x 2m).
tangerines Gestaltungsentwurf für Snoozebox.com kreierte ein hochwertiges Hotelgasterlebnis auf sehr begrenzten Raum (3,6 m x 2 m x 2 m).

02 **British Airways' Club World**
In the second generation of Club World, by angling the pair of seats at two degrees to the centre line of the aircraft and making the arms drop, the bed was made 25 per cent wider within the same footprint.
In der zweiten Club-World-Generation wurde das Sitzpaar in einem Winkel von 2 Grad zur Mittellinie des Flugzeugs positioniert; die Armlehnen lassen sich herunterklappen. Dadurch wurde das Bett innerhalb der gleichen Grundfläche um 25 Prozent verbreitert.

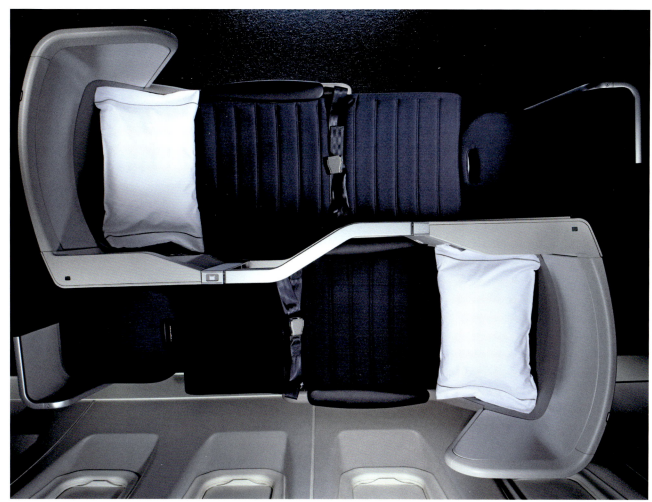

02

"My work philosophy is about challenging preconceptions and creating breakthrough change, improving lives and generating wealth."
„Meine Arbeitsphilosophie besteht darin, vorgefasste Meinungen zu hinterfragen und bahnbrechende Veränderungen zu erzielen, die das Leben verbessern und Wohlstand erzeugen."

The Red Dot Award has been uncovering the best designs for 60 years now. Which product innovations would you like to see in the next 60 years, and why?
I would like to see more applications of modern technologies and a greater focus on using modern technological advances to develop cheaper medical products, for instance creating low-cost portable equipment for developing countries.

What do you take special notice of when you are assessing a product as a jury member?
When assessing a new car entry for instance, I look at the interior first, taking great interest in the detailing and finishes before looking at the exterior. For me it is important that both have been given equal importance and attention by the design team.

What significance does winning an award in a design competition have for the designer?
It is a good feeling to know that your peers have judged you worthy of it, because a Red Dot carries such prestige within the design and commercial community.

Der Red Dot Award ermittelt seit 60 Jahren die besten Gestaltungen. Welche Produktinnovationen würden Sie sich für die nächsten 60 Jahre wünschen, und warum?
Ich würde gerne mehr Anwendungen moderner Technologien und ein größeres Augenmerk auf moderne technologische Fortschritte sehen, um günstigere Medizinprodukte zu entwickeln, wie zum Beispiel bei der Herstellung von kostengünstigen, tragbaren Geräten für Entwicklungsländer.

Worauf achten Sie besonders, wenn Sie ein Produkt als Juror bewerten?
Wenn ich etwa ein Auto bewerte, schaue ich zuerst auf die Innenausstattung, wobei mich die Details und die Verarbeitung besonders interessieren. Erst dann inspiziere ich das Äußere. Für mich ist es wichtig, dass beide die gleiche Aufmerksamkeit vom Designteam erhalten haben.

Welche Bedeutung hat die Auszeichnung in einem Designwettbewerb für den Designer?
Es ist ein gutes Gefühl zu wissen, dass dich deine Fachkollegen als dessen würdig erachten, da ein Red Dot ein so großes Ansehen in der Design- und Geschäftswelt genießt.

01

Robin Edman
Sweden
Schweden

Robin Edman has been the chief executive of SVID, the Swedish Industrial Design Foundation, since 2001. After studying industrial design at Rhode Island School of Design he joined AB Electrolux Global Design in 1981 and parallel to this started his own design consultancy. In 1989, Robin Edman joined Electrolux North America as vice president of Industrial Design for Frigidaire and in 1997, moved back to Stockholm as vice president of Electrolux Global Design. Throughout his entire career he has worked towards integrating a better understanding of users, their needs and the importance of design in society at large. His engagement in design related activities is reflected in the numerous international jury appointments, speaking engagements, advisory council and board positions he has held. Robin Edman served on the board of Icsid from 2003 to 2007, the last term as treasurer. Since June 2015, he is the president of BEDA (Bureau of European Design Associations).

Robin Edman ist seit 2001 Firmenchef der SVID, der Swedish Industrial Design Foundation. Nach seinem Industriedesign-Studium an der Rhode Island School of Design kam er 1981 zu AB Electrolux Global Design. Zeitgleich startete er seine eigene Unternehmensberatung für Design. 1989 wechselte Edman zu Electrolux North America als Vizepräsident für Industrial Design für Frigidaire und kehrte 1997 als Vizepräsident von Electrolux Global Design nach Stockholm zurück. Während seiner gesamten Karriere hat er daran gearbeitet, ein besseres Verständnis für Nutzer zu entwickeln, für deren Bedürfnisse und die Wichtigkeit von Design in der Gesellschaft insgesamt. Sein Engagement in designbezogenen Aktivitäten weist sich durch zahlreiche Jurierungsberufungen aus sowie durch Rednerverpflichtungen und Positionen in Gremien sowie Beratungsausschüssen. Von 2003 bis 2007 war Robin Edman Mitglied im Vorstand des Icsid, in der letzten Amtsperiode als Schatzmeister. Seit Juni 2015 ist er Präsident von BEDA (Bureau of European Design Associations).

02

"My work philosophy is to create a better place to live in for as many people as possible ... and to have fun while doing so."

„Meine Arbeitsphilosophie besteht darin, für so viele Menschen wie möglich einen besseren Lebensraum zu schaffen ... und dabei Spaß zu haben."

Which product area will Sweden have most success with in the next ten years?
Assuming the definition of "product" follows the now common view of both goods and services, I would like to stress the enormous potential in the public sector. The areas ranging from the development of the health care sector to regional and local innovation will include an array of design-focused competences that will transform our future society.

The Red Dot Award has been uncovering the best designs for 60 years now. Which product innovations would you like to see in the next 60 years?
A combination of solutions that will secure the continuation of humankind on earth and decrease our footprint on earth.

What significance does winning an award in a design competition have for the manufacturer?
It means recognition and is a great marketing tool that gives a seal of credibility and drives internal innovation.

In welchem Produktbereich wird Schweden in den nächsten zehn Jahren den größten Erfolg haben?
Voraussetzend, dass sich die Definition des Begriffs „Produkt" auf dessen zurzeit gängige Ansicht als Güter und Dienstleistungen bezieht, möchte ich das enorme Potenzial im öffentlichen Sektor hervorheben. Die Bereiche von der Entwicklung der Gesundheitsversorgung bis hin zu regionalen und lokalen Innovationen werden ein breites Spektrum designbezogener Kompetenzen beinhalten, das unsere Gesellschaft zukünftig verändern wird.

Der Red Dot Award ermittelt seit 60 Jahren die besten Gestaltungen. Welche Produktinnovationen würden Sie sich für die nächsten 60 Jahre wünschen?
Eine Kombination aus Lösungen, die das Fortbestehen der Menschheit auf der Erde sichern und unseren ökologischen Fußabdruck auf der Erde verringern.

Welche Bedeutung hat die Auszeichnung in einem Designwettbewerb für den Hersteller?
Sie bedeutet Anerkennung und ist ein großartiges Marketingwerkzeug, das ein Siegel der Glaubwürdigkeit darstellt und die interne Motivation antreibt.

01

02

Hans Ehrich
Sweden
Schweden

Hans Ehrich, born 1942 in Helsinki, Finland, has lived and been educated in Finland, Sweden, Germany, Switzerland, Spain and Italy. From 1962 to 1967, he studied metalwork and industrial design at the University College of Arts, Crafts and Design (Konstfackskolan), Stockholm. In 1965, his studies as a designer took him to Turin and Milan. With Tom Ahlström, he co-founded and became a director of A&E Design AB in 1968, a company which he still heads as director. From 1982 to 2002, he was managing director of Interdesign AB, Stockholm. Hans Ehrich has designed for, among others, Alessi, Anza, ASEA, Cederroth, Colgate, Fagerhults, Jordan, RFSU, Siemens, Turn-O-Matic and Yamagiwa. His work has been exhibited at many international exhibitions and collections and he has received numerous awards.

Hans Ehrich, 1942 in Helsinki, Finnland, geboren, lebte und lernte in Finnland, Schweden, Deutschland, der Schweiz, Spanien und Italien. Von 1962 bis 1967 studierte er Metall- und Industriedesign am University College of Arts, Crafts and Design (Konstfackskolan) in Stockholm. 1965 führten ihn Studien als Designer nach Turin und Mailand. 1968 war er zusammen mit Tom Ahlström Gründungsdirektor von A&E Design AB, Stockholm, das er heute immer noch als Direktor leitet, und von 1982 bis 2002 war er geschäftsführender Direktor von Interdesign AB, Stockholm. Hans Ehrich gestaltete u. a. für Alessi, Anza, ASEA, Cederroth, Colgate, Fagerhults, Jordan, RFSU, Siemens, Turn-O-Matic und Yamagiwa. Seine Arbeiten sind in zahlreichen internationalen Ausstellungen und Sammlungen vertreten und er wurde vielfach ausgezeichnet.

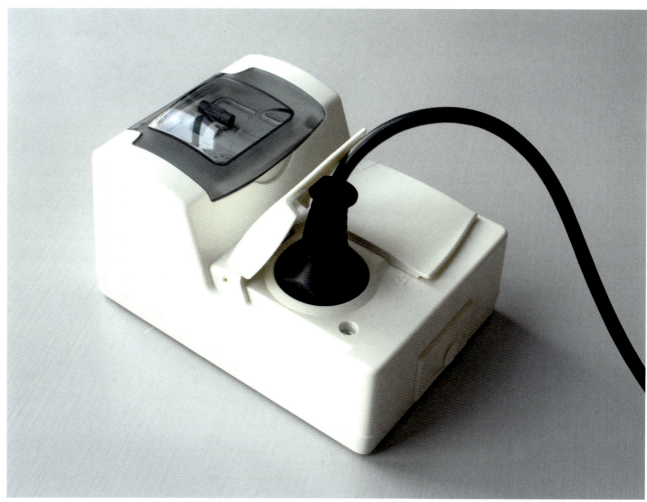

03

"My work philosophy has always
been: create for tomorrow with
a foothold on yesterday."
„Meine Arbeitsphilosophie war stets:
Gestalte für morgen mit einem Halt
in der Vergangenheit."

**This year you have donated the complete collection
from your 50 years of work to the Swedish
National Museum. Which motto do you think
would be most fitting for this collection?**
"Handmade models and prototypes, approximately
400 objects, which represent half a century of
successful Swedish product design."

**Which of the projects in your lifetime are you
particularly proud of?**
I am particularly proud of the "Stockholm II" folding
stool for museums, which is in use worldwide.

**What impressed you most during the Red Dot
judging process?**
The high design standards, the good functionality
and the appealing material quality of many of the
submitted products.

**What is the advantage of commissioning an
external designer compared to having an in-house
design team?**
An external designer often has a broader spectrum
of professional expertise with more influences than
an in-house-designer.

Sie haben in diesem Jahr die gesamte Sammlung
Ihrer 50 Berufsjahre dem schwedischen National-
museum überlassen. Unter welches Motto möchten
Sie diese Sammlung am liebsten stellen?
„Handgefertigte Modelle und Prototypen, etwa 400
Objekte, die ein halbes Jahrhundert erfolgreiches
schwedisches Produktdesign repräsentieren."

Auf welches Projekt in Ihrem Leben sind Sie
besonders stolz?
Auf den weltweit verbreiteten Museumsklapphocker
„Stockholm II".

Was hat Sie bei der Red Dot-Jurierung am meisten
beeindruckt?
Das hohe Gestaltungsniveau, die gute Funktionalität
und die ansprechende Materialqualität einer Vielzahl
der eingereichten Produkte.

Worin liegt der Vorteil, einen externen Designer
zu beauftragen, im Vergleich zu einem Inhouse-
Designteam?
Ein externer Designer verfügt meistens über eine
breiter gefächerte Berufskompetenz mit mehr Einfalls-
winkeln als ein Inhouse-Designer.

01

Joachim H. Faust
Germany
Deutschland

Joachim H. Faust, born in 1954, studied architecture at the Technical University of Berlin, the Technical University of Aachen, as well as at Texas A&M University (with Professor E. J. Romieniec), where he received his Master of Architecture in 1981. He worked as a concept designer in the design department of Skidmore, Owings & Merrill in Houston, Texas and as a project manager in the architectural firm Faust Consult GmbH in Mainz. From 1984 to 1986, he worked for KPF Kohn, Pedersen, Fox/Eggers Group in New York and as a project manager at the New York office of Skidmore, Owings & Merrill. In 1987, Joachim H. Faust took over the management of the HPP office in Frankfurt am Main. Since 1997, he has been managing partner of the HPP Hentrich-Petschnigg & Partner GmbH + Co. KG in Düsseldorf. He also writes articles and gives lectures on architecture and interior design.

Joachim H. Faust, 1954 geboren, studierte Architektur an der TU Berlin und der RWTH Aachen sowie – bei Professor E. J. Romieniec – an der Texas A&M University, wo er sein Studium 1981 mit dem Master of Architecture abschloss. Er war Entwurfsarchitekt im Design Department des Büros Skidmore, Owings & Merrill, Houston, Texas, sowie Projektleiter im Architekturbüro der Faust Consult GmbH in Mainz. Anschließend arbeitete er im Büro KPF Kohn, Pedersen, Fox/Eggers Group in New York und war Projektleiter im Büro Skidmore, Owings & Merrill in New York. 1987 übernahm Joachim H. Faust die Leitung des HPP-Büros in Frankfurt am Main und ist seit 1997 geschäftsführender Gesellschafter der HPP Hentrich-Petschnigg & Partner GmbH + Co. KG in Düsseldorf. Er ist zudem als Autor tätig und hält Vorträge zu Fachthemen der Architektur und Innenarchitektur.

02

"My work philosophy is:
to your own self be true."
„Meine Arbeitsphilosophie lautet:
Bleibe dir selbst treu."

What piece of advice has been useful in your youth or at the beginning of your professional career?
To face big things with composure, and small ones with close attention.

What will a single-family home look like in the year 2050?
Modular, prefabricated, interactive, adaptive, but hopefully also personal with human-centred standards and materials.

The Red Dot Award has been uncovering the best designs for 60 years now. Which product innovation would you like to see in the next 60 years, and why?
Intrinsic value is one of the highest goals in the design of a product or building. This means continuity in the best sense of a "classic".

What has impressed you most during the Red Dot judging process?
The wide range of products and the sensitivity for material conformity have increased every year.
And with that my enthusiasm to take part.

Welcher Ratschlag hat Sie in Ihrer Jugend oder frühen Berufskarriere weitergebracht?
Großen Dingen begegnet man mit Gelassenheit, kleinen mit besonderer Aufmerksamkeit.

Wie wird ein Einfamilienhaus im Jahr 2050 aussehen?
Modular, vorfabriziert, interaktiv, adaptiv, aber hoffentlich auch privat mit menschlich angemessenem Maßstab und Material.

Der Red Dot Award ermittelt seit 60 Jahren die besten Gestaltungen. Welche Produktinnovation würden Sie sich für die nächsten 60 Jahre wünschen, und warum?
Werthaltigkeit ist eines der höchsten Ziele in der Gestaltung eines Produkts bzw. Bauwerks. Das bedeutet Kontinuität im besten Sinne des „Klassikers".

Was hat Sie bei der Red Dot-Jurierung am meisten beeindruckt?
Die Vielfalt der Produkte und die Sensibilität für Materialkonformität sind in jedem Jahr gestiegen. Damit natürlich auch meine Begeisterung, mitzuwirken.

01

Hideshi Hamaguchi
USA / Japan
USA / Japan

Hideshi Hamaguchi graduated with a Bachelor of Science in chemical engineering from Kyoto University. Starting his career with Panasonic in Japan, Hamaguchi later became director of the New Business Planning Group at Panasonic Electric Works, Ltd. and then executive vice president of Panasonic Electric Works Laboratory of America, Inc. In 1993, he developed Japan's first corporate Intranet and also led the concept development for the first USB flash drive. Hideshi Hamaguchi has over 15 years of experience in defining strategies and decision-making, as well as in concept development for various industries and businesses. As Executive Fellow at Ziba Design and CEO at monogoto, he is today considered a leading mind in creative concept and strategy development on both sides of the Pacific and is involved in almost every project this renowned business consultancy takes on. For clients such as FedEx, Polycom and M-System he has led the development of several award-winning products.

Hideshi Hamaguchi graduierte als Bachelor of Science in Chemical Engineering an der Kyoto University. Seine Karriere begann er bei Panasonic in Japan, wo er später zum Direktor der New Business Planning Group von Panasonic Electric Works, Ltd. und zum Executive Vice President von Panasonic Electric Works Laboratory of America, Inc. aufstieg. 1993 entwickelte er Japans erstes Firmen-Intranet und übernahm zudem die Leitung der Konzeptentwicklung des ersten USB-Laufwerks. Hideshi Hamaguchi verfügt über mehr als 15 Jahre Erfahrung in der Konzeptentwicklung sowie Strategie- und Entscheidungsfindung in unterschiedlichen Industrien und Unternehmen. Als Executive Fellow bei Ziba Design und CEO bei monogoto wird er heute als führender Kopf in der kreativen Konzept- und Strategieentwicklung auf beiden Seiten des Pazifiks angesehen und ist in nahezu jedes Projekt der renommierten Unternehmensberatung involviert. Für Kunden wie FedEx, Polycom und M-System leitete er etliche ausgezeichnete Projekte.

02

"My work philosophy is:
all I need is less."

„Meine Arbeitsphilosophie lautet:
Alles, was ich brauche, ist weniger."

What impressed you most during the Red Dot judging process?

This year still felt like the end of a transitional period – a time to resolve some of the critical tensions that have emerged out of the massive changes of technology, consumer experience, and business models in the past ten years.

What challenges do you see for the future in design?

The challenge is finding the sweet spot between what resonates with the consumer and what is true to the brand.

Do you see a correlation between the design quality of a company's products and the economic success of this company?

I see a strong correlation between them. If a company has a good design in all three phases of consumer interaction – attract, engage and extend – it should directly impact its success.

Was hat Sie bei der Red Dot-Jurierung am meisten beeindruckt?

Dieses Jahr fühlte sich immer noch wie das Ende einer Übergangszeit an – eine Zeit, um einen Teil der kritischen Spannungen, die aus den massiven Veränderungen der Technologie, Konsumentenerfahrung und Geschäftsmodelle während der letzten zehn Jahre resultieren, aufzulösen.

Welche zukünftigen Herausforderungen sehen Sie im Designbereich?

Die Herausforderung besteht darin, das richtige Verhältnis zwischen dem, was beim Konsumenten Anklang findet, und dem, was der Wahrheit der Marke entspricht, zu finden.

Sehen Sie einen Zusammenhang zwischen der Designqualität, die sich in den Produkten eines Unternehmens äußert, und dem wirtschaftlichen Erfolg dieses Unternehmens?

Ich sehe eine starke Korrelation zwischen beiden. Wenn ein Unternehmen gute Gestaltung für alle drei Phasen der Interaktion mit dem Konsumenten – auffallen, einnehmen, ausbauen – bietet, dann sollte sich das unmittelbar auf seinen Erfolg auswirken.

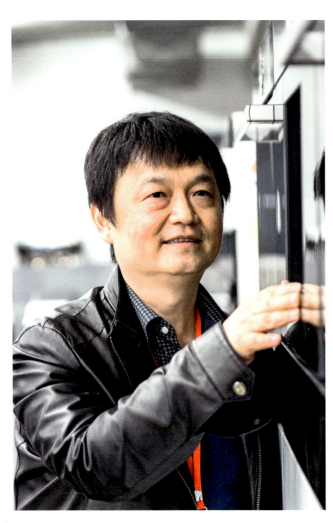

01

Prof. Renke He
China
China

Professor Renke He, born in 1958, studied civil engineering and architecture at Hunan University in China. From 1987 to 1988, he was a visiting scholar at the Industrial Design Department of the Royal Danish Academy of Fine Arts in Copenhagen and, from 1998 to 1999, at North Carolina State University's School of Design. Renke He is dean and professor of the School of Design at Hunan University and is also director of the Chinese Industrial Design Education Committee. Currently, he holds the position of vice chair of the China Industrial Design Association.

Professor Renke He wurde 1958 geboren und studierte an der Hunan University in China Bauingenieurwesen und Architektur. Von 1987 bis 1988 war er als Gastprofessor für Industrial Design an der Royal Danish Academy of Fine Arts in Kopenhagen tätig, und von 1998 bis 1999 hatte er eine Gastprofessur an der School of Design der North Carolina State University inne. Renke He ist Dekan und Professor an der Hunan University, School of Design, sowie Direktor des Chinese Industrial Design Education Committee. Er ist derzeit zudem stellvertretender Vorsitzender der China Industrial Design Association.

01/02 Fashion product designs inspired by the Dong nationality's traditional textile patterns, developed in the project New Channel Design & Social Innovation Summer Camp 2014, at the School of Design of Hunan University, China.
Modeprodukt-Designs inspiriert von den traditionellen Stoffmustern der Dong-Nationalität, entwickelt im Projekt „New Channel Design & Social Innovation Summer Camp 2014" an der School of Design der Hunan University, China.

02

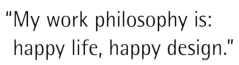

"My work philosophy is:
happy life, happy design."
„Meine Arbeitsphilosophie ist:
Ein glückliches Leben führt
zu glücklicher Gestaltung."

Which country do you consider to be a pioneer in product design?
The USA, because companies like Apple combine technologies with service design, business model and interaction design in order to create brand new products for the global markets.

What are the main challenges in a designer's everyday life?
Finding a balance between business and social responsibility in design.

What significance does winning an award in a design competition have for the designer?
It is the ultimate recognition of a designer's professional skills and reputation in this competitive world.

What significance does winning an award in a design competition have for the manufacturer?
A design award is a wonderful ticket to success in the marketplace.

Welche Nation ist für Sie Vorreiter im Produktdesign?
Die USA. Unternehmen wie Apple vereinen Technologien mit Servicedesign, Geschäftsmodell und Interaction Design, um so brandneue Produkte für den globalen Markt zu entwerfen.

Was sind große Herausforderungen im Alltag eines Designers?
Das Gleichgewicht zwischen Geschäft und sozialer Verantwortung im Design zu finden.

Welche Bedeutung hat die Auszeichnung in einem Designwettbewerb für den Designer?
Sie ist die beste Bestätigung für die Fertigkeiten und den Ruf eines Designers in unserer wettbewerbsorientierten Welt.

Welche Bedeutung hat die Auszeichnung in einem Designwettbewerb für den Hersteller?
Eine Auszeichnung in einem Designwettbewerb ist eine wunderbare Fahrkarte zum Markterfolg.

01

Prof. Herman Hermsen
Netherlands
Niederlande

Professor Herman Hermsen, born in 1953 in Nijmegen, Netherlands, studied at the ArtEZ Institute of the Arts in Arnhem from 1974 to 1979. Following an assistant professorship, he began his career in teaching in 1985. Since 1979, he is an independent jewellery and product designer. Until 1990, he taught product design at the Utrecht School of the Arts (HKU), after which time he returned to Arnhem as lecturer at the Academy. Hermsen has been professor of product and jewellery design at the University of Applied Sciences in Düsseldorf since 1992. He gives guest lectures at universities and colleges throughout Europe, the United States and Japan, and began regularly organising specialist symposia in 1998. He has also served as juror for various competitions. Herman Hermsen has received numerous international awards for his work in product and jewellery design, which is shown worldwide in solo and group exhibitions and held in the collections of renowned museums, such as the Cooper-Hewitt Museum, New York; the Pinakothek der Moderne, Munich; and the Museum of Arts and Crafts, Kyoto.

Professor Herman Hermsen, 1953 in Nijmegen in den Niederlanden geboren, studierte von 1974 bis 1979 am ArtEZ Institute of the Arts in Arnheim und ging nach einer Assistenzzeit ab 1985 in die Lehre. Seit 1979 ist er unabhängiger Schmuck- und Produktdesigner. Bis 1990 unterrichtete er Produktdesign an der Utrecht School of the Arts (HKU) und kehrte anschließend nach Arnheim zurück, um an der dortigen Hochschule als Dozent zu arbeiten. Seit 1991 ist Hermsen Professor für Produkt- und Schmuckdesign an der Fachhochschule Düsseldorf; er hält Gastvorlesungen an Hochschulen in ganz Europa, den USA und Japan, organisiert seit 1988 regelmäßig Fachsymposien und ist Juror in verschiedenen Wettbewerbsgremien. Für seine Arbeiten im Produkt- und Schmuckdesign, die weltweit in Einzel- und Gruppenausstellungen präsentiert werden und sich in den Sammlungen großer renommierter Museen befinden – z. B. Cooper-Hewitt Museum, New York, Pinakothek der Moderne, München, und Museum of Arts and Crafts, Kyoto –, erhielt Herman Hermsen zahlreiche internationale Auszeichnungen.

02

"My work philosophy is: if 'less or more' is wanted, it should at least contribute to the poetry."

„Meine Arbeitsphilosophie ist: Wenn ‚weniger oder mehr' gewünscht wird, sollte es wenigstens zur Poesie beitragen."

What can laypeople learn, when they look at award-winning products in design museums?
They can learn that design is not a direct translation of emotions, but that a lot of thought has gone into which design language can aesthetically communicate the perception of the product's function, and how this perception has developed in different cultures and eras.

Which of the projects in your lifetime are you particularly proud of?
I have created designs which I believe provide a poetic answer; they communicate the essence of a concept, such as my lamp "Charis" for Classicon.

Which nation to you consider to be a pioneer in product design?
I have come to increasingly appreciate Scandinavian design, because the designers have developed a successful combination of design language, function, zeitgeist, innovation and high-quality manufacturing.

Was können Laien lernen, wenn sie in Design-museen hervorragend gestaltete Produkte betrachten?
Dass Gestaltung keine direkte Umsetzung von Emotionen ist, dass aber genau überlegt wurde, welche Formensprache die Sichtweise auf die Produktfunktion ästhetisch kommunizieren kann. Und wie sich diese Sichtweisen in den unterschiedlichen Kulturen und Epochen entwickelten.

Auf welches Projekt in Ihrem Leben sind Sie besonders stolz?
Ich habe natürlich Entwürfe, von denen ich denke, mit der gestalterischen Umsetzung eine poetische Antwort gefunden zu haben, die die Essenz des Konzeptes kommuniziert; z. B. meine Lampe „Charis" für Classicon.

Welche Nation ist für Sie Vorreiter im Produktdesign?
Immer mehr schätze ich das skandinavische Design, denn die Designer entwickelten ein gelungenes Zusammenspiel aus Formensprache, Funktion, Zeitgeist, Innovation und qualitativ sehr guter Herstellung.

01

Prof. Carlos Hinrichsen
Chile
Chile

Professor Carlos Hinrichsen, born in 1957, graduated as an industrial designer in Chile in 1982 and earned his master's degree in engineering in Japan in 1991. Currently, he is dean of the Faculty of Engineering and Business at the Gabriela Mistral University in Santiago. From 2007 to 2009 he was president of Icsid and currently serves as senator within the organisation. He has since been heading research projects in the areas of innovation, design and education, and in 2010 was honoured with the distinction "Commander of the Order of the Lion of Finland". From 1992 to 2010 he was director of the School of Design Duoc UC, Chile and from 2011 to 2014 director of the Duoc UC International Affairs. He has led initiatives that integrate trade, engineering, design, innovation, management and technology in Asia, Africa and Europe and is currently design director for the Latin American Region of Design Innovation. Since 2002, Carlos Hinrichsen has been an honorary member of the Chilean Association of Design. Furthermore, he has been giving lectures at various conferences and universities around the world.

Professor Carlos Hinrichsen, 1957 geboren, erlangte 1982 seinen Abschluss in Industriedesign in Chile und erhielt 1991 seinen Master der Ingenieurwissenschaft in Japan. Aktuell ist er Dekan der Fakultät „Engineering and Business" an der Universität Gabriela Mistral in Santiago. Von 2007 bis 2009 war er Icsid-Präsident und ist heute Senator innerhalb der Organisation. Seither leitet er Forschungsprojekte in den Bereichen Innovation, Design sowie Erziehung und wurde 2010 mit der Auszeichnung „Commander of the Order of the Lion of Finland" geehrt. Von 1992 bis 2010 war er Direktor der School of Design Duoc UC in Chile und von 2011 bis 2014 Direktor der Duoc UC International Affairs. In Asien, Afrika und Europa leitete Carlos Hinrichsen Initiativen, die Handel, Ingenieurwesen, Design, Innovation, Management und Technologie integrieren, und ist aktuell Designdirektor der Latin American Region of Design Innovation. Seit 2002 ist er Ehrenmitglied der Chilean Association of Design. Außerdem hält er Vorträge auf Konferenzen und in Hochschulen weltweit.

01 Foguita
Kerosene heater, designed for
Compañía Tecno Industrial CTI in
1993: it is still in production in
Chile and sold in local and regional
markets – currently the company
is part of Electrolux Chile.
Ölheizgerät, entworfen 1993 für
Compañía Tecno Industrial CTI:
Das Gerät wird immer noch in Chile
hergestellt und lokal und regional
vertrieben – derzeit gehört die Firma
zu Electrolux Chile.

02 ZERO/ONE
The initiative explores new scenarios
of mobility for Milan, Italy, based on
different time horizons (2010 to 2025)
and proposals for sustainable innovation with regard to transport, logis-
tics and usability.
Die Initiative erforscht neue Szenarien
der Mobilität für Mailand, Italien, auf-
bauend auf verschiedenen Zeithorizon-
ten (2010 bis 2025) sowie Vorschläge
für nachhaltige Innovation in Bezug auf
Transport, Logistik und Benutzerfreundlichkeit.

02

"My work philosophy is well
described by the following quote:
'If you can imagine it, you can
create it. If you can dream it, you
can become it.'"
„Meine Arbeitsphilosophie lässt sich
gut mit folgendem Zitat beschreiben:
‚Wenn du es dir vorstellen kannst,
kannst du es kreieren. Wenn du
davon träumen kannst, kannst du
es werden.'"

When did you first think consciously about design?
I realised as a child that design contributes to human
happiness, and I have seen this insight confirmed over
the years.

**What impressed you most during the Red Dot
judging process?**
This time I saw how products offer realisations of the
desires and dreams of the users, as well as those of
the designers and producers. In product categories,
design quality and innovation are playing a key role
in turning technological innovations into good and
useful solutions. For this the Red Dot Award is like
a mirror, always reflecting what is going on in the
current design industry and market. It reveals the
prevalent trends and enables us to foresee other
potential trends, all of which have opened an unpre-
cedented field of knowledge and expectations.

**Wann haben Sie das erste Mal bewusst über
Design nachgedacht?**
Als Kind habe ich erkannt, dass Design dazu beiträgt,
dass sich Menschen glücklich fühlen. Und diese
Erkenntnis hat sich über die Jahre hinweg bestätigt.

**Was hat Sie bei der Red Dot-Jurierung am meisten
beeindruckt?**
Dieses Mal habe ich gesehen, wie Produkte Umset-
zungen der Wünsche und Träume sowohl der
Nutzer als auch der Designer und Hersteller bieten.
In den Produktkategorien spielen Designqualität
und -innovation eine zentrale Rolle dabei, technische
Neuerungen in gute und nützliche Lösungen zu über-
führen. Daher ist der Red Dot Award wie ein Spiegel,
der wiedergibt, was momentan in der Designbranche
und auf dem Markt passiert. Er offenbart die vor-
herrschenden Trends und ermöglicht uns, andere po-
tenzielle Trends vorherzusehen, die alle zusammen
einen neuartigen Bereich an Wissen und Erwartungen
eröffnet haben.

01

Tapani Hyvönen
Finland
Finnland

Tapani Hyvönen graduated in 1974 as an industrial designer from the present Aalto University School of Arts, Design and Architecture. He founded the design agency "Destem Ltd." in 1976 and was co-founder of ED-Design Ltd. in 1990, one of Scandinavia's largest design agencies. He has served as CEO and president of both agencies until 2013. Since then, he has been a visiting professor at, among others, Guangdong University of Technology in Guangzhou and Donghua University in Shanghai, China. His many award-winning designs are part of the collections of the Design Museum Helsinki and the Cooper-Hewitt Museum, New York. Tapani Hyvönen was an advisory board member of the Design Leadership Programme at the University of Art and Design Helsinki from 1989 to 2000, and a board member of the Design Forum Finland from 1998 to 2002, as well as the Icsid from 1999 to 2003 and again from 2009 to 2013. He has been a jury member in many international design competitions, a member of the Finnish-Swedish Design Academy since 2003 and a board member of the Finnish Design Museum since 2011.

Tapani Hyvönen graduierte 1974 an der heutigen Aalto University School of Arts, Design and Architecture zum Industriedesigner. 1976 gründete er die Design-agentur „Destem Ltd." und war 1990 Mitbegründer der ED-Design Ltd., einer der größten Designagenturen Skandinaviens, die er beide bis 2013 als CEO und Präsident leitete. Seitdem lehrt er als Gastprofessor u. a. an der Guangdong University of Technology in Guangzhou und der Donghua University in Shanghai, China. Seine vielfach ausgezeichneten Arbeiten sind in den Sammlungen des Design Museum Helsinki und des Cooper-Hewitt Museum, New York, vertreten. Tapani Hyvönen war von 1989 bis 2000 in der Bera-tungskommission des Design Leadership Programme der University of Art and Design Helsinki, von 1998 bis 2002 Vorstandsmitglied des Design Forum Finland sowie von 1999 bis 2003 und von 2009 bis 2013 des Icsid. Er ist international als Juror tätig, seit 2003 Mitglied der Finnish-Swedish Design Academy und seit 2011 Vorstandsmitglied des Finnish Design Museum.

02

"My work philosophy is to be open to new ideas."

„Meine Arbeitsphilosophie ist es, für neue Ideen offen zu sein."

What advice did you find helpful in your younger years or in the early days of your career?
Be curious, look and study different aspects seriously and don't fall in love with the first idea that comes.

Are there any designers who are role models for you?
A designer who puts good design, function, aesthetics and structure in balance. Alvar Aalto's beautiful, simple and functional designs are something I appreciate.

The Red Dot Award has been uncovering the best designs for 60 years now. Which product innovation would you like to see in the next 60 years?
Technology will become invisible. We will have products similar to the ones from now but they will feature a new kind of intelligence.

What significance does winning an award in a design competition have for the manufacturer?
It challenges the company to invest in design.

Welcher Ratschlag hat Sie in Ihrer Jugend oder frühen Berufskarriere weitergebracht?
Neugierig zu sein, sich ernsthaft verschiedene Aspekte anzuschauen und zu studieren und sich nicht in die erstbeste Idee, die daherkommt, zu verlieben.

Gibt es Designer, die Ihnen als Vorbilder dienen?
Designer, die gute Gestaltung, Funktion, Ästhetik und Konstruktion in Einklang bringen. Alvar Aaltos schöne, schlichte und funktionale Gestaltungen schätze ich durchaus.

Der Red Dot Award ermittelt seit 60 Jahren die besten Gestaltungen. Welche Produktinnovation würden Sie sich für die nächsten 60 Jahre wünschen?
Technologie wird unsichtbar werden. Wir werden ähnliche Produkte wie die heutigen haben, aber sie werden eine neue Art von Intelligenz aufweisen.

Welche Bedeutung hat die Auszeichnung in einem Designwettbewerb für den Hersteller?
Es fordert das Unternehmen dazu heraus, in Gestaltung zu investieren.

01

Guto Indio da Costa
Brazil
Brasilien

Guto Indio da Costa, born in 1969 in Rio de Janeiro, studied product design and graduated from the Art Center College of Design in Switzerland in 1993. He is design director of Indio da Costa A.U.D.T., a consultancy based in Rio de Janeiro, which develops architectural, urban planning, design and transportation projects. It works with a multidisciplinary strategic-creative group of designers, architects and urban planners, supported by a variety of other specialists. Guto Indio da Costa is a member of the Design Council of the State of Rio de Janeiro, former Vice President of the Brazilian Design Association (Abedesign) and founder of CBDI (Brazilian Industrial Design Council). He has been active as a lecturer and a contributing writer to different design magazines and has been a jury member of many design competitions in Brazil and abroad.

Guto Indio da Costa, geboren.1969 in Rio de Janeiro, studierte Produktdesign und machte 1993 seinen Abschluss am Art Center College of Design in der Schweiz. Er ist Gestaltungsdirektor von Indio da Costa A.U.D.T., einem in Rio de Janeiro ansässigen Beratungsunternehmen, das Projekte in Architektur, Stadtplanung, Design- und Transportwesen entwickelt und mit einem multidisziplinären, strategisch-kreativen Team aus Designern, Architekten und Stadtplanern sowie mit der Unterstützung weiterer Spezialisten operiert. Guto Indio da Costa ist Mitglied des Design Councils des Bundesstaates Rio de Janeiro, ehemaliger Vize-Präsident der brasilianischen Designvereinigung (Abedesign) und Gründer des CBDI (Industrial Design Council Brasilien). Er ist als Lehrbeauftragter aktiv, schreibt für verschiedene Designmagazine und ist als Jurymitglied zahlreicher Designwettbewerbe in und außerhalb Brasiliens tätig.

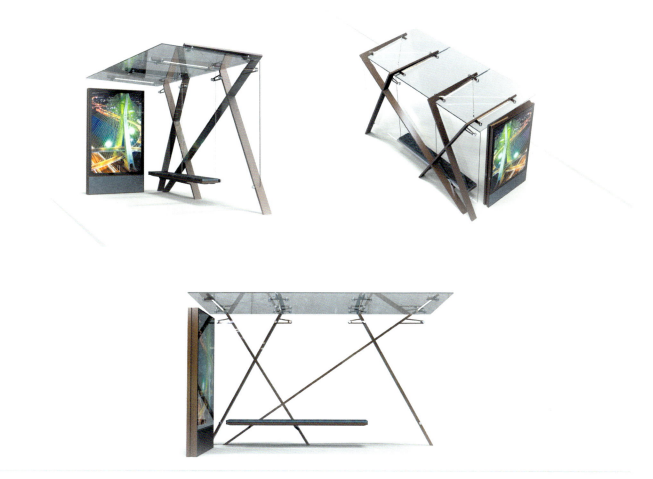

02

"My work philosophy focuses on how to please and cleverly surprise the user not only through aesthetics but also through functional and technical innovations."

„Meine Arbeitsphilosophie konzentriert sich darauf, wie man Benutzer erfreuen und geschickt überraschen kann, nicht nur durch Ästhetik, sondern auch durch funktionelle und technische Innovationen."

Which product design area will Brazil have most success with in the next ten years?
Considering the world's urgent need for a more eco-friendly industrial production and that Brazil's vast and diversified natural resources could lead to the research and development of eco-friendly materials, Brazilian design has the opportunity to play a leading role in this new eco-production revolution.

What are your sources of inspiration?
People. Observing the way people behave, the way people work, live or enjoy life.

The Red Dot Award has been uncovering the best designs for 60 years now. Which product innovations would you like to see in the next 60 years, and why?
I would love to see innovations that lead to zero-footprint production, where waste is easily transformed into resources. Designers can play a leading role in this transformation.

In welchen Produktbereichen wird Brasilien in den nächsten zehn Jahren den größten Erfolg haben?
In Anbetracht dessen, dass die Welt ein dringendes Bedürfnis nach einer umweltfreundlicheren industriellen Herstellung hat und dass Brasiliens riesige Vorkommen an verschiedenen Rohstoffen zur Erforschung und Entwicklung umweltfreundlicher Materialien dienen können, hat brasilianisches Design die Gelegenheit, eine führende Rolle in dieser Revolution hin zu einer neuen, umweltfreundlichen Produktion zu spielen.

Was sind Ihre Inspirationsquellen?
Menschen. Zu beobachten, wie sie sich verhalten, wie sie arbeiten, leben oder das Leben genießen.

Der Red Dot Award ermittelt seit 60 Jahren die besten Gestaltungen. Welche Produktinnovationen würden Sie sich für die nächsten 60 Jahre wünschen, und warum?
Innovationen hin zu einer umweltneutralen Produktion, in der Abfälle einfach wieder in Rohstoffe umgewandelt werden. Designer können bei dieser Umwandlung eine führende Rolle spielen.

01

Prof. Cheng-Neng Kuan
Taiwan

Taiwan

In 1980, Professor Cheng-Neng Kuan earned a Master's degree in Industrial Design (MID) from the Pratt Institute in New York. He is currently a full professor and the vice president of Shih-Chien University, Taipei, Taiwan. With the aim of developing a more advanced design curriculum in Taiwan, he founded the Department of Industrial Design, in 1992. He served as department chair until 1999. Moreover, Professor Kuan founded the School of Design in 1997 and had served as the dean from 1997 to 2004 and as the founding director of the Graduate Institute of Industrial Design from 1998 to 2007. He had also held the position of the 16th chairman of the Board of China Industrial Designers Association (CIDA), Taiwan. His fields of expertise include design strategy and management as well as design theory and creation. Having published various books on design and over 180 research papers and articles, he is an active member of design juries in his home country and internationally. He is a consultant to major enterprises on product development and design strategy.

1980 erwarb Professor Cheng-Neng Kuan einen Master-Abschluss in Industriedesign (MID) am Pratt Institute in New York. Derzeit ist er ordentlicher Professor und Vizepräsident der Shih-Chien University in Taipeh, Taiwan. 1992 gründete er mit dem Ziel, einen erweiterten Designlehrplan zu entwickeln, das Department of Industrial Design in Taiwan. Bis 1999 war Professor Kuan Vorsitzender des Instituts. Darüber hinaus gründete er 1997 die School of Design, deren Dekan er von 1997 bis 2004 war. Von 1998 bis 2007 war er Gründungsdirektor des Graduate Institute of Industrial Design. Zudem war er der 16. Vorstandsvorsitzende der China Industrial Designers Association (CIDA) in Taiwan. Seine Fachgebiete umfassen Designstrategie, -management, -theorie und -kreation. Neben der Veröffentlichung verschiedener Bücher über Design und mehr als 180 Forschungsarbeiten und Artikel ist er aktives Mitglied von Designjurys in seiner Heimat sowie auf internationaler Ebene. Zudem ist er als Berater für Großunternehmen im Bereich Produktentwicklung und Designstrategie tätig.

02

"My work philosophy is to bridge the known and unknown, familiarity and strangeness and to make new and good things happen continuously."
„Meine Arbeitsphilosophie besteht darin, eine Brücke zwischen Bekanntem und Unbekanntem, Familiärem und Fremdem zu schlagen und ständig neue und gute Dinge zu verwirklichen."

What challenges will designers face in the future?
In response to the interaction of technology (Internet, IoT, Big Data, 3D printing), entrepreneurship (crowdfunding) and micro lifestyles (cross-culture, cross-age, cross-region), designers will not only have to discover new ways of thinking and working, but also face creative challenges from individuals without a design background.

What do you take special notice of when you are assessing a product as a jury member?
As a language, design has to integrate all criteria; however, what concerns me most in terms of the degree of integration is the expressive uniqueness and the exquisite qualities of a product.

What is the advantage of commissioning an external designer compared to having an in-house design team?
It can broaden the vision for the design with regards to creativity and through the differentiation of designs improves the discovery and interpretation of a brand's DNA from different perspectives.

Was sind Herausforderungen für Designer in der Zukunft?
Als Reaktion auf die Interaktion zwischen Technologie (Internet, Internet der Dinge, Big Data, 3D-Drucken), Unternehmergeist (Crowdfunding) und Mikro-Lebensstilen (kultur-, alters- und regionenübergreifend) werden Designer nicht nur neue Arten des Denkens und Arbeitens entdecken müssen, sondern sich auch kreativen Herausforderungen von Individuen ohne Gestaltungshintergrund stellen müssen.

Worauf achten Sie besonders, wenn Sie ein Produkt als Juror bewerten?
Als eine Sprache muss Design alle Kriterien integrieren. Was mir jedoch am wichtigsten bei dem Grad der Integration ist, sind die Einzigartigkeit des Ausdrucks und die hervorragenden Qualitäten eines Produktes.

Worin liegt der Vorteil, einen externen Designer zu beauftragen, im Vergleich zu einem Inhouse-Designteam?
Es kann die Vision des Designs kreativ erweitern und verbessert durch Differenzierung der Entwürfe die Entdeckung und Interpretation der Marken-DNA aus verschiedenen Perspektiven.

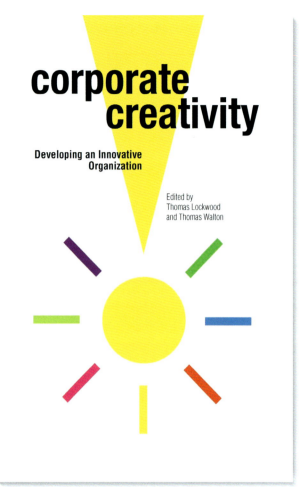

01

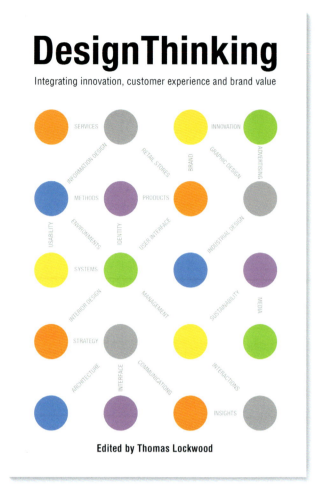

02

Dr Thomas Lockwood
USA
USA

Thomas Lockwood is the author of books on design thinking, design management, corporate creativity and design strategy. He has a PhD, works in design management and is regarded as a pioneer in the integration of design and innovation practices with business as well as for founding international design and user-experience organisations. He was guest professor at Pratt Institute in New York City, is a Fellow at the Royal Society of the Arts in London and a frequent design award judge. Lockwood is a founding partner of Lockwood Resource, an international recruiting firm specialising in design leadership recruiting. From 2005 to 2011, he was president of DMI, the Design Management Institute, a non-profit research association in Boston. From 1995 to 2005, he was design director at Sun Microsystems and StorageTek, and creative director at several design and advertising firms for a number of years. In addition, he manages a blog about design leadership at lockwoodresource.com.

Thomas Lockwood ist Autor von Büchern zu den Themen Design Thinking, Designmanagement, Corporate Creativity und Designstrategie und als Doktor der Philosophie im Designmanagement tätig. Er gilt als Vordenker für die Integration der Design- und Innovationspraxis ins Geschäftsleben sowie für die Gründung internationaler Design- und User-Experience-Organisationen. Er war Gastprofessor am Pratt Institute in New York City, ist Mitglied der Royal Society of the Arts in London und regelmäßig Juror bei Designwettbewerben. Lockwood ist Gründerpartner von Lockwood Resource, einer internationalen Rekrutierungsfirma, die sich auf die Anwerbung von Führungspersonal für den Designbereich spezialisiert hat. Von 2005 bis 2011 war er Präsident des DMI (Design Management Institute), einer gemeinnützigen Forschungsorganisation in Boston. Von 1995 bis 2005 war er Designdirektor bei Sun Microsystems und StorageTek und über mehrere Jahre hinweg Kreativdirektor in verschiedenen Design- und Werbefirmen. Darüber hinaus führt er den Blog lockwoodresource.com über Mitarbeiterführung im Designsektor.

01 Corporate Creativity (2008) explores
    how to develop more creativity and
    innovation power at the individual,
    team and organisational levels.
    Corporate Creativity (2008) untersucht,
    wie sich Kreativität und Innovationskraft
    auf individueller, Team- und Organisati-
    onsebene steigern lassen.

02 Design Thinking (2010) presents best
    practice cases of design thinking
    methods applied to innovation, brand
    building, service design and customer
    experience.
    Design Thinking (2010) präsentiert
    Best-Practice-Fallstudien gestalterischer
    Denkansätze für Innovation, Marken-
    bildung, Service-Design und Kunden-
    erfahrung.

"My work philosophy is to learn, grow, and change. When we learn, we grow. When we grow, we change. I choose to embrace this process."

„Meine Arbeitsphilosophie ist zu lernen, zu wachsen und zu verändern. Wenn wir lernen, wachsen wir. Wenn wir wachsen, verändern wir uns. Ich habe entschieden, mir diesen Prozess zu eigen zu machen."

In your opinion, which decade brought the greatest progress in the design sector?
I think the years from 2006 to 2011, the Great Recession, were fabulous for design. Because many CEOs realised that recovery would not just mean going back to business as usual. These CEOs gradually realised that design is the connection between their innovation ideas and new customers.

The Red Dot Award has been uncovering the best designs for 60 years now. Which product innovations would you like to see in the next 60 years?
I would love to see innovations in food distribution, in energy development, in health, and in equality of wealth.

What impressed you most during the Red Dot judging process?
I love the process: to review all the entries by myself first, and then to review them again with the team of judges for discussions, which are very insightful.

Welches Jahrzehnt hat Ihrer Meinung nach den größten Fortschritt im Designbereich gebracht?
Ich glaube, die Jahre von 2006 bis 2011, die große Rezession, waren fabelhaft für Design. Vielen CEOs wurde bewusst, dass ein Aufschwung nicht einfach bedeuten würde, dass alles wie gehabt weitergeht. Diese CEOs haben nach und nach gemerkt, dass Design die Verbindung zwischen ihren Innovationsideen und neuen Kunden darstellt.

Der Red Dot Award ermittelt seit 60 Jahren die besten Gestaltungen. Welche Produktinnovationen würden Sie sich für die nächsten 60 Jahre wünschen?
Ich würde gerne mehr Innovationen in der Nahrungsmittelverteilung, der Energieentwicklung, im Gesundheitsbereich und im Angleichen des Wohlstands sehen.

Was hat Sie bei der Red Dot-Jurierung am meisten beeindruckt?
Ich liebe den Prozess: alle Einreichungen zuerst allein zu begutachten, um sie dann erneut zusammen mit dem Jurorenteam anzuschauen und sehr erkenntnisreiche Diskussionen zu führen.

01

Lam Leslie Lu
Hong Kong
Hongkong

Lam Leslie Lu received a Master of Architecture from Yale University in Connecticut, USA in 1977, and was the recipient of the Monbusho Scholarship of the Japanese Ministry of Culture in 1983, where he conducted research in design and urban theory in Tokyo. He is currently the principal of the Hong Kong Design Institute and academic director of the Hong Kong Institute of Vocational Education. Prior to this, he was head of the Department of Architecture at the University of Hong Kong. Lam Leslie Lu has worked with, among others, Cesar Pelli and Associates, Hardy Holzman Pfeiffer and Associates, Kohn Pedersen Fox and Associates and Shinohara Kazuo on the design of the Centennial Hall of the Institute for Technology Tokyo. Moreover, he was visiting professor at Yale University and the Delft University of Technology as well as assistant lecturer for the Eero Saarinen Chair at Yale University. He also lectured and served as design critic at major international universities such as Columbia, Cambridge, Delft, Princeton, Yale, Shenzhen, Tongji, Tsinghua and the Chinese University Hong Kong.

Lam Leslie Lu erwarb 1977 einen Master of Architecture an der Yale University in Connecticut, USA, und war 1983 Monbusho-Stipendiat des japanischen Kulturministeriums, an dem er die Forschung in Design und Stadttheorie in Tokio leitete. Derzeit ist er Direktor des Hong Kong Design Institutes und akademischer Direktor des Hong Kong Institute of Vocational Education. Zuvor war er Leiter des Architektur-Instituts an der Universität Hongkong. Lam Leslie Lu hat u. a. mit Cesar Pelli and Associates, Hardy Holzman Pfeiffer and Associates, Kohn Pedersen Fox and Associates und Shinohara Kazuo am Design der Centennial Hall des Instituts für Technologie Tokio zusammengearbeitet, war Gastprofessor an der Yale University und der Technischen Universität Delft sowie Assistenz-Dozent für den Eero-Saarinen-Lehrstuhl in Yale. Er hielt zudem Vorträge und war Designkritiker an großen internationalen Universitäten wie Columbia, Cambridge, Delft, Princeton, Yale, Shenzhen, Tongji, Tsinghua und der chinesischen Universität Hongkong.

01 Centennial Hall of
the Tokyo Institute of
Technology
Die Centennial Hall
des Instituts für
Technologie Tokio

"My work philosophy is to advance knowledge and talent through calculated strategies and research while drawing in fresh thinking."

„Meine Arbeitsphilosophie besteht darin, Wissen und Talent durch wohl kalkulierte Strategien und Forschung voranzutreiben und gleichzeitig frische Denkansätze einzubeziehen."

Which of the projects in your lifetime are you particularly proud of?
I spent a period of time working with Shinohara Kazuo on the Centennial Hall building. The depth of thought and the level of experimentation in space, form and structure was the furthest I ever got in the design of a building. The interactive experience of designing as a duo set the standard for all my designs to follow.

What significance does winning an award in a design competition have for the designer?
The need to be recognised by one's peers is paramount and maybe more important than public acclaim. We need it not for our ego but for our soul.

What is the advantage of commissioning an external designer compared to having an in-house design team?
Change and new conceptions can only come about with the ebb and flow of talent, either through internal deployment or external injection.

Auf welches Ihrer bisherigen Projekte sind Sie besonders stolz?
Ich habe eine Zeit lang mit Shinohara Kazuo am Centennial-Hall-Gebäude zusammengearbeitet. Die Tiefe der Gedanken und das Niveau der Experimente in Bezug auf Raum, Form und Struktur waren die extremste Arbeit, die ich je bei der Gestaltung eines Gebäudes geleistet habe. Die interaktive Erfahrung des Gestaltens zu zweit hat Maßstäbe für alle meine nachfolgenden Entwürfe gesetzt.

Welche Bedeutung hat die Auszeichnung in einem Designwettbewerb für den Designer?
Die Notwendigkeit, von seinen Partnern anerkannt zu werden, ist entscheidend und womöglich wichtiger als öffentliche Anerkennung. Wir brauchen sie nicht für unser Ego, sondern für unsere Seele.

Worin liegt der Vorteil, einen externen Designer zu beauftragen, im Vergleich zu einem Inhouse-Designteam?
Veränderung und neue Ideen können nur mit der Ebbe und Flut an Talenten entstehen, entweder durch internen Personaleinsatz oder durch Zuführung von außen.

01

Wolfgang K. Meyer-Hayoz
Switzerland
Schweiz

Wolfgang K. Meyer-Hayoz studied mechanical engineering, visual communication and industrial design and graduated from the Stuttgart State Academy of Art and Design. The professors Klaus Lehmann, Kurt Weidemann and Max Bense have had formative influence on his design philosophy. In 1985, he founded the Meyer-Hayoz Design Engineering Group with offices in Winterthur/Switzerland and Constance/Germany. The design studio offers consultancy services for national as well as international companies in the five areas of design competence: design strategy, industrial design, user interface design, temporary architecture and communication design, and has received numerous international awards. From 1987 to 1993, Wolfgang K. Meyer-Hayoz was president of the Swiss Design Association (SDA). He is a member of the Association of German Industrial Designers (VDID), Swiss Marketing and the Swiss Management Society (SMG). Wolfgang K. Meyer-Hayoz also serves as juror on international design panels and supervises Change Management and Turnaround projects in the field of design strategy.

Wolfgang K. Meyer-Hayoz absolvierte Studien in Maschinenbau, Visueller Kommunikation sowie Industrial Design mit Abschluss an der Staatlichen Akademie der Bildenden Künste in Stuttgart. Seine Gestaltungsphilosophie prägten die Professoren Klaus Lehmann, Kurt Weidemann und Max Bense. 1985 gründete er die Meyer-Hayoz Design Engineering Group mit Büros in Winterthur/Schweiz und Konstanz/Deutschland. Das Designstudio bietet Beratungsdienste für nationale wie internationale Unternehmen in den fünf Designkompetenzen Designstrategie, Industrial Design, User Interface Design, temporäre Architektur und Kommunikationsdesign und wurde bereits vielfach ausgezeichnet. Von 1987 bis 1993 war Wolfgang K. Meyer-Hayoz Präsident der Swiss Design Association (SDA); er ist Mitglied im Verband Deutscher Industrie Designer (VDID), von Swiss Marketing und der Schweizerischen Management Gesellschaft (SMG). Wolfgang K. Meyer-Hayoz engagiert sich auch als Juror internationaler Designgremien und moderiert Change-Management- und Turnaround-Projekte im designstrategischen Bereich.

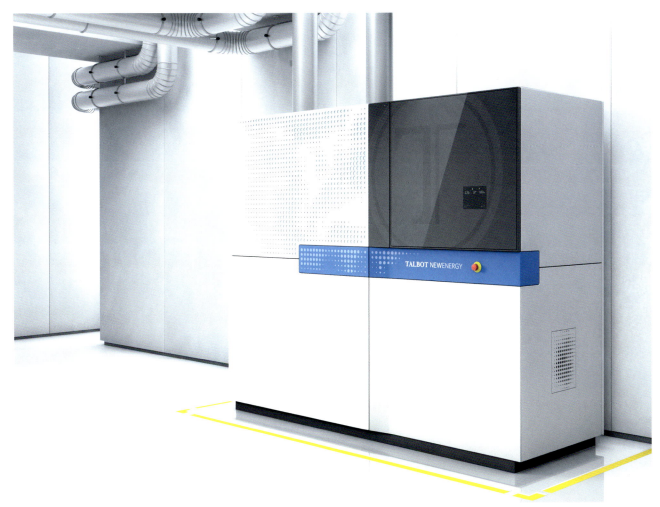

02

"My work philosophy is to see
technological innovations and
societal changes as opportunities
for the development of better
products."

„Meine Arbeitsphilosophie lautet,
die technologischen Innovationen
und gesellschaftlichen Veränderungen als Chancen für die Entwicklung
besserer Produkte zu begreifen."

**You have been a Red Dot juror for many years.
What significance does being appointed to the jury
have for you?**
I see it is an expression of the appreciation for my
work as a designer, and I appreciate the exchange
with international colleagues from all areas of design.

**What advice has been helpful in your youth or at
the beginning of your professional career?**
The advice to always be open to new developments,
be they technical or societal, and to fathom what
they mean for new design approaches.

**What do you take special notice of when you are
assessing a product as a juror?**
In the field of medical technology products in
particular, the aspects of safety, clarity of function,
manufacturing processes, hygienic properties and
stability of value are important criteria.

**What significance does winning an award in
a design competition have for the designer?**
A competition gives designers the opportunity to
compete internationally and thus get to know their
skills better.

**Sie sind langjähriger Red Dot-Juror. Welchen Wert
hat eine Berufung in diese Jury für Sie?**
Sie ist für mich Ausdruck der Wertschätzung meiner
gestalterischen Arbeit, und ich selbst schätze den
Austausch mit den internationalen Kollegen aus allen
Sparten des Designs.

**Welche Ratschläge haben Sie in Ihrer Jugend oder
frühen Berufskarriere weitergebracht?**
Immer offen zu sein für neue Entwicklungen, seien es
technische oder gesellschaftliche, und auszuloten, was
dies für neue Designansätze bedeutet.

**Worauf achten Sie besonders, wenn Sie ein
Produkt als Juror bewerten?**
Speziell in den Produktsortimenten der Medizintechnik
sind für mich die Aspekte Sicherheit, Sinnfälligkeit,
Herstellungsprozesse, Hygieneeigenschaften und
Wertbeständigkeit wichtige Kriterien.

**Welche Bedeutung hat die Auszeichnung in einem
Designwettbewerb für den Designer?**
Ein Wettbewerb gibt Designern die Gelegenheit, sich
international zu messen und so ihre Qualifikation
besser kennenzulernen.

01

Prof. Jure Miklavc
Slovenia
Slowenien

Professor Jure Miklavc graduated in industrial design from the Academy of Fine Arts in Ljubljana, Slovenia, and has nearly 20 years of experience in the field of design. Miklavc started his career working as a freelance designer, before founding his own design consultancy, Studio Miklavc. Studio Miklavc works in the fields of product design, visual communications and brand development and is a consultancy for a variety of clients from the industries of light design, electronic goods, user interfaces, transport design and medical equipment. Sports equipment designed by the studio has gained worldwide recognition. From 2013 onwards, the team has been working for the prestigious Italian motorbike manufacturer Bimota. Designs by Studio Miklavc have received many international awards and have been displayed in numerous exhibitions. Jure Miklavc has been involved in design education since 2005 and is currently a lecturer and head of industrial design at the Academy of Fine Arts and Design in Ljubljana.

Professor Jure Miklavc machte seinen Abschluss in Industrial Design an der Academy of Fine Arts and Design in Ljubljana, Slowenien, und verfügt über nahezu 20 Jahre Erfahrung im Designbereich. Er arbeitete zunächst als freiberuflicher Designer, bevor er sein eigenes Design-Beratungsunternehmen „Studio Miklavc" gründete. Studio Miklavc ist in den Bereichen Produktdesign, visuelle Kommunikation und Markenentwicklung sowie in der Beratung zahlreicher Kunden der Branchen Lichtdesign, elektronische Güter, Benutzeroberflächen, Transport-Design und medizinisches Equipment tätig. Die von dem Studio gestalteten Sportausrüstungen erfahren weltweit Anerkennung. Seit 2013 arbeitet das Team für den angesehenen italienischen Motorradhersteller Bimota. Studio Miklavc erhielt bereits zahlreiche Auszeichnungen sowie Präsentationen in Ausstellungen. Seit 2005 ist Jure Miklavc in der Designlehre tätig und aktuell Dozent und Head of Industrial Design an der Academy of Fine Arts and Design in Ljubljana.

01 Alpina ESK Pro
Racing boots for cross-country
skiing with a third of global market
share and countless podium finishes for top athletes at the highest
level of racing.
Racingschuhe für den Skilanglauf,
die ein Drittel des Weltmarktes
beherrschen und unzählige Podiums-
gewinne für Topathleten in der
höchsten Racingkategorie vorweisen
können.

02 Helmet for Bimota SA
It is part of the redesign of the
identity and strategy, new commu-
nication materials, fair stand design
and merchandise products.
Helm für Bimota SA
Er ist Teil der Neugestaltung der
Identität und Strategie, der neuen
Kommunikationsmaterialien, Messe-
standgestaltung und Handelswaren.

02

"My work philosophy is: quality
over quantity."
„Meine Arbeitsphilosophie lautet:
Qualität über Quantität."

**You are involved in design education. What
fascinates you about this job?**
The exchange between the practical knowledge that
I gain in my studio and the theoretical knowledge
that is based on the Academy. Mixing those influences
enriches both parts of my professional life.

When did you first think consciously about design?
When I was in primary school, I spent all the time
enhancing, modifying and redesigning my toys. This
sounds like just playing, but I added electric motors,
new metal parts or just enhanced their finish.

**What impressed you most during the Red Dot
judging process?**
All jury members took the task very seriously and
displayed a high level of professionalism. It is
a privilege and remarkably easy to work with so
many renowned experts. The work is quite intense
and focused, but enjoyable.

**Sie sind in der Designlehre tätig. Was fasziniert
Sie an dieser Aufgabe?**
Der Austausch zwischen praktischem Wissen, das ich
in meinem Designstudio erwerbe, und dem theore-
tischen Wissen, das auf der Akademie vermittelt wird.
Diese Einflüsse zu vermischen, bereichert beide
Bereiche meines beruflichen Lebens.

**Wann haben Sie das erste Mal bewusst über
Design nachgedacht?**
Als ich in der Grundschule war, verbrachte ich meine
gesamte Zeit damit, mein Spielzeug zu verbessern, zu
verändern und umzugestalten. Das klingt nach bloßem
Spielen, aber ich baute auch elektrische Motoren ein,
neue Teile aus Metall oder veränderte einfach die
Lackierung.

**Was hat Sie bei der Red Dot-Jurierung am meisten
beeindruckt?**
Alle Mitglieder der Jury nahmen die Aufgabe sehr ernst
und zeigten sich sehr professionell. Es ist ein Privileg
und bemerkenswert einfach, mit so vielen berühmten
Experten zusammenzuarbeiten. Die Arbeit ist ziemlich
intensiv und konzentriert, aber unterhaltsam.

01

Prof. Ron A. Nabarro
Israel
Israel

Professor Ron A. Nabarro is an industrial designer, strategist, entrepreneur, researcher and educator. He has been a professional designer since 1970 and has designed more than 750 products to date in a wide range of industries. He has played a leading role in the emergence of age-friendly design and age-friendly design education. From 1992 to 2009, he was a professor of industrial design at the Technion Israel Institute of Technology, where he founded and was the head of the graduate programme in advanced design studies and design management. Currently, Nabarro teaches design management and design thinking at DeTao Masters Academy in Shanghai, China. From 1999 to 2003, he was an executive board member of Icsid and now acts as a regional advisor. He is a frequent keynote speaker at conferences, has presented TEDx events, has lectured and led design workshops in over 20 countries and consulted to a wide variety of organisations. Furthermore, he is co-founder and CEO of Senior-touch Ltd. and design4all. The principle areas of his research and interest are design thinking, age-friendly design and design management.

Professor Ron A. Nabarro ist Industriedesigner, Stratege, Unternehmer, Forscher und Lehrender. Seit 1970 ist er praktizierender Designer, gestaltete bisher mehr als 750 Produkte für ein breites Branchenspektrum und spielt eine führende Rolle im Bereich des altersfreundlichen Designs und dessen Lehre. Von 1992 bis 2009 war er Professor für Industriedesign am Technologie-Institut Technion Israel, an dem er das Graduiertenprogramm für fortgeschrittene Designstudien und Designmanagement einführte und leitete. Aktuell unterrichtet Nabarro Designmanagement und Design Thinking an der DeTao Masters Academy in Shanghai, China. Von 1999 bis 2003 war er Vorstandsmitglied des Icsid, für den er aktuell als regionaler Berater tätig ist. Er ist ein gefragter Redner auf Konferenzen, hat bei TEDx-Veranstaltungen präsentiert, hielt Vorträge und Workshops in mehr als 20 Ländern und beriet eine Vielzahl von Organisationen. Zudem ist er Mitbegründer und Geschäftsführer von Senior-touch Ltd. und design4all. Die Hauptbereiche seiner Forschung und seines Interesses sind Design Thinking, altersfreundliches Design und Designmanagement.

02

"My work philosophy is to improve the lives of ageing people by gathering leading companies, entrepreneurs, designers and technologists in order to educate and innovate together."

„Meine Arbeitsphilosophie ist die Verbesserung des Lebens älterer Menschen durch Innovation und Erziehung in Zusammenarbeit mit führenden Firmen, Unternehmern, Designern und Technologen."

What impressed you most during the Red Dot judging process?
The most important aspect of the jury process was the way jury members treat the process, in particular the respect for each work presented and the commitments to the high standards of adjudication.

Do you see a correlation between the design quality of a company's products and the economic success of this company?
In most cases this "formula" could work, still it is important to acknowledge the importance of marketing and not put everything on the designers' shoulders. We also can see designs that have been ahead of their time and totally failed, although the design was brilliant.

Which project would you like to realise one day?
At this stage of my life I find more interest in sharing my professional experience and life experience with young designers and design students.

Was hat Sie bei der Red Dot-Jurierung am meisten beeindruckt?
Der wichtigste Aspekt während der Jurierung war die Art, wie die Jurymitglieder dem Prozess begegnet sind, insbesondere der Respekt für jede einzelne Arbeit und die Hingabe an die hohen Jurierungsstandards.

Sehen Sie einen Zusammenhang zwischen der Designqualität, die sich in den Produkten eines Unternehmens äußert, und dem wirtschaftlichen Erfolg dieses Unternehmens?
In den meisten Fällen funktioniert diese „Formel", dennoch ist es wichtig, auch die Rolle des Marketings zu erkennen und nicht alles auf die Schultern der Gestalter zu laden. Wir kennen auch Gestaltungen, die ihrer Zeit voraus waren, aber komplett gescheitert sind, obgleich die Gestaltung brillant war.

Welches Projekt würden Sie gerne einmal realisieren?
In dieser Phase meines Lebens interessiert es mich mehr, meine professionelle Erfahrung und Lebenserfahrung mit jungen Designern und Designstudenten zu teilen.

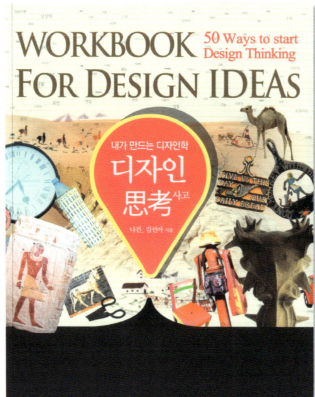

01

Prof. Dr. Ken Nah
Korea
Korea

Professor Dr Ken Nah graduated with a Bachelor of Arts in industrial engineering from Hanyang University, South Korea, in 1983. He deepened his interest in human factors/ergonomics by earning a master's degree from Korea Advanced Institute for Science and Technology (KAIST) in 1985 and he gained a PhD from Tufts University in 1996. In addition, Ken Nah is also a USA Certified Professional Ergonomist (CPE). He was the dean of the International Design School for Advanced Studies (IDAS) and is currently professor of design management as well as director of the Human Experience and Emotion Research (HE.ER) Lab at IDAS, Hongik University, Seoul. From 2002 he was the director of the International Design Trend Center (IDTC). Ken Nah was the director general of "World Design Capital Seoul 2010". Alongside his work as a lecturer he is also the vice president of the Korea Association of Industrial Designers (KAID), the Ergonomics Society of Korea (ESK) and the Korea Institute of Design Management (MIDM), as well as the chairman of the Design and Brand Committee of the Korea Consulting Association (KCA).

Professor Dr. Ken Nah graduierte 1983 an der Hanyang University in Südkorea als Bachelor of Arts in Industrial Engineering. Sein Interesse an Human Factors/Ergonomie vertiefte er 1985 mit einem Master-Abschluss am Korea Advanced Institute for Science and Technology (KAIST) und promovierte 1996 an der Tufts University. Darüber hinaus ist Ken Nah ein in den USA zertifizierter Ergonom (CPE). Er war Dekan der International Design School for Advanced Studies (IDAS) und ist aktuell Professor für Design Management sowie Direktor des „Human Experience and Emotion Research (HE.ER)"-Labors an der IDAS, Hongik University, Seoul. Von 2002 an war er Leiter des International Design Trend Center (IDTC). Ken Nah war Generaldirektor der „World Design Capital Seoul 2010". Neben seiner Lehrtätigkeit ist er Vizepräsident der Korea Association of Industrial Designers (KAID), der Ergonomics Society of Korea (ESK) und des Korea Institute of Design Management (MIDM) sowie Vorsitzender des „Design and Brand"-Komitees der Korea Consulting Association (KCA).

01 **Workbook for Design Ideas**
The Korean and Chinese
version on 50 ways to start
design thinking
Die koreanische und chinesische Version über 50 Wege, mit
Design Thinking zu beginnen

"My work philosophy is to do my best in all areas and every moment of every day, since time never stops and opportunity never waits."

„Meine Arbeitsphilosophie besteht darin, in allen Bereichen und in jedem Moment mein Bestes zu geben, da die Zeit ständig voran-schreitet und sich Gelegenheiten nicht zweimal bieten."

What motivates you to get up in the morning?
My question every morning before getting up is: "What if today were my last day?" This question motivates me to get back to work and use my time and energy in the best possible way.

When did you first think consciously about design?
It was in the winter of 1987, when I had to decide what majors to choose at university. Reading books and articles, I instantly fell in love with the words "Human Factors", defined as designing for people and optimising living and working conditions. Not only physical and physiological characteristics, but also psychological ones are important in "design". Since then, design has been everything to me!

What do you take special notice of when you are assessing a product as a jury member?
I pay attention to the "balance" between form and function. The product should also be well balanced between logic and emotion.

Was motiviert Sie, morgens aufzustehen?
Die Frage, die ich mir jeden Morgen vor dem Aufstehen stelle, ist: „Was wäre, wenn heute mein letzter Tag wäre?" Sie motiviert mich, wieder an die Arbeit zu gehen und meine Zeit und Energie optimal zu nutzen.

Wann haben Sie das erste Mal bewusst über Design nachgedacht?
Es war im Winter 1987, als ich an der Universität meine Hauptfächer wählen musste. Beim Lesen vieler Bücher und Artikel hatte ich mich sofort in die Worte „Human Factors" verliebt. Human Factors bedeutet, etwas für Menschen zu gestalten und die Lebens- und Arbeitsverhältnisse zu optimieren. Nicht nur physische und physiologische, sondern auch psychologische Eigenschaften sind wichtig in der Gestaltung. Seither ist Design mein Ein und Alles!

Worauf achten Sie besonders, wenn Sie ein Produkt als Juror bewerten?
Ich achte auf das Gleichgewicht zwischen Form und Funktion. Zudem sollte das Produkt auch zwischen Logik und Emotion gut ausgewogen sein.

01

Prof. Dr. Yuri Nazarov
Russia
Russland

Professor Dr Yuri Nazarov, born in 1948 in Moscow, teaches at the National Design Institute in Moscow where he is also provost. As an actively involved design expert, he serves on numerous boards, for example as president of the Association of Designers of Russia, as a corresponding member of Russian Academy of Arts, and as a member of the Russian Design Academy. Yuri Nazarov has received a wide range of accolades for his achievements: he is a laureate of the State Award of the Russian Federation in Literature and Art as well as of the Moscow Administration' Award, and he also has received a badge of honour for "Merits in Development of Design".

Professor Dr. Yuri Nazarov, 1948 in Moskau geboren, lehrt am National Design Institute in Moskau, dessen Rektor er auch ist. Als engagierter Designexperte ist er in zahlreichen Gremien des Landes tätig, zum Beispiel als Präsident der Russischen Designervereinigung, als korrespondierendes Mitglied der Russischen Kunstakademie sowie als Mitglied der Russischen Designakademie. Für seine Verdienste wurde Yuri Nazarov mit einer Vielzahl an Auszeichnungen geehrt. So ist er Preisträger des Staatspreises der Russischen Föderation in Literatur und Kunst sowie des Moskauer Regierungspreises und besitzt zudem das Ehrenabzeichen für „Verdienste in der Designentwicklung".

01 Roly-Poly
Mascot for the Russian
Federation's pavilion of the
World Expo Milano 2015 real-
ised with Vitaly Stavitcky.
Maskottchen für den rus-
sischen Pavillon auf der
Weltausstellung Expo Milano
2015, umgesetzt mit Vitaly
Stavitcky.

"My work philosophy ranges from functional designs for social regional projects and low-income individuals to engage in joint work with young designers and to integrate Russian design on an international scale."

„Meine Arbeitsphilosophie reicht von betont funktionalen Gestaltungen für soziale regionale Projekte wie für Geringverdienende bis zur Anbindung junger Designer sowie des russischen Designs an internationales Niveau."

What is, in your opinion, the significance of design quality in the product categories you evaluated?
The significance of design quality lies in confirming the usability and safety of the products.

What are the important criteria for you as a juror in the assessment of a product?
Creative ideas and the quality of their realisation.

What impressed you most during the Red Dot judging process?
The most outstanding for me was the mutual understanding of and similarities between our viewpoints.

Do you see a correlation between the design quality of a company's products and the economic success of this company?
It depends on how we interpret economic success. If we talk about profit it may simply be due to a hot commodity. But real design quality always implies taking into account consumer preferences.

Wie schätzen Sie den Stellenwert der Designqualität in den von Ihnen beurteilten Produktkategorien ein?
Der Stellenwert der Designqualität liegt darin, die Bedienbarkeit und Sicherheit der Produkte zu bekräftigen.

Worauf achten Sie als Juror, wenn Sie ein Produkt bewerten?
Auf kreative Ideen und die Qualität ihrer Umsetzung.

Was hat Sie bei der Red Dot-Jurierung am meisten beeindruckt?
Das Hervorstechendste für mich waren das gegenseitige Verständnis und die Ähnlichkeiten unserer Standpunkte.

Sehen Sie einen Zusammenhang zwischen der Designqualität, die sich in den Produkten eines Unternehmens äußert, und dem wirtschaftlichen Erfolg dieses Unternehmens?
Das hängt davon ab, wie man wirtschaftlichen Erfolg interpretiert. Sprechen wir von Profit, mag das schlicht an einem „heißen" Produkt liegen. Aber echte Designqualität impliziert immer die Berücksichtigung der Vorlieben der Verbraucher.

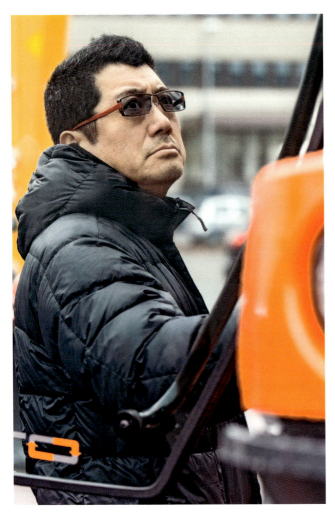

01

Ken Okuyama
Japan
Japan

Ken Kiyoyuki Okuyama, industrial designer and CEO of KEN OKUYAMA DESIGN, was born 1959 in Yamagata, Japan, and studied automobile design at the Art Center College of Design in Pasadena, California. He has worked as a chief designer for General Motors, as a senior designer for Porsche AG, and as design director for Pininfarina S.p.A., being responsible for the design of Ferrari Enzo, Maserati Quattroporte and many other automobiles. He is also known for many different product designs such as motorcycles, furniture, robots and architecture. KEN OKUYAMA DESIGN was founded in 2007 and provides business consultancy services to numerous corporations. Ken Okuyama also produces cars, eyewear and interior products under his original brand. He is currently a visiting professor at several universities and also frequently publishes books.

Ken Kiyoyuki Okuyama, Industriedesigner und CEO von KEN OKUYAMA DESIGN, wurde 1959 in Yamagata, Japan, geboren und studierte Automobildesign am Art Center College of Design in Pasadena, Kalifornien. Er war als Chief Designer bei General Motors, als Senior Designer bei der Porsche AG und als Design Director bei Pininfarina S.p.A. tätig und zeichnete verantwortlich für den Ferrari Enzo, den Maserati Quattroporte und viele weitere Automobile. Zudem ist er für viele unterschiedliche Produktgestaltungen wie Motorräder, Möbel, Roboter und Architektur bekannt. KEN OKUYAMA DESIGN wurde 2007 als Beratungsunternehmen gegründet und arbeitet für zahlreiche Unternehmen. Ken Okuyama produziert unter seiner originären Marke auch Autos, Brillen und Inneneinrichtungsgegenstände. Derzeit lehrt er als Gastprofessor an verschiedenen Universitäten und publiziert zudem Bücher.

02

"My design philosophy reads:
modern, simple, timeless."
„Meine Designphilosophie lautet:
Modern, schlicht, zeitlos."

What trends have you noticed in the field of "Vehicles" in recent years?
Driving performance is no longer a sales point. Transport design should propose not only mobility but also a new lifestyle. A car's design reflects its owner's character and lifestyle more than ever.

Do you see a correlation between the design quality of a company's products and the economic success of this company?
The correlation is the result of a clear vision and the teamwork that made it happen, plus the personalities of individual team members.

What are the important criteria for you as a juror in the assessment of a product?
A juror has to determine a product's value to society and the market. Therefore, an objective view and wide ranging knowledge of technology, materials, manufacturing, etc. are necessary.

Welche Trends konnten Sie im Bereich „Fahrzeuge" in den letzten Jahren ausmachen?
Fahr-Performance ist kein Verkaufsargument mehr. Im Segment „Transport" sollte Gestaltung nicht nur auf Mobilität abzielen, sondern auch auf einen neuen Lebensstil. Das Design eines Autos spiegelt mehr denn je den Charakter und Lebensstil seines Besitzers wider.

Sehen Sie einen Zusammenhang zwischen der Designqualität, die sich in den Produkten eines Unternehmens äußert, und dem wirtschaftlichen Erfolg dieses Unternehmens?
Diese Wechselwirkung ist das Ergebnis einer klaren Vision und der ihr zugrunde liegenden Teamarbeit – plus der Persönlichkeiten der einzelnen Teammitglieder.

Worauf achten Sie als Juror, wenn Sie ein Produkt bewerten?
Ein Juror muss den Wert bestimmen, den ein Produkt für die Gesellschaft und den Markt hat. Daher sind eine objektive Sichtweise und eine große Bandbreite an Wissen über Technik, Werkstoffe, Herstellung etc. notwendig.

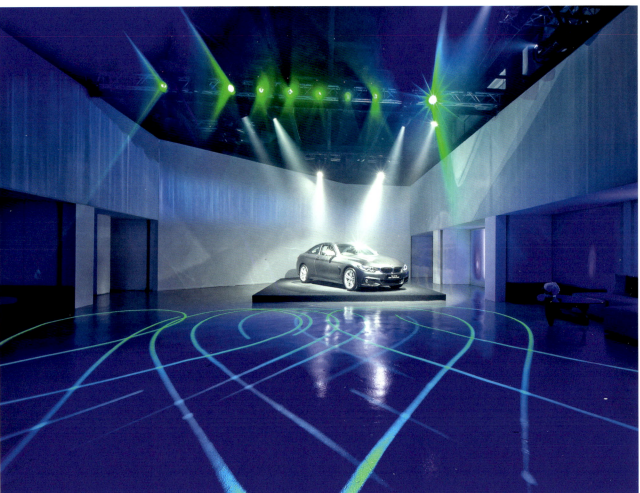

01

Simon Ong
Singapore
Singapur

Simon Ong, born in Singapore in 1953, graduated with a master's degree in design from the University of New South Wales and an MBA from the University of South Australia. He is the group managing director and co-founder of Kingsmen Creatives Ltd., a leading communication design and production group with 18 offices across the Asia Pacific region and the Middle East. Kingsmen has won several awards, such as the President's Design Award, SRA Best Retail Concept Award, SFIA Hall of Fame, Promising Brand Award, A.R.E. Retail Design Award and RDI International Store Design Award USA. Simon Ong is actively involved in the creative industry as chairman of the design group of Manpower, the Skills & Training Council of Singapore Workforce Development Agency. Moreover, he is a member of the advisory board of the Singapore Furniture Industries Council and School of Design & Environment at the National University of Singapore, Design Business Chamber Singapore and Interior Design Confederation of Singapore.

Simon Ong, geboren 1953 in Singapur, erhielt einen Master in Design der University of New South Wales und einen Master of Business Administration der University of South Australia. Er ist Vorstandsvorsitzender und Mitbegründer von Kingsmen Creatives Ltd., eines führenden Unternehmens für Kommunikationsdesign und Produktion mit 18 Geschäftsstellen im asiatisch-pazifischen Raum sowie im Mittleren Osten. Kingsmen wurde vielfach ausgezeichnet, u. a. mit dem President's Design Award, SRA Best Retail Concept Award, SFIA Hall of Fame, Promising Brand Award, A.R.E. Retail Design Award und RDI International Store Design Award USA. Simon Ong ist als Vorsitzender der Designgruppe von Manpower, der „Skills & Training Council of Singapore Workforce Development Agency", aktiv in die Kreativindustrie involviert, ist unter anderem Mitglied des Beirats des Singapore Furniture Industries Council, der School of Design & Environment an der National University of Singapore, des Design Business Chamber Singapore und der Interior Design Confederation of Singapore.

02

"My work philosophy is 'less is more'
in everything we do. Having less
'quantity' or details enables us to
have more time to focus on quality
and what matters."

„Meine Arbeitsphilosophie ist
‚Weniger ist mehr', in allem, was wir
tun. Die Reduktion von Quantität
oder Details gibt uns mehr Zeit, uns
auf die Qualität und das Wesentliche
zu konzentrieren."

What motivates you to get up in the morning?
Humour or having a good laugh. In our rapidly
developing world, we are often so caught up in
our work that we tend to forget the finest thing
in life – that is to laugh.

**In your opinion, which decade brought the greatest
progress in the design sector?**
Affordable computer software in the 1990s opened up
vast opportunities to go beyond traditional processes
in design thinking. Designs could be "tested" through
walk-through imaging – saving time and resources for
prototyping.

What challenges will designers face in the future?
Designers will have to think ahead and look at what
lies beyond sustainable design.

Was motiviert Sie, morgens aufzustehen?
Etwas zum Lachen zu haben. In unserer schnelllebigen
Welt sind wir oft so stark in unsere Arbeit verstrickt,
dass wir dazu neigen, das Beste im Leben zu vergessen –
und zwar zu lachen.

**Welches Jahrzehnt hat Ihrer Meinung nach den
größten Fortschritt im Designbereich gebracht?**
Erschwingliche Computer-Software in den 1990er
Jahren eröffnete riesige Möglichkeiten, über die
Grenzen traditioneller Prozesse im Design Thinking
hinauszugehen. Entwürfe konnten durch Computer-
simulationen „getestet" werden, um so Zeit und Ressourcen bei der Herstellung von Prototypen zu sparen.

**Vor welchen Herausforderungen werden Designer
in der Zukunft stehen?**
Designer werden vorausdenken und ausmachen
müssen, was nach nachhaltiger Gestaltung kommt.

01

Prof. Martin Pärn
Estonia
Estland

Professor Martin Pärn, born in Tallinn in 1971, studied industrial design at the University of Industrial Arts Helsinki (UIAH). After working in the Finnish furniture industry he moved back to Estonia and undertook the role of the ambassadorial leader of design promotion and development in his native country. He was actively involved in the establishment of the Estonian Design Centre and continues directing the organisation as chair of the board. Martin Pärn founded the multidisciplinary design office "iseasi", which creates designs ranging from office furniture to larger instruments and from small architecture to interior designs for the public sector. Having received many awards, Pärn begun in 1995 with the establishment and development of design training in Estonia and is currently head of the Design and Engineering's master's programme, a joint initiative of the Tallinn University of Technology and the Estonian Academy of Arts, which aims, among other things, to create synergies between engineers and designers.

Professor Martin Pärn, geboren 1971 in Tallinn, studierte Industriedesign an der University of Industrial Arts Helsinki (UIAH). Nachdem er in der finnischen Möbelindustrie gearbeitet hatte, ging er zurück nach Estland und übernahm die Funktion des leitenden Botschafters für die Designförderung und -entwicklung des Landes. Er war aktiv am Aufbau des Estonian Design Centres beteiligt und leitet seither die Organisation als Vorstandsvorsitzender. Martin Pärn gründete das multidisziplinäre Designbüro „iseasi", das ebenso Büromöblierung wie größere Instrumente, „kleine Architektur" oder Interior Designs im öffentlichen Sektor gestaltet. Vielfach ausgezeichnet, startete Pärn 1995 mit der Entwicklung und dem Ausbau der Designlehre in Estland und ist heute Leiter des Masterprogramms Design und Engineering, einer gemeinsamen Initiative der Tallinn University of Technology und der Estonian Academy of Arts, u. a. mit dem Ziel, durch den Zusammenschluss Synergien von Ingenieuren und Designern zu erreichen.

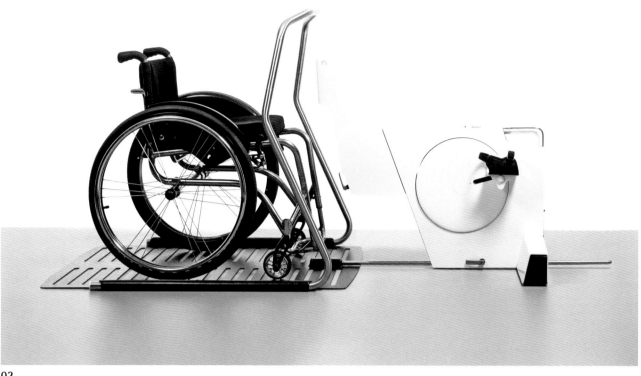

02

"My work philosophy is to search for something that is obvious, but yet unnoticed."

„Meine Arbeitsphilosophie besteht darin, nach etwas zu suchen, das offensichtlich, aber bisher unbemerkt geblieben ist."

What motivates you to get up in the morning?
Sun or, in short, life itself. It is full of new and unseen miracles I do not want to miss.

What challenges will designers face in the future?
The challenge of shifting the focus from fast consumer success towards long-lasting effectiveness and sustainability.

What do you take special notice of when you are assessing a product as a jury member?
I am looking more for the new "Why?" than the new "How?".

What significance does winning an award in a design competition have for the designer?
It means they have gained the respect of their colleagues, and thus it is a matter of honour. The real credits have to be earned in the field.

Was motiviert Sie, morgens aufzustehen?
Die Sonne oder, kurz gesagt, das Leben an sich. Es ist voller neuer und unbemerkter Wunder, die ich nicht verpassen möchte.

Vor welchen Herausforderungen werden Designer in der Zukunft stehen?
Vor der Herausforderung, das Augenmerk vom schnellen Markterfolg auf langlebige Leistungsfähigkeit und Nachhaltigkeit zu verschieben.

Worauf achten Sie besonders, wenn Sie ein Produkt als Juror bewerten?
Ich suche mehr nach einem neuen „Warum?" als nach einem neuen „Wie?".

Welche Bedeutung hat die Auszeichnung in einem Designwettbewerb für den Designer?
Es bedeutet, dass der Designer den Respekt seiner Kollegen gewonnen hat, was daher eine Frage der Ehre ist. Die echten Auszeichnungen müssen in der Praxis verdient werden.

01

02/03

Dr Sascha Peters
Germany
Deutschland

Dr Sascha Peters is founder and owner of the agency for material and technology HAUTE INNOVATION in Berlin. He studied mechanical engineering at the RWTH Aachen, Germany, and product design at the ABK Maastricht, Netherlands. He wrote his doctoral thesis at the University of Duisburg-Essen, Germany, on the complex of problems in communication between engineering and design. From 1997 to 2003, he led research projects and product developments at the Fraunhofer Institute for Production Technology IPT in Aachen and subsequently became head of the Design Zentrum Bremen. Sascha Peters is author of various specialised books on sustainable raw materials, smart materials, innovative production techniques and energetic technologies. He is a leading material expert and trend scout for new technologies. Since 2014, he has been an advisory board member of the funding initiative "Zwanzig20 – Partnerschaft für Innovation" (2020 – Partnership for innovation) by order of the German Federal Ministry of Education and Research.

Dr. Sascha Peters ist Gründer und Inhaber der Material- und Technologieagentur HAUTE INNOVATION in Berlin. Er studierte Maschinenbau an der RWTH Aachen und Produktdesign an der ABK Maastricht. Seine Doktorarbeit schrieb er an der Universität Duisburg-Essen über die Kommunikationsproblematik zwischen Engineering und Design. Von 1997 bis 2003 leitete er Forschungsprojekte und Produktentwicklungen am Fraunhofer-Institut für Produktionstechnologie IPT in Aachen und war anschließend bis 2008 stellvertretender Leiter des Design Zentrums Bremen. Sascha Peters ist Autor zahlreicher Fachbücher zu nachhaltigen Werkstoffen, smarten Materialien, innovativen Fertigungsverfahren und energetischen Technologien und zählt zu den führenden Materialexperten und Trendscouts für neue Technologien. Seit 2014 ist er Mitglied im Beirat der Förderinitiative „Zwanzig20 – Partnerschaft für Innovation" im Auftrag des Bundesministeriums für Bildung und Forschung.

04

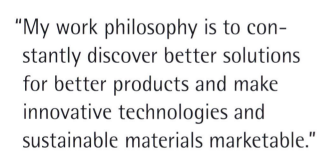

"My work philosophy is to constantly discover better solutions for better products and make innovative technologies and sustainable materials marketable."
„Meine Arbeitsphilosophie ist, beständig bessere Lösungen für bessere Produkte zu finden und innovative Technologien und nachhaltige Materialien marktfähig zu machen."

What impressed you most during the Red Dot judging process?
The jury with its experts from so many different countries. This network of opinions emerging from different cultural backgrounds is the basis for global acceptance and the success of the Red Dot Award.

What properties must a new material have to convince you of its outstanding quality?
With regards to our resources becoming ever scarcer, the issues we are faced with concerning waste disposal and the challenges resulting from a growing world population, I judge the development of a material as outstanding when it opens up the possibility for sustainable use and leaves a particularly small ecological footprint.

Which material development from the last hundred years has had the biggest influence on today's world?
Plastics. They enable the creation of almost limitless properties. However, most of them are not bio-degradable.

Was hat Sie bei der Red Dot-Jurierung am meisten beeindruckt?
Die Herkunft der Jury mit Experten aus den unterschiedlichsten Ländern. Dieses Netzwerk aus Meinungen verschiedener kultureller Einflüsse ist die Grundlage für die globale Akzeptanz und den Erfolg des Red Dot Awards.

Welche Eigenschaften muss ein neues Material vorweisen, um Sie von seiner herausragenden Qualität zu überzeugen?
Mit Blick auf die knapper werdenden Ressourcen, die Probleme, die wir mit der Abfallentsorgung haben, und die Herausforderungen, die sich durch die wachsende Weltbevölkerung ergeben, bewerte ich Materialentwicklungen als herausragend, wenn sie eine Möglichkeit zu einer nachhaltigen Nutzung aufzeigen und einen besonders geringen ökologischen Fußabdruck offenbaren.

Welche Materialentwicklung der letzten hundert Jahre hat den größten Einfluss auf die heutige Zeit?
Kunststoffe. Ihre Qualitäten lassen sich nahezu beliebig einstellen. Ihr Großteil ist jedoch nicht biologisch abbaubar.

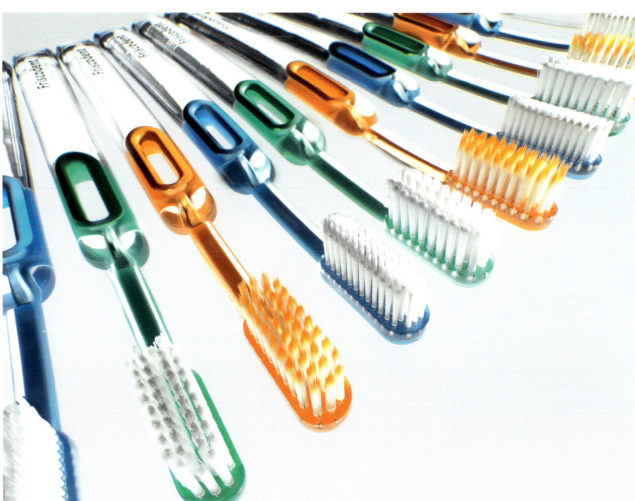

01

Oliver Stotz
Germany
Deutschland

Oliver Stotz, born in 1961 in Stuttgart, Germany, studied industrial design at the University Essen, Germany and at the Royal College of Art in London. As the founder of his own studio "stotz-design.com" in Wuppertal, Germany, he has more than 20 years of experience in the fields of industrial and corporate design. His studio has created established brands such as Proseat in the automotive sector and Blomus in the field of glass, porcelain and ceramic. With his eight-strong team, Stotz advises companies in various industries regarding the implementation and realisation of new design concepts. Since 2010, Oliver Stotz has been a board member of the foundation "Mia Seeger Stiftung", which promotes young designers after graduation with the "Mia Seeger Preis" award. In addition, he is a lecturer at the design department of the University Wuppertal. His design achievements have received several awards and have been on display in numerous exhibitions.

Oliver Stotz, 1961 in Stuttgart geboren, studierte Industriedesign an der Universität Essen und am Royal College of Art in London. Als Gründer des Studios „stotz-design.com" mit Sitz in Wuppertal kann er inzwischen auf mehr als 20 Jahre Berufserfahrung in den Bereichen Industrial und Corporate Design zurückgreifen. In seinem Studio entstanden bereits etablierte Marken wie Proseat im Automotive-Sektor oder auch Blomus im Bereich Glas, Porzellan, Keramik. Mit seinem achtköpfigen Team berät Stotz national wie international Unternehmen aus unterschiedlichsten Branchen bei der Implementierung und Umsetzung von neuen Designkonzepten. Seit 2010 ist Oliver Stotz Vorstandsmitglied der „Mia Seeger Stiftung", die junge Designerinnen und Designer nach ihrem Studienabschluss mit dem „Mia Seeger Preis" fördert. Darüber hinaus ist er als Dozent an der Universität Wuppertal im Fachbereich Design tätig. Seine Designleistungen wurden vielfach ausgezeichnet und in zahlreichen Ausstellungen präsentiert.

01 **Medical toothbrush
for the global market**
Commissioned by
Sunstar Group
Medizinische Zahnbürste
für den Weltmarkt
Im Auftrag der Sunstar Group

02 **Proseat automotive brand**
Corporate design for the
European market for Recticel
Woodbridge
Automotive-Marke Proseat
Corporate Design für den
europäischen Markt für
Recticel Woodbridge

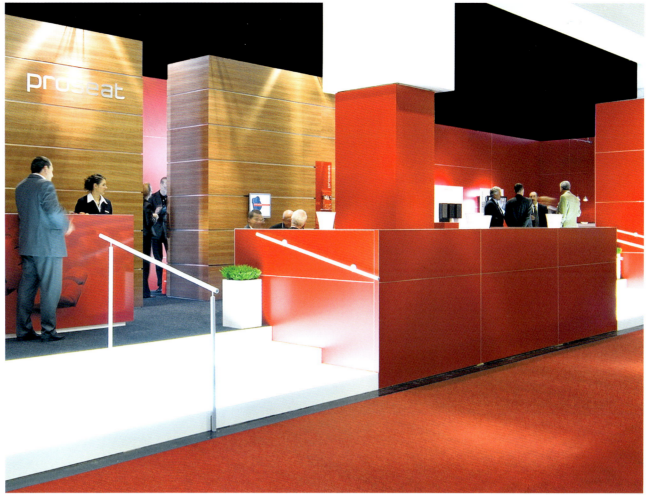

02

"My work philosophy is:
a good designer is always
ahead of his time."

„Meine Arbeitsphilosophie ist:
Ein guter Designer ist immer
seiner Zeit voraus."

In which product industry will Germany be most successful in the next ten years?
Germany is currently technology leader in the field of transportation and will achieve more successes in this field. However, networked thinking of the global players is a prerequisite for this to happen.

What are the main challenges in a designer's everyday life?
Keeping it simple is always the main challenge.

The Red Dot Award has been uncovering the best designs for 60 years now. Which product innovation would you like to see in the next 60 years, and why?
Driven by the thought that in future we will have to deal with the digitisation of many processes, as a result of which more and more things will become virtual, I wish to see materiality as product innovation, that haptic qualities become an enrichment of things.

What do you take special notice of when you are assessing a product as a jury member?
Whether it is an original design.

Mit welcher Produktbranche wird Deutschland in den nächsten zehn Jahren am erfolgreichsten sein?
Im Bereich Transportation ist Deutschland Technologieführer und wird Erfolge erzielen können. Voraussetzung ist allerdings, dass die Global Player vernetzt denken.

Was sind große Herausforderungen im Alltag eines Designers?
Keep it simple! Das ist immer wieder die große Aufgabe.

Der Red Dot Award ermittelt seit 60 Jahren die besten Gestaltungen. Welche Produktinnovation würden Sie sich für die nächsten 60 Jahre wünschen, und warum?
Von dem Gedanken getrieben, sich zukünftig mit der Digitalisierung vieler Prozesse auseinandersetzen zu müssen, in deren Folge immer mehr Dinge virtuell werden, wünsche ich mir bei den Produkten als Innovation die Materialität, das Haptische als konkret erlebbare Bereicherung der Dinge.

Worauf achten Sie besonders, wenn Sie ein Produkt als Juror bewerten?
Darauf, ob die Eigenständigkeit gewährleistet ist.

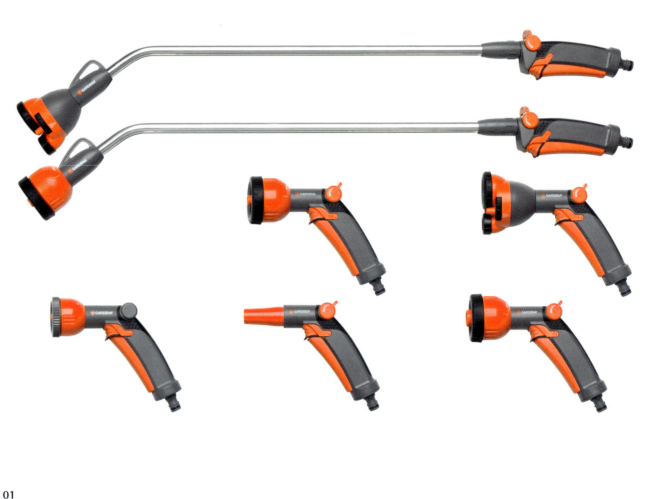

01

Aleks Tatic
Germany / Italy
Deutschland / Italien

Aleks Tatic, born 1969 in Cologne, Germany, is product designer and founder of Tatic Designstudio in Milan, Italy. After his studies at the Art Center College of Design in the USA and Switzerland, he specialised in his focal areas, sports and lifestyle products, in various international agencies in London and Milan. Afterwards, he guided the multiple award-winning Italian design studio Attivo Creative Resource to international success, leading the agency for 12 years. Together with his multicultural team of designers and product specialists, he today designs and develops – amongst others – sailing yachts, sporting goods, power tools, FMCGs and consumer electronics for European and Asian premium brands. Aleks Tatic lectures practice-oriented industrial design and innovation management at various European universities and seminars.

Aleks Tatic, geboren 1969 in Köln, ist Produktdesigner und Gründer der Agentur Tatic Designstudio in Mailand. Nach seinem Studium am Art Center College of Design in den USA und der Schweiz hat er sich zunächst in verschiedenen internationalen Büros in London und Mailand auf sein Schwerpunktgebiet Sport- und Lifestyleprodukte spezialisiert. Danach führte er zwölf Jahre lang das mehrfach ausgezeichnete italienische Designbüro Attivo Creative Resource zu internationalem Erfolg. Heute gestaltet und entwickelt er mit seinem multikulturellen Team von Designern und Produktspezialisten u. a. Segeljachten, Sportgeräte, Hobby- und Profiwerkzeuge, FMCGs und Unterhaltungselektronik für europäische und asiatische Premiummarken. Aleks Tatic unterrichtet an verschiedenen europäischen Hochschulen und Seminaren praxisorientiertes Industriedesign und Innovationsmanagement.

02

"My work philosophy is:
it has to be fun (for my
colleagues and me)!"

„Meine Arbeitsphilosophie ist:
Es muss uns (meinen Kollegen
und mir) Spaß machen!"

Which of the projects in your lifetime are you particularly proud of?
That would be our first sailing yacht, which we designed and then promptly won the "Boat of the Year" award for, even though we were new to the industry. This is the perfect proof that you can design any product even without specialising in the industry.

The Red Dot Award has been uncovering the best designs for 60 years now. Which product innovation would you like to see in the next 60 years, and why?
I would love to have someone launch Scotty's beamer in the market. This would cut travel times to our clients significantly.

What significance does winning an award in a design competition have for the manufacturer?
Many of our clients want to enter the Red Dot Award, because it is often hard for their clients – consumers or buying decision makers – to judge the design quality. Awards such as the Red Dot, as internationally recognised quality seals, are excellent indicators for this purpose.

Auf welches Projekt in Ihrem Leben sind Sie besonders stolz?
Auf unsere erste Segeljacht, die wir als komplett Branchenfremde gestaltet und dann prompt die Auszeichnung „Boat of the Year" gewonnen haben – der perfekte Beweis dafür, dass man jedes Produkt auch ohne Branchenspezialisierung gestalten kann.

Der Red Dot Award ermittelt seit 60 Jahren die besten Gestaltungen. Welche Produktinnovation würden Sie sich für die nächsten 60 Jahre wünschen, und warum?
Ich würde mich sehr freuen, wenn endlich jemand Scottys „Beamer" auf den Markt bringen würde. Die Reisezeiten zu unseren Kunden würden sich endlich deutlich verringern.

Welche Bedeutung hat die Auszeichnung in einem Designwettbewerb für den Hersteller?
Viele unserer Kunden möchten gerne beim Red Dot Award mitmachen. Denn für ihre Kunden – Verbraucher oder Kaufentscheider – ist die Designqualität oft schwierig zu beurteilen. Auszeichnungen wie der Red Dot als international anerkanntes Qualitätssiegel bilden hierfür herausragende Gradmesser.

01

Nils Toft
Denmark
Dänemark

Nils Toft, born in Copenhagen in 1957, graduated as an architect and designer from the Royal Danish Academy of Fine Arts in Copenhagen in 1985. He also holds a Master's degree in Industrial Design and Business Development. Starting his career as an industrial designer, Nils Toft joined the former Christian Bjørn Design in 1987, an internationally active design studio in Copenhagen with branches in Beijing and Ho Chi Minh City. Within a few years, he became a partner of CBD and, as managing director, ran the business. Today, Nils Toft is the founder and managing director of Designidea. With offices in Copenhagen and Beijing, Designidea works in the following key fields: communication, consumer electronics, computing, agriculture, medicine, and graphic arts, as well as projects in design strategy, graphic and exhibition design.

Nils Toft, geboren 1957 in Kopenhagen, machte seinen Abschluss als Architekt und Designer 1985 an der Royal Danish Academy of Fine Arts in Kopenhagen. Er verfügt zudem über einen Master im Bereich Industrial Design und Business Development. Zu Beginn seiner Karriere als Industriedesigner trat Nils Toft 1987 bei dem damaligen Christian Bjørn Design ein, einem international operierenden Designstudio in Kopenhagen, das mit Niederlassungen in Beijing und Ho-Chi-Minh-Stadt vertreten ist. Innerhalb weniger Jahre wurde er Partner bei CBD und leitete das Unternehmen als Managing Director. Heute ist Nils Toft Gründer und Managing Director von Designidea. Mit Büros in Kopenhagen und Beijing operiert Designidea in verschiedenen Hauptbereichen: Kommunikation, Unterhaltungselektronik, Computer, Landwirtschaft, Medizin und Grafikdesign sowie Projekte in den Bereichen Designstrategie, Grafik- und Ausstellungsdesign.

02

"My work philosophy is: design is
a language that carries your brand
and tells the story of how good
your products are."

„Meine Arbeitsphilosophie lautet:
Design ist eine Sprache, die deine
Marke transportiert und erzählt, wie
gut deine Produkte sind."

When did you first think consciously about design?
I was around seven or eight years old when my parents introduced me to arts and design. I remember how I, in my mind, redesigned the interior of my playmates' homes, when I saw how different they looked compared to mine.

What motivates you to get up in the morning?
I have always looked at life as a big apple and I can't wait to get up in the morning to take the next bite.

What impressed you most during the Red Dot judging process?
The vast number of exceptional entries and the fantastic discussions with the other jury members.

What significance does winning an award in a design competition have for the designer?
It is a tribute to the uniqueness of their talent and a reminder not to take talent for granted.

Wann haben Sie das erste Mal bewusst über Design nachgedacht?
Ich war ungefähr sieben oder acht Jahre alt, als mich meine Eltern an Kunst und Design heranführten. Ich erinnere mich daran, wie ich in meinem Kopf die Inneneinrichtung der Häuser meiner Spielkameraden umgestaltete, als ich sah, wie stark sie sich von meiner unterschied.

Was motiviert Sie, morgens aufzustehen?
Ich habe das Leben schon immer als einen großen Apfel betrachtet und kann es kaum erwarten, morgens aufzustehen und den nächsten Bissen zu nehmen.

Was hat Sie bei der Red Dot-Jurierung am meisten beeindruckt?
Die riesige Anzahl bemerkenswerter Einreichungen und die fantastischen Diskussionen mit den anderen Jurymitgliedern.

Welche Bedeutung hat die Auszeichnung in einem Designwettbewerb für den Designer?
Sie zollt der Einzigartigkeit seines Talentes Tribut und erinnert daran, Talent nicht als selbstverständlich anzusehen.

01

Prof. Danny Venlet
Belgium
Belgien

Professor Danny Venlet was born in 1958 in Victoria, Australia and studied interior design at Sint-Lukas, the Institute for Architecture and Arts in Brussels. Back in Australia in 1991, Venlet started to attract international attention with large-scale interior projects such as the Burdekin hotel in Sydney, and Q-bar, an Australian chain of nightclubs. His design projects range from private mansions, lofts, bars and restaurants all the way to showrooms and offices of large companies. The interior projects and the furniture designs of Danny Venlet are characterised by their contemporary international style. He says that the objects arise from an interaction between art, sculpture and function. These objects give a new description to the space in which they are placed – with respect, but also with relative humour. Today, Danny Venlet teaches his knowledge to students at the Royal College of the Arts in Ghent.

Professor Danny Venlet wurde 1958 in Victoria, Australien, geboren und studierte Interior Design am Sint-Lukas Institut für Architektur und Kunst in Brüssel. Nachdem er 1991 wieder nach Australien zurückgekehrt war, begann er, mit der Innenausstattung großer Projekte wie dem Burdekin Hotel in Sydney und der Q-Bar, einer australischen Nachtclub-Kette, internationale Aufmerksamkeit zu erregen. Seine Designprojekte reichen von privaten Wohnhäusern über Lofts, Bars und Restaurants bis hin zu Ausstellungsräumen und Büros großer Unternehmen. Die Innenausstattungen und Möbeldesigns von Danny Venlet sind durch einen zeitgenössischen, internationalen Stil ausgezeichnet und entspringen, wie er sagt, der Interaktion zwischen Kunst, Skulptur und Funktion. Seine Objekte geben den Räumen, in denen sie sich befinden, eine neue Identität – mit Respekt, aber auch mit einer Portion Humor. Heute vermittelt Danny Venlet sein Wissen als Professor an Studenten des Royal College of the Arts in Gent.

02

"My work philosophy is to create objects that, while embodying the theory of relativity, leave us no other choice than to act differently."

„Meine Arbeitsphilosophie besteht darin, Objekte zu kreieren, die – obwohl sie die Relativitätstheorie verkörpern – uns keine andere Wahl lassen, als anders zu handeln."

Which design project was the greatest intellectual challenge for you?
The Easyrider for Bulo, because my briefing consisted of the two opposing concepts of "relaxation/office", which gave birth to, if I may say so, a great innovation.

Which country do you consider to be a pioneer in product design?
Of course Belgium, due to its excellent training, its history and companies which are willing to take the design path.

What is the advantage of commissioning an external designer compared to having an in-house design team?
The biggest advantage is that an external designer is free-spirited and not conditioned by the constraints that an in-house designer might have encountered working in a company. In order to innovate you need to be able to look beyond!

Welches Ihrer bisherigen Designprojekte war die größte Herausforderung für Sie?
Der Easyrider für Bulo, da mein Briefing aus den zwei gegensätzlichen Konzepten „Entspannung/Büro" bestand, aus denen eine, wenn ich das einmal so sagen darf, großartige Innovation hervorging.

Welche Nation ist für Sie Vorreiter im Produktdesign?
Natürlich Belgien, aufgrund seiner hervorragenden Ausbildungsmöglichkeiten, seiner geschichtlichen Vergangenheit und seiner Unternehmen, die gewillt sind, auf Design zu setzen.

Worin liegt der Vorteil, einen externen Designer zu beauftragen, im Vergleich zu einem Inhouse-Designteam?
Der größte Vorteil ist, dass ein externer Designer über mehr geistige Freiheit verfügt und nicht von den Beschränkungen konditioniert ist, denen ein Inhouse-Designer in einer Firma unterworfen ist. Um Innovationen zu kreieren, muss man die Möglichkeit haben, über den Tellerrand hinauszuschauen!

01

Günter Wermekes
Germany
Deutschland

Günter Wermekes, born in Kierspe, Germany in 1955, is a goldsmith and designer. After many years of practice as an assistant and head of the studio of Professor F. Becker in Düsseldorf, he founded his own studio in 1990. He attracted great attention with his jewellery collection "Stainless steel and brilliant", which he has been presenting at national and international fairs since 1990. His designs are also appreciated by renowned manufacturers such as BMW, Rodenstock, Niessing or Tecnolumen, for which he designed, among other things, accessories, glasses, watches and door openers. In exhibitions and lectures around the world he has illustrated his personal design philosophy, which is: "Minimalism means reducing things more and more to their actual essence, and thus making it visible." He has also implemented this motto in his design of the Red Dot Trophy. In 2014, Günter Wermekes was one of ten finalists in the design competition for the medal of the Tang Prize, the "Asian Nobel Prize". His works have won numerous prizes and are part of renowned collections.

Günter Wermekes, 1955 geboren in Kierspe, ist Gold-schmied und Designer. Nach langjähriger Tätigkeit als Assistent und Werkstattleiter von Professor F. Becker in Düsseldorf gründete er 1990 sein eigenes Studio. Große Aufmerksamkeit erregte er mit seiner Schmuck-kollektion „Edelstahl und Brillant", die er ab 1990 auf nationalen und internationalen Messen präsentierte. Sein Design schätzen auch namhafte Hersteller wie BMW, Rodenstock, Niessing oder Tecnolumen, für die er u. a. Accessoires, Brillen, Uhren und Türdrücker ent-warf. In Ausstellungen und Vorträgen weltweit ver-deutlicht er seine persönliche Gestaltungsphilosophie, dass „Minimalismus bedeutet, Dinge so auf ihr Wesen zu reduzieren, dass es dadurch sichtbar wird". Diesen Leitspruch hat er auch in seinem Entwurf der Red Dot Trophy umgesetzt. 2014 war Günter Wermekes einer von zehn Finalisten des Gestaltungswettbewerbs für die Preismedaille des „Tang Prize", des „Asiatischen Nobelpreises". Seine Arbeiten wurden mehrfach prämiert und befinden sich in bedeutenden Sammlungen.